Compendium

featuring all the content from the following books

Scion

First published 2008

A CIP catalogue record for this book is available from the British Library.

ISBN 978 1 904842 58 3

Scion Publishing Limited
Bloxham Mill, Barford Road, Bloxham, Oxfordshire OX15 4FF
www.scionpublishing.com

Typeset by Phoenix Photosetting, Chatham, Kent, UK
Printed by Cromwell Press Group, Trowbridge, UK

Contents

Catch Up Biology

Catch Up Chemistry

Catch Up Maths and Stats

Biology

for the medical sciences

Philip Bradley and **Jane Calvert**

Faculty of Medical Sciences, The Medical School,
Newcastle University, Framlington Place,
Newcastle upon Tyne, UK

First published 2006

A CIP catalogue record for this book is available from the British Library.

ISBN 1 904842 32 1

Scion Publishing Limited
Bloxham Mill, Barford Road, Bloxham, Oxfordshire OX15 4FF
www.scionpublishing.com

Important Note from the Publisher

The information contained within this book was obtained by Scion Publishing Limited from sources believed by us to be reliable. However, while every effort has been made to ensure its accuracy, no responsibility for loss or injury whatsoever occasioned to any person acting or refraining from action as a result of information contained herein can be accepted by the authors or publishers.

Typeset by Phoenix Photosetting, Chatham, Kent, UK
Printed by Biddles Ltd, King's Lynn, UK, www.biddles.co.uk

Contents

Preface

Students entering university courses in the medical or biomedical sciences have a wide range of different qualifications and knowledge. According to the route of entry, different students will have covered topics in varying levels of detail. This short text aims to provide an overview of some of the important concepts that will help a student to understand and gain maximum benefit from their university course.

The book takes a hierarchical approach, starting with an introduction to the molecules of life, moving on to consider cells and their functions, how cells are assembled into tissues and ultimately the various systems of the body. Biology is a huge subject so this text selects material that will be most useful to students studying courses related to medicine and the medical sciences.

It is important to be aware of the position of humans in relation to other life on the planet. Life can be divided up into prokaryotes and eukaryotes, which differ markedly in the properties of their cells (see Chapter 6). The prokaryotes include two major domains – the bacteria and the archaea, a detailed discussion of which is beyond the scope of this book. The eukaryotes can be divided into kingdoms, as shown in the diagram.

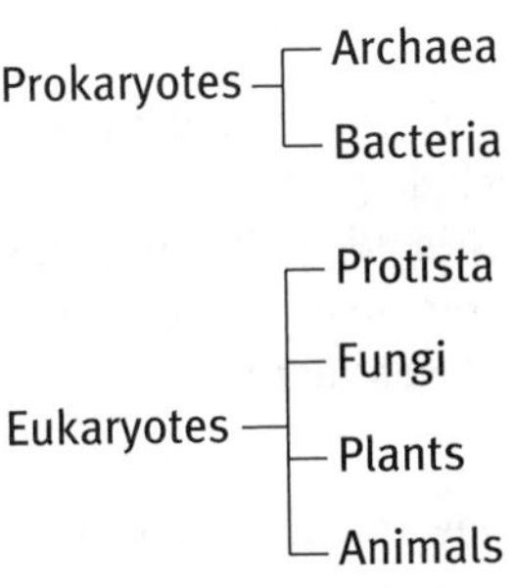

Classification of humans within the animal kingdom

Phylum	Chordata
Class	Mammalia
Order	Primates
Family	Hominidae
Genus	Homo
Species	*Homo sapiens*

Humans, of course, belong to the animal kingdom but this contains many thousands of species and so we further classify humans as shown above. Biological classification groups together organisms according to degrees of similarity and attempts to reflect evolutionary relationships. The material covered in the first part of the book is relevant to the whole of biology, as the molecules of life are, to a very large extent, shared across the kingdoms. As we go further into the book the material covered becomes more selective. Humans are mammals and, while the focus here is on human biology, most of the information covered will be true for other mammalian species.

Biology is a fascinating subject because it tells us how our bodies work and helps us to understand what can go wrong in disease. It is also a subject that has progressed in leaps and bounds as new technologies have allowed us to analyse the processes of life at ever more sophisticated levels. We hope that you will continue to be excited by this science and share your enthusiasm with others.

Philip Bradley and Jane Calvert
Newcastle, May 2006

Acknowledgements

We would like to thank all those who have helped and advised in relation to the material in this book, including both reviewers and our academic colleagues. Most particularly, we would like to thank Austin Diamond and Monica Hughes for their comments on sections of the book. We would also like to thank Julie Alexander for her patient help and support.

Most of all we are grateful to all our many students over the years who have taught us far more than we taught them.

1 Water and life

Basic concepts:
Water makes up approximately 60% of the human body. Its molecular structure allows it to act as a solvent for many of the other key molecules which enable cells to function and life to be maintained. An understanding of the distribution of water in the body, the composition of the various fluid compartments and the control of the movement of water between compartments is crucial to understanding many basic life processes.

1.1 The properties of water

Water is essential for life. The cells of living organisms are composed of around 70% water and many of the reactions essential to life occur in an aqueous environment. The chemical properties of water make it a particularly suitable medium for supporting life. Water is a *polar* molecule, which is to say it has an uneven distribution of charge (Fig. 1.1).

Figure 1.1 A water molecule showing distribution of charge

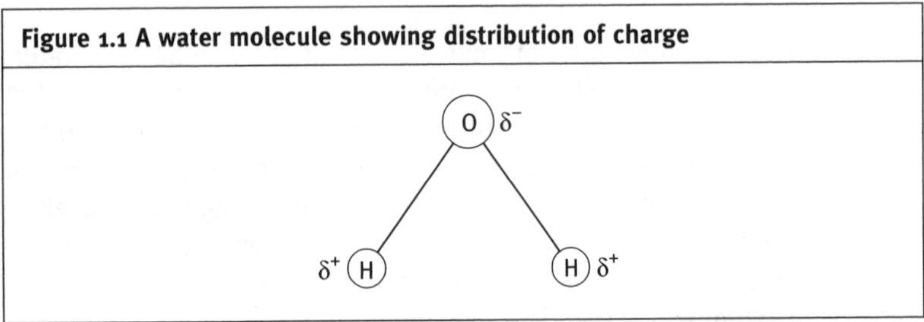

This means that it is able to interact with other polar and charged groups. Molecules or groups that interact with water are described as *hydrophilic*, whereas non-polar groups are described as *hydrophobic*.

Virtually all the molecules of life are based around the element carbon. These include:

- sugars and polysaccharides
- amino acids and proteins
- nucleotides and nucleic acids
- lipids

Polysaccharides, proteins and nucleic acids are very large molecules, termed *macromolecules*, and are polymers of sugars, amino acids and nucleotides respectively. Biological macromolecules contain both hydrophilic groups (such as OH, NH_2 and COOH) and hydrophobic groups (for example hydrocarbons) and the relative amounts of these influence solubility (for further information see Section 3.7 in *Catch Up Chemistry*).

Interactions with water play an important part in determining the structure of these biological molecules. Generally speaking, hydrophilic groups tend to be exposed on the surface of a molecule or structure from where they are able to interact with water molecules. In contrast, hydrophobic groups tend to orientate themselves towards the inside of the molecule or structure where they interact with each other forming hydrophobic bonds. Interactions between hydrophobic chains of fatty acids allow the formation of cell membranes (see Chapter 4). Other molecules that are associated with membranes, such as proteins, often have hydrophobic regions which are inserted into the membrane to form an anchor.

Water is also very important as a medium of transport and forms the basis of blood. Gases dissolve in water, and this is important in allowing oxygen to be taken to cells and carbon dioxide to be removed.

1.2 Water in the human body

Approximately 60% of the weight of the human body is water – thus a 60 kg person will contain approximately 36 litres of water. Within the body the water is distributed between three main compartments. The bulk of body water (65%) is contained in the cytoplasm of cells and is known as *intracellular fluid.* Most of the remaining extracellular fluid is divided into the *interstitial fluid* (25%) which bathes the cells and the *plasma* (7.5%) which is contained within the blood vessels of the circulatory system. The remaining 2.5% of fluid is known as *transcellular fluid* and includes, for example, the water in the bladder and the contents of the gastrointestinal tract.

Intracellular fluid is separated from interstitial fluid by the plasma membrane of the cell (see Chapter 6). The ionic composition of these two compartments is dramatically different. The extracellular fluid has a similar composition to seawater and contains approximately 140 mmol Na^+ and 110 mmol Cl^-. Extracellular fluid also contains significant levels of bicarbonate ions. By contrast, intracellular fluid contains high levels of K^+ (approximately 160 mmol compared with 4 mmol in extracellular fluid) and low levels of Na^+ (10 mmol). The intracellular negative charge is provided not by Cl^- but by proteins, bicarbonate and phosphate ions.

The concentration gradients of Na^+ and K^+ across cell membranes form the basis of many physiological processes (see Chapters 8 and 20). Ions contained within body fluids are known as *electrolytes.*

A general rule which applies when considering the ionic balance of any one compartment is that it should contain equivalent positive and negative charges (determined by the relative numbers of cations and anions). Each compartment is said to be *electroneutral.* This has significance when considering the movement of ions across membranes because, wherever possible, the body strives to ensure that movement of positively charged cations is accompanied by an equivalent negative charge in anions. When this does not happen electrical potentials are generated across membranes and this forms the basis of the function of excitable tissues (see Chapter 14).

The two components of extracellular fluid are separated from each other by the capillary wall. In most capillaries this is freely permeable to the movement of ions and small organic molecules but does not allow the passage of proteins. Thus under normal circumstances interstitial fluid contains no protein whereas both plasma and intracellular fluid are protein rich.

Clinical example: Dehydration

On a hot day a runner may lose up to 2 litres per hour in sweat. For a normal individual a loss of water constituting more than 3% of body weight (about 2 litres) may lead to the early stages of clinical dehydration and cause feelings of light-headedness and disorientation. Further water loss will affect the ability of cells to function and may lead to death due to shock caused by low blood volume. This is why it is particularly important for fun runners to ensure that they take on plenty of water when competing in marathons and other long-distance races.

1.3 Test yourself

1. Where in a biological macromolecule would hydrophobic groups generally be found?
2. What are the three main compartments in which body water is distributed?
3. What is the main cation of: (a) extracellular fluid; (b) intracellular fluid?
4. Organic molecules are based around which element?
5. Which key component of plasma does not normally pass across the capillary wall?

2 Proteins

Basic concepts:
Proteins are macromolecules assembled as a sequence of amino acids. There are twenty different amino acids, giving rise to a wide range of possible proteins. According to the particular amino acid sequence, proteins will adopt different three-dimensional structures. Proteins are present in all cells and can perform many roles, including as structural elements and as enzymes. It is important to understand how the amino acid sequence of proteins can determine the properties of different proteins, and also how these properties can be altered by external factors such as the binding of another molecule or the addition of a phosphate group.

2.1 Introduction

Proteins are a highly diverse and important group of molecules, central to life. Proteins are biological macromolecules and are polymers of *amino acids.*

Figure 2.1 General structure of an amino acid

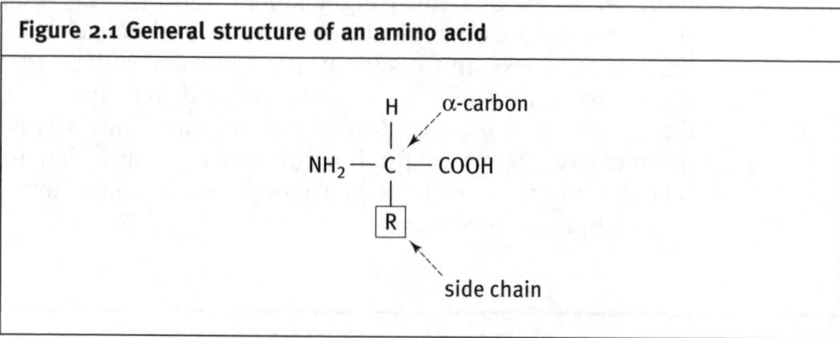

Amino acids contain an amino group and a carboxylic acid group (Fig. 2.1), both attached to an alpha carbon atom. Also attached to the alpha carbon is a side chain, which is different in different amino acids (Fig. 2.2). Side chains have their properties too – some carry a positive or negative charge, some are polar and others are hydrophobic (they prefer not to be in contact with water). The different properties of the side chains are important in determining the structure and function of proteins. There are twenty different amino acids that are found in proteins. Because these can occur in different orders and combinations, this leads to a very large number of possible protein structures.

Figure 2.2 Examples of different kinds of amino acids

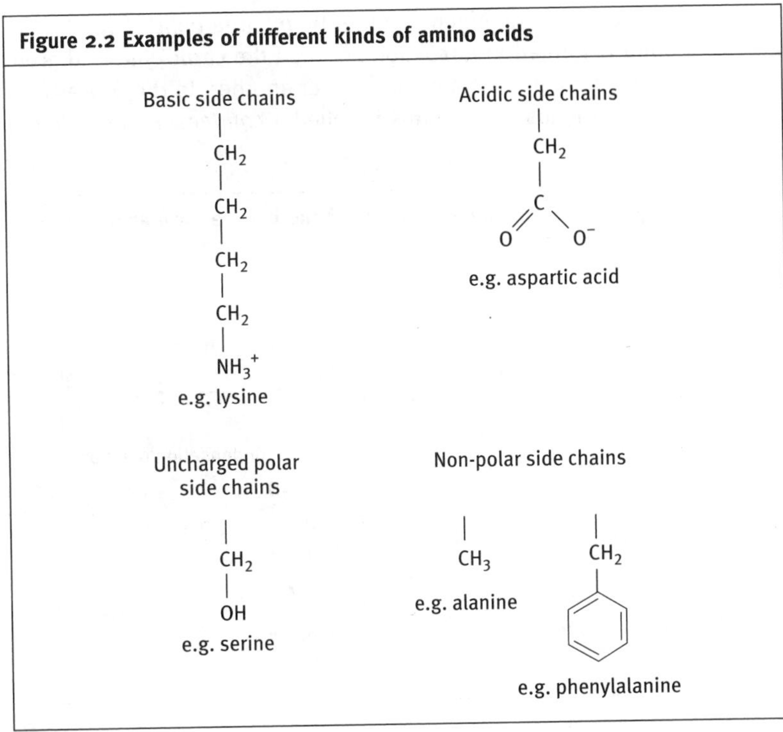

Amino acids can exist as different isomers, depending upon the arrangement of the groups attached to the alpha carbon. Isomers are defined as 'two or more different compounds with the same chemical formula but different structures and characteristics'. The alpha carbon in an amino acid participates in four covalent bonds forming a tetrahedral arrangement, and mirror image forms can exist, called *enantiomers*. The different enantiomers are described by the letters D and L. All amino acids occurring in proteins are L-isomers.

Figure 2.3 D and L forms of amino acids

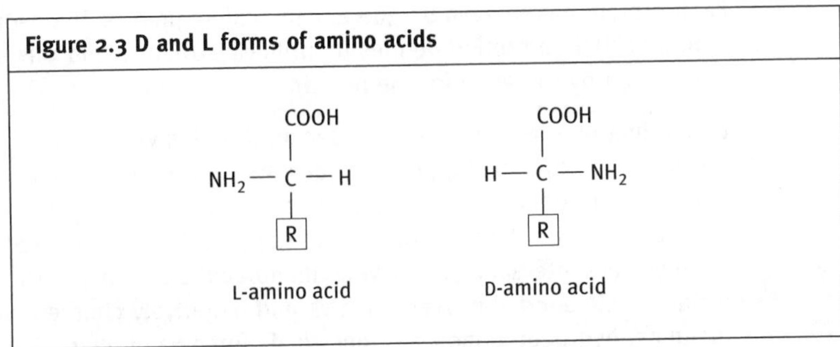

Amino acids are joined together by *peptide bonds* (Fig. 2.4). A peptide bond is formed in a reaction between the carboxylic acid group of one amino acid and the amino group of another. In the process, a molecule of water is lost and so this is called a *condensation reaction.*

Figure 2.4 Formation of a peptide bond between two amino acids

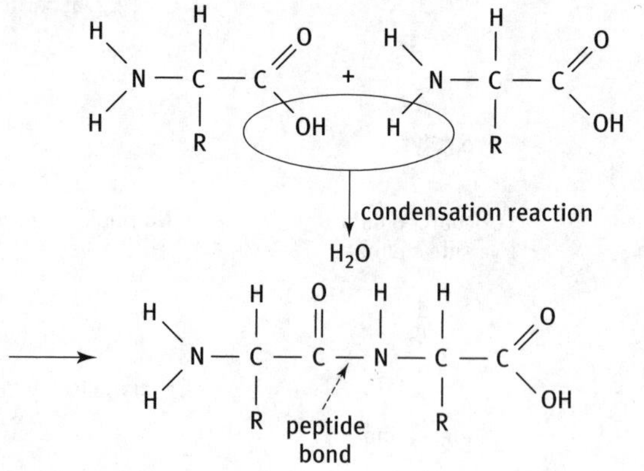

The amino acids at each end of a protein molecule participate in only one peptide bond, hence they have either a free NH_2 group or a free COOH group. The end of the polypeptide chain with a free amino group is called the *N-terminus*, and the end with the free carboxyl group is called the *C-terminus.*

2.2 Primary structure

Each protein has its own unique amino acid sequence. The sequence of amino acids in a protein defines its primary structure and this sequence is encoded by the gene for the protein.

Depending on the amino acid sequence, proteins will, under physiological conditions, preferentially adopt a particular folded structure, or *conformation* (see the sections on secondary and tertiary structure below). The conformation of the protein is maintained by non-covalent interactions involving amino acid side chains. These include ionic bonds between positive and negatively charged amino acid residues, hydrogen bonds, van der Waals forces and hydrophobic interactions (see *Catch Up Chemistry* for further information on these). Hydrophobic interactions are particularly important as they bring together non-polar amino acid side chains and ensure that these are not exposed to water. Following the initial synthesis of proteins within cells (see Chapter 5) their folding into secondary and tertiary structures is aided by the presence of other proteins called *molecular chaperones.*

2.3 Secondary structures

Some common patterns of folding occur. Two classical protein folds that recur in many different proteins are the *alpha helix* and the *beta sheet* (Fig. 2.5). Both of these structures depend on interactions between groups in the polypeptide backbone and they can be found in a wide range of protein molecules.

Alpha helix

This is a structure in which the polypeptide chain twists on itself in a highly regular manner. The structure is stabilised by hydrogen bonding between the C=O group of one amino acid residue and the NH group of the residue four amino acids further along the primary sequence. In this way every C=O and NH, as well as being involved in the covalent peptide bond, also participates in hydrogen bonding. This rule confers particular dimensions on the alpha helix – each turn of the helix represents 3.6 amino acid residues. Alpha helices normally assume a right-handed or clockwise twist as this is energetically more favourable. The amino acid side chains are exposed on the outside of the helix. Some amino acids, in particular proline, tend to disrupt an alpha helical structure and are known as *helix breakers.*

Two or more alpha helices can intertwine to form a *superhelix.* Such superhelices are found in proteins such as keratin, the major constituent of hair.

Beta sheet

The beta sheet, like the alpha helix, is a structure which is maintained by hydrogen bonding between C=O and NH groups, but in this case these interactions occur between adjacent strands. The amino acid side chains protrude alternately above and below the plane of the sheet. Where the adjacent strands run in a similar orientation this is said to be a *parallel* beta sheet; where the adjacent strands lie in opposing orientations the sheet is said to be *anti-parallel.* The beta-sheet structure is found in many important proteins, including antibodies.

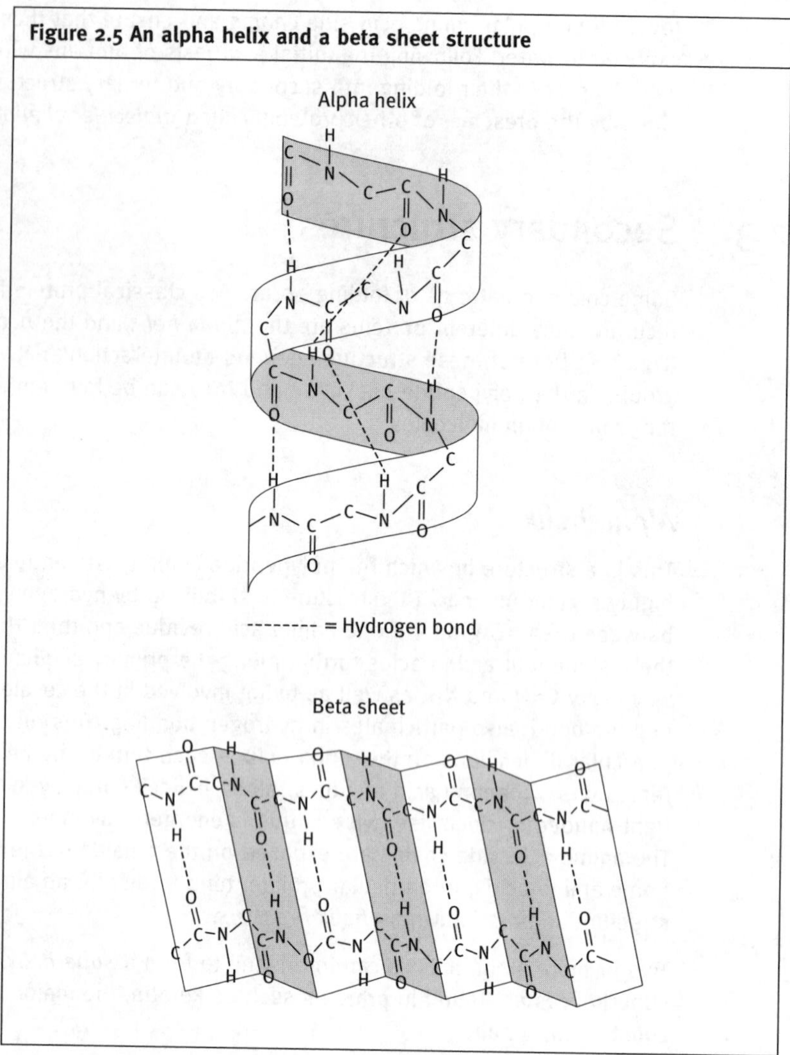

Figure 2.5 An alpha helix and a beta sheet structure

Collagen triple helix

A third type of structure is found in collagen, an important structural protein of connective tissues. Collagen has a triple-helical structure, in which three amino acid chains are wound around each other. This structure is allowed because of the particular primary sequence of collagen and related proteins – every third amino acid residue is a glycine. Because glycine (Gly) has the smallest side chain (H) this allows the chains to interact closely to form the triple helix. Collagen-like proteins are also rich in the amino acid proline and show a repeat sequence of Gly–X–Y. Although X and Y can be any amino acid, these positions are most commonly taken by proline and hydroxyproline.

2.4 Tertiary structure

In aqueous solution, all proteins fold in such a way as to internalise those amino acid side chains that are hydrophobic and to ensure that those that are exposed to the solvent are charged or polar residues. Many alpha helices and beta-sheet structures are found to be *amphipathic* – this means that the hydrophobic side chains tend to occur on one side of the helix or beta sheet and the polar and charged residues on the other. In this way the protein can fold to ensure that hydrophobic surfaces interact with each other, and not with solvent.

Clinical example: BSE and CJD

It is crucial to cellular function that proteins fold into their correct conformation. Some diseases are associated with the presence of abnormally folded proteins. The prion disease bovine spongiform encephalopathy (BSE) and its human equivalent Creutzfeldt–Jakob disease (CJD) are caused by the prion protein PrP, which differs from a normal cellular protein by only a few amino acids. The prion protein, in which α-helices are replaced with β-sheets, assumes an abnormal conformation which is resistant to digestion by proteases (the enzymes involved in the breakdown of proteins). Moreover, in the presence of the prion protein, the normal cellular form of the protein is induced to adopt a similar conformation. The presence of these abnormally folded proteins is associated with degeneration of the brain, dementia and death.

2.5 Quaternary structure

A protein can comprise a single polypeptide chain or multiple polypeptides which are associated by covalent and/or non-covalent bonds. A common type of covalent bond is the *disulphide bridge*, which can occur between cysteine residues, either on the same or on different polypeptide chains. A protein which is made up of two chains is called a *dimer*; proteins comprising three, four and five chains are similarly called *trimers*, *tetramers* and *pentamers* respectively. Where a dimer consists of two identical polypeptide chains this is said to be a *homodimer*. Where the chains are different this is called a *heterodimer*. Haemoglobin, the main protein found in red blood cells, is an example of a tetrameric protein, comprising two identical alpha chains and two identical beta chains.

2.6 Domains

Proteins tend to be folded up into subunits called *domains*. Within a protein a single polypeptide chain can contribute one or more domains. The domains can be structurally similar or quite different from each other. Often a particular property of a protein (for example the ability to bind to another molecule or ligand) can be attributed to one domain of the protein. At the genetic level each domain is likely to be encoded by a separate exon (see Chapter 5).

Clinical example: Domains in antibodies

Antibodies provide an example of proteins with multiple domains. Antibodies are an important molecular defence against infection. They are multifunctional molecules – firstly, they are able to recognise and bind to other molecules (antigens) with a high degree of specificity; secondly, they can interact with cells and other molecules of the immune system to allow the elimination of an infection. Antibodies are made up of two heavy chains and two light chains. Within each polypeptide chain there are multiple domains, and each domain has a structure comprising two beta sheets. The N-terminal domains are responsible for binding to the antigen, and the domains towards the C-terminus of the heavy chains interact with cellular receptors on phagocytes (see Chapter 23).

2.7 Functions of proteins

Proteins carry out many functions within the body. They may form the structural elements of tissues, important for mechanical support. They can act as transporters to carry other molecules from one location to another. They can also act as hormones, which are chemical messengers that carry information from one part of the body to another. Proteins are present in cell membranes, where they may act as receptors to alert the cell to the presence of molecules in its environment. Proteins also play an important role in defence against infection.

One of the key roles of proteins is to act as *enzymes*. Enzymes are found in both intracellular and extracellular locations within the body and catalyse the various chemical reactions on which life depends. The table shows examples of the functions carried out by proteins.

Function	Example	Role
Structural	Collagen	Provides tensile strength in connective tissues
Transporter	Haemoglobin	Transports oxygen from the lungs to the various tissues of the body
Hormone	Insulin	Controls the concentration of blood glucose
Receptor	Acetylcholine receptor	Present on muscle cells where it binds acetylcholine leading to muscle contraction
Defence against infection	Antibodies	Bind to an infectious agent and allow its destruction by cells and molecules of the immune system
Enzyme	Trypsin	Breaks down food proteins in the intestine so that they can be absorbed into the blood
Movement	Myosin	Present in muscle where it forms part of the contractile mechanism

2.8 Conformational change

In order to be fully functional, proteins must be folded correctly – they must be in their correct *conformation.* Protein conformation can be disrupted under a number of conditions, including extremes of pH, high temperature and in the presence of detergents. The function of proteins can also be regulated under physiological conditions by altering their conformation. Conformational change can be induced in a number of ways, for example by the binding of a ligand or by covalent modification of protein molecules. *Phosphorylation* is a form of covalent modification of proteins that is commonly used to regulate the activity of enzymes. The process of phosphorylation is itself mediated by enzymes called *protein kinases.* Another way in which protein conformation can be regulated is by *cleavage.* Some enzymes are synthesised as inactive precursors which only become active when they are cleaved. One such enzyme is trypsin which is found in the intestine where it plays a role in breaking down the proteins in food. The enzyme is initially synthesised in the pancreas as an inactive precursor called trypsinogen, which only becomes active when a peptide bond is cleaved. The active enzyme can then catalyse the breakdown of further trypsinogen molecules generating more active enzyme.

2.9 Test yourself

1. What are the subunits called that polymerise to form proteins?
2. Which two functional groups participate in the formation of a peptide bond?
3. What is the primary structure of a protein?
4. In an alpha helix

   (a) where are the amino acid side chains located?

   (b) How many amino acid residues are there per turn of the helix?

   (c) What sorts of bonds occur between every fourth peptide bond?
5. What is a protein called that comprises: (a) two identical polypeptide chains; (b) two non-identical polypeptide chains?

3 Carbohydrates

Basic concepts:
Carbohydrates range in structure from simple sugar molecules, such as glucose, to polymers of sugar molecules, such as those that constitute the rigid wall around all plant cells. Carbohydrates are a major source of energy for the body and so knowledge of their basic structure is very important in understanding energy metabolism. Carbohydrates can be attached to other molecules, including proteins and lipids, and are present on the surface of cells where they can act as recognition molecules.

3.1 Introduction

Carbohydrates, as the name suggests, are molecules made up from carbon and water (hydrogen and oxygen). Carbohydrates serve a number of important functions. They are a major source of energy for life. They are also important structural molecules in many organisms. Additionally, carbohydrates can be attached to other biological molecules, such as proteins, and in doing so modify the properties of the molecule. Probably the most abundant organic molecule on earth is the carbohydrate *cellulose*, a major constituent of plant cell walls. Sugars are small carbohydrates and these can be joined together to form *oligosaccharides* and *polysaccharides*. Different sugars can be joined together in many different ways to give rise to a very diverse array of structures.

3.2 Monosaccharides

The simplest sugars are monosaccharides, which have the general formula $(CH_2O)_n$. According to the value of *n*, sugars are described as *trioses*, *pentoses*, *hexoses* etc. An important sugar in energy metabolism is *glucose*. This is a hexose with the formula $C_6H_{12}O_6$. Monosaccharides such as glucose can occur as mirror image D and L isomers (*enantiomers*), similar to amino acids (see Fig. 3.1). D or L refers specifically to the arrangement of groups at the asymmetric carbon furthest from the aldehyde group (carbon 5 in glucose).

Figure 3.1 D and L glucose

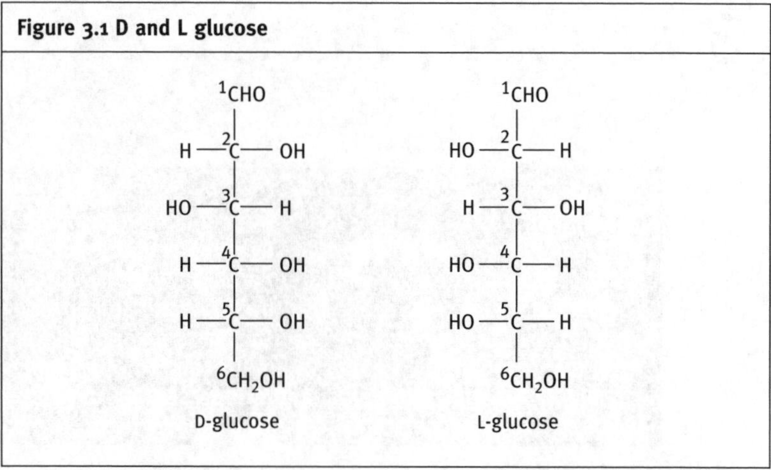

Other sugars, including mannose and galactose, share the same chemical formula as glucose but are not mirror image forms (Fig. 3.2). Glucose frequently assumes a ring form, as shown in Fig. 3.2, by the reaction of the aldehyde group on carbon 1 with the hydroxyl group on carbon 5. The glucose ring can also be found in either alpha or beta form, according to the arrangement of the OH group attached to the first carbon atom. In alpha glucose the OH group is below the plane of the ring, whereas in beta glucose it is above the plane of the ring. This minor structural difference has important consequences for the properties of the molecules and their polymers. Starch is a polymer of alpha glucose molecules and cellulose is a polymer of beta glucose. Note that humans have digestive enzymes that can degrade starch but are unable to digest cellulose.

Sugars may also contain additional groups. Glucosamine is a derivative of glucose in which an amino (NH_2) group replaces one of the OH groups. The amino group frequently has an additional acetyl group ($COCH_3$) attached to it, forming N-acetyl glucosamine. Another important glucose derivative is glucuronic acid, which contains a carboxylic acid group. Sugar derivatives such as N-acetyl-glucosamine and glucuronic acid are present in glycosaminoglycans, which are an important component of the extracellular matrix (see Chapter 13).

3.3 Glycosidic bond

When sugars join together to form oligosaccharides or polysaccharides this creates a *glycosidic bond.* A glycosidic bond is formed by the

Figure 3.2 (a) Glucose and mannose in open chain form; (b) Glucose in ring form

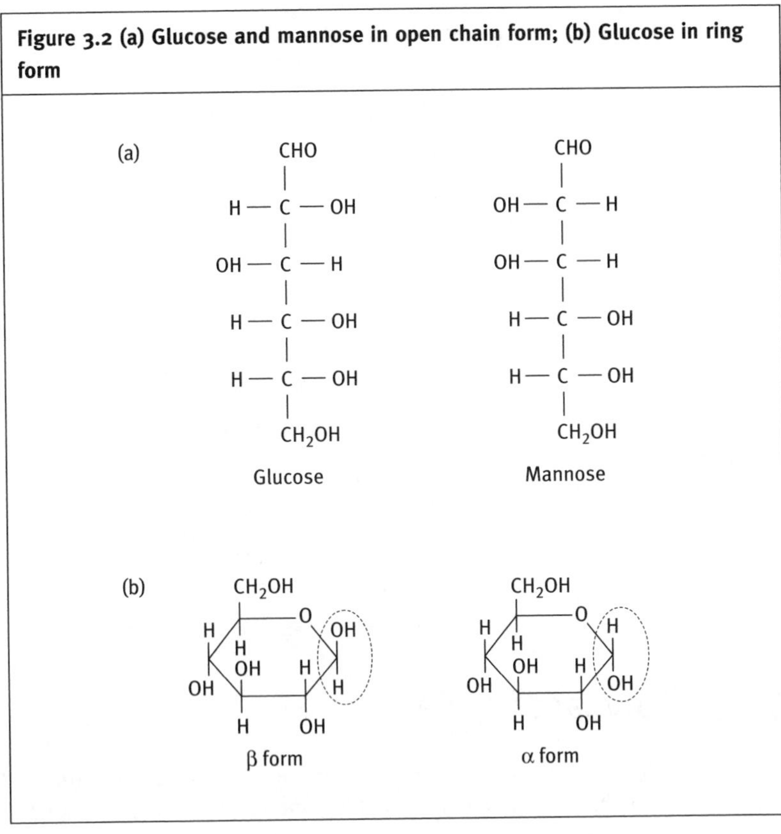

interaction of an OH group on one sugar with an OH group on another sugar. This involves the elimination of water and is therefore a condensation reaction. Enzymes that can break down glycosidic bonds are termed *glycosidases*. In comparison to the peptide bond, which is always formed between the NH_2 and COOH groups on the alpha carbon of an amino acid, there is much greater diversity in the way in which glycosidic bonds can form. Different carbon atoms can be involved, and an individual sugar can participate in multiple glycosidic bonds to give rise to branched structures. The bond is described according to the particular carbon atoms involved and whether the hydroxyl groups are in the alpha or beta position. In starch the glucose units are joined by α-1,4 linkages (Fig. 3.3), which means that the reaction involves the OH group on carbon 1 of one sugar reacting with the OH group on carbon 4 of the next. Glycosidases are highly specific for a particular type of glycosidic linkage.

Figure 3.3 Formation of an α-1,4 glycosidic bond

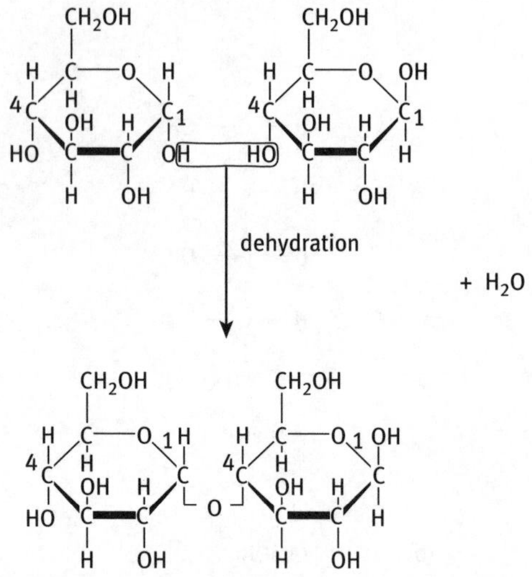

dehydration

+ H_2O

3.4 Polysaccharides

Polysaccharides consist of many monosaccharide subunits covalently linked together. They may involve one or more different sugar units and can be either linear or branched in structure. Starch is a storage polysaccharide of plants that plays an important role in the human diet. Starch has two components – *amylose*, which is a linear polymer of glucose joined by α-1,4 linkages, and *amylopectin*, which is similar but also contains branches due to additional α-1,6 linkages occurring at points along the chain (Fig. 3.4). The enzyme that digests starch is called *amylase* and is present in saliva and in the small intestine.

Figure 3.4 α-1,4 and α-1,6 linkages in glycogen

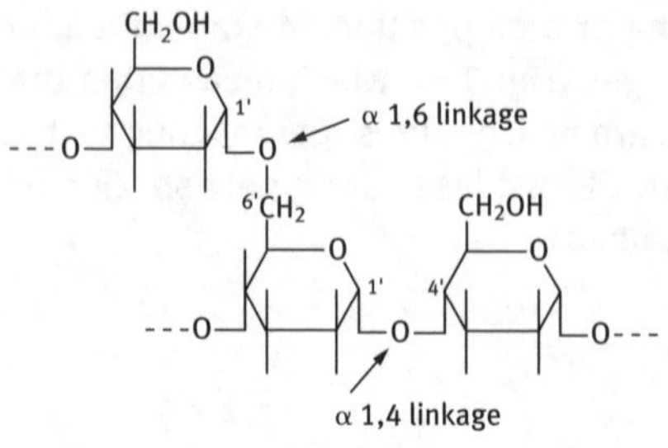

α 1,6 linkage

α 1,4 linkage

Mammals store carbohydrates in the form of *glycogen* and this can be found in the liver and in muscle. Glycogen is similar in structure to amylopectin but is more branched.

Cellulose is an unbranched polymer of glucose linked by β-1,4 glycosidic bonds and is an important structural component of the cell walls of plants. Humans do not have an enzyme that can hydrolyse this type of linkage and cannot use cellulose as an energy source. However, cellulose does play an important role in digestive function as a source of roughage, or insoluble matter, helping to maintain the consistency of the faeces and to stimulate mucus secretion as waste matter passes through the large intestine. Ruminants, such as cattle and sheep, are able to make use of cellulose because they harbour bacteria that secrete the enzyme cellulase in a part of their digestive tract known as the rumen.

3.5 Glycoconjugates

Other molecules may also have sugars attached to them – a process known as *glycosylation*. Many proteins are glycosylated, particularly those that are found in cell membranes or that are secreted from the cell. Glycoproteins contain one or more oligosaccharides attached to the protein backbone, either at the hydroxyl group of serine or threonine residues (when they are described as O-linked) or via the amide group of asparagine (N-linked). The relative amount of carbohydrate in glycoproteins varies enormously, from merely 1 or 2% to over 70%. Mucin is a very large and heavily glycosylated molecule that forms a major constituent of mucus and contributes to its gel-like properties. Mucin contains a very high proportion of carbohydrate in the form of O-linked oligosaccharides. Lipids may also be glycosylated.

Clinical example: ABO blood groups

The human ABO blood group antigens are oligosaccharides present on glycoproteins and glycolipids on red blood cells. Blood group specificity is defined by the oligosaccharide component. Enzymes called glycosyl transferases are required for the synthesis of oligosaccharides. These are highly specific and whether an individual produces the blood group oligosaccharide O, A or B will depend on which glycosyl transferases they have inherited from their parents. If an individual with type O blood were to receive a transfusion of type A blood then their immune system would react against the A oligosaccharide, destroying the transfused cells and leading to serious consequences.

3.6 Functions of carbohydrates

Carbohydrates serve a range of biological functions, several of which have been referred to above. They are an important part of the diet, providing a ready source of energy. In the form of polysaccharides such as starch and glycogen they play a role in storage. They are also important structural components of cells (e.g. cellulose) and influence the mechanical properties of tissues or secretions (e.g. glycosaminoglycans, mucins).

Sugars are also important *recognition molecules*. Within the cell, the carbohydrates on glycoproteins play a role in ensuring that these molecules are transferred to the appropriate cellular compartments. Carbohydrates present on cell surface glycoproteins and glycolipids also interact with proteins called *lectins*. This type of interaction plays an important role in the trafficking of white blood cells to sites of infection (see Chapter 23). Oligosaccharides also play an important role in the recognition of an egg by a sperm (see Chapter 10).

3.7 Test yourself

1. What is the chemical formula for glucose?
2. What is the name of the bond that joins sugars together in a polysaccharide?
3. What is the enzyme that degrades starch called?
4. What are the enzymes that synthesise oligosaccharides?
5. What polymer is used to store carbohydrates in animals?

4 Lipids

Basic concepts:
Lipids are commonly called fats or waxes. They are generally insoluble in water. They include phospholipids, which have a hydrophilic head region and a hydrophobic tail. Lipids are an important constituent of cell membranes where their properties determine membrane permeability. Lipids can also act as long-term energy stores in the body.

4.1 Introduction

Lipids are molecules of varying sizes and structures that tend to be insoluble in water. Lipids are a major constituent of biological membranes, where they form *bilayers*. Lipids are also an important source of energy yielding approximately twice the energy value of carbohydrates on a weight-for-weight basis.

4.2 Fatty acids

Fatty acids consist of a long hydrocarbon chain with a terminal carboxyl (COOH) group. Fatty acids vary in the length of their carbon chain and in the degree of saturation – that is, how many hydrogen atoms are bound to the various carbon atoms (see Fig. 4.1 for examples). Fatty acids with the maximum number of hydrogen atoms are described as *saturated*. *Unsaturated* fatty acids have fewer hydrogen atoms attached to the carbon chain and contain double bonds between two or more of the

Figure 4.1 (a) Stearic acid – a saturated fatty acid found in animal fat; (b) Oleic acid – an unsaturated fatty acid found in olive oil

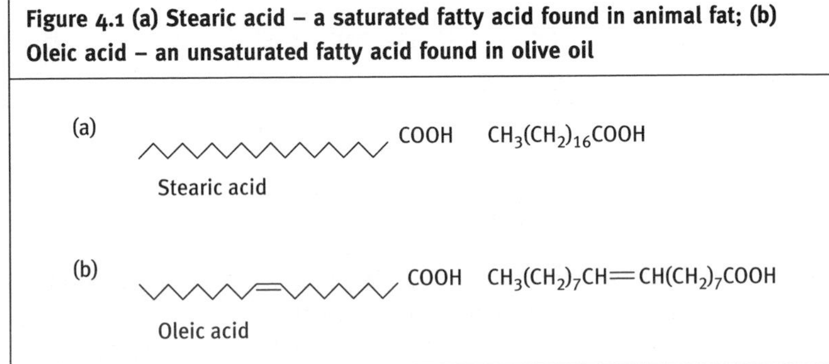

carbon atoms. Those that have multiple double bonds are *polyunsaturated.* The degree of saturation affects the physical properties, with unsaturated fats tending to have a lower melting point than saturated fats. Polyunsaturated fats in the diet are believed to be beneficial because they help to lower the levels of cholesterol in the blood and reduce the risk of coronary heart disease.

4.3 Triglycerides and phospholipids

The carboxylic acid group on a fatty acid can react with a hydroxyl group on the three-carbon alcohol glycerol to form a *glyceride.* Triglycerides (Fig. 4.2) consist of three fatty acid molecules attached to one glycerol molecule. This is the main form in which fat is stored in the body and provides a highly efficient energy reserve.

In *phosphoglycerides*, which are a major class of lipid in cell membranes, two of the OH groups of glycerol have a fatty acid molecule attached, while the other OH is attached via a phosphate group to a polar head group. The head group varies and gives phospholipids their names. Commonly occurring phospholipids include phosphatidyl serine, phosphatidyl choline and phosphatidyl inositol, in which the head groups are serine, choline and inositol respectively. Another type of phospholipid is based not on glycerol but on the molecule sphingosine.

Figure 4.2 General structure of a triglyceride (a) and a phospholipid (b)

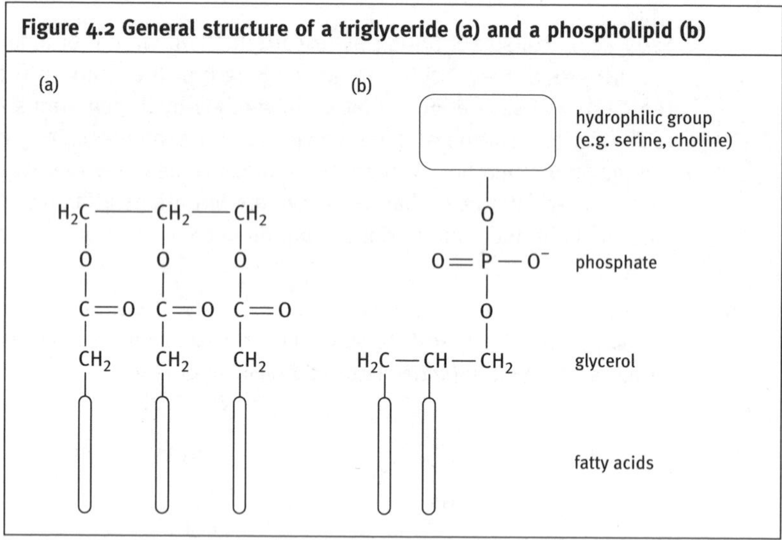

The fatty acid chains of phospholipids (Fig. 4.2) are hydrophobic, whereas the polar head groups are hydrophilic, making phospholipids *amphipathic.* In an aqueous environment these molecules will tend to

Figure 4.3 A lipid bilayer

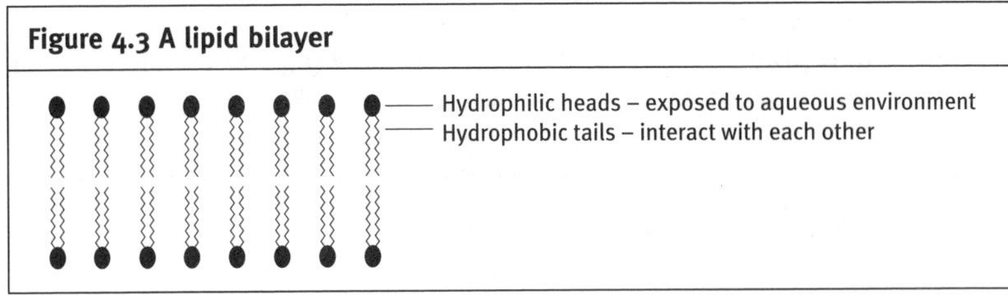

arrange themselves so that the hydrophobic fatty acid chains interact with each other and the head groups are exposed to water. These interactions give rise to the phospholipid bilayer that is characteristic of cell membranes (see Chapter 6 and Fig. 4.3).

4.4 Cholesterol

Another important lipid found in mammalian cell membranes is *cholesterol*. Cholesterol has a very different structure from the fatty acid type lipids described above. Instead of having long hydrocarbon chains, cholesterol is based on a structure containing four linked hydrocarbon rings with a short hydrocarbon chain. Cholesterol is found in varying proportions in different membranes and plays an important role in determining membrane properties such as fluidity and permeability.

Clinical example: Familial hypercholesterolaemia

Cholesterol absorbed from the diet circulates in the bloodstream bound to carrier proteins in a complex known as a low density lipoprotein (LDL). It is taken into cells after binding to a specific LDL receptor on the cell surface. An inherited defect in the LDL receptor can lead to failure to internalise cholesterol with a consequent raised level of cholesterol in the blood. Because of their raised blood cholesterol levels, affected individuals will have a greater likelihood of developing heart attacks and strokes. In individuals with two copies of the defective LDL receptor gene, circulating cholesterol is so high that they are likely to die in their early thirties.

4.5 Functions of lipids

The discussion so far has focused on lipids as a primary constituent of cell membranes. Lipids have other important biological functions. Triglycerides can be stored in *adipose tissue* (fat) to provide a reserve of energy. Fatty tissues also play a role in thermal insulation. Lipids are an important component of the myelin sheath, which surrounds neurons and provides electrical insulation, allowing the effective transmission of nerve impulses.

4.6 Test yourself

1. What is the term used to describe fatty acids in which the maximum number of hydrogen atoms are attached to the carbon chain?
2. What is the name of the three-carbon alcohol that occurs in phospholipids such as phosphatidyl choline?
3. How many fatty acid chains are present in a molecule such as phosphatidyl choline?
4. What is the term used to describe molecules, such as phospholipids, which have a hydrophilic and a hydrophobic region?
5. What is the name of the tissue used to store fat in mammals?

5 Nucleic acids and genes

Basic concepts:
DNA is the basic molecule of life. A DNA molecule consists of two strands of nucleic acid which are capable of replication. A chromosome consists of a long strand of DNA with associated proteins and each human cell contains 46 chromosomes. Each chromosome carries a series of genes, which are sections of DNA in which the nucleic acid sequence codes for the production of a specific protein. The genetic code is based on triplets of nucleotides, which specify which amino acid goes into a particular position in a protein molecule. The structure and function of an individual cell is determined by the genes which are active in that cell. An understanding of how genes are turned on and off in specific cells is necessary to appreciate the diversity found within individuals.

5.1 Introduction

Nucleic acids hold the key to life. They are the stuff that *genes* are made of – the macromolecules containing the information that defines species and individuals. There are two main nucleic acids: *DNA*, or deoxyribonucleic acid, and *RNA*, or ribonucleic acid. In most organisms, including humans, genetic information is encoded within DNA but is then copied (transcribed) into RNA before being translated into proteins.

Nucleic acids (see Fig. 5.1), like proteins and polysaccharides, are polymers made up of a large number of subunits joined together. The subunits of DNA and RNA are called *nucleotides*. DNA and RNA are synthesised by copying from existing DNA or RNA templates using enzymes called *polymerases*. In order for a cell to divide it must first undergo DNA synthesis to produce a new copy of the genetic material. In order to produce proteins, cells must first transcribe the genes in the DNA into RNA.

5.2 DNA and RNA

The nucleotides that make up both DNA and RNA comprise three parts – a five-carbon sugar (*r*ibose in RNA, *d*eoxyribose in DNA), a phosphate group and an organic base. Four different bases are used in DNA – adenine, thymine, guanine and cytosine. In RNA, uracil replaces thymine. Adenine and guanine contain two joined rings and are called *purines*; cytosine, thymine and uracil contain a single ring and are called *pyrimidines*. The backbone of the nucleic acid is formed by the sugars joined through phosphate groups. The OH on the third carbon of one sugar is covalently bound to the phosphate group, which in turn forms a bond with the OH on the fifth carbon of the next sugar. This is called a 3′–5′ *phosphodiester* bond. At one end of the chain a phosphate group is attached to the 5′-carbon of the sugar, and this is called the 5′-end – the other end has a free 3′-OH group and this is called the 3′-end.

Figure 5.1 (a) Basic structure of a nucleic acid; (b) Structure of the five bases present in DNA and RNA

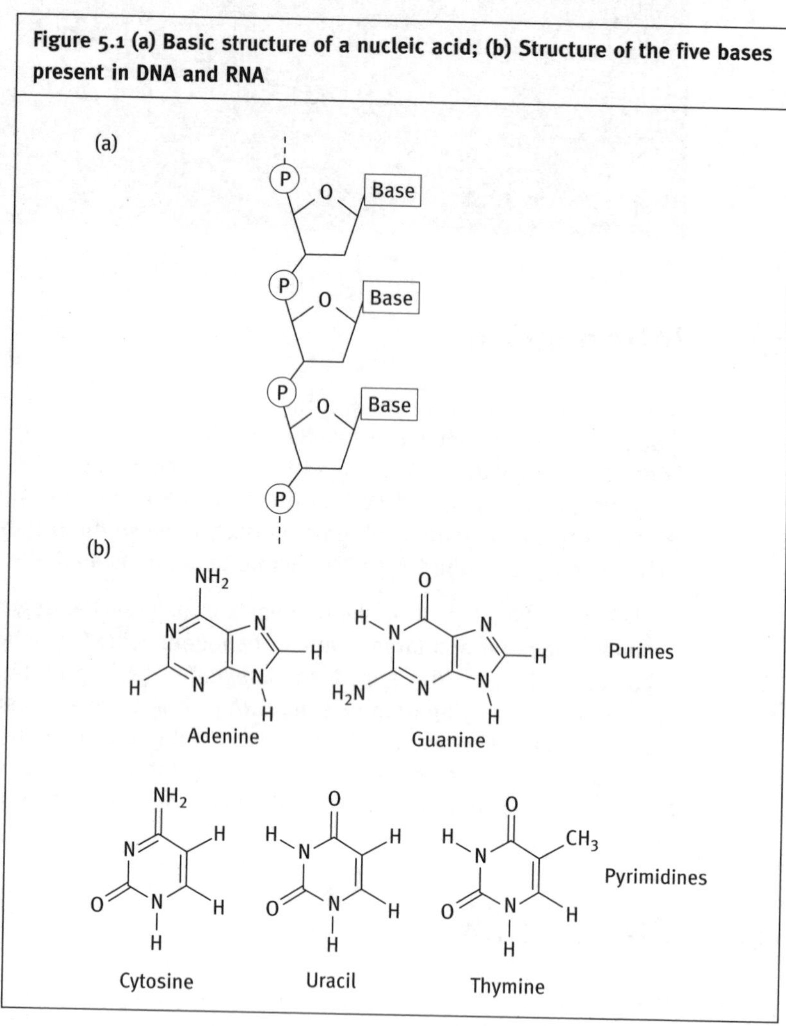

The bases protrude from the sugar–phosphate backbone of the molecule, and the key to the function of DNA and RNA is the ability of certain bases to interact non-covalently, by means of hydrogen bonds, to form *base pairs*. Adenine has the ability to pair with thymine (or uracil in RNA) and guanine has the ability to pair with cytosine. The DNA molecule is double-stranded and the two strands are complementary. Where adenine occurs on one strand of the molecule thymine is found on the other strand where it can form a base pair. Similarly guanine occurs in the opposite position to cytosine. This allows for the two strands of DNA to interact in a highly regular fashion with the sugar–phosphate backbone on the outside of the molecule and the paired bases on the inside. The double-stranded DNA forms a double helix (Fig. 5.2), with ten bases per turn of the helix. The base pairing only works if the two chains run in opposite directions – one chain lies 5′–3′ and the other lies 3′–5′.

Figure 5.2 The DNA double helix showing base pairing between strands

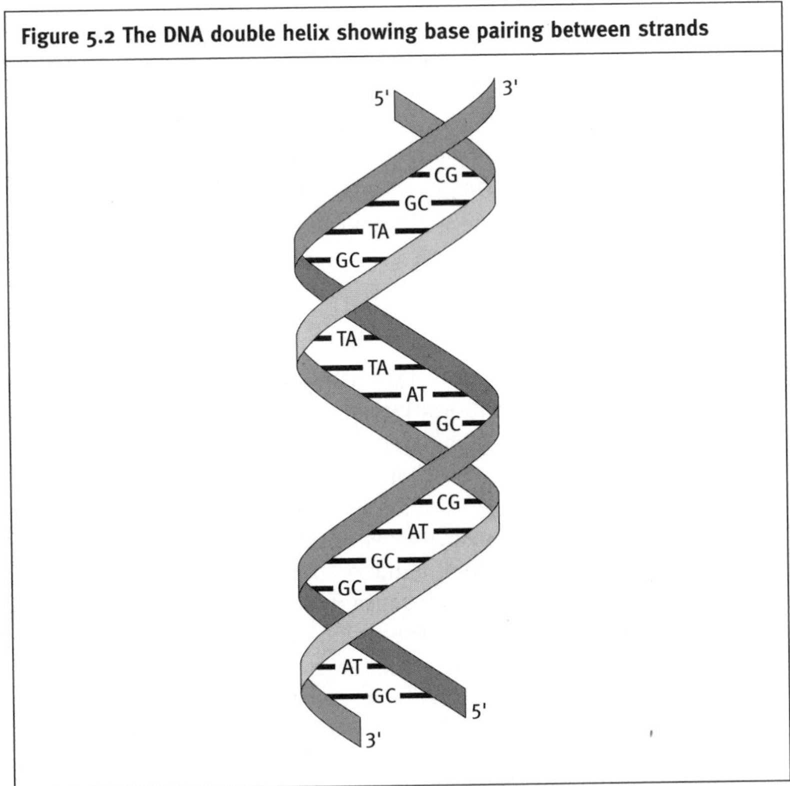

RNA is a single-stranded molecule, though it may fold back on itself in such a way as to allow internal pairing of some of the bases. There are three main types of RNA within a cell – *messenger* RNA (mRNA), *transfer* RNA (tRNA) and *ribosomal* RNA (rRNA). Each of these has a particular role in protein synthesis.

5.3 DNA synthesis

In order for cells to divide they must first reproduce their DNA. DNA is synthesised in a process called *semi-conservative replication*. The two strands of the double helix become separated from each other, and then each acts as a template for the synthesis of a new and complementary strand. When synthesis is complete, each daughter cell inherits DNA in which one strand is derived from the original DNA and the other is newly synthesised. Accuracy is crucially important in DNA synthesis and so mechanisms exist to correct any errors that might arise. Several enzymes are involved in DNA synthesis – *helicases* separate the two strands prior to replication; *polymerases* allow extension of the new strand by the addition of nucleotides. A third type of enzyme, called a DNA *ligase*, is also required. This is because DNA synthesis proceeds continuously on only one of the two strands – on the other strand DNA is made in small fragments which must subsequently be joined together. This is the function of the ligase.

5.4 RNA synthesis

Like DNA, RNA is also synthesised by polymerases in a template-dependent fashion. DNA forms the template for RNA synthesis. Synthesis of RNA is initiated at sites within the DNA known as *promoters* – these are characterised by particular sequences of bases found before the start of genes. The promoters of most genes have a characteristic sequence known as the TATA box located about 30 bases upstream of the site of initiation of RNA synthesis. The TATA box is so called because of its sequence (typically TATAAA). As with DNA, RNA synthesis also requires the action of helicases to open up the DNA so that the polymerase can gain access to the gene. In addition to the RNA polymerase, other proteins, known as *basal transcription factors*, are required for transcription to be initiated in eukaryotes. The TATA box plays a key role in allowing the assembly of the various proteins required for transcription of DNA by the RNA polymerase. RNA is then produced by the sequential addition of nucleotides complementary to those in the DNA, guided by the rules of base pairing. An important difference between DNA synthesis and RNA synthesis is that in the latter case only one of the two strands of DNA is copied.

5.5 The genetic code

It had been established by the middle of the twentieth century that the genetic information in an organism was contained in DNA. Further experiments established that genes encoded proteins, giving rise to the 'one-gene one-polypeptide hypothesis'. The discovery of the genetic code provided the vital link that connected DNA and proteins.

Central to the genetic code is the existence of *codons.* A codon is a sequence of three bases that specify a particular amino acid. Each base is part of only one codon, and so the code is said to be *non-overlapping.* A single codon can specify only one amino acid, and the same codon always encodes the same amino acid, regardless of the species, so the code is said to be *universal* (although it should be noted that there are some exceptions to this rule, for example in mitochondrial DNA). As there are four different bases this gives rise to a possible 64 ($4 \times 4 \times 4$) codons. There are only 20 amino acids and for most of these there is more than one codon, so the code is said to be *degenerate.* For example, the amino acid tyrosine is coded for by the codons UAC and UAU. Some codons (UAA, UAG, UGA) do not specify an amino acid but instead provide a signal to terminate protein synthesis – these are called *stop codons.* There is also a *start codon* (AUG, which codes for methionine) which provides the signal to allow translation to be initiated.

5.6 Protein synthesis

For proteins to be synthesised, a DNA gene must be transcribed into a messenger RNA which is then translated into protein. This flow of information from DNA → RNA → protein is sometimes called the *central dogma* (Fig. 5.3).

Figure 5.3 Flow of information from DNA to RNA to a protein

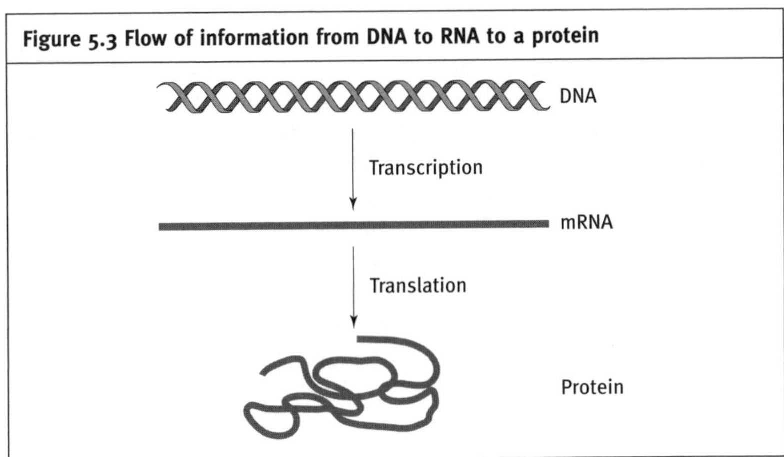

Clinical example: HIV breaks the central dogma

When the central dogma was proposed it was believed that the flow of information could occur in only one direction – from DNA to RNA to protein. Subsequently, viruses were discovered that were able to copy RNA into DNA by means of an enzyme called *reverse transcriptase*. These are called *retroviruses* and include the human immunodeficiency virus HIV, which is responsible for AIDS. The genetic information in HIV is contained within RNA. In infected cells this is then copied into DNA under the action of reverse transcriptase. The double-stranded DNA that is produced can then insert itself into the DNA of the host cell. The DNA copy of the viral RNA can subsequently be transcribed to produce mRNA which is then translated to produce new viral proteins. Because reverse transcriptase is specific to the virus and not required for the functioning of the host cells, reverse transcriptase has been the target of some of the antiviral drugs that have been developed to treat HIV infection.

Transcription of genes proceeds as described above. Once mRNA has been produced this then becomes associated with the *ribosomes*, which are the site of protein synthesis. Ribosomes are particles made up of protein and RNA and are responsible for catalysing the formation of peptide bonds between amino acids as proteins are synthesised.

As protein synthesis proceeds, the ribosome moves along the mRNA. Amino acids, the precursors of proteins, are picked up by adaptor molecules of tRNA. For each amino acid there is a particular tRNA that is able to bind only that amino acid. The tRNA also contains a site known as the *anticodon*, which comprises three bases complementary to the codon for that amino acid. This allows tRNA molecules to align themselves with codons within the mRNA so that this is translated into the correct sequence of amino acids (Fig. 5.4).

Amino acids are added sequentially to a growing polypeptide chain under the influence of the catalytic activity of the ribosome. Interestingly, this activity seems to be associated not with ribosomal protein enzymes but with rRNA, giving rise to the concept of a *ribozyme*, or catalytic RNA. This *peptidyl transferase* activity allows the formation of peptide bonds between the NH_2 group of an incoming amino acid on a tRNA molecule and the free COOH group of the preceding amino acid. After formation of the peptide bond the tRNA can be released and reused.

Figure 5.4 Translation of mRNA into protein

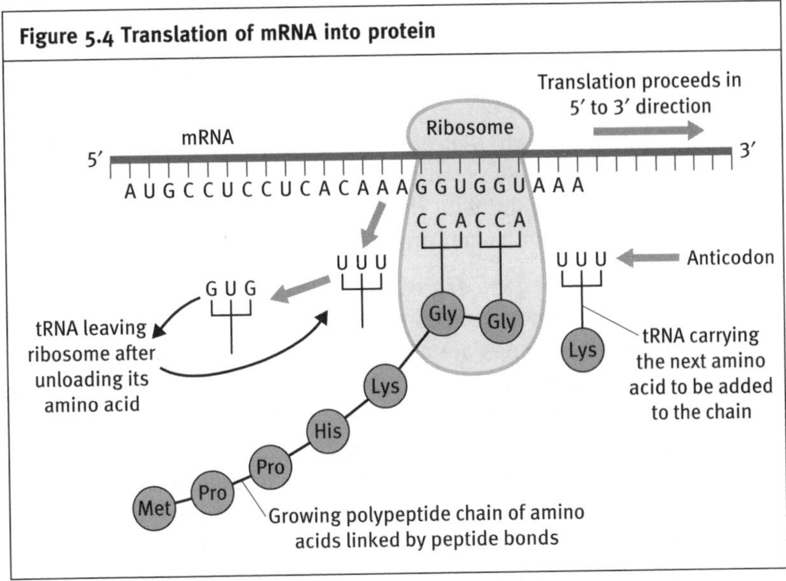

5.7 Introns and exons

In mammals, and other higher organisms, genes do not consist of continuous sequences of DNA but of coding sequences which are expressed (*exons*) separated by non-coding intervening sequences (*introns*). Both the exons and introns of a gene are transcribed into RNA and then the introns are removed. Within proteins an individual domain may be encoded by an exon.

5.8 Regulation of gene expression

Cells need to regulate the expression of their genes for various reasons. Some proteins need to be expressed under particular environmental conditions, and it would be a waste of valuable resources to produce these when they are not needed. Additionally, in complex multicellular organisms, different cells perform specialised functions and so the repertoire of proteins they express will vary accordingly. Within any particular cell some genes will be expressed *constitutively* (all the time), regardless of the environment, and some will be expressed *inducibly* when their protein products are needed.

Regulation of gene expression involves a number of mechanisms. In bacteria, genes encoding related proteins in a pathway are clustered together into *operons*. Regulation of expression of the genes within an operon can be influenced by proteins that bind to the DNA and affect accessibility to the RNA polymerase. The classic example of this is the *lac* operon, described by Jacob and Monod in 1960. The lac operon encodes genes that enable cells to use the substrate lactose as an energy source. In the absence of lactose the genes are not expressed because a *repressor* protein is bound to a region of the operon called the *operator*, where it prevents the RNA polymerase from binding. In the presence of lactose, an *inducer* binds to the repressor molecule, reducing its affinity for the operator. The inducer is not lactose itself but allolactose, an isomer produced from lactose.

Regulation of gene expression in eukaryotes is more complex. DNA is present in chromosomes where it is bound tightly to proteins known as *histones*. Histones are an important structural component of chromosomes and play a role in the regulation of gene expression by regulating the accessibility of promoters. In this way they can determine whether or not transcription may begin. Histones are clustered in groups of eight (octamers) and each octamer is entwined in DNA to form a *nucleosome*. Histones can be covalently modified in such a way as to alter their affinity for DNA. Acetylation of histones leads to reduced affinity for DNA so that it becomes more accessible.

The importance of the TATA box in initiating transcription has been noted above. Additionally there may be other sites upstream of this that can bind transcription factors and enhance or repress expression of the gene. Sequences found within DNA called *enhancers* also play an important role in regulating transcription. Enhancers can be found at a distance of 100s–1000s of base pairs from a gene that they influence, but can interact with the relevant promoter region by the looping of DNA. Enhancers bind to specific regulatory proteins which facilitate transcription. The production of regulatory proteins able to bind to particular enhancers is specific to certain types of differentiated cells and this ensures that cells make the range of proteins appropriate to their particular function.

A simple example of the regulation of gene expression in mammals is provided by the response that is seen to steroid hormones (see Chapter 16). These molecules are hydrophobic and so they pass through cell membranes and bind to intracellular receptor molecules. These molecules, once the hormone has bound, are then able to move into the nucleus where they bind to special sites in the DNA to activate transcription of selected genes (Fig. 5.5).

Figure 5.5 Regulation of gene expression by a steroid hormone

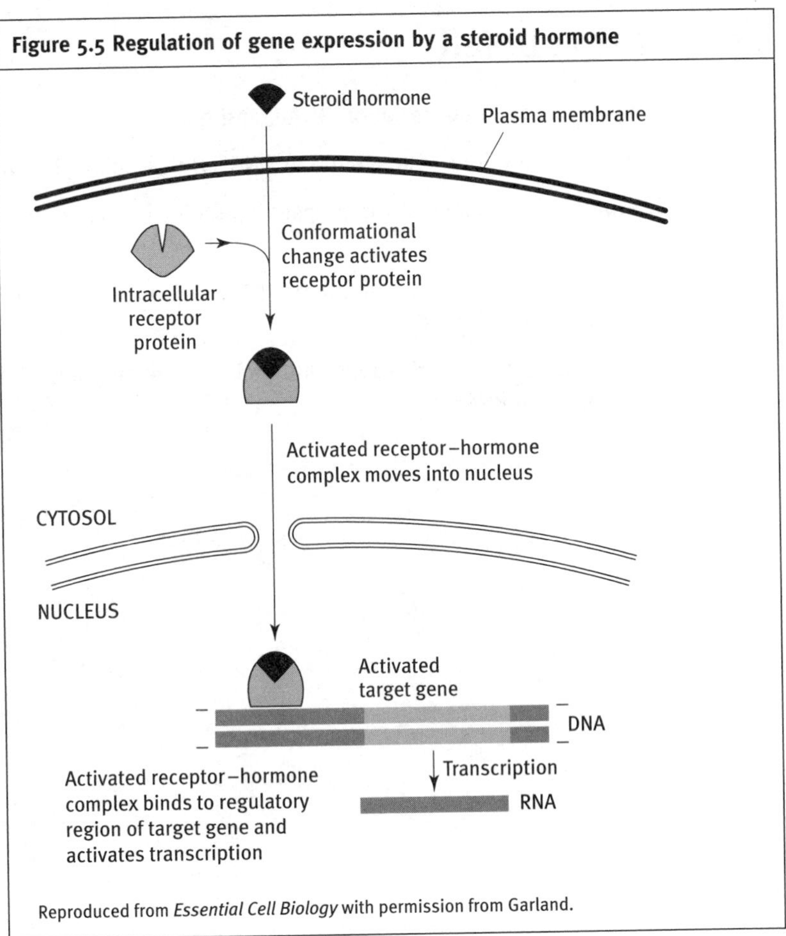

Reproduced from *Essential Cell Biology* with permission from Garland.

Clinical example: Enhancers, oncogenes and cancer

Cells of the specific immune system called B lymphocytes are responsible for the production of immunoglobulins (antibodies), which are important proteins in defence against infection. B lymphocytes are the only cells able to synthesise immunoglobulins, and the production of these proteins is under the regulation of an immunoglobulin enhancer. In some types of cancer affecting B cells, a cellular gene called *myc* is brought into the vicinity of the immunoglobulin enhancer following a chromosomal translocation, when a piece of the chromosome containing the myc gene is moved to the chromosome containing the immunoglobulin genes. The resulting over-expression of myc leads to unregulated proliferation of the cells giving rise to cancer.

5.9 Test yourself

1. What does the abbreviation DNA stand for?
2. What are the four bases found in DNA? Which pairs with which?
3. What is a sequence of three bases that specifies an amino acid called?
4. What term is used to describe the DNA sequence at which RNA synthesis is initiated?
5. What are the molecules that bind amino acids and match them to codons in mRNA?

6 The cell

Basic concepts:
The cell is the basic structural unit of life. In eukaryotic cells the genetic material is contained within the nucleus. Multicellular organisms consist of assemblies of specialised cells organised into tissues and organs. Each cell is surrounded by a plasma membrane which regulates its interactions with the external environment. Within each cell, basic functions such as respiration, protein synthesis and excretion are compartmentalised into internal structures known as organelles. Knowledge of the functions of these organelles is a prerequisite for understanding the most basic concepts of cell biology.

6.1 Introduction

All living organisms are made of *cells*. Cells are membrane-bound units in which many of the reactions essential to life occur. The simplest organisms consist of a single cell and are sometimes referred to as *unicellular*. Single-celled organisms may be described as either *prokaryotic* or *eukaryotic*. Eukaryotic cells are characterised by having a nucleus and have a more complex cellular architecture than prokaryotic cells. More complex, *multicellular* organisms including plants, animals and fungi are also eukaryotes. Bacteria are prokaryotes and their cells lack a nucleus and tend to be smaller than eukaryotic cells.

6.2 Eukaryotic cells

Membranes

All cells are surrounded by a plasma *membrane* (Fig. 6.1). Membranes consist of a phospholipid bilayer. Phospholipids are amphipathic molecules, with a hydrophilic head group and hydrophobic fatty acid chains (see Chapter 4). These are arranged so that the hydrophilic head groups are exposed on the surface (either external or internal to the cell) and the fatty acid chains are shielded from the aqueous environment and interact with each other.

Figure 6.1 The fluid mosaic model of the cell membrane

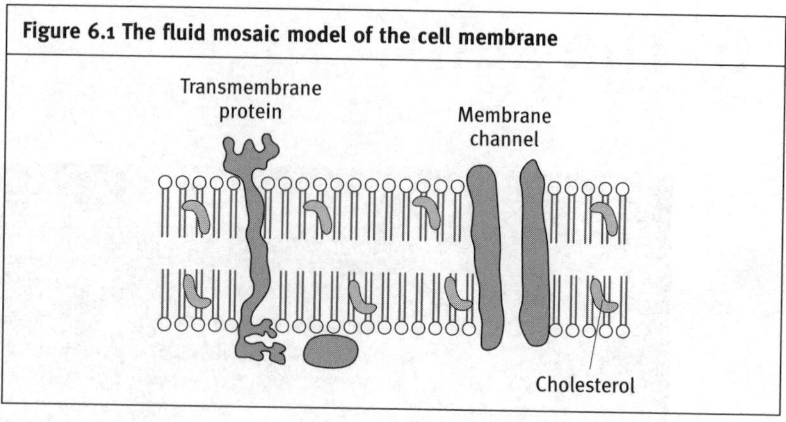

In addition to phospholipids, membranes also contain other lipids such as cholesterol, and proteins which can contribute up to 45% of the mass of a membrane. Membranes are sometimes described as a fluid mosaic (Fig. 6.1) because the proteins and lipids can freely diffuse in the plane of the membrane.

In eukaryotic cells there are subcellular structures called *organelles* (Fig. 6.2), which may also be surrounded by membranes. Different organelles have different functions within the cell and allow reactions to take place in separate compartments. Some reactions take place on membranes and are catalysed by enzymes that are membrane-bound.

The nucleus

The *nucleus* is the organelle that contains the genetic material in the form of *chromosomes*. Chromosomes consist of DNA and associated proteins. The nucleus is surrounded by a double membrane, which separates it from the cytoplasm of the cell. The nuclear membrane contains pores that allow material to move between the nucleus and cytoplasm. It is within the nucleus that DNA and RNA synthesis take place. Following transcription, mRNA leaves the nucleus via the pores in the nuclear membrane and enters the cytoplasm, where it becomes associated with ribosomes and is translated into protein.

Cytoplasm

The part of the cell that is not the nucleus is referred to as the *cytoplasm*. Many of the reactions of life occur in the cytoplasm. The cytoplasm is the principal site of protein synthesis (see Chapter 5) and this is carried out by ribosomes. Ribosomes may be found either free in the cytoplasm or attached to membranes of the endoplasmic reticulum.

Figure 6.2 Eukaryotic cell with organelles – a typical secretory cell

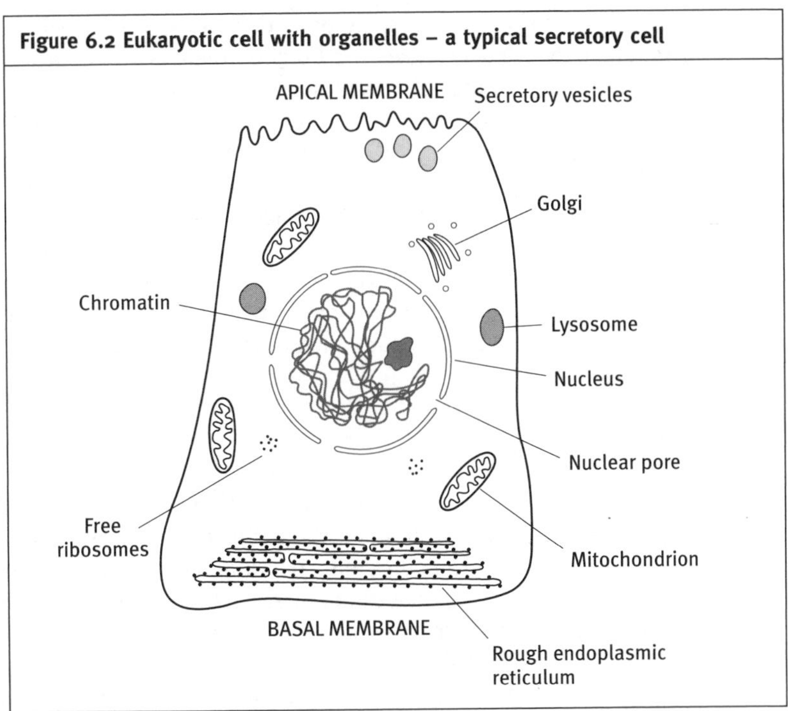

Endoplasmic reticulum

The *endoplasmic reticulum* (ER) consists of a collection of interconnected membrane-bound sacs. These are the sites of production of new membrane, and of proteins destined for secretion from the cell. ER that has ribosomes attached to it is described as rough endoplasmic reticulum. Proteins are synthesised by ribosomes attached to the cytoplasmic face of the ER membrane and then transported into the lumen of the ER. Within the ER, proteins have sugars added to them – this is described as a post-translational modification – and they also become correctly folded.

ER which is not specialised for protein synthesis is known as smooth endoplasmic reticulum. This can serve a variety of functions including calcium storage, synthesis of steroid hormones and detoxification.

Golgi apparatus

The *Golgi apparatus* is a stack of membrane-bound sacs. After synthesis in the ER, proteins destined for secretion are transported to the Golgi where they may undergo further post-translational modification. Proteins leaving the Golgi are directed to the appropriate destination, which may be the cell surface, a lysosome or a secretory vesicle. Membrane lipids are similarly transported via the Golgi on their way to being incorporated into the cell membrane.

Lysosomes

Lysosomes are membrane-bound vesicles that contain a range of hydrolytic enzymes, including proteases. The main function of lysosomes in a cell is to provide a site of intracellular digestion. Within phagocytic cells of the immune system (see Chapter 23) lysosomes play an important role in the destruction of infectious agents that have been taken into the cell.

Mitochondria

Mitochondria (singular *mitochondrion*) are organelles found within the cytoplasm of the cell. Like the nucleus these are surrounded by a double membrane. The outer mitochondrial membrane contains large pores which allow free movement of small to medium sized organic molecules. The inner mitochondrial membrane is extensively folded and is the site of ATP production in the process known as oxidative phosphorylation (Chapter 7). In the centre of the mitochondrion is a region known as the mitochondrial matrix which contains the enzymes necessary for the citric acid cycle as well as a circular strand of mitochondrial DNA, which encodes for some mitochondrial proteins.

Clinical example: Mitochondrial myopathy

Mitochondrial myopathy is a condition in which an inherited defect in one of the proteins involved in oxidative phosphorylation leads to a gradual loss of muscle function. It is thought to affect muscle cells in particular because of their high energy demands. The affected genes are usually in the mitochondrial DNA and so it is a condition that is passed from a mother to her children. Only the daughters of an affected individual can then pass it on to their children. The muscle damage is often seen first in the muscles around the hips and sufferers will complain of muscle pain and difficulty in getting up from a seated position.

The cytoskeleton

Within the cell is a framework of protein filaments that make up the *cytoskeleton*. The cytoskeleton is responsible for maintaining a cell's shape, mechanical strength and directing the movement of organelles. It also plays an important role in the process of cell division by *mitosis* (see Chapter 9). In phagocytic cells that can ingest material from the surroundings, the cytoskeleton is also important in the process of phagocytosis.

The cytoskeleton is made up of three types of filaments. *Microfilaments* consist primarily of the protein *actin*, which is also found in muscle, and are found in highest density just beneath the plasma membrane. *Intermediate filaments* are made up of multiple rod-like proteins and extend across the cytoplasm providing mechanical strength. *Microtubules* consist of the protein tubulin, and these play an important role in mitosis and internal transport through the cytoplasm.

The cytoskeleton is a dynamic structure, in which the filaments assemble and disassemble as required.

6.3 Cell specialisation

Although all eukaryotic cells have the same basic components they can become specialised (differentiated) to allow them to perform their particular functions. For example, red blood cells do not have a nucleus but are filled with the protein haemoglobin, which allows them to transport oxygen to tissues. Nerve cells have a cell body that contains the nucleus and an extended projection, called an axon, along which nerve impulses are conveyed. Cells whose main function is to secrete proteins may have a very well-developed rough endoplasmic reticulum, as in the case of plasma cells that secrete antibody molecules. Sperm cells have a long tail or flagellum that allows them to swim up the reproductive tract – the tail contains abundant mitochondria to provide the necessary energy. Fat cells are specialised for storage of fat, and contain a large lipid globule surrounded by only a thin rim of cytoplasm. Cells in the gut, which function to absorb nutrients, have a greatly increased surface area due to the presence of many short finger-like projections called *microvilli.* Cells in the airways have hair-like protrusions, called *cilia*, which allow them to move a stream of mucus along the surface of the airway.

Plant cell specialisations

Unlike animal cells, plant cells are surrounded by a rigid cell wall made up of the polysaccharide cellulose. Plant cells also have a special type of organelle, the *chloroplast*, containing the green pigment chlorophyll. Chloroplasts are essential for the process of photosynthesis in which plants convert carbon dioxide and water into glucose using light as a source of energy.

6.4 Prokaryotic cells

All plants, animals and fungi are eukaryotes. In contrast, bacteria are prokaryotes. Prokaryotic cells are smaller than eukaryotic cells and, by definition, do not have a nucleus. Their DNA consists of a single circular strand which lies free in the cytoplasm. Prokaryotes also do not have most of the membrane-bound organelles described for eukaryotic cells. They do have ribosomes, which are smaller than those of eukaryotes, and these are found only in free form as there is no endoplasmic reticulum. Bacteria, like plant cells, are surrounded by a rigid cell wall. In bacteria this is made up of a substance called proteoglycan, which contains carbohydrate and protein.

6.5 Test yourself

1. Which membrane-bound organelles contain hydrolytic enzymes?
2. Where in the cell is energy mainly produced?
3. In animal cells, apart from the nucleus, which membrane-bound organelles contain DNA?
4. What protein occurs in microfilaments?
5. Which cytoplasmic particles, found in both prokaryotic and eukaryotic cells, play a central role in protein synthesis?

7 Energy metabolism

Basic concepts:
Cellular metabolism is driven by the energy stored in the ATP molecule. Without energy, cells will stop functioning and it is therefore important to understand the factors which regulate its production. ATP is produced by using the energy derived from the breakdown of complex organic molecules. To break down these molecules fully requires the presence of oxygen, which is combined with the carbon and hydrogen to form carbon dioxide and water. This process is known as aerobic respiration and takes place within the mitochondria.

7.1 Introduction

We derive the energy necessary for life from the various foods we eat. Ultimately the energy stored in food is derived from the sun through the actions of plants which utilise special organelles known as chloroplasts to combine carbon dioxide and water into carbohydrates. In order for us to release the energy stored in foodstuffs they must be broken down. Energy released during these reactions is transferred to carrier molecules of ATP (*adenosine triphosphate*). In ATP the third phosphate group is bound by a high-energy bond (Fig. 7.1), and this energy can be released as required by the conversion of ATP into ADP (*adenosine diphosphate*).

Figure 7.1 Structure of ATP

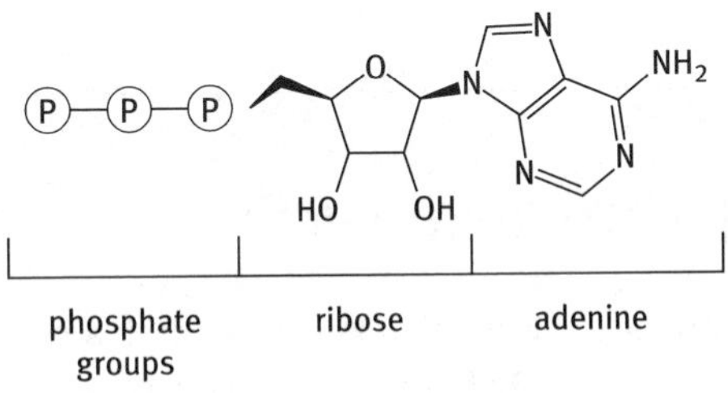

P P P O N N NH2 N N HO OH

phosphate groups | ribose | adenine

The breakdown of glucose gives rise to large quantities of ATP. Glucose derived from food is broken down in a step-by-step process into carbon dioxide and water. The reaction uses oxygen and can be represented as:

$$C_6H_{12}O_6 + 6O_2 \rightarrow 6CO_2 + 6H_2O.$$

This does not occur as a single chemical reaction but in an ordered series of reactions in which the energy from glucose is progressively released and stored in carrier molecules.

The first series of reactions occurs in the cytoplasm and is described as *glycolysis.* In glycolysis the six-carbon sugar glucose is broken down into two molecules of the three-carbon compound *pyruvate.* This occurs as a series of ten separate reactions, each requiring its own enzyme. Such a series of reactions, in which the product of each reaction forms the substrate for the next, is called a *metabolic pathway.* The principle is illustrated in Fig. 7.2. For every molecule of glucose that is converted to pyruvate there is an associated release of energy in the form of two molecules of ATP. Additionally glycolysis leads to the production of NADH from NAD^+ (nicotine adenine dinucleotide). NADH serves as a carrier of high-energy electrons and their energy is subsequently used to generate more ATP (see below).

Figure 7.2 Representation of a metabolic pathway

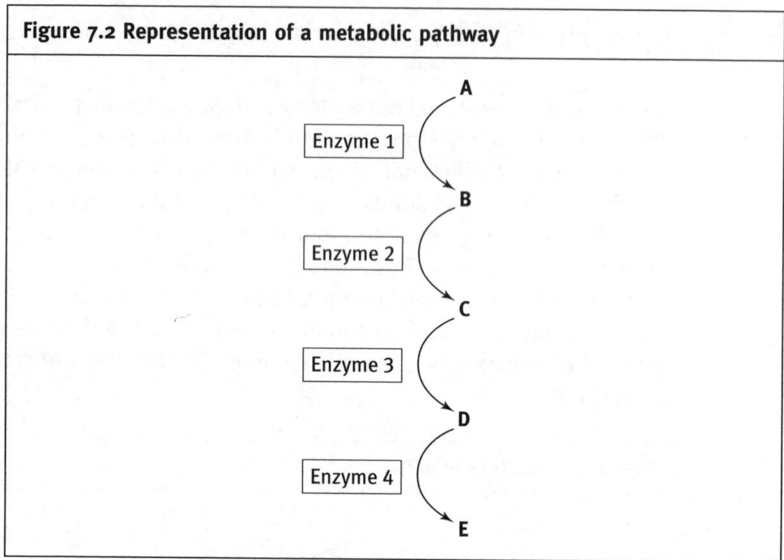

7.2 The citric acid cycle

The citric acid cycle (also known as Kreb's cycle or the tricarboxylic acid (TCA) cycle) is a metabolic pathway whose enzymes are located in the mitochondrial matrix. It generates CO_2 and NADH.

Pyruvate produced as a result of glycolysis in the cytoplasm is transported to the mitochondria, where it is combined with coenzyme A (CoA) to form a molecule called acetyl CoA, with the release of carbon dioxide and the production of another molecule of NADH. The acetyl group is transferred from acetyl CoA to a four-carbon molecule called oxaloacetate to produce the six-carbon molecule citric acid and the CoA is released. Citric acid then forms the starting point of a series of reactions in which the two carbons derived from the acetyl group of acetyl CoA are oxidised into carbon dioxide. The end product of these reactions is oxaloacetate, which can combine with more molecules of acetyl CoA so that the cycle continues. As a result of the citric acid cycle more NADH is produced.

7.3 Release of energy from fats and proteins

So far the discussion has focused on the release of energy from carbohydrates, particularly glucose. Fats are also a major source of energy and these are digested in the gut to release fatty acids. Acetyl CoA can be produced in the cell from fatty acids and this then enters the citric acid cycle as described above. Similarly proteins are digested into amino acids and these can be further broken down to feed the citric acid cycle and generate energy.

Clinical example: Beriberi

Deficiency of vitamin B1, also known as thiamine, leads to a condition called beriberi, which mainly affects the nervous system. Symptoms may include numbness, pain and paralysis. Thiamine is required for the action of the enzyme pyruvate dehydrogenase, which is involved in converting pyruvate produced from glucose in glycolysis into acetyl CoA, which can then enter the citric acid cycle. Why is the nervous system selectively affected by this? Unlike other cells of the body which can also use fats and proteins, it appears that nerve cells depend on glucose as their main source of fuel. So when pyruvate dehydrogenase is inactive this leads to neurological problems.

7.4 Oxidative phosphorylation

The high-energy electrons in the NADH, generated as a result of glycolysis and the citric acid cycle, are then used to generate more ATP. The electrons are transferred in an electron transport chain down a series of carriers. During this process their energy is released and used to pump hydrogen ions (H^+) out of the mitochondrial matrix into the space between the inner and outer mitochondrial membranes to create a concentration gradient. This gradient is then used by the ATP synthase enzyme to produce ATP. This enzyme is located in the inner mitochondrial membrane and as hydrogen ions pass back down their concentration gradient through a narrow channel in the enzyme the energy of their flow is used to drive the production of ATP. This can be viewed as the molecular equivalent of the way in which water flowing over a waterwheel is used to generate electricity. This process is termed *oxidative phosphorylation* (Fig. 7.3). Ultimately the electrons and H^+ ions are transferred to oxygen to produce water in an overall reaction that can be described by the equation

$$NADH + \tfrac{1}{2}O_2 + H^+ \rightarrow NAD^+ + H_2O$$

Figure 7.3 Oxidative phosphorylation in the mitochondrion

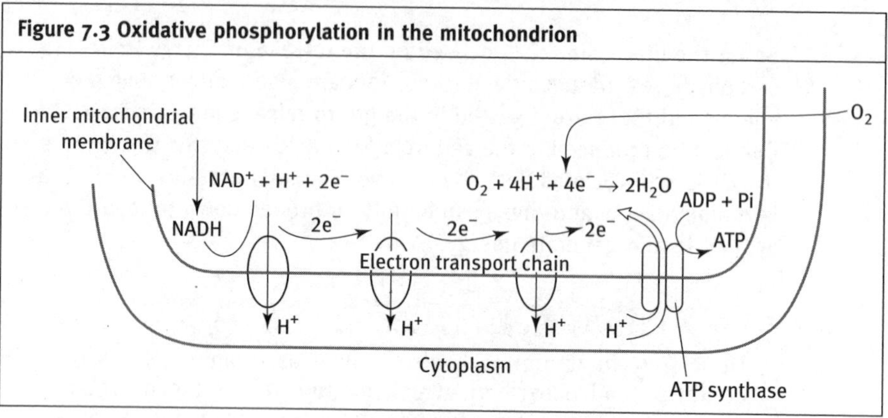

Overall each molecule of glucose generates about 30 molecules of ATP. Two molecules of ATP are produced directly from glycolysis, but the majority of the ATP is produced in the process of oxidative phosphorylation using the NADH generated in glycolysis and the citric acid cycle (Fig. 7.4).

7.5 Anaerobic respiration

The complete breakdown of glucose as described above requires oxygen. Under *anaerobic* conditions, where oxygen is deficient, *glycolysis* is the main source of ATP. Anaerobic conditions can arise in muscle tissues as a result of oxygen depletion in exercise. In this case the pyruvate produced in glycolysis is converted into lactic acid. This process uses electrons donated by NADH, which is converted back to NAD^+.

Figure 7.4 Relationship between glycolysis, the citric acid cycle and oxidative phosphorylation

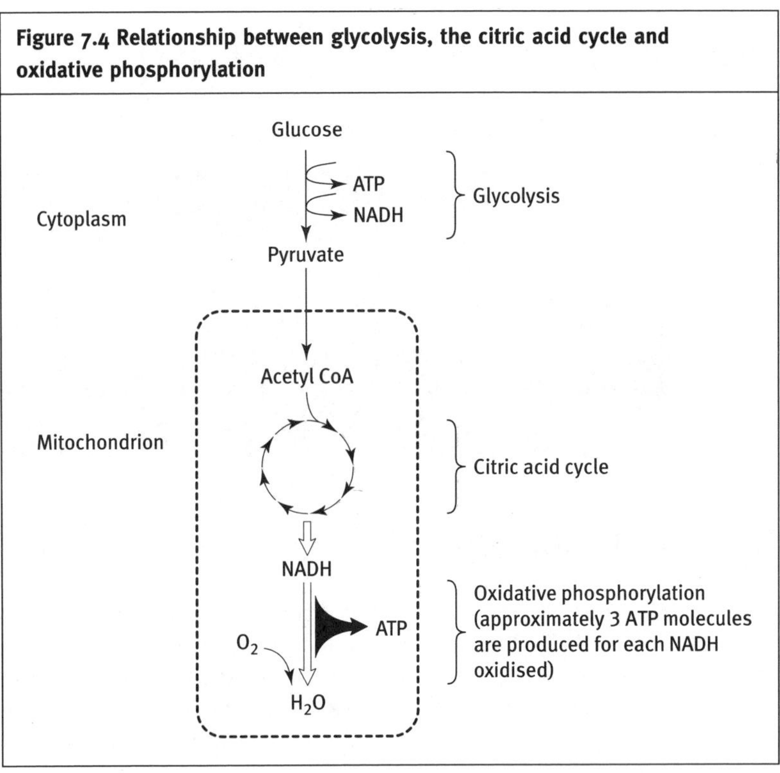

7.6 Test yourself

1. What three-carbon compound is produced from glucose as a result of glycolysis?
2. What does the abbreviation ATP stand for?
3. Where in the cell does the citric acid cycle occur?
4. What term is used to describe the process whereby a proton gradient across the inner mitochondrial membrane is used to drive the synthesis of ATP?
5. In the absence of oxygen into what is pyruvate converted?

8 Membrane transport

Basic concepts:
The cell membrane is semi-permeable, which means that it will permit some substances to pass through it more easily than others. Cell functions depend on the passage of water, ions and metabolites into and out of the cell. This can be achieved through simple diffusion or with the help of specific proteins in the membrane. Understanding the mechanisms of membrane transport is essential for understanding many basic functions of the body.

8.1 Introduction

All cells need to exchange material with the extracellular fluid which bathes them. This may include nutrients, oxygen, carbon dioxide, waste products, hormones, ions or other cellular secretions. The plasma membrane is selectively permeable to these various substances and the mechanisms by which they enter or leave the cell will depend on their permeability.

Small molecules such as O_2, CO_2 and urea are freely diffusible across the plasma membrane. Their transport follows the laws of diffusion in that the rate of movement depends primarily on the concentration difference between the inside and outside of the cell. Gaseous exchange in the alveoli of the lung (see Chapter 19) is an example of this mechanism. Lipid-soluble molecules such as steroid hormones will also pass freely through the plasma membrane.

Other molecules are not able to diffuse across the cell membrane and are transported by specialised mechanisms involving membrane-associated transport proteins. Where molecules are moving down a concentration gradient this process is described as facilitated diffusion or passive transport. In order to move molecules against a concentration gradient – from areas of low concentration to areas of high concentration – energy is required and a process of active transport is involved.

8.2 Osmosis

Perhaps surprisingly, given the hydrophobic nature of the centre of the lipid bilayer, water can easily pass through the plasma membrane. This is due to the small size and lack of overall charge of the water molecule. The movement of water across plasma membranes is achieved by a process known as *osmosis* (Fig. 8.1) which occurs when two basic conditions are satisfied:

- there must be a higher concentration of particles of solute on one side of the membrane than the other;
- the membrane must be more permeable to water than to the solute particles.

If these conditions are met then water will move from an area of low solute concentration to one of high solute concentration.

The *osmolarity* of a solution is a measure of the number of particles dissolved in it and is determined by the molarity of the solute (see section 4.3 in *Catch Up Chemistry*) multiplied by the number of particles into which it dissociates. Osmolarity is expressed in the units osmol/l. Thus a 1 molar solution of sucrose has an osmolarity of 1 osmol/l as each sucrose molecule in solution is a single particle. However, a 1 molar solution of NaCl has an osmolarity of 2 osmol/l because in solution each molecule of NaCl dissociates into the two osmotically active particles Na^+ and Cl^-. Both intracellular and extracellular fluids in the body have an osmolarity of about 290 mosmol/l and an important function of homeostasis (see Chapter 15) is to maintain this value. Solutions that have the same osmolarity are said to be *iso-osmolar*.

An alternative way of expressing osmolarity is to use the term *osmolality*. This is a measure of the number of particles of solute per mass of solvent rather than its volume (i.e. the units are mosm/kg H_2O). At standard temperature and pressure for an aqueous solution containing only one solute at low concentration, the values for osmolarity and osmolality will be identical. This is because the volume occupied by the solute is negligible and a litre of solution will contain 1 kg of H_2O. However, in the measurement of osmolality in biological fluids such as plasma, it is necessary to take into account the volume occupied by other solutes such as protein. Thus 1 litre of plasma will weigh 1 kg but will actually only contain 930 g of H_2O (total plasma protein = approx 70 g). When calculating osmolality of Na^+ in a litre of plasma we calculate not as if this litre was equal to 1 kg of H_2O but instead as if it were 930 g H_2O. This means that a solute with a calculated osmolarity in plasma of 100 mosmol/l will have a higher calculated osmolality since it is actually distributed in less water. **The terms osmolarity and osmolality are often used interchangeably and in most circumstances this lack of precision has little functional significance.**

Solutions that have the same osmolality are said to be iso-osmolal. A solution with a lower osmolality than another is hypo-osmolal and one with a higher osmolality is hyperosmolal. A difference in osmolality between two solutions which leads to the movement of water is described as the effective osmolality or tonicity. The movement of water into a compartment generates osmotic pressure. A difference in effective osmolality of 1 mosmol/kg is equivalent to a pressure of 19.3 mmHg. If cells which are surrounded only by a plasma membrane are placed into pure water osmosis will cause the movement of water into the cells, raise their internal pressure, and they will swell until they burst, a process known as lysis (Fig. 8.1).

The internal pressure generated by the movement of water into plant cells (which have a rigid cellulose cell wall and therefore cannot swell) is known as turgor and helps to prevent the leaves and stems of plants from going floppy. (Think what happens to cut flowers if they are left out of water!)

Figure 8.1 Osmosis and its effects on cells

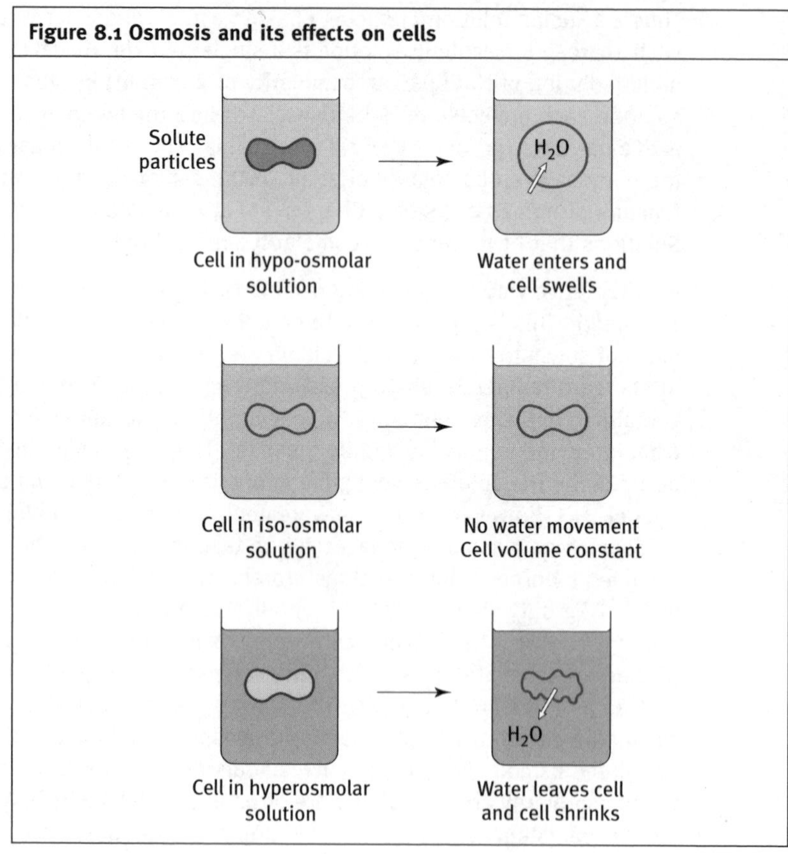

The movement of water between the various compartments of the body is controlled predominantly by the movement of ions. For example, the secretion of water from the cells of the epithelium lining the airways in order to reduce the stickiness of the mucus which coats them is achieved by moving sodium ions and chloride ions out of the airway cells. This increases the extracellular solute concentration and water follows by osmosis.

The movement of water back into capillaries from the interstitial fluid is also due to osmosis, but in this case the key particles are the proteins within the capillary. In many tissues the capillary walls are as freely permeable to ions as they are to water and so the ions exert no osmotic influence – the first condition required for osmosis to occur is not satisfied. However, the proteins are too large to move from the capillary and so are osmotically active and because the interstitial fluid contains no protein there is a concentration gradient of solute which draws water into the capillary (see *Oncotic pressure* in Chapter 18).

8.3 Facilitated diffusion

Many substances need to be transported into and out of cells but have low permeability in the cell membrane and so will not readily diffuse across. This is true of all ions and of larger organic molecules such as glucose. Where such substances have a concentration gradient across the membrane then their transport can occur through a process known as *facilitated diffusion* – also known as *passive transport*. This can occur with the assistance of either *carrier molecules* or *channels*.

Carrier molecules are trans-membrane proteins which bind the molecule to be transported on the side of the membrane where it is at high concentration and then undergo a conformational change which moves the molecule through the membrane before releasing it on the other side. The carrier protein then returns to its original shape and is ready to transport another molecule. Some carriers only move one molecule or ion through the membrane and are known as uniports. Others transport two different molecules or ions in opposite directions across the membrane and are known as antiports.

Channels are small protein-lined openings in the membrane which permit the movement through the membrane of a molecule down its concentration gradient. Channels have two key properties in that they are regulated and selective. Channels may exist in either open or closed states and the probability of being in one of those states depends on external factors. Thus some channels will tend to open when a signal molecule, such as a neurotransmitter, binds to the proteins which make up the channel. Other channels are voltage sensitive and will open when the membrane potential (see Chapter 14) moves above a certain threshold. The size of the channel opening and the charge on the proteins which make up the channel lining will determine which molecules may pass through. Many channels are only large enough to permit the passage of ions and will be selective for particular cations or anions such as Na^+ and Cl^-.

8.4 Active transport

In both diffusion and facilitated diffusion, substances move across the membrane down their concentration gradient. It is also possible to use energy directly or indirectly to move substances up their concentration gradient.

In primary *active transport* the energy derived from the breakdown of ATP to ADP is used directly to pump substances across a membrane against their concentration gradients. One of the most widespread of these pumps is the Na^+/K^+ ATPase. This protein complex pumps three Na^+ ions out of the cell and two K^+ ions into the cell. This helps to maintain the high extracellular Na^+ and high intracellular K^+ levels which are characteristic of eukaryotic cells. This process is so important to cell function that approximately 30% of all ATP generated is used to fuel this pump. Other pumps important to cellular function are those that pump Ca^{2+} ions out of the cytoplasm of excitable tissues once the depolarising event which has triggered Ca^{2+} entry into the cytoplasm is complete.

Clinical example: Cholera

Infection by the bacterium *Vibrio cholerae* causes the disease cholera which is characterised by watery diarrhoea and potentially fatal dehydration. The bacteria release a protein toxin which enters cells in the intestine and causes the opening of channels in their membranes. This allows Cl^- ions to flow from the cells into the lumen of the gut. This movement of ions draws water with it by osmosis, leading to diarrhoea and producing dehydration in the rest of the body. This is potentially fatal. The most effective way of rehydrating the body is to drink a mixture of glucose and salts dissolved in water. This mixture activates co-transporters in the gut wall, which take up glucose in the presence of Na^+ ions. The inward movement of Na^+ ions and glucose moves water by osmosis back from the lumen of the gut into the bloodstream and reverses the effects of the toxin.

Secondary active transport depends on the maintenance by the Na^+/K^+ ATPase of an Na^+ ion gradient between the outside of the cell and the inside. The movement of Na^+ ions into the cell down their concentration gradient is then coupled with the transport by a carrier protein (co-transporter or symport) of a substance into the cell against its concentration gradient. This mechanism is used in the transport of sugars, amino acids, nucleotides and other essential nutrients into cells. The Na^+ ions that enter the cell are immediately recycled out of the cell by the Na^+/K^+ ATPase and the Na^+ ion gradient is thus maintained to drive further transport activity.

8.5 Exocytosis and endocytosis

Transfer of very large molecules, such as proteins, across the cell membrane is achieved not by the transport of individual molecules using carrier proteins but by the mechanisms of *exocytosis* and *endocytosis*. In the process of exocytosis, proteins which are destined for export from the cell are packaged in the Golgi apparatus into small vesicles. These vesicles then move towards the area of the cell membrane from which they are to be secreted and are held next to the membrane by linker proteins. When the appropriate signal is received a Ca^{2+}-mediated mechanism triggers the fusion of the vesicle with the cell membrane and the contents of the vesicle are emptied into the interstitial fluid. Exocytosis is used for the storage and release of hormones into the bloodstream (see Chapter 16) and digestive enzymes into the lumen of the gut (see Chapter 20).

Endocytosis is the reverse of exocytosis and involves the binding of substances to be taken into the cell by receptors on the cell surface. Once the receptors are occupied then the membrane is drawn in and a vesicle is pinched off. Such vesicles are most frequently directed to the lysosome where the ingested material is broken down.

8.6 Test yourself

1. What is the osmolarity of a 1 molar solution of $CaCl_2$?
2. True or false: passive transport can only occur down a concentration gradient?
3. The movement of amino acids into a cell against their concentration gradient is coupled with the movement of which cation into the cell?
4. What is the name of the enzyme responsible for maintaining high extracellular concentrations of Na^+ ions and high intracellular concentrations of K^+ ions?
5. Is urea osmotically active in the body?

9 Cell division and mitosis

Basic concepts:
The growth of organisms and the replacement of cells both depend on the formation of new cells. These arise by a process of division, either from differentiated cells within a tissue or from stem cells. In order to divide, cells have to make a copy of each of their chromosomes so that the daughter cells can be genetically identical to the parent cell. The sequence of events by which a cell reproduces itself is known as the cell cycle. Understanding the cell cycle and its regulation allows us to understand what may go wrong when uncontrolled cell division produces tumours.

9.1 Introduction

For a multicellular organism to grow and for tissues to be regenerated requires an increase in the number of cells. This takes place through an orderly process of events involving the duplication of the genetic material and separation of this into two new daughter cells.

9.2 Cell cycle

The stages through which a cell proceeds as it divides are described as the *cell cycle* (Fig. 9.1). Replication of the cell's genetic material, the DNA, is described in Chapter 5 and this occurs during the S-phase of the cell cycle. At the end of the S-phase a cell therefore has twice its normal amount of DNA, with two copies of each chromosome. Each copy is known as a *chromatid* and the two chromatids are joined by a structure known as the *centromere.* During the M-phase of the cell cycle this genetic material is carefully segregated into two new daughter cells, each of which inherits one copy of each chromosome; a process known as mitosis (Fig. 9.2).

After the M-phase and before the cell enters the S-phase and starts to replicate its DNA again there is a period described as G1. Similarly after the S-phase and before the M-phase there is a second period called G2. G1 and G2 provide checkpoints where the cell can decide whether or not to continue on to the next phase of the cell cycle. Together G1, S and G2 are described as *interphase.* G1 is also important as the period during which cell growth occurs. One of the factors which would prevent

Figure 9.1 The cell cycle

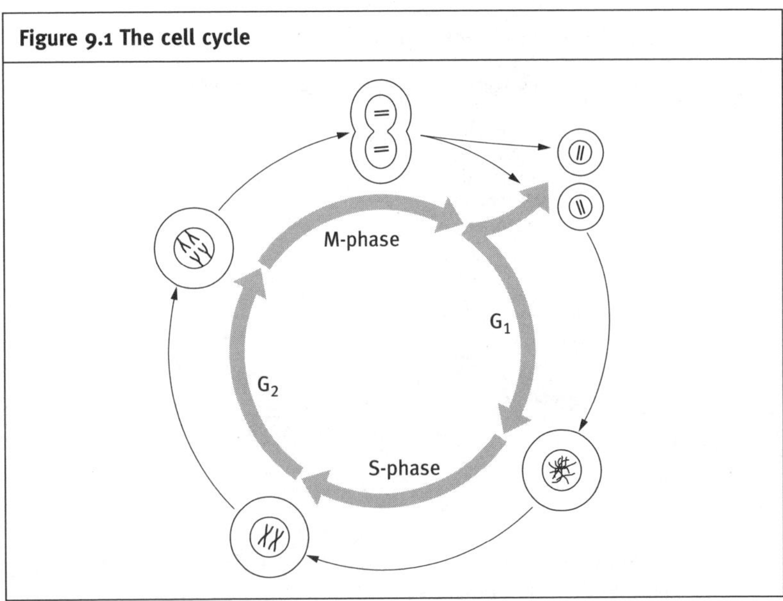

the cell entering the S-phase and proceeding through a further mitotic division is small size. Thus, cells spend longer in G1 in malnourished individuals and if this occurs during development it can permanently reduce the eventual cell population of organs such as the brain.

The M-phase can be subdivided into six phases, each of which is characterised by particular events.

- In *prophase* the chromosomes condense and the *mitotic spindle* starts to form in the cytoplasm. The mitotic spindle is made up of various microtubules and, in animal cells, involves an organelle called the *centrosome*, which duplicates and gives rise to the two poles of the spindle.
- In *prometaphase* the nuclear membrane starts to break down so that the chromosomes can now become attached to the mitotic spindle. The sister chromatids of the duplicated chromosomes become attached to the spindle via the centromere.
- In *metaphase* the chromosomes arrange themselves in the centre of the cell equidistant from each pole of the spindle.
- In *anaphase* the sister chromatids now separate, forming two daughter chromosomes, and these are pulled towards the poles of the spindle (one of each pair going to each pole).
- There are now two identical sets of chromosomes within the cell, one at each end. In *telophase* a nuclear envelope is formed around each of the two sets of chromosomes

Figure 9.2 The phases of mitosis

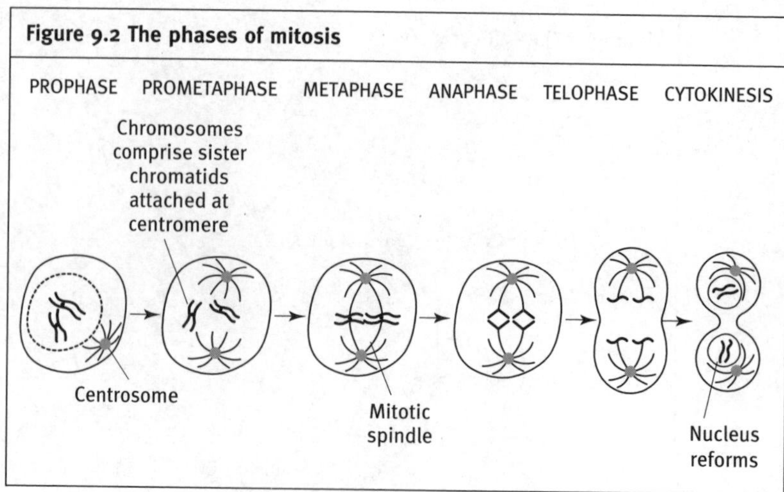

- Finally the cytoplasm divides to give rise to two daughter cells: this is known as *cytokinesis.* Chromosomes within the two daughter cells now de-condense and cell division is complete.

9.3 Control of cell division

Within the body, cells in different tissues tend to divide at very different rates. Cells in the gut are dividing rapidly to replace cells that are constantly being lost. In contrast, nerve cells are essentially non-dividing cells. Cell division is strictly regulated by a series of molecular controls and if these go wrong then uncontrolled cell division can occur, leading to the development of cancers.

The cell cycle is safeguarded by *checkpoints.* The first checkpoint comes at the end of G1 – at this point cells are allowed to enter the S-phase (DNA synthesis) only if conditions in the environment of the cell are appropriate and the DNA is not damaged. The second checkpoint occurs at the end of G2 and only allows the cell to enter the M-phase if the DNA has replicated correctly. The third checkpoint occurs during the M-phase and checks that all of the chromosomes have been correctly arranged on the spindle during metaphase before they separate during anaphase. Together these checkpoints ensure that cell division only occurs under appropriate conditions and that each daughter cell inherits a complete and undamaged copy of the genetic material. These controls depend on enzymes called *cyclin-dependent kinases,* which phosphorylate proteins that regulate the cell cycle.

Cell division is also regulated by factors in the external environment. In mammals there is a group of proteins called *growth factors* that bind to cell surface receptors and stimulate cells to divide. Growth factors can either be broadly specific for a range of tissues or act on a very narrow range of cell types. Erythropoietin is an example of a growth factor that is highly specific and stimulates the production of red blood cells (see Chapter 18). This property has been exploited by some athletes who have used erythropoietin as a performance-enhancing drug.

Many human cells are limited in the number of rounds of cell division they can undertake. As they divide, material is progressively lost from the ends of the chromosomes (called *telomeres*). This shortening of the telomeres ultimately prevents further cell division occurring. This is thought to provide protection against unregulated cell growth and the development of cancer.

Clinical example: Cancer and p53

Many human cancers are associated with mutations in a gene called *p53*. The p53 protein plays an important role in the cell cycle at the checkpoint between G1 and entry into the S-phase. If DNA damage is detected this leads to the expression of p53 which then blocks DNA replication until the damage has been repaired. Mutations leading to a loss of p53 function will allow cells with damaged DNA to divide and to accumulate further mutations. These mutations, by disrupting the functions of other important genes, can lead to the unregulated cell division that is characteristic of cancer. The p53 gene is therefore described as a *tumour suppressor gene.*

9.4 Cell division and differentiation

As cells divide they give rise to more cells of the same type. For example, a dividing epithelial cell will give rise to further epithelial cells. These *differentiated* cells maintain their properties due to the regulation of gene expression (see Chapter 5) which ensures that a set of proteins is produced that is characteristic of that cell type.

When they reach their fully differentiated state some cells lose the ability to undergo further cell division and are said to be *terminally differentiated.* An example of this is red blood cells which have no nucleus and can no longer divide. As red blood cells die they are replaced from *stem cells*, which give rise to the various cells of the blood. Stem cells are not differentiated but when they divide they give rise both to differentiated cells (e.g. red blood cells) and more stem cells. The stem cells in various tissues in an adult have a limited potential to give rise to different cell types. For example the haematopoietic stem cells that occur in the bone marrow give rise to red blood cells, white blood cells and platelets.

Researchers are very interested in the potential for using stem cells to replace damaged tissues in certain diseases. Of particular interest are the stem cells that can be found in early embryos (*embryonic stem cells*). These have the potential to give rise to any cell type. In the future, it is hoped that embryonic stem cells might be used to replace damaged tissues and restore normal function in patients with a range of diseases.

9.5 Test yourself

1. What three phases of the cell cycle comprise interphase?
2. What is the name given to the structure formed from microtubules that moves duplicated chromosomes to opposite poles of the cells?
3. What is the region called in a mitotic chromosome where the two sister chromatids are attached to each other and to the spindle?
4. What name is given to the regions at the ends of chromosomes which tend to shorten as cells undergo successive rounds of cell division?
5. What are the enzymes called that regulate progression through the cell cycle by phosphorylating other proteins?

10 Reproduction

Basic concepts:
The purpose of reproduction is to ensure the survival of the species by the production of new individuals. Asexual reproduction results in offspring that are identical to the parent. Sexual reproduction leads to the mixing of genes from two parents to form a unique organism. Because evolution depends on variability within a population it is important that during the production of the egg and sperm there is a random allocation of genetic material from each of the parents. This randomisation, which occurs during meiosis, ensures that every egg and sperm is genetically unique and that no two individuals born of the same parents will be identical. Understanding this process allows us to appreciate how individuality is created at the genetic level and how a species becomes equipped to face multiple, unknown challenges.

10.1 Introduction

An essential characteristic of living organisms is the ability to reproduce. Reproduction can be achieved in a number of different ways, some of which are more complex than others. Single-celled organisms such as bacteria tend to reproduce by cell division. Such reproduction is described as *asexual* (not involving sex). In asexual reproduction only a single parent is required and the progeny are identical to the parent. Asexual reproduction is also seen in more complex, multicellular organisms. Plants have a variety of means of asexual reproduction involving, for example, the production of tubers in potatoes, bulbs in flowering plants such as daffodils, runners in strawberries etc. Some animals can also reproduce asexually, for example the jellyfish.

Most plants and animals use sexual means of reproducing themselves, either instead of or in addition to asexual reproduction. *Sexual reproduction* involves two parents and the offspring inherit genes from both (Fig. 10.1). The genes are inherited in various combinations so that the offspring will be different from each other and from the two parents. This increased variability is thought to be advantageous in allowing a species to adapt to the environmental conditions.

10.2 Sexual reproduction

Gametogenesis

Organisms that reproduce through sexual means are *diploid*, that is to say each cell contains two copies of each chromosome – one inherited from the mother (*maternal*) and the other from the father (*paternal*). In humans there are 23 pairs of chromosomes. The two chromosomes of a pair are said to be *homologous* and each carries the same set of genes, which means that each gene is present in two copies.

In order to reproduce sexually the organism must produce *germ cells* or *gametes*. Gametes are *haploid*, which is to say they carry only a single set of chromosomes. In order to produce haploid gametes, cells in the reproductive organs must undergo a special type of cell division called *meiosis*. Because meiosis leads to the production of cells which have only half as many chromosomes as the cell from which they are derived this is sometimes called *reductive division*. The germ cells of the male are described as *spermatozoa*, or *sperm*, and the germ cells of the female are called *ova*, or *eggs*. The process through which eggs and sperm are produced is called *gametogenesis*. In the process of fertilisation, male and female gametes fuse to form a zygote, which is now diploid, having acquired one set of chromosomes from each gamete.

Figure 10.1 Sexual reproduction results in offspring with genetic material derived from both parents

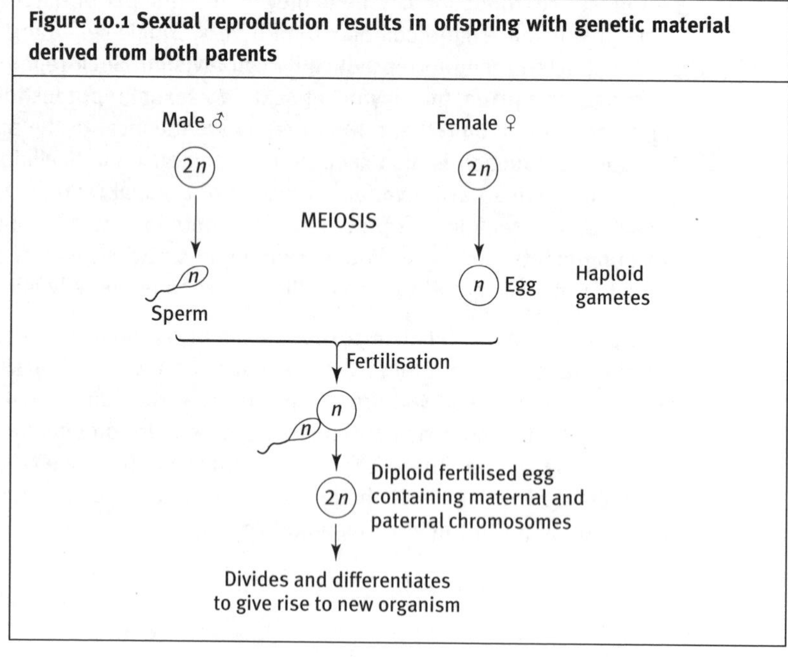

Meiosis

The process of meiosis shares some similarities with mitosis but also shows some important differences. DNA is duplicated in mitosis and meiosis but in the latter there are two, rather than just one, rounds of cell division with the result that each cell inherits half the DNA of the parent cell.

At the start of meiosis, as in mitosis, the chromosomes condense, the nuclear membrane breaks down and the spindle appears. DNA undergoes replication resulting in the production of two sister chromatids that stay closely associated with each other. This stage is described as *prophase I*. The next event is unique to meiosis and is not observed in mitosis. Each replicated chromosome comes together with its homologous pair (Fig. 10.2). This pairing of homologous chromosomes is essential for the next stage because it allows each daughter cell to inherit one member of the pair.

Figure 10.2 Pairing of homologous chromosomes during metaphase in meiosis compared with metaphase in mitosis

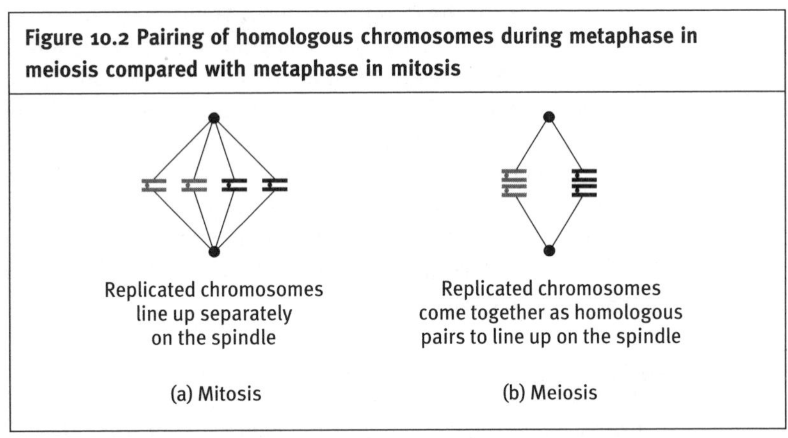

Pairing of homologous chromosomes also allows for *recombination* to occur. In the process of recombination, or *crossing-over* (Fig. 10.3), there is a reciprocal exchange of material between homologous chromosomes. Thus the paternal chromosome in a pair donates some material to the maternal chromosome and acquires in exchange some of the maternal chromosome. Crossing-over requires the formation of a connection between two non-sister chromatids and this is known as a *chiasma* (plural *chiasmata*). On average each pair of human chromosomes will form 2–3 chiasmata. This phase is described as *anaphase I*.

Figure 10.3 Crossing-over between homologous chromosomes at meiosis

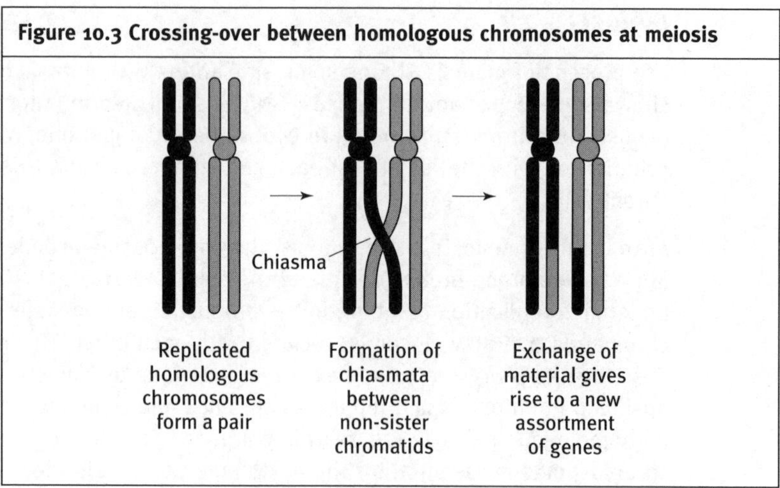

The next step is for the duplicated homologous chromosomes (*homologues*), each consisting of two chromatids, to separate. These are pulled to opposite poles of the spindle and then cell division occurs so that each daughter cell inherits a set of duplicated chromosomes. The daughter cells now enter into a second round of cell division. Spindle formation occurs in each daughter cell (*prophase II*) and then the chromosomes line up on the equator (*metaphase II*). At this point, the sister chromatids now separate and are pulled to opposite poles. This stage is described as *anaphase II* and is followed by reformation of the nuclear membrane and cell division.

Clinical example: Down syndrome

Errors sometimes occur during meiosis such that the chromosomes separate incorrectly – this is known as *non-disjunction* – and can result in the production of gametes that lack a particular chromosome or have two copies. Such an error occurs in Down syndrome, which results from an extra chromosome 21 (trisomy 21). Down syndrome arises when one of the gametes, usually the egg, acquires two copies of chromosome 21 so that after fertilisation the resulting embryo has three copies of this chromosome. This results in abnormal development and affected children show learning disabilities and characteristic facial features. The condition is also associated with an increased risk of infection and, in some cases, heart problems.

Down syndrome occurs more commonly in the children of older mothers and can be detected by examination of fetal cells following amniocentesis.

In this way a diploid cell undergoes one round of DNA replication and two rounds of cell division resulting in four gametes, each with half the DNA of the original cell. Each gamete has a full set of chromosomes (23 in humans) with a mixture of maternal and paternal homologues. The random shuffling of maternal and paternal homologues together with the crossing-over that occurs in the first cell division leads to enormous genetic variability between gametes.

10.3 Fertilisation

Fertilisation involves the fusion of two haploid gametes to produce a diploid zygote. Spermatozoa, the male gametes, are produced in the *testes* (singular *testis*; see Chapter 21) in the process of *spermatogenesis*. A sperm is a very small specialised cell comprising a head, where the nucleus is located, and a tail to enable motility. There is also a special organelle at the tip of the head called the *acrosome*, which is derived from the Golgi apparatus and is involved in fertilisation. Mitochondria in the tail provide the energy to allow the sperm to swim in search of the egg.

Eggs (*oocytes*) are produced in the *ovaries* (see Chapter 21) in the process of *oogenesis*. These are large spherical cells and when they are shed from the ovaries they are surrounded by a special coating called the *zona pellucida* and a layer of cells from the ovarian follicle in which they developed.

For fertilisation to occur a sperm must contact the egg and then penetrate the outer layer of cells and the zona pellucida before fusing with the plasma membrane of the egg. Sperm are produced in large numbers and several can bind to an egg but it is crucial that only one sperm is allowed to fertilise the egg. To ensure this, once fertilisation has occurred an increase in intracellular Ca^{2+} ions leads to changes in the zona pellucida that prevent further sperm from penetrating. Within the fertilised egg the two nuclei must then fuse to form a diploid nucleus and the process of *embryogenesis* leading to the development of a new individual can begin. The cells divide and then implant themselves into the wall of the *uterus*, where the developing embryo can obtain nutrients from the mother's blood via the *placenta* (see Chapter 21). In humans the period from fertilisation to birth is 40 weeks (the *gestation period*).

10.4 Reproductive and therapeutic cloning

The term *clone* refers to a group of cells or organisms that originate from and are identical to a single precursor cell or organism. Asexual reproduction leads to the production of clones with an identical genetic make-up to the parent. Cloning has also been achieved in mammals by artificial means, and the best-known example of such *reproductive cloning* is Dolly the sheep. Reproductive cloning is achieved by taking an unfertilised egg, removing the nucleus and replacing it with the nucleus of a diploid cell from another animal. In the case of Dolly the nucleus was taken from an epithelial cell in the mammary gland. The egg is then allowed to develop in culture where it may give rise to an early embryo, which can then be implanted in the uterus of a foster mother, where it develops into a cloned animal. The animal will be genetically identical to the animal from which the nucleus was taken.

Therapeutic cloning is still under development. It uses a similar approach to give rise to embryonic stem cells of a defined genetic make-up. In this case, instead of the early embryo being implanted in a foster mother, it is used instead as a source of embryonic stem cells (see Chapter 9). Because these cells have the potential to differentiate into different tissue types there is hope that they could be used to help replace damaged tissues in a wide range of different diseases.

10.5 Test yourself

1. What is the general term used to describe the haploid germ cells produced in sexual reproduction?
2. What type of cell division results in the production of haploid cells?
3. What name is given to the points on the chromatids at which crossing-over occurs?
4. What is the layer surrounding the egg that must be penetrated by the sperm in order for fertilisation to occur?
5. Therapeutic cloning aims to produce what type of cells in culture?

11 Inheritance

Basic concepts:
Genes provide the information that determines the overall appearance and characteristics of an organism. The laws of inheritance allow us to predict how the genes inherited from a father and a mother will interact to produce the complex pattern of features that make up an individual. Some inherited traits, such as eye colour, are explained by the action of a few genes; others, such as personality, are the product of many genes and also of non-genetic, environmental factors. Understanding how inheritance works is essential in understanding the basis of genetic disease.

11.1 Introduction

It is widely recognised that offspring have a tendency to resemble their parents. This is because the genetic material, DNA, is passed from one generation to the next. A diploid organism such as a human being inherits one set of chromosomes from each parent and the genes that are inherited as part of the chromosomes account for many characteristics of the offspring.

Variation within a population can be due to either genes or environment, or a combination of both. For example, tall parents will pass on the genes for tallness to their children – however, if the children are poorly nourished this can restrict their height. Some characteristics that are inherited are due to single genes while other characteristics represent the combined effects of multiple genes (these traits are said to be *polygenic*).

11.2 Definition of terms

The two copies of a gene that are inherited from the two parents are called *alleles*. For any gene, if the two alleles that are inherited are identical then that individual is said to be *homozygous*. If the two alleles that are inherited are different then the individual is *heterozygous*. The position occupied by a gene on a chromosome is described as a genetic *locus*.

The set of genes that is inherited by an individual is described as the *genotype*. The expressed characteristics (for example height, hair colour, blood group) constitute the *phenotype*. In a heterozygous individual one allele can be *dominant* while the other allele is *recessive* – in this case the phenotype represents the dominant allele. An example of dominance is seen in rhesus blood groups, where individuals who have either one or two copies of the gene for rhesus protein are rhesus positive, and only individuals who are homozygous for the recessive allele are rhesus negative. Some genes can be expressed co-dominantly, for example human cells have molecules on their membranes called *MHC* molecules. There are a number of different MHC alleles in human populations (such genes are said to be highly *polymorphic)* and so individuals are frequently heterozygous at these loci. A heterozygous individual will express on their cells the MHC proteins corresponding to the genes inherited from both the father and the mother.

11.3 Mutations

The primary source of genetic variation is *mutation*. A mutation manifests as a change in DNA. Many mutations are *silent*, because they occur in non-coding regions or because the change does not affect the coding sequence. Other mutations will cause changes to the protein encoded by the mutated genes. It is only mutations that occur in the sex cells of an organism that will be inherited. Mutations in somatic cells will usually pass unnoticed, although sometimes they can have serious effects, e.g. giving rise to cancers.

Gene mutations can affect a single base pair (*point mutations*) or multiple base pairs. There are three main types of gene mutation:

- *substitution* of one nucleotide for another
- *deletion* mutations that remove nucleotides
- *insertion* mutations where nucleotides are added.

Chromosomal mutations are also seen and Down syndrome, which is due to trisomy of chromosome 21, where the sufferers have three copies of chromosome 21 instead of the normal two, is an example (see Chapter 10).

Mutations occur spontaneously at a low rate but can be accelerated by agents such as radiation and exposure to certain chemicals.

11.4 Mendelian inheritance

Gregor Mendel was an Austrian monk whose work on plant breeding in the nineteenth century forms the basis of the science of genetics.

Mendel's breeding experiments allowed him to deduce and formulate a number of genetic 'rules'. Mendel observed that inheritance was due to discrete particles – what we now call genes – and that these are passed unchanged from one generation to the next. Different genes encode for different characteristics and each organism has a pair of genes – in modern terminology, alleles – encoding a particular characteristic. The alleles of these genes *segregate* during gametogenesis so that each gamete has only one copy of a particular gene. If the genes reside on different (non-homologous) chromosomes then their alleles show *independent assortment* – i.e. the inheritance of particular alleles at one genetic locus does not affect the inheritance of alleles at another locus.

11.5 Monohybrid inheritance

One of the characteristics studied by Mendel was the inheritance of height in pea plants. In pea plants height may show discontinuous variation, with peas being either tall or short (dwarf). The gene that determines this characteristic has two alleles – T, the allele for tallness, and t, the allele for dwarfness. Tall peas that breed true (i.e. always produce tall peas) when crossed with each other are homozygous for T (TT), while dwarf peas that breed true are homozygous for t (tt).

When these true-breeding tall peas and dwarf peas are crossed with each other, all of the offspring are tall. This is explained in Fig. 11.1.

Figure 11.1 Outcome of crossing plants homozygous for the tall (T) and dwarf (t) alleles

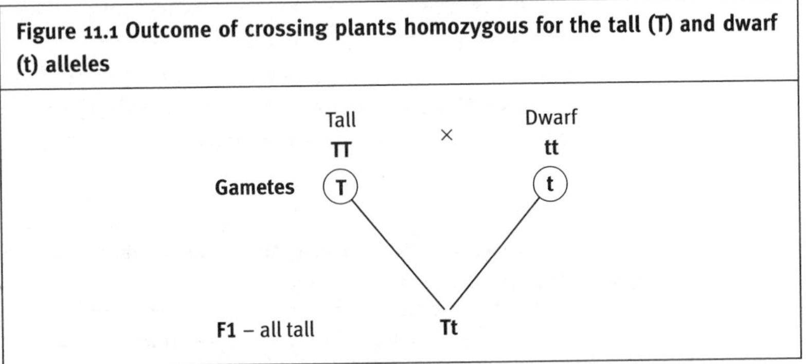

All the offspring are heterozygous. The fact that they are all tall tells us that T is dominant to t. The offspring of this cross are described as the *F1* (standing for first filial) generation.

If a further cross is now performed between the F1 offspring, the pattern shown in Fig. 11.2 emerges.

Figure 11.2 Outcome of crossing F1 generation plants heterozygous for the tall (T) and dwarf (t) alleles

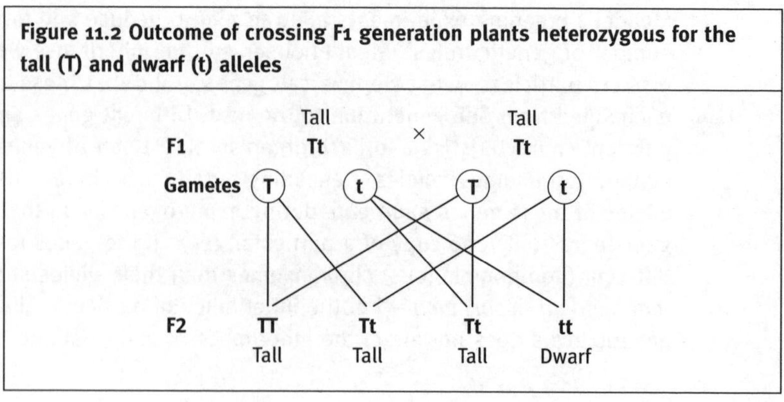

Each F1 plant produces equal numbers of gametes containing either the T or t alleles. Random combination of gametes produces the next (F2) generation where the offspring comprise approximately 75% tall plants and 25% dwarf plants. The three possible genotypes, TT, Tt and tt will be represented in a ratio of roughly 1:2:1.

11.6 Dihybrid inheritance

Mendel also studied the simultaneous inheritance of two genes encoding different characteristics. In this case the genes were for height (tall or dwarf) and for flower colour (purple or white). When true-breeding tall purple-flowered plants were crossed with true-breeding dwarf white-flowered plants, all of the progeny were tall with purple flowers. We already know that the tall allele is dominant to dwarf – this experiment further shows that the allele for purple flowers is dominant to that for white flowers. The cross is illustrated in Fig. 11.3.

Figure 11.3 Outcome of cross between tall purple and dwarf white plants to illustrate dihybrid inheritance

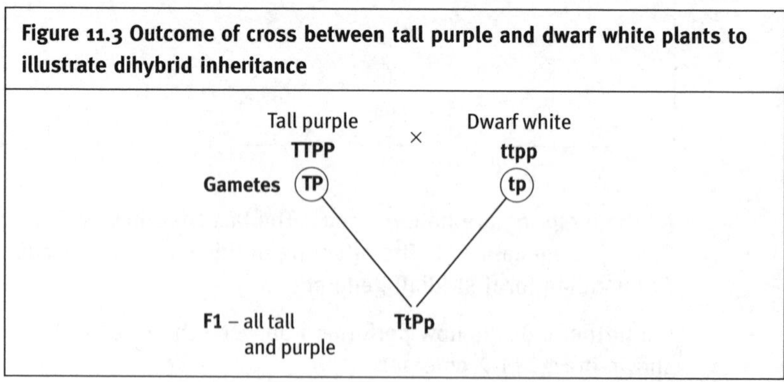

The F1 plants are therefore heterozygous at both loci, and because the two genes are located on different chromosomes and their alleles assort independently, gametes produced by F1 plants can be of four types – TP, Tp, tP or tp – produced in equal numbers. When the F1 plants are crossed, a random combination of the four gametes from each parent produces the pattern shown in Fig. 11.4.

Figure 11.4 Outcome of cross between F1 generation plants heterozygous for tall, dwarf, purple and white

F1 **TtPp** **TtPp**

Gametes (TP) (Tp) (tP) (tp) (TP) (Tp) (tP) (tp)

F2

	TP	**Tp**	**tP**	**tp**
TP	**TTPP** Tall, purple	**TTPp** Tall, purple	**TtPP** Tall, purple	**TtPp** Tall, purple
Tp	**TTPp** Tall, purple	**TTpp** Tall, white	**TtPp** Tall, purple	**Ttpp** Tall, white
tP	**TtPP** Tall, purple	**TtPp** Tall, purple	**ttPP** Dwarf, purple	**ttPp** Dwarf, purple
tp	**TtPp** Tall, purple	**Ttpp** Tall, white	**ttPp** Dwarf, purple	**ttpp** Dwarf, white

In other words, the following phenotypes are observed:

- 9 tall purple-flowered
- 3 tall white-flowered
- 3 dwarf purple-flowered
- 1 dwarf white-flowered.

This shows that the two characteristics, height and flower colour, behave independently of each other, consistent with Mendel's law of independent assortment. It is worth noting that in the above cross the only plants whose genotype can be deduced from the phenotype are the dwarf white plants. Plants that are either tall or purple could be either homozygous or heterozygous for these genes because the alleles for these characters are dominant.

11.7 Linkage

Some pairs of genes are observed to disobey the law of independent assortment and tend to be inherited together. This is because they are located on the same chromosome, and such genes are said to be *linked*. Because crossing-over between homologous chromosomes occurs at meiosis (see Chapter 10) genes that are far enough apart on the same chromosome may also assort independently because of the high likelihood of recombination events occurring in the intervening DNA. Genes that lie close together are much less likely to behave independently as the probability of recombination is much lower. In this way, by observing how often recombination occurs between a pair of genes, we can estimate their relative distance from each other on the chromosome to give a *linkage map*. Such mapping has been important in helping to identify genes associated with several human diseases.

11.8 Autosomal and sex-linked genes

Human beings have 23 pairs of chromosomes. One pair is known as the *sex chromosomes*, the other 22 pairs being described as *autosomes*. The sex chromosomes are called X and Y. Females have two X chromosomes and males have an X and a Y chromosome. The Y chromosome is much smaller than X and contains only a small number of genes. Genes that are found on the sex chromosomes are said to be sex-linked. For a gene on the X chromosome, a female may be either homozygous or heterozygous. However, a male who inherits only a single X chromosome will have only one copy of the gene. In the case of a recessive gene it is therefore more likely that the trait will be expressed in males. This is the case in red/green colour blindness and for some sex-linked diseases such as haemophilia.

11.9 Genetic fingerprinting

Human beings, with the exception of identical twins, are genetically unique. Genetic fingerprinting takes advantage of this fact to help identify criminals from forensic evidence and for paternity testing.

Genetic fingerprinting makes use of the fact that our DNA contains repetitive DNA sequences – short tandem repeats or STRs. These are present at multiple sites in the DNA and the number of repeats at each site varies between individuals. The technique allows scientists to determine the number of repeats at any given locus in a sample of DNA. By performing this analysis at several loci it is possible to say if a particular sample matches the DNA in another sample with a very high degree of certainty.

11.10 Evolution by natural selection

Each organism inherits a set of characteristics specified by genes acquired from its parents. However, populations can undergo changes in some of these characteristics over time. Such gradual changes are described as the process of *evolution.*

In the late nineteenth century two biologists, Darwin and Wallace, proposed a theory by which such evolutionary change might occur. This was the theory of *evolution by natural selection.* According to Darwin and Wallace, within a population of organisms there is genetic diversity. The alleles of some genes confer characteristics that are better than others in adapting the organism to the particular environment. Those organisms that carry these alleles will, on average, be more likely to survive and reproduce themselves so that these alleles are passed on to the next generation. In this way the population will change over time such that the frequency of such alleles increases. This idea is often referred to as *survival of the fittest* and the environmental factors that favour some genes over others give rise to *selective pressure.* Those organisms that survive are said to have a *selective advantage.* Cumulative changes can eventually give rise to organisms that are genetically so different from the original population that they would be unable to breed with them, and a new *species* has been produced.

Point mutations are one source of diversity that can lead to evolutionary change. Additionally there is evidence that new genes can arise through a process in which existing genes are duplicated, and the two copies of the gene can then diversify though independent mutations. An example of evolution through gene duplication is seen in the pigment genes in the eye that are responsible for colour vision. Colour vision in humans depends on genes for three visual pigments – red, green and blue. In contrast, New World monkeys have only two visual pigment genes – blue and red/green.

Major chromosomal changes can also contribute to evolution. Whereas humans have 23 pairs of chromosomes, chimpanzees and gorillas have 24. It appears that one of the large human chromosomes (chromosome 2) may have derived from two smaller chromosomes that are present in the apes but not in humans.

Evolution usually occurs over long periods of time. In this way all life on earth is envisaged to have evolved from a common ancestor over billions of years. However, where selection pressures are very high, evolutionary change can be observed over much shorter time frames. This is illustrated by the growth in antibiotic resistance that is being observed in a number of bacterial species. As the use of antibiotics has increased, so has the frequency of bacteria with genes conferring resistance to these drugs. One such antibiotic-resistant organism that is currently a cause of extreme concern is MRSA, which stands for methicillin-resistant *Staphylococcus aureus*. *Staphylococcus aureus* is a bacterium that can cause a range of problems from superficial abscesses to life-threatening conditions such as pneumonia and septicaemia. The antibiotic of choice for treating *Staphylococcus aureus* infections used to be penicillin. However, strains started to appear that were resistant to this drug and so in 1960 a new antibiotic, methicillin, was introduced. Within a very short space of time methicillin-resistant strains started to appear and MRSA is now a significant cause of hospital-acquired infections that are very difficult to treat. Fortunately there is an alternative antibiotic that can be used called vancomycin. Worryingly (but not surprisingly), strains of *Staphylococcus aureus* have now started to appear which are also vancomycin-resistant.

Clinical example: Sickle cell anaemia

Sickle cell anaemia is an inherited disease caused by a mutation in the haemoglobin gene. The mutation, which leads to a single amino acid substitution, when present in the homozygous form causes red blood cells to 'sickle' or change shape at low oxygen concentrations, which leads to blockage of blood vessels. The mutation is autosomal recessive – homozygous individuals have the disease whereas heterozygous individuals are carriers.

Interestingly, the mutation is found in higher frequencies in populations that originate from areas where malaria is endemic. The reason for this is that heterozygous carriers of the sickle cell mutation show an increased resistance to the malaria parasite, which infects red blood cells. This is an important lesson for us – it may seem desirable to eliminate from the population alleles that appear to be deleterious, but we should remember the sickle cell story and the advantage conferred by this disease-associated gene.

11.11 Test yourself

1. How is an individual described who has two identical alleles at a particular locus?

2. What type of mutation results in the addition of nucleotides to a coding sequence?
3. What can be deduced about two alleles that do *not* show independent assortment?
4. Which sex chromosomes are present in: (a) human females; (b) human males?
5. What is the term used to describe the benefit conferred on individuals by a genotype that enables them to cope better with environmental challenges?

12 Epithelial tissues

Basic concepts:
Epithelial cells form the covering of the external and the internal surfaces of the body. They also form glands. They provide the interface between the organs of the body and their external and internal environments. Epithelia are vital in ensuring that the internal environment is maintained in a state consistent with life. They perform a range of important transport functions. Understanding the role of epithelia is vital to an understanding of the body's interaction with its environment.

12.1 Introduction

Collections of cells that are specialised to perform a specific function are called *tissues*. The organ systems of the body are made up of four basic tissues – *epithelia*, *connective tissue*, *muscle* and *nervous tissue*.

An *epithelium* (plural *epithelia*) is a continuous layer of cells covering an internal or external body surface. The *epidermis* of the skin, which can be up to 200 cells thick, is an epithelium, as is the delicate single-cell lining of the alveoli of the lungs. Glands, which are the secretory organs of the body, are also derived from epithelial tissue.

12.2 Classification

Epithelia can be classified as *simple* or *stratified*. A simple epithelium consists of one layer of cells and a stratified epithelium has more than one layer of cells. In addition, epithelia are further described according to the shape of the cells in their upper layer. Three basic shapes are described – *squamous* (flattened), *cuboidal* and *columnar* (Fig. 12.1).

Epithelia have many functions and these are reflected in their cellular organisation as is summarised in the table below.

Classification	Function	Location
Simple squamous	Facilitate diffusion and transport from one compartment of the body to another	Alveoli of lungs Capillaries Kidney glomerulus
Simple cuboidal	Often associated with transport of ions across cells	Kidney tubule Thyroid follicle
Simple columnar	Secretion and absorption	Small and large intestine Pancreas
Stratified cuboidal	Lining of ducts	Sweat gland duct
Stratified squamous	Protection against wear and tear	Skin Vagina Oral cavity

Figure 12.1 Types of epithelia

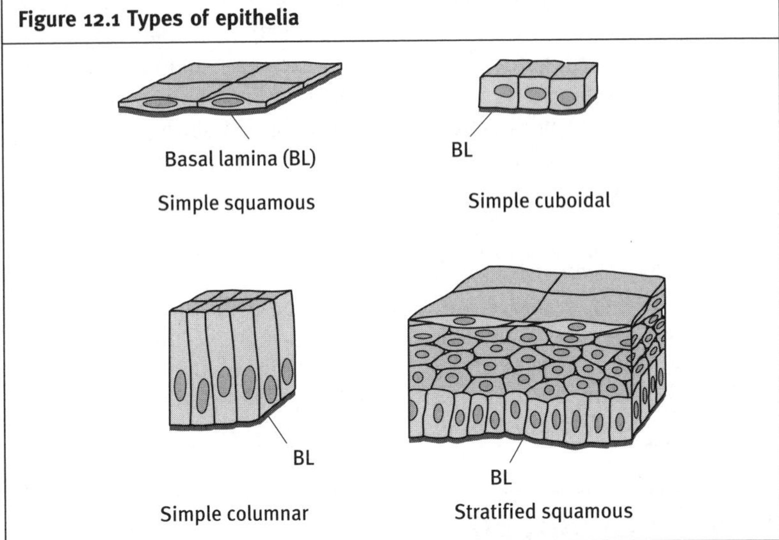

12.3 Adhesion

It is important that epithelial cells maintain a coherent sheet-like structure by being able to adhere both to neighbouring cells and to the underlying tissues. The majority of epithelial cells sit on a thin protein sheet known as the *basal lamina*. The basal lamina is secreted by the epithelial cells themselves and is a fine meshwork of proteins, including collagen (see Chapter 2). This extracellular meshwork provides attachment both for the epithelial cells above and for the connective tissue below and is important in linking epithelial sheets to the underlying connective tissues.

The interactions between adjacent epithelial cells depend on three main types of intercellular junctions.

- *Desmosomes* are adhesions between adjacent cells characterised by dense circular protein plaques inside the cell membrane. These act as an anchor point for extracellular protein bridges and as a point of attachment within the cell for cytoskeletal proteins. Desmosomes also link the epithelial cells to the basal lamina and thus anchor the epithelial sheet in place.
- *Tight junctions* are areas where the cell membranes of adjacent cells are held close together by shared trans-membrane proteins. They act to restrict the movement of ions and other substances between epithelial cells and thus help to prevent epithelia being 'leaky'.
- *Gap junctions* are small pores formed between adjacent cells that permit the free movement of water, ions and small molecules such as glucose and amino acids. Gap junctions allow cells in an epithelium to act together in response to external stimuli. In stratified epithelia gap junctions allow the passage of metabolic substrates to cells that are not in direct contact with the blood supply. Gap junctions are also found in excitable tissues such as cardiac muscle (see Chapter 14).

Clinical example: Bullous pemphigus

This is a rare autoimmune disease in which the body develops antibodies against proteins of the desmosomes in skin. This results in the epidermis of the skin lifting away from the underlying connective tissue and leads to the formation of blisters. The disease can be life-threatening.

12.4 Test yourself

1. How many layers of cells occur in a simple epithelium?
2. How are epithelia containing multiple layers of flattened cells described?
3. What is the term used to describe the protein sheet that underlies an epithelium?
4. What type of junctions in epithelia are characterised by shared membrane proteins between adjacent cells?
5. What type of junctions link an epithelium to its basal lamina?

13 Connective tissues

Basic concepts:
Connective tissues, such as bone and cartilage, are key components of the skeleton. As such they are essential in locomotion. Other connective tissues are important structural elements of many organs and some allow the body to store energy in the form of fat. Connective tissues are all composed of the same basic elements but vary in how these are put together. Understanding the structure–function relationships in connective tissues is the key to understanding their diverse roles.

13.1 Introduction

There are a wide variety of types of *connective tissue* in the body – including adipose tissue, tendon, cartilage and bone – as well as the specialised tissues of the teeth – enamel and dentine. All connective tissues have the same basic elements in that they consist of specialised cells surrounded by an *extracellular matrix*. The matrix is made up of a hydrated glycosaminoglycan gel in which are embedded protein fibres. The matrix is secreted and maintained by the cells which sit within it. In mineralised connective tissues, such as bone, the matrix also contains crystals of calcium phosphate which provide rigidity.

13.2 Glycosaminoglycans

These are unbranched polysaccharide chains with repeating disaccharide units. There are a number of different *glycosaminoglycans* (GAGs) depending on the composition of the disaccharide. Some GAGs are also sulphated. Individual GAGs are assembled onto a protein core (like the teeth of a comb) to form a larger structure known as a *proteoglycan*. These proteoglycans are in turn linked together to form a supermolecule known as an *aggrecan*. All GAGs carry significant negative charge which draws osmotically-active cations (predominantly Na^+) into the matrix which in turn attract water and create a hydrated gel. The length and composition of the individual GAGs within the aggrecan molecule will determine the total negative charge and thus the amount of water drawn in. This will determine the turgidity (degree of swelling) of the extracellular matrix. A connective tissue such as cartilage has a high water content and is thus able to resist compression.

13.3 Fibres

The fibrous components of the connective tissue matrix comprise principally *collagen* and *elastin.*

Collagen

Collagen provides tensile strength and flexibility. The presence of parallel bundles of collagen fibres in a tendon provides a good example of these properties. As discussed in Chapter 2, collagen is a fibrous protein assembled from three helical α-chains. There are up to 25 collagen α-chains coded for in the genome and the collagen type is determined by the precise combination of the α-chains from which it is made. The collagen type which is found in most major connective tissues, with the exception of cartilage, is Type 1 and this is made up of two α_1 and one α_2 chains. Once the individual triple helical collagen molecules have been secreted into the extracellular matrix they begin to self-assemble into fibrils and then into fibres. A collagen fibre can be up to 10 μm in diameter. Not all collagen types assemble into fibres – for example, Type IV collagen, which is found in the basal lamina, assembles as a fine meshwork.

Clinical example: Scurvy

Scurvy is a condition once associated with sailors on long voyages and is characterised by weakness, bleeding under the skin, teeth falling out and ultimately death. It is caused by a lack of fresh fruit leading to vitamin C deficiency which results in collagen fibre assembly malfunction. The reason for this is that the assembly of collagen molecules into fibres depends on the presence of the amino acid 4-hydroxyproline in the collagen molecule. The production of this hydroxylated amino acid depends on the presence of vitamin C.

Elastic fibres

Elastic fibres have a composite structure consisting of a core of a coiled and cross-linked protein, elastin, surrounded by a fine meshwork of another protein, fibrillin. Elastic fibres allow connective tissues to be distorted and then to return to their original shape. The walls of blood vessels contain elastic fibres which allow them to expand and then recoil as the blood pulses through them (see Chapter 18).

Other matrix proteins

In addition to collagen and elastin, the extracellular matrix of connective tissues contains a wide variety of glycoproteins with diverse functions, which include regulating the assembly of the other components of the matrix, acting as binding sites for adhesion molecules on cell surfaces, acting as molecular markers for cell migration pathways and inhibiting unwanted calcification.

13.4 Cells

A range of cell types are found within connective tissues. The cells responsible for the secretion of the extracellular matrix are known as '...blast' cells and are further described according to the nature of the connective tissue they produce. Thus *fibroblasts* secrete fibrous connective tissues such as tendon and ligament while *chondroblasts* lay down cartilage and *osteoblasts* deposit bone.

Once these cells have deposited a matrix around themselves they become the mature cells of the connective tissue and are responsible for the subsequent repair and maintenance of the matrix. Death of these cells would result in resorption of the matrix. The mature cells are known as '...cytes' – as in *fibrocytes, adipocytes* (in adipose tissue), *chondrocytes* and *osteocytes.* In fibrous connective tissues both fibrocytes and stem cells within the connective tissue retain the capacity to divide and will do so in response to injury or insult. This can lead to scarring (*fibrosis*) if excessive cellular activity with consequent deposition of collagen fibres into the extracellular matrix occurs.

Connective tissue matrices also play host to many other cell types:

- *mast cells* – which secrete heparin and histamine and are important in producing localised inflammation in response to injury and in certain types of allergic reaction;
- *plasma cells* – which are responsible for antibody production;
- *macrophages* – which are blood-derived scavenger cells.

13.5 Test yourself

1. What substance of the extracellular matrix is responsible for drawing in water to form a gel?
2. What protein provides strength and flexibility to the extracellular matrix?
3. What name is given to the cells responsible for laying down cartilage?
4. What are the mature cells in cartilage called?
5. What is the function of mast cells in connective tissues?

14 Excitable tissues

Basic concepts:
Muscle and nerve are known as excitable tissues. This is because in response to stimuli they can rapidly alter the disposition of ions across their membranes, leading to changes in electrical potential. These electrical changes can be transmitted along the membranes of excitable cells. In muscle cells this will trigger contraction. Nerve cells use electrical changes to transmit information and are capable of exciting adjacent nerve cells to generate a network in which cooperative activity can lead to higher cognitive processes such as thought. Understanding the basic processes by which electrical changes are generated is important in understanding the basis of thought, senses and movement.

14.1 Introduction

Muscle and *nerve* are both described as *excitable tissues*. This relates to the fact that both types of tissue have membranes which maintain a significant electric potential difference between the inside and outside of the cell. This is normally expressed in millivolts, with the outside of the cell considered as being at 0 mV. The inside of the cell is negative with respect to the outside and in excitable tissues normally lies between −60 and −80 mV. This potential difference (known as the *resting membrane potential*) can be altered rapidly by the movement of ions across the membrane, and when this results in the inside of the cell becoming less negative is known as *depolarisation* or *excitation*.

14.2 Membrane potential

In order to understand how excitable tissues work it is useful to understand how the potential difference across a cell membrane is generated. In order to explain this we will use a hypothetical situation in which we consider only a limited number of the charged molecules which are found within cells and in the extracellular fluid.

As explained in Chapter 1, the extracellular fluid (ECF) has high levels of Na^+ ions and the intracellular fluid (ICF) has high levels of K^+ ions. Consider a system at equilibrium in which the ECF and ICF contain equal numbers of ions and have equal numbers of positive and negative charges – and thus are electrically neutral. In this situation there is no difference in charge across the membrane. If we were now to open in the membrane an ion channel selective for K^+ ions then the movement of this ion through the channel would depend on two forces. The first would be a force generated by the concentration gradient for K^+ ions across the membrane where internally the K^+ ion is at a high concentration (160 mmol) and externally it is at a low concentration (4 mmol). Thus simple diffusion would tend to cause K^+ ions to move out of the cell (see Fig. 14.1). However, every K^+ ion that moves out of the cell will create an electrical imbalance across the membrane since a positive ion moving out will leave a spare negative ion behind which cannot cross the membrane because there are no channels open for it to use. The developing difference in charge created by leakage of K^+ ions out of the cell will make the movement of further K^+ ions increasingly difficult as they are drawn back by the negative charge. The outward force generated by the concentration gradient is opposed by a developing electrical gradient. Ultimately these two forces will reach equilibrium and a membrane potential will be established at which net inward and outward movement of K^+ ions is balanced.

Figure 14.1 Generation of the resting membrane potential

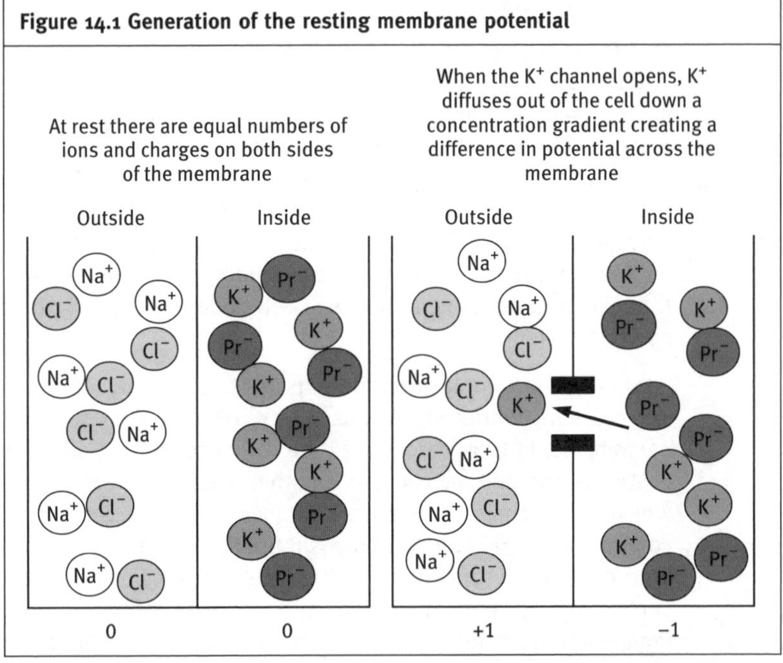

For a single ion, like K^+, this is known as the *equilibrium potential* (E_k) and is described by an adaptation of the Nernst equation (below) in which *R* is the gas constant, *F* is the Faraday constant, *T* is the temperature (°Kelvin), Z is the valence (charge) of the ion and the terms in square brackets are the concentrations of K^+ in the ECF, $[K^+]_e$ and the ICF, $[K^+]_i$:

$$E_k = \frac{RT}{ZF} \log \frac{[K^+]_e}{[K^+]_i}$$

When this equation is applied to the physiological values for K^+ it gives an answer of −75 mV. That this value is remarkably close to the actual resting membrane potential of excitable tissues is because the only major ion which is free to move at rest in these cells is K^+ for which there exists a specific leak channel. Movement of the other ions contributes very little to the resting membrane potential.

14.3 Muscle

Muscle cells are capable of contraction either spontaneously or in response to external stimuli. There are three main types of muscle in the body – *skeletal muscle*, *cardiac muscle* and *smooth muscle*.

Skeletal muscle

This is the type of muscle which is found in the musculoskeletal system and is responsible for voluntary movement. An individual skeletal muscle (such as the biceps in the arm) is attached at both ends to a bone via a *tendon* (see Chapter 24). Contraction of the muscle results in the movement of one bone relative to the other. The cells of skeletal muscle are elongated, multinucleated tubular structures known as *myocytes* or muscle fibres. Each muscle fibre is innervated centrally, in an area known as the neuromuscular junction, by the terminal branch of the axon of a motor neuron whose cell body is located in the central nervous system.

Internally, the cytoplasm of the muscle fibre is filled with structures known as *myofibrils* which consist of repeated units known as *sarcomeres*. In turn, sarcomeres are composed of microfilaments of the proteins *actin* and *myosin* and it is the interaction of these proteins which is responsible for contraction (see Fig. 14.2). Each sarcomere contains a central bundle of myosin filaments which are overlapped at both ends by an array of actin filaments. Each myosin filament is surrounded by, and can interact with, six actin filaments. The myosin filament is assembled from a number of myosin molecules, each of which contains a rod-like domain and a flexible head which is capable of binding actin and breaking down ATP. The actin filaments at both ends of the sarcomere insert into a protein band known as the *Z-line*. Thus a sarcomere runs from one Z-line to another. The part of the sarcomere containing the thicker myosin filaments tends to stain more darkly when viewed down the microscope and is known as the *A-band*. The area of the sarcomere containing only actin is known as the *I-band*. The alternating dark and light bands give skeletal muscle a striped appearance – hence it is sometimes referred to as *striated muscle*.

Figure 14.2 Structure of actin and myosin filaments

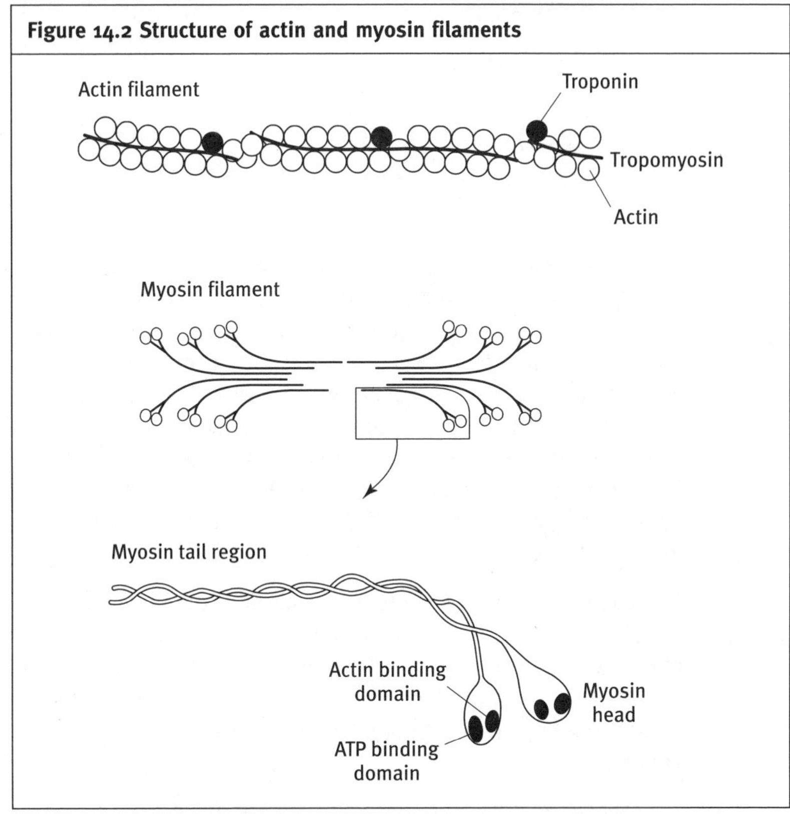

The process of muscle contraction depends on the interaction between myosin and actin filaments. Basically, the extent to which the actin and myosin filaments overlap is increased during contraction and this results in the shortening of the sarcomere. The mechanism by which this occurs is outlined below and in Fig. 14.3.

Nerve impulses arriving at the neuromuscular junction trigger the release of the neurotransmitter acetylcholine (ACh). Binding of ACh to specific receptors on the muscle membrane causes the opening of Na^+ channels, leading to the influx of Na^+ ions and depolarisation of the muscle cell at the neuromuscular junction.

Figure 14.3 Muscle contraction

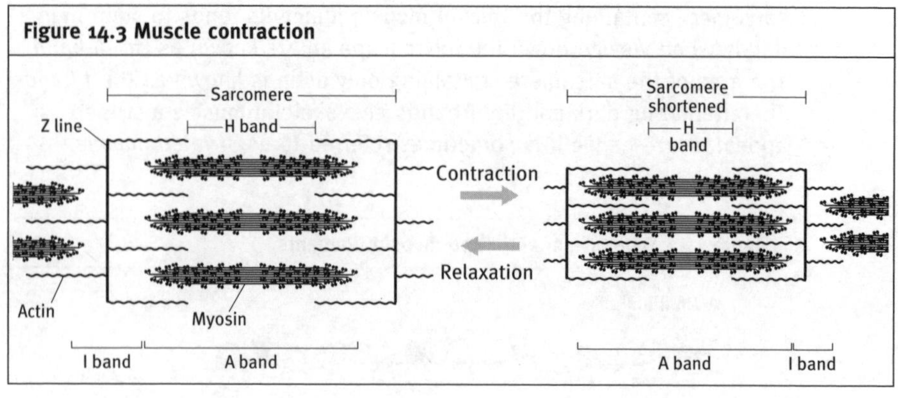

The depolarisation spreads away from the neuromuscular junction, across the surface of the muscle fibre as voltage-dependent Na^+ channels adjacent to the initial site of depolarisation are triggered to open. The depolarisation is also transmitted through a system of invaginations of the cell membrane (*t-tubules*) which run into the centre of the muscle fibre. The t-tubules are closely related to expansions of smooth endoplasmic reticulum (SER) within the cytoplasm of the muscle fibre. The depolarisation of the t-tubules triggers the opening of voltage-dependent Ca^{2+} channels in the membrane of the SER allowing Ca^{2+} ions to enter the cytoplasm of the muscle fibre.

The Ca^{2+} ions bind to a protein known as troponin which, in conjunction with a second protein tropomyosin, lies along the actin filament and blocks access to the myosin binding site on the actin molecules. Ca^{2+} binding to troponin induces a conformational change which exposes the myosin binding sites. The myosin heads then cross-link with the actin filament.

The myosin acts as an ATPase and breaks down ATP to ADP. The energy from this reaction is used to cause a change in the angle of the myosin head such that the actin filament is drawn towards the centre of the A-band.

The myosin releases the actin and replaces the ADP it now has bound to its head with ATP. The angle of the head changes back to its resting conformation.

If Ca^{2+} ions are still present then these events are repeated and contraction proceeds. Contraction is terminated by the active removal of Ca^{2+} ions from the cytoplasm of the muscle fibre back into the SER. This occurs when depolarisation at the neuromuscular junction has been terminated by the breakdown of ACh by the enzyme acetylcholinesterase.

The force of the contraction generated by the shortening of the sarcomeres is transmitted through their attachment to the cell membrane by a set of proteins, including linker proteins known as the *dystrophins*, and from there to the fibres of the connective tissue in which the muscle fibres are embedded. These connective tissue fibres connect with more densely packed fibres of the tendon or ligament at the end of the muscle, and thus force is transmitted to the bone.

Clinical example: Muscular dystrophy

Muscular dystrophy is a wasting disease of muscle which affects young boys. It is an X-linked genetic disease caused by a mutation in the gene for a dystrophin. The improper attachment of myofilaments to the cell membrane of muscle fibres results in the membranes of the muscle fibres gradually becoming damaged and the cells degenerating. The effects of the mutation are cumulative and so those affected may have normal muscle function in infancy but become progressively weaker and they normally die in adolescence.

Cardiac muscle

Cardiac muscle is a specialised form of muscle found only in the heart. It consists of a network of branched cells joined end-to-end by a specialised junctional complex known as the *intercalated disc*. This structure consists of alternating segments of desmosal-like cell adhesions and gap junctions. The gap junctions allow the free movement of ions from one cardiac muscle cell to another, and thus enable cardiac muscle to operate as an electrical *syncytium* in which depolarisation of one cell leads to depolarisation and contraction of all connected cells.

Internally, cardiac muscle cells contain myofibrils similar to those found in skeletal muscle and thus cardiac muscle also has a striated appearance. The mechanism of contraction of cardiac muscle cells is similar to that in skeletal muscle depending on the internal spread of depolarisation through t-tubules and the release of Ca^{2+} ions which bind to troponin. Cardiac muscle cells are able to depolarise spontaneously and this property is used by pacemaker cells in the heart wall to generate rhythmic beating (see Chapter 18).

Smooth muscle

Smooth muscle is normally found in the walls of internal organs – such as the gastrointestinal tract, the bladder and the uterus – and is responsible for slow and sustained involuntary movements. The cells of smooth muscle (Fig. 14.4) are spindle shaped and in most tissues electrical continuity is provided through the presence of connecting gap junctions. Internally the basic unit of contraction is the sarcomere but there are significant differences in terms of both the organisation of the sarcomeres and the mechanism of contraction between smooth muscle and that described above for skeletal and cardiac muscle.

Sarcomeres within smooth muscle cells run diagonally across the muscle cell and are attached at both ends to a structure known as a *dense body* rather than into a Z-line. Contraction of the sarcomere results in a smooth muscle cell becoming shorter and fatter. Most smooth muscle cells are in a permanently semicontracted state and this resting level of contraction is known as *muscle tone*. Smooth muscle contraction can then be regulated either upwards or downwards leading to further contraction or to relaxation.

Figure 14.4 Smooth muscle cell

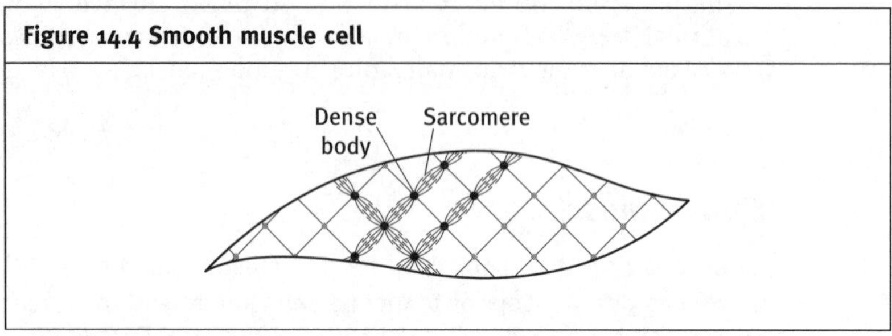

Contraction in smooth muscle cells, as in striated muscle, is regulated by the levels of intracellular calcium. Contraction is initiated by the entry of Ca^{2+} ions into the cytoplasm, either through regulated Ca^{2+} channels in the cell membrane or by release from intracellular stores. The cytoplasmic Ca^{2+} ions bind to the protein *calmodulin* to produce an activated Ca^{2+}-calmodulin complex. This interacts with the inactive cytoplasmic protein *myosin light chain kinase* (MLCK) to convert it into active MLCK. The MLCK phosphorylates a component of the myosin head enabling it to interact with actin. At this point the myosin converts ATP to ADP and the energy released is used to adjust the angle of the myosin head to slide the actin filament towards the centre of the sarcomere. If Ca^{2+} ion levels remain high the contraction may be maintained, or if Ca^{2+} ion levels fall then relaxation may ensue.

14.4 Nerve

The nervous system is made up of two main cell types – *neurons* and *glia*. The neurons are the main excitable cells within the nervous system and the glia act predominantly as supporting cells functioning to myelinate the axons of neurons and to regulate the ionic and nutrient content of the local microenvironment. The nervous system will be considered in detail in Chapter 17 and in this section only the structure of neurons and their basic electrical properties will be considered.

Neurons

A neuron has three main components as shown in Fig. 14.5. There is a *cell body*, which contains the nucleus, from which arise two structures – *dendrites* and an *axon*. Information in the nervous system is represented as patterns of electrical activity within its neurons. The nervous system is constructed to allow the flow of information from one set of neurons to the next, and the neurons are organised into complex networks in which each cell may receive information from up to 10 000 other cells. This transmission of information occurs at specialised structures known as *synapses*, which are formed between the axon of one neuron and the dendrites of the next. The dendrites of a neuron form an extensively branched network arising from the cell body, and the membrane of the dendrites is capable of being depolarised at multiple points as the result of activity at synapses (see below). The depolarisation of the dendrites spreads passively towards the cell body, which acts to sum the activity in all the dendrites which arise from it. Depending on the level of summed activity arriving at the cell body, a wave of depolarisation known as an *action potential* may be initiated, and then transmitted along the axon to its terminal. Arrival of the action potential at the axon terminal triggers release of a chemical

Figure 14.5 A neuron

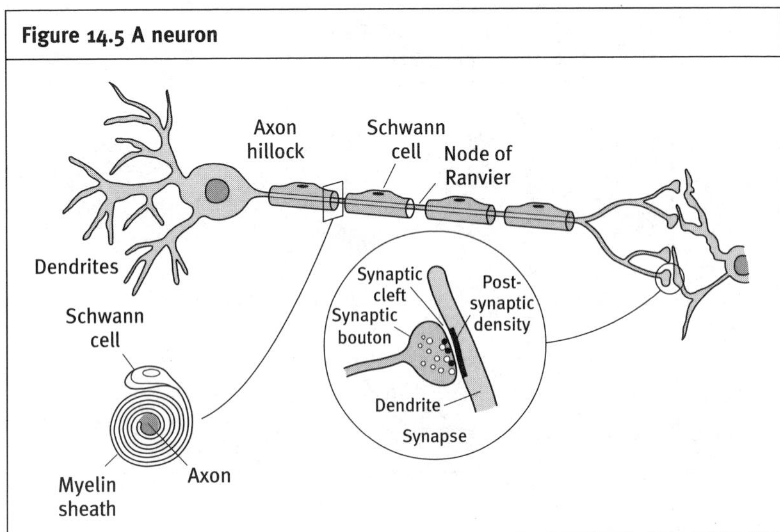

neurotransmitter which acts at the synapse to depolarise the adjacent dendritic membrane (see below).

Action potential

The action potential is a wave of depolarisation which can be transmitted along an axon. It is an all-or-nothing phenomenon, meaning that it has the same amplitude regardless of both the magnitude of the stimulus and wherever it is recorded along an axon. An action potential may be represented graphically as shown in Fig. 14.6. This indicates what would be measured if one recorded the membrane potential at a fixed point along the axon with a recording electrode placed inside the axon. The vertical axis is voltage (the membrane potential) and the horizontal axis is time. At rest (point *A*) the membrane is at a fixed resting membrane potential of –70 mV. As a result of local depolarisation the membrane is gradually depolarised (*B*) until it reaches a threshold point (*C*) of around –50 mV. At this point, large numbers of voltage-sensitive Na^+ channels in the axonal membrane open and Na^+ ions flood into the cytoplasm of the axon down their concentration gradient. This rapid influx begins to depolarise the membrane. Because Na^+ is now the major ion moving across the membrane, the Nernst equation can be applied for Na^+ and the membrane potential heads rapidly (*D*) towards the Na^+ equilibrium potential, which is +50 mV. However, before the membrane can reach the Na^+ equilibrium potential, the depolarisation triggers the opening of K^+ channels in the membrane. This results both in the efflux of K^+ ions and the closure of the Na^+ channels. Now K^+ is the major ion moving across the membrane (*E*) and so the membrane potential heads back

Figure 14.6 Action potential

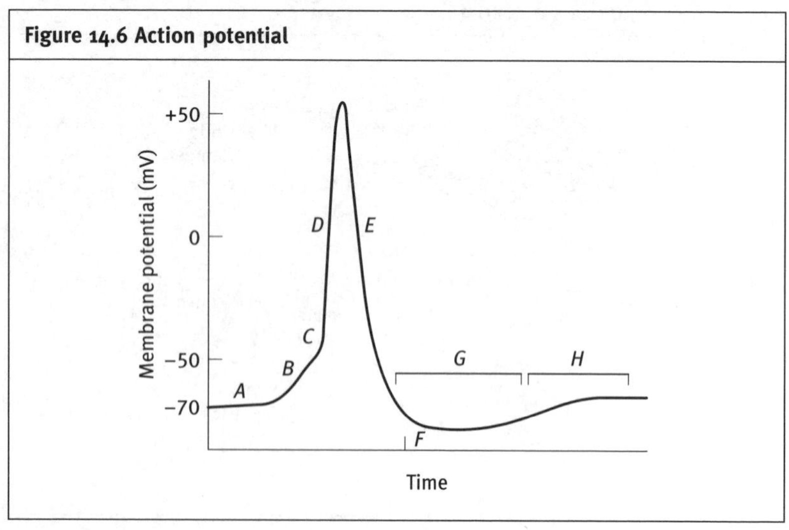

towards the K^+ ion equilibrium potential of –75 mV. At point *F* the membrane hyperpolarises (goes below its resting membrane potential) as the K^+ ion equilibrium potential is approached. The K^+ channels then close and the activation of the Na/K ATPase restores the ionic balance across the membrane. During this period of hyperpolarisation there is an initial phase (*G*) during which the Na^+ channels are rendered incapable of reopening and thus no further action potentials may be generated in this region of membrane. This is known as the *absolute refractory period.* This is functionally important because action potentials normally travel only away from the cell body. This wave of depolarisation is propagated along the axon by depolarisation in one region triggering the opening of Na^+ channels in the adjacent membrane regions. Because the membrane which has just been depolarised is refractory, it cannot undergo further depolarisation and thus the action potential can move in only one direction. Following the absolute refractory period there is a further period during which the membrane is slightly hyperpolarised (*H*) and thus further away from threshold value and more difficult to depolarise again. This is known as the *relative refractory period.*

For most neurons, information is coded in the frequency and pattern of action potentials which they transmit.

Myelination

Conduction of an action potential along an axon is a relatively slow process, depending as it does on the sequential opening of Na^+ channels. In thin unmyelinated axons the rate of conduction can be as slow as 2 m/s. Given that reaction times between receiving a visual cue and responding with a finger movement can be as fast as 0.2 s and that the distance from the brain to the finger tip is at least 1 m, it is clear that most axons are able to conduct action potentials at a much faster rate. Indeed the fastest axons in our body have a conduction velocity of 100 m/s. Conduction rates are enhanced because axons are surrounded by a *myelin sheath* (see Fig. 14.5). This is formed from the cell membrane of a specialised glial cell. In the brain these cells are known as *oligodendroglia* and in peripheral nerves myelination is carried out by *Schwann cells.* A myelin sheath consists of a number of concentric layers of the cell membrane of the glial cell wrapped around the axon. The membranes within the myelin sheath contain specialised lipids and associated proteins such as myelin basic protein. A basal lamina lies between the myelin sheath and the axonal membrane.

Because it is predominantly lipid, the myelin sheath acts as an electrical insulator. Along the length of an axon the myelin sheath is not continuous and at the points between adjacent glial cells the axonal membrane is uncovered at areas known as the *nodes of Ranvier.* These nodes are the only areas of the axon where depolarisation may occur. During transmission of an action potential along a myelinated axon, depolarisation at one node triggers immediate depolarisation at the next node and the action potential leaps from node to node down the axon. This is known as *saltatory conduction* and greatly enhances both the speed and efficiency of conduction by reducing the number of channel opening events required to get from one end of the axon to the other.

Clinical example: Multiple sclerosis

MS is a disease in which the myelin sheath of axons is destroyed. It is caused by the immune system attacking the myelin sheath by directing an immune response against components such as myelin basic protein. The resultant demyelination produces significant impairment of function in the affected pathways. Individuals with MS may suffer from loss of vision, loss of motor function or loss of peripheral sensation. The disease is particularly devastating because of its random nature – there is no way of knowing when or where it will strike and which functions will be affected. At present there is no cure and the only treatments are drugs which suppress immune function, such as corticosteroids.

Synaptic transmission

Although each neuronal cell body gives rise to only a single axon this may branch many times at its termination and thus make contact with many dendrites. The contact between an axon terminal and a dendrite is known as a *synapse* (see Fig. 14.5). There is a small swelling at the end of the axon which is filled with many vesicles containing a chemical neurotransmitter. When the action potential arrives at the end of the axon it triggers the opening of voltage-dependent Ca^{2+} channels and the subsequent influx of Ca^{2+} ions triggers vesicle fusion with the presynaptic membrane (Fig. 14.6). Each vesicle then releases into the synaptic cleft a small amount of neurotransmitter known as a *quantum.* Vesicle recycling occurs in that once a vesicle has released its contents it is internalised back into the axon terminal and refilled with neurotransmitter. On the dendritic side of the synaptic cleft are membrane receptors that bind the neurotransmitter. Some neurotransmitters, such as acetylcholine, are known as excitatory and open Na^+ channels in the postsynaptic membrane. The effect of this is to produce small depolarisations. Other neurotransmitters, such as gamma-aminobutyric acid (GABA), are inhibitory and these open Cl^- channels, tending to oppose depolarisation. The net effect of these events at all the synapses contacting one neuron may lead to the generation of an action potential at the cell body. Termination of the interaction between the transmitter and the receptor is achieved either by enzymatic breakdown of the transmitter in the synaptic cleft or by the re-uptake of the transmitter into the axon and its internal recycling.

Clinical example: Depression and SSRIs

Serotonin is a neurotransmitter which acts to modify the activity of neurons in the brain, which are important for our control of mood. It is known that one of the causes of depression is a reduced level of serotonin within certain areas of the brain. A class of antidepressant drugs known as selective serotonin re-uptake inhibitors (SSRIs) block the mechanism which takes serotonin back up into the axon terminal and thus increases the amount and duration of action of the serotonin present in the synaptic cleft. This helps to alleviate the symptoms of depression.

14.5 Test yourself

1. Which cation, found in higher concentration in the cytoplasm than in the extracellular fluid, is principally responsible for maintaining the potential difference across a cell membrane?
2. What two proteins present in the microfilaments of muscle cells are mainly responsible for muscle contractions?
3. What neurotransmitter released at the neuromuscular junction causes the depolarisation of the muscle cell membrane?
4. What are the three main components of a neuron?
5. What is the function of the myelin sheath?

15 Homeostasis

Basic concepts:
Homeostasis is the regulatory process by which the body maintains a constant internal environment. For each parameter to be regulated it is important to have a sensor mechanism which allows its current status to be monitored. The current status is then checked against a predetermined 'ideal state' and deviations from this will initiate a response that will move the measured parameter back towards the 'ideal'. The response may be achieved via hormonal or neural mechanisms. An understanding of homeostasis underpins an understanding of much of physiology.

15.1 Introduction

Homeostasis is the set of processes by which the body strives to maintain a constant internal environment. This principally involves regulation of the composition of the extracellular fluid compartment. The key properties that are regulated are *temperature*, *blood pressure* and *volume*, *plasma metabolite levels*, *plasma oxygen levels* and *plasma osmolarity*. A constant internal environment is needed to ensure that cells can function normally. For example, if plasma potassium levels are not maintained within very tight limits then membrane potentials in excitable tissues are affected. This can rapidly result in heart failure and death.

Most homeostatic mechanisms in the body are based around a feedback loop in which the levels of the substance or parameter to be regulated are constantly monitored. Fluctuations of these levels away from a preset norm are detected and signals passed to an effector mechanism which adjusts the levels back into the normal range. The signalling mechanisms for both the feedback and effector arms of the system can be either hormonal or neural. Once the levels are back within the normal range the effector mechanism is shut down.

In many systems, constant levels are achieved by the action of two opposing mechanisms – one to elevate the parameter and one to depress it. Homeostasis often results in oscillations of any given parameter about the set level because feedback results in constant adjustments up and down. An effective homeostatic mechanism will minimise these fluctuations.

15.2 Regulation of plasma glucose

Homeostasis of plasma glucose offers an example of a mechanism which relies predominantly on hormonal feedback loops (Fig. 15.1). Following a meal, absorption of glucose from the small intestine results in a rapid rise of plasma glucose. Prolonged elevation of plasma glucose can harm the body in a number of ways, including damage to small blood vessels in the retina and kidney. It is thus important that plasma glucose levels are maintained within a normal range of about 4.5–5.6 mmol/l.

Detection of plasma glucose levels is achieved by beta cells located within the pancreas gland. The beta cells contain vesicles which store the protein hormone insulin. Glucose enters the beta cells and triggers the release of the insulin into the bloodstream. The release of insulin is stimulated at plasma glucose levels above 5 mmol/l. Insulin travels in the bloodstream to its target tissues, which are primarily the liver and skeletal muscle. Insulin binds to its receptor and enhances the uptake of glucose into the cells of these tissues. Once plasma glucose levels have fallen to within the normal range then insulin release is shut down and glucose uptake is down-regulated.

If plasma glucose levels fall below the normal range then a second type of cell in the pancreas, the alpha cell, is stimulated to produce the hormone glucagon. This acts on liver to stimulate the breakdown of glycogen into glucose which is then released into the bloodstream.

Figure 15.1 Feedback loop in the control of blood glucose

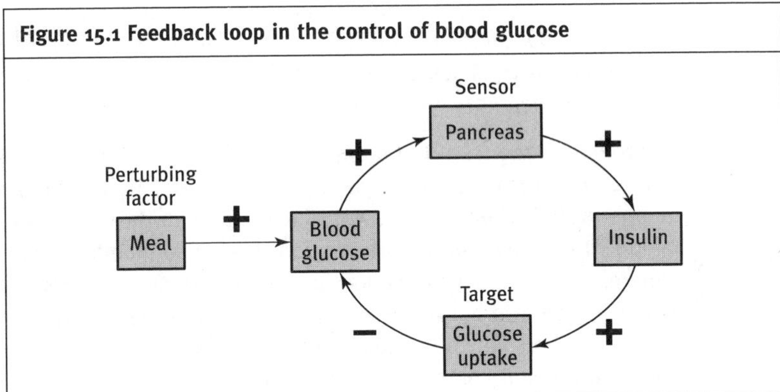

Clinical example: Diabetes

Diabetes is characterised by a loss of control over plasma glucose resulting in increased levels of glucose in the blood (hyperglycaemia).

In Type 1 diabetes, which particularly affects adolescents, the body's own immune system destroys the beta cells in the pancreas. Thus the body is no longer able to respond to increases in plasma glucose by secreting insulin, and meals are followed by rapid rises in plasma glucose, the osmotic and metabolic effects of which can lead to coma.

In Type II diabetes, which tends to affect older obese patients, the target cells in liver and muscle lose their ability to respond to insulin. This is the result of a desensitisation of insulin receptors. Patients with Type II diabetes show chronically elevated plasma glucose levels. In the long term this can damage small blood vessels leading to retinal degeneration and blindness and to loss of blood flow to the legs and feet causing death of tissues in the feet and their eventual amputation.

15.3 Thermoregulation

Maintenance of a constant internal temperature is vital for the correct functioning of body systems. If body temperature rises above about 41°C then proteins can start to denature and metabolic systems break down. Below 32°C metabolic processes slow down so much that energy production is negligible and cell functions are seriously impaired. *Thermoregulation* is primarily achieved through the actions of the autonomic nervous system (see Chapter 17) and provides an example of a neural homeostatic mechanism.

Neurons in a region of the brain called the hypothalamus (see Fig. 15.2) act as the primary temperature regulatory centre. They monitor core temperature by being sensitive to fluctuations in the temperature of the blood which flows through the region. They also monitor peripheral temperature by receiving neural input from thermoreceptors in the skin, which respond to both cold and warmth. Within the hypothalamus there is a set of cells which increase their firing rate in response to a drop in temperature and another set which respond to a rise in temperature. In response to a change in temperature, there are a number of mechanisms (neural, metabolic and behavioural) which the body can utilise to alter heat generation and to regulate heat loss at the periphery.

As temperature falls, an increase in the activity of the sympathetic nervous system (see Chapter 17) leads to constriction (narrowing) of blood vessels near the body surface, causing a diversion of blood flow away from the vessels immediately underneath the epidermis of the skin – thus reducing heat loss from the surface. In addition, the muscles which are attached to hair follicles are stimulated to contract, causing hair to stand on end. In furry animals this leads to an increase in the thickness of the layer of air trapped in the hair – thus improving insulation. The principal effect in humans is to produce 'goosebumps'. Skeletal muscles are stimulated to contract and relax repeatedly, causing shivering. This increases their heat production as a result of increased metabolic activity. The thyroid and adrenal glands (see Chapter 16) are stimulated to release thyroid hormone and adrenaline respectively. Both of these hormones act to increase metabolic activity in organs such as the liver and thus generate heat. In addition, behavioural programmes are initiated which lead to simple behaviours such as huddling up and to complex adaptive behaviours such as seeking shelter and putting on additional clothing.

Newborn babies have an additional thermoregulatory mechanism in that they possess a specialised adipose tissue known as brown fat.

Figure 15.2 Temperature regulation

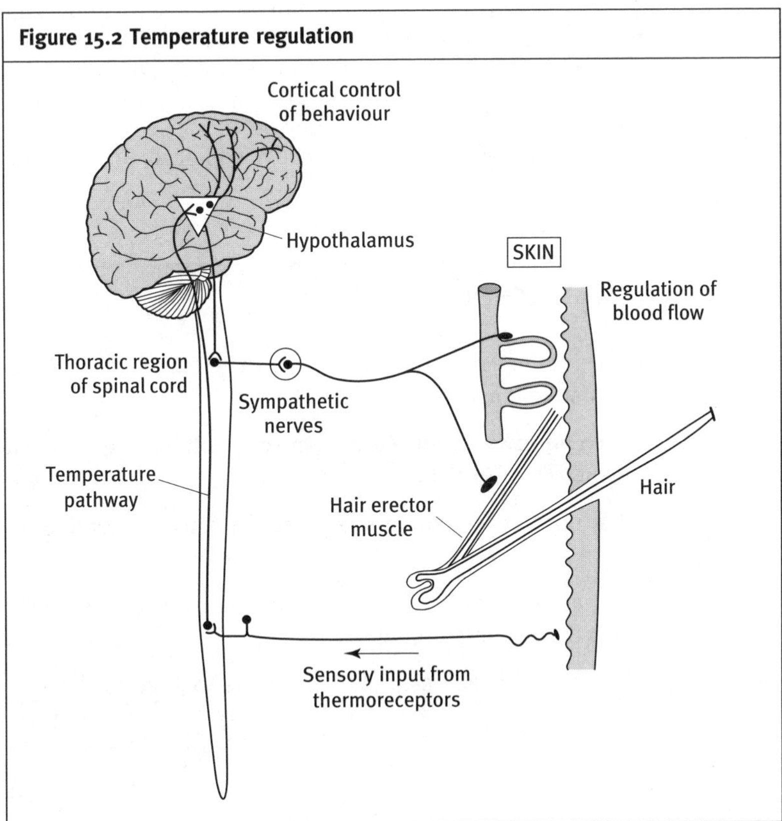

Activation of this tissue leads to the expression within mitochondria of a protein which uncouples the products of the citric acid cycle from the electron transport chain (see Chapter 7) with the result that the energy normally used to generate ATP is converted directly to heat.

As temperature rises, peripheral blood vessels dilate, resulting in an increase in blood flow near to the skin and consequent heat loss through radiation. Activation of sweat glands causes an increased secretion of sweat onto the skin surface where evaporation causes further heat loss. Activation of the muscles associated with hair follicles is reduced and so hairs will lie flat and close to the skin surface, encouraging heat loss through convection and evaporation. The production of thyroid hormone and adrenaline is also reduced with the resultant slowing of metabolic rates. Heat-reducing behaviours such as shade seeking and clothes shedding are induced.

Clinical example: Fever

Fever is a response by the body to help fight infection. The raised temperature improves the performance of the cells of the immune system and impairs the ability of microorganisms to replicate. Toxins from the microorganisms induce the production of a protein called IL-1 from cells of the immune system. IL-1 acts within the blood vessels of the hypothalamus to stimulate production of prostaglandin E2, which acts to reduce the firing of warm sensitive neurons and to initiate mechanisms which will lead to a rise in body temperature.

15.4 Test yourself

1. Which hormone is produced by the pancreatic beta cells in response to a rise in plasma glucose?
2. Which hormone produced by pancreatic alpha cells stimulates the breakdown of glycogen into glucose?
3. What region of the brain acts as the primary temperature regulatory centre?
4. What mechanism in humans acts to reduce heat loss from the skin when the temperature falls?
5. Which specialised tissue do infants use to generate heat?

16 The endocrine system

Basic concepts:
The endocrine system acts to regulate a variety of functions in the body by the release of chemical messengers into the bloodstream. These messengers, or hormones, act on their target cells by binding to receptor molecules on the cell surface or in the cytoplasm. The binding of hormone to receptor triggers a cascade of cellular events which ultimately leads to a change in cell function. This can be achieved by enzyme activation, channel modification or changes in gene expression. Understanding the ways in which hormones exert their effects on target cells forms the basis of understanding many regulatory processes in the body.

16.1 Introduction

Hormones are traditionally defined as substances which are produced by cells within *endocrine glands* and which act on cells or tissues at a distance. Endocrine glands are collections of secretory cells which release their hormonal secretions into the bloodstream. Hormones interact with receptor molecules located either on the surface or within the cytoplasm of the target cell. They can produce immediate metabolic effects, as in the case of the stimulation of glucose uptake into liver and muscle by insulin (see Chapter 15). They can also have long-lasting effects on gene expression and cellular differentiation, as in the role of testosterone in the development of sexual characteristics. The *endocrine system* is defined as the set of organs or glands which produce hormones and includes the hypothalamus, pituitary, adrenal gland, thyroid gland, gonads, parathyroid glands and the pancreatic islets. Hormones are central to the body's homeostatic mechanisms and their production is often regulated via negative feedback loops (see Chapter 15).

It is now recognised that almost every cell in the body produces signal molecules which are released into the extracellular fluid, and which may influence other cells either nearby or at a distance. Many of these signal molecules are not described as hormones but may be known by other names such as cytokines, modulators or transmitters. An example of a non-classical hormonal system is the production of leptin by adipose tissue. Levels of this hormone are related to levels of stored fats and it acts on the brain to suppress appetite.

16.2 Types of hormone

There are three main classes of hormone – *steroid*, *polypeptide* and *amino acid* – but other molecules, such as glucose, may also be effective in cell signalling.

Steroid family hormones

Examples are testosterone, oestrogen and corticosterone. These are small molecules synthesised primarily from cholesterol which are released immediately following synthesis. They are hydrophobic and circulate in the blood bound to carrier proteins. They diffuse readily through cell membranes and act on intracellular receptors, which then bind to DNA and alter gene expression. In general they have slow, longlasting effects.

Polypeptide or protein hormones

These consist of chains containing between 3 and 332 amino acids. They are often synthesised as inactive precursors and stored prior to modification and release. They act on cell surface receptors then via second messenger systems (see section 16.4). Examples are insulin and gastrin.

Amino acid hormones

Examples are thyroid hormone, noradrenaline and adrenaline. These are all derived from the amino acid tyrosine and are stored within their cells of production for immediate release. They have a variety of modes of action. Thyroid hormone acts through an intracellular receptor whereas the others act through cell surface receptors.

16.3 Cell signal receptors

The function of receptors is to act as the first stage in the process of converting the presence of the signal molecule into a cellular response. This process is known as *signal transduction*. There are two main classes of receptor – *cell surface and intracellular.*

Cell surface receptors

These are proteins which span the cell membrane and whose extracellular domain operates as a specific binding site for the signal molecule. The binding of the signal molecule to the binding site initiates a change in the receptor which alters the activity of the intracellular domain of the receptor. Changes to the intracellular domain then trigger an intracellular cascade of reactions which ultimately lead to the cellular response. Some receptors may be linked directly to channels and binding of the signal molecule will cause a change in the open–closed configuration of the channel, often leading to an influx of Ca^{2+} ions.

Intracellular receptors

These receptors are located in the cytoplasm or nucleus of the cell. Their active DNA-binding site is normally blocked by the presence of an inhibitory protein. They respond to lipid-soluble hormones which can pass directly through the cell membrane. Binding of the hormone causes a conformational change which releases the inhibitory protein and exposes the DNA-binding site. The receptor then acts as a transcription factor to initiate gene expression (see above and Chapter 5).

16.4 Second messenger systems

The binding of the signal molecule to a cell surface receptor may be a transient event lasting a few milliseconds. This may often lead to a cellular response which lasts minutes or hours. This process of signal amplification is achieved by the activation of second messenger systems. Second messengers are molecules whose levels are raised within the cytoplasm as a result of the interaction between the signal molecule and the receptor, and which, in turn, interact with intracellular proteins to produce a cellular response. A simple example of this principle can be seen in the regulation of smooth muscle contraction (see Chapter 14) where Ca^{2+} ions, which enter the cytoplasm either from outside the cell or from intracellular stores, interact with a calcium-binding protein which, in turn, interacts with the proteins of the contractile apparatus of the cell. An initial, extremely rapid depolarisation event can lead to a slow and sustained contraction.

G-protein linked receptors

G-proteins are small multi-unit proteins which are found on the cytoplasmic surface of the plasma membrane in close association with the intracellular domain of receptors. In the inactive state they bind guanosine diphosphate (GDP). Binding of the signal molecule to the receptor leads to a change in the conformation of the G-protein, which now exchanges GDP for GTP (guanosine triphosphate). This leads to detachment of the G-protein from the receptor. The G-protein then diffuses in the plane of the cell membrane until it binds to a target protein. This may result in either the modification of an ion channel or the activation of an enzyme. The activated enzyme then produces a second messenger molecule that enters the cytoplasm and interacts with its target proteins leading to a cellular response. Eventually the G-protein converts the bound GTP into GDP and is restored to its inactive state.

An example of this mechanism is seen in the kidney (see Chapter 22) where antidiuretic hormone (ADH) acts through a G-protein-linked receptor to activate the membrane enzyme adenylate cyclase. This enzyme converts ATP into a second messenger molecule, cyclic AMP (cAMP). This then acts through a cascade of protein intermediaries to cause the insertion of water channels into the cell membrane of the collecting duct – this promotes the re-uptake of water, preventing excessive fluid loss.

Tyrosine kinase receptors

A number of receptors, of which the insulin receptor is an example, have an intracellular domain which acts as a tyrosine kinase. A tyrosine kinase is an enzyme which adds phosphate groups to the amino acid tyrosine in other proteins. Phosphorylation alters the conformation of the recipient protein and, in most instances, causes its activation (see Chapter 2). The initial target proteins that are phosphorylated are often themselves kinases which, in turn, activate a further subset of proteins. This cascade effect leads to signal amplification, both in time (phosphorylated proteins may stay active for hours) and in amplitude (one receptor will activate many thousands of target proteins). In the case of the insulin receptor, phosphorylation of proteins in the cell may lead to the insertion of a glucose transporter into the cell membrane and to the activation of key proteins in the glucose→glycogen metabolic pathway.

16.5 Endocrine glands

The hypothalamus and the pituitary gland provide central control of a number of endocrine organs (Fig. 16.1). There is a hierarchy of systems in which the hypothalamus operates at the highest level and in which feedback to each level from the levels below acts to regulate hormonal production (see Fig. 16.3). The hypothalamus is a region of the brain which contains groups of neurons whose function is to provide homeostatic control over important bodily functions. For example, the hypothalamus regulates plasma osmolarity, appetite and sexual drive. In addition, the hypothalamus provides a link between higher consciousness and emotional responses through its regulation of the autonomic nervous system (see Chapter 17) and controls stress reactions through regulation of cortisol production by the adrenal gland.

Figure 16.1 The endocrine glands

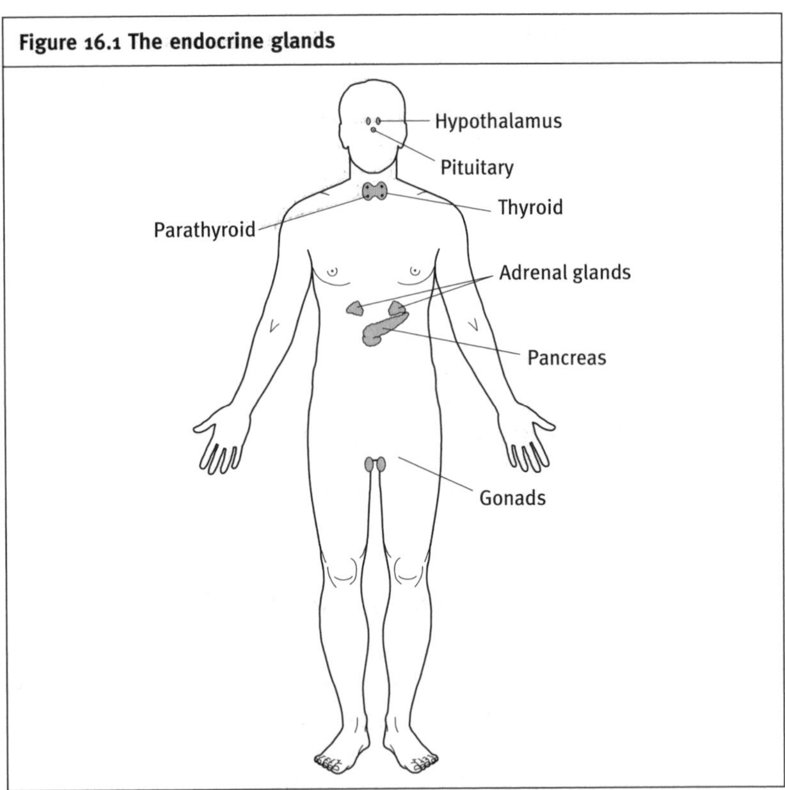

16.6 Hypothalamo-pituitary axis

Suspended beneath the hypothalamus and connected to it by the pituitary stalk is the *pituitary gland*. The pituitary gland is divided into anterior and posterior lobes. The anterior pituitary contains a collection of endocrine cells known as '...troph' cells. They are called this because the primary function of their secretions is to stimulate endocrine glands and other tissues elsewhere in the body. The '...troph' cells produce several protein hormones – *thyroid stimulating hormone, growth hormone, adrenocorticotrophic hormone, luteinising hormone, follicle stimulating hormone*, and *prolactin*.

Thyroid stimulating hormone (TSH)

Produced from thyrotrophs, TSH acts to stimulate the thyroid gland to produce the thyroid hormones T_3 and T_4. These are essential in normal development and promote increases in the basal metabolic rate of cells. Thyroid hormones act at nuclear receptors to alter gene expression. Lack of thyroid hormone will produce growth and mental retardation in the child and a sense of tiredness in the adult.

Growth hormone (GH)

Produced from somatotrophs, GH acts on multiple target tissues, including liver and muscle, to promote cell growth and enhance cellular metabolism. GH can act either directly or by stimulating the release of growth factors such as insulin-like growth factor-1.

> **Clinical example: Acromegaly**
>
> Excessive production of growth hormone following puberty, often as the result of a tumour of the anterior pituitary, results in excessive growth of bone and soft tissues. It is characterised by a thickening of the bones of the skull and by growth of skin. There is not normally an increase in height as the growth zones at the ends of long bones are sealed by this time. An individual with acromegaly will have a larger than normal head with coarse features.

Adrenocorticotrophic hormone (ACTH)

Produced from corticotrophs, ACTH acts on the adrenal gland to stimulate the production of the glucocorticoid hormone, cortisol. Cortisol has multiple effects including mobilisation of glucose from intracellular stores, suppression of the immune system, anti-inflammatory activity, reduction in bone production and effects on the central nervous system. Levels of cortisol are raised in chronic stress and analogues of cortisol are used as anti-inflammatory drugs.

Luteinising hormone (LH) and follicle stimulating hormone (FSH)

Produced from gonadotrophs, these hormones are the primary regulators of the ovary in females and the testis in males. In females, FSH promotes development of the ovarian follicle which surrounds the ovum and stimulates cells in the ovary to produce oestradiol. LH triggers ovulation and supports the formation of the corpus luteum, which is necessary for the development of the lining of the uterus in readiness for the implantation of the fertilised ovum. In the male, LH stimulates Leydig cells in the testis to produce testosterone, while FSH stimulates Sertoli cells to produce a number of factors which support sperm production.

Prolactin (PRL)

Produced by lactotrophs, PRL promotes the development of the breasts during pregnancy, initiates milk production and helps to maintain milk production once it has been established.

All of the '...troph' cells in the anterior pituitary are directly under the control of the hypothalamus. Hormones produced in the hypothalamus, and known as releasing hormones, are secreted into a special capillary network which connects the hypothalamus to the anterior pituitary. These hormones, such as gonadotrophin releasing hormone (GnRH) which acts on the gonadotroph cells, are often released in a pulsatile fashion and are subject to negative feedback from the hormones whose production they stimulate.

16.7 Posterior pituitary

The posterior pituitary is part of the *neuroendocrine* system. Neurons with their cell bodies in two nuclei in the hypothalamus send their axons down the pituitary stalk where they release hormones into capillaries in the posterior pituitary (Fig. 16.2). One of these nuclei

Figure 16.2 The pituitary and hypothalamus

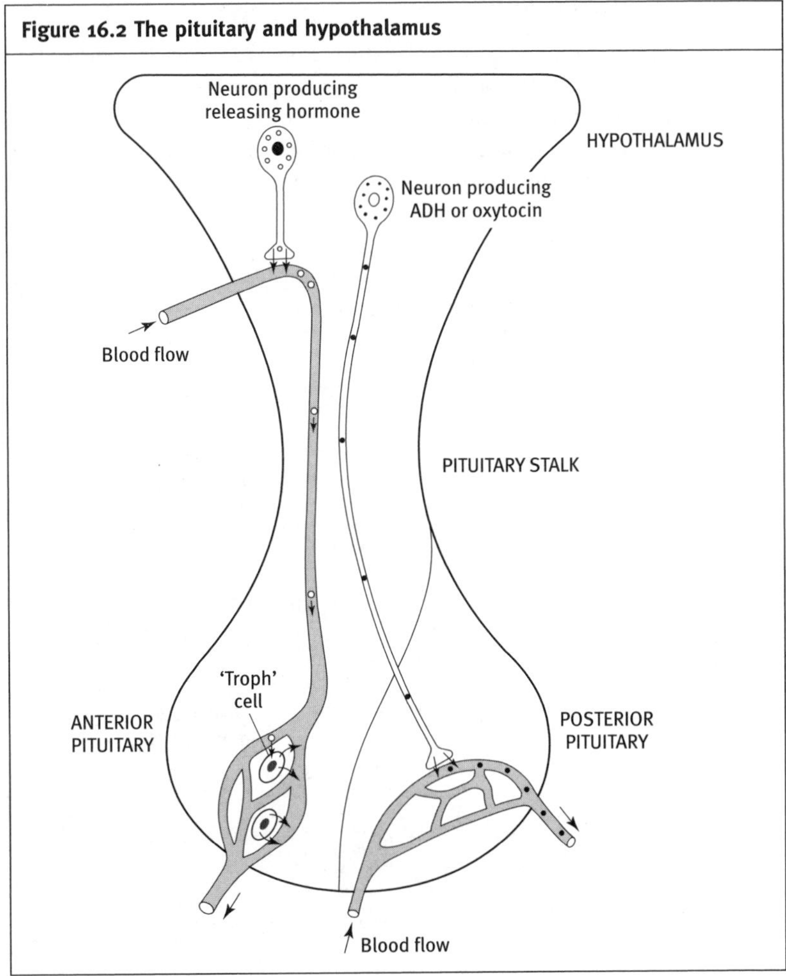

produces the antidiuretic hormone (ADH) which acts on the collecting duct of the kidney to increase water reabsorption (see above). The other nucleus produces oxytocin which stimulates contraction of the uterus during the act of giving birth and milk release during breastfeeding.

16.8 Other endocrine glands

Parathyroid glands

There are four parathyroid glands located on the back of the thyroid gland. They produce parathyroid hormone (PTH) in response to falling plasma Ca^{2+} ion levels. The actions of PTH are to promote Ca^{2+} ion reabsorption in the kidney and to stimulate Ca^{2+} ion release from bone. This increases plasma calcium.

Endocrine pancreas

Within the pancreas are clusters of cells known as *pancreatic islets.* These contain α- and β-cells. The β-cells detect plasma glucose levels and respond to rising plasma glucose by secreting insulin. Insulin triggers the uptake and storage of glucose in liver and muscle cells. Glucagon, which is produced by the α-cells, acts to promote glucose

Figure 16.3 Growth hormone as an example to illustrate the hierarchy of systems and feedback in the endocrine system

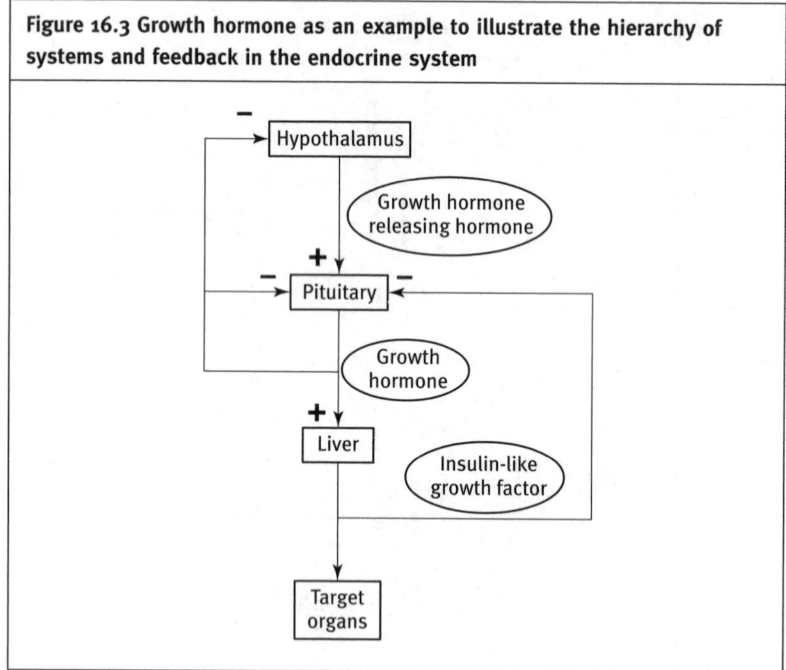

production and release from the liver and is secreted at high levels following fasting.

Adrenal gland

In addition to cortisol the adrenal gland is responsible for the production of at least two other important hormones. The first is aldosterone which is involved in the regulation of extracellular fluid volume; it does this by promoting Na^+ ion, and thus water, reabsorption in the kidney (see Chapter 22). The second is adrenaline (epinephrine) which is produced in the centre of the adrenal gland by cells which are the equivalent of postganglionic neurons in the sympathetic nervous system. Adrenaline is released directly into the bloodstream and integrates the 'fight or flight' reaction (see Chapter 17) throughout the body.

16.9 Test yourself

1. Which family of hormones are small lipid molecules that act on intracellular receptors?
2. Which hormone is responsible for the 'fight or flight' response?
3. What nucleotide is bound to G-proteins when they are in the active state?
4. From which hormone, produced by the adrenal glands in response to stress, are many anti-inflammatory drugs derived?
5. List: (a) six hormones produced by the anterior pituitary gland; (b) two hormones released from the posterior pituitary gland.

17 The nervous system

Basic concepts:
Nerve cells act to coordinate information received from multiple sources and to transmit this information to other cells. This ability of nerve cells to integrate information lies at the core of the functioning of the nervous system. Sets of nerve cells are organised into functional units which deal with specific types of information. Some may process sensory information, while others handle motor functions. Part of the nervous system is important in homeostasis. Understanding the basic structural patterns which underlie the organisation of the nervous system is important in appreciating its complexity.

17.1 Introduction

The nervous system is a complex assembly of nerve cells (see Chapter 14) and supporting glial cells which functions to control and coordinate all aspects of behaviour. The nervous system is responsible for higher thought processes, coordination of movement, integration of sensory information, regulation of levels of consciousness, control of emotion and memory storage as well as control of essential bodily functions such as breathing.

17.2 Structure of nervous system

The nervous system comprises the *brain*, *spinal cord* and *peripheral nerves*. The brain is contained within the *cranial cavity* of the skull, and the spinal cord runs within the *vertebral column* (backbone) – see Fig. 17.1. Nerves run out from both the brain (*cranial nerves*) and the spinal cord (*spinal nerves*). *Motor neurons* convey information from the brain and spinal cord to muscles. *Sensory neurons* carry information from the sense organs to the spinal cord and brain.

An important functional classification is into the central nervous system (CNS) and the peripheral nervous system (PNS). The CNS includes all those neurons whose cell bodies, dendrites and axons lie entirely within the brain and spinal cord. The PNS includes neurons whose cell bodies and/or axons lie outside the brain and spinal cord. The differentiation into CNS and PNS is particularly important in terms of the ability to recover following damage, since regrowth of damaged axons can occur

Figure 17.1 The brain and spinal cord

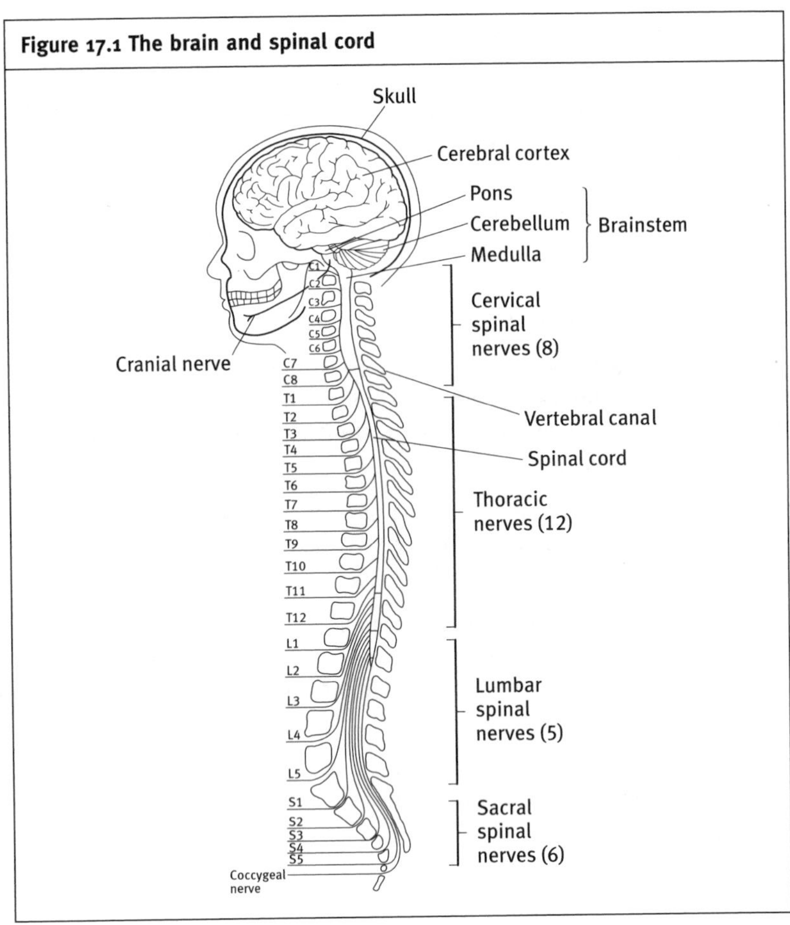

in the PNS but not within the CNS. This has profound implications for patients who suffer damage to the nervous system.

Peripheral nerves are composed of the myelinated axons of motor and sensory neurons. Within the CNS, grey matter and white matter can be distinguished. *Grey matter* contains predominantly cell bodies and dendrites of neurons. Collections of nerve cells are known as *nuclei* or *ganglia* depending on their location. *White matter* contains myelinated axons running in tracts which connect one set of neurons to another.

17.3 The brain

The brain integrates information received from a variety of sources and produces a coordinated response. The brain can be divided into distinct components – the *cerebral cortex*, the *midbrain* and the *brainstem*.

Cerebral cortex

This is a folded sheet of neurons which covers the outer surface of the two hemispheres of the brain. The cerebral cortex is responsible for higher cognitive functions such as speech, thought and memory. The cerebral cortex is divided into four lobes (Fig. 17.2) each of which is specialised for particular functions.

- The *frontal lobe* is involved in planning movement and in regulation of emotions. On the left-hand side of the brain the frontal lobe contains a region responsible for speech production.
- The *parietal lobe* receives sensory information regarding pain, touch and temperature from the entire body and also acts to integrate this with other sensory information such as vision, hearing, taste and smell.
- The *occipital lobe* at the back of the brain is the area into which visual information is received and where processing of this information takes place.
- The *temporal lobe* contains areas associated with emotion and the regulation of emotional responses. It enables the formation of memory. The temporal lobe also receives hearing and smell sensory input.

Figure 17.2 The cerebral cortex

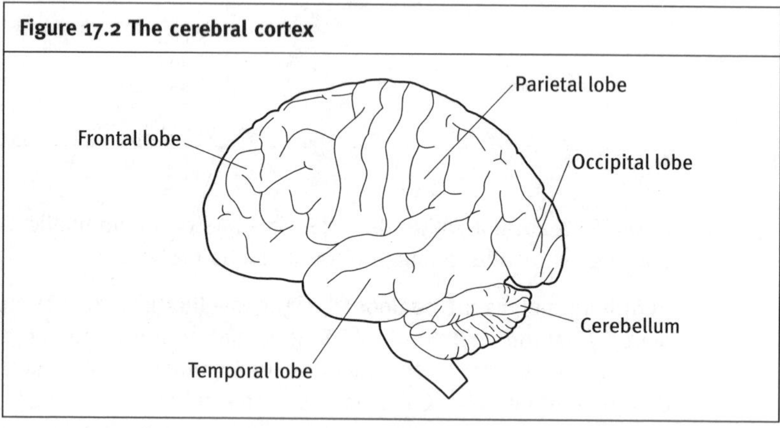

The two hemispheres of the brain are connected by a white matter bridge known as the *corpus callosum* which ensures integration between the two sides. In general, the right-hand side of the brain receives sensory input from and sends motor output to the left side of the body, and the left-hand side of the brain controls the right side of the body.

Midbrain

This region contains important sets of nuclei which act to link the cerebral cortex with the brainstem and spinal cord. The *thalamus* contains a number of nuclei which relay sensory information towards the cortex. The *basal ganglia* are involved in the planning of movement. Also in the midbrain is the *hypothalamus* which contains nuclei involved in homeostasis and endocrine functions (see Chapters 15 and 16).

Brainstem

This comprises the *pons*, *medulla* and *cerebellum*. The pons and medulla contain important nuclei associated with sensory and motor functions of the cranial nerves (such as movement of the eyes and tongue, taste, control of facial expression and chewing, and touch, pain and temperature sensation from the face) as well as key nuclei involved in respiratory and cardiovascular regulation. The cerebellum ensures that movements are carried out smoothly and helps to regulate balance.

17.4 The spinal cord

The primary function of the spinal cord is to convey information from the brain to the rest of the body, and from the rest of the body to the brain. The spinal cord gives rise to 31 pairs of spinal nerves (Fig. 17.1). Each spinal nerve is formed from the fusion of dorsal and ventral roots which arise from the spinal cord (Fig. 17.3). The dorsal roots carry sensory axons into the spinal cord. The ventral roots convey motor axons away from the spinal cord. Motor axons originate from motor neurons whose cell bodies lie within the ventral horn of the central grey matter of the spinal cord. The motor neurons are controlled by input from the cerebral cortex and from brainstem nuclei, and their axons will eventually terminate at neuromuscular junctions on skeletal muscle.

17.5 Peripheral nervous system

The peripheral nervous system is divided into *somatic* and *autonomic* divisions. The somatic division contains sensory fibres from the skin and special sense organs such as the eye and the ear. It supplies motor innervation to skeletal muscle. The autonomic nervous system is involved in the control of glands, the heart and visceral smooth muscle.

Figure 17.3 The spinal cord – front view

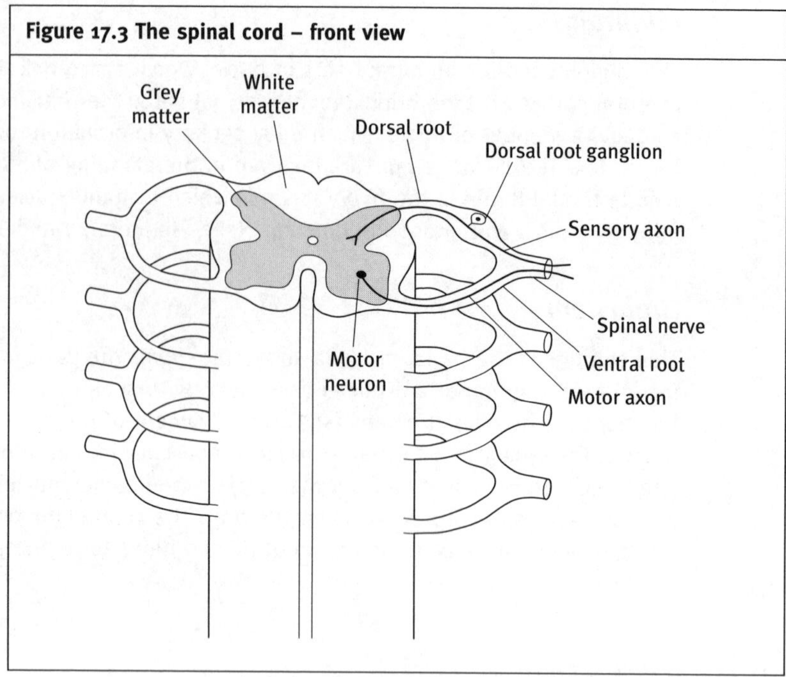

Somatic division

Many spinal nerves, once they have emerged from the vertebral column, join up with spinal nerves from higher and lower levels of the spinal cord to form nerve *plexuses.* In these plexuses, axons from various spinal nerves join together to form a single peripheral nerve. This nerve will then run through the body to reach its target regions. As an example, the *sciatic nerve* is formed from the fusion of five spinal nerves which emerge from the lower regions of the vertebral column. The sciatic nerve runs down the back of the leg and supplies the muscles of the back of the leg above the knee and all the muscles of the leg below the knee. Sensory nerve endings in the skin and in muscles and joints of the leg give rise to axons which run back in the sciatic nerve to enter the spinal cord through the dorsal roots of the individual spinal nerves which comprise it. The cell bodies of these sensory axons are clustered in special ganglia associated with the dorsal roots.

Autonomic division

The autonomic nervous system is divided into *sympathetic* and *parasympathetic* divisions. Both divisions tend to innervate smooth muscle and glandular tissue and in many instances have opposite effects on their target tissues (Fig. 17.4).

Figure 17.4 Innervation of the heart and GI tract by the autonomic nervous system

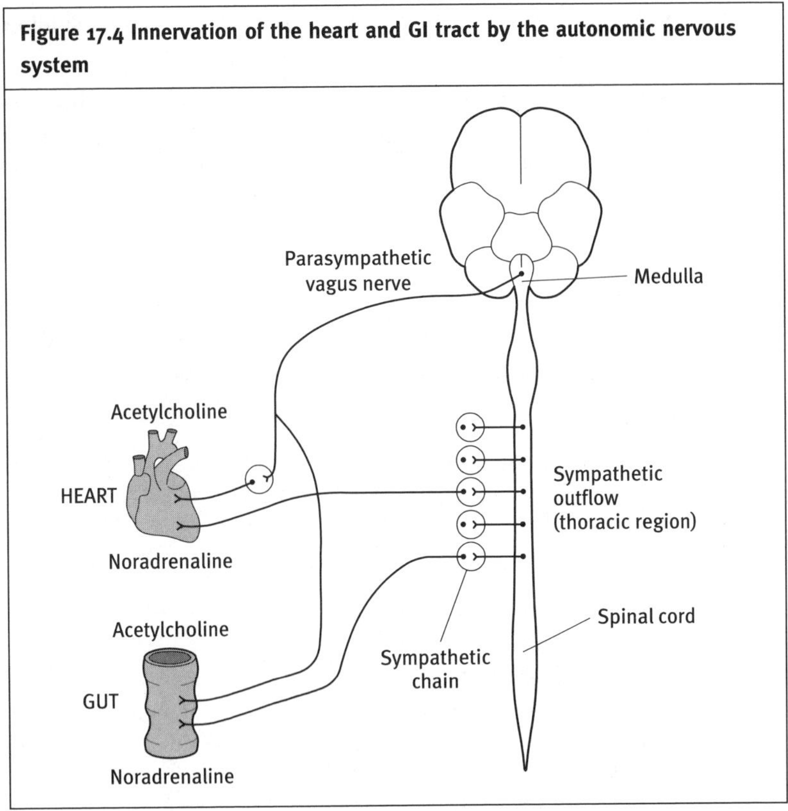

Sympathetic nerves contain axons originating from a series of ganglia which lie alongside the vertebral column. These ganglia are controlled by axons which arise from neurons in the thoracic or lumbar regions of the spinal cord. Sympathetic nerves use *noradrenaline* as their neurotransmitter at the target organ. They produce physiological changes which characterise the 'fight or flight' reaction. These include:

- increased heart rate and force of contraction of the heart
- dilation of the arteries to the heart
- decreased blood flow to the skin
- increased blood flow to muscles
- dilation of the airways
- sweating
- piloerection (hair stands on end)
- reduced blood flow to the GI tract
- production of a viscous saliva

- stimulation of the adrenal gland to release adrenaline into the bloodstream.

Parasympathetic nerves contain axons which originate from the brainstem or the sacral region of the spinal cord. They synapse with neurons in ganglia very close to the target organs and the axons of these ganglionic neurons use acetylcholine as their neurotransmitter at the target organ. The vagus nerve, which originates in the brainstem, is the main parasympathetic nerve in the body and supplies all thoracic and abdominal organs.

The physiological changes produced by the parasympathetic system can be characterised as 'rest and digest' and include:

- decreased heart rate
- increased blood flow to the GI tract
- increased acid production in the stomach
- increased production of a watery enzyme-rich saliva
- constriction of the airways.

Clinical example: Stroke

Blockage of arteries bringing blood to the brain causes the death of neurons in the area of the brain supplied. This is known as a stroke and leads to a loss of function in the parts of the body controlled by the affected cells. A common site for stroke is the area supplied by the middle cerebral artery which includes large sections of the frontal and parietal lobes of the cerebral cortex. This leads to loss of motor and sensory functions on the opposite side of the body and, if the lesion is in the left hemisphere of the brain, to a loss of the ability to speak.

17.6 Test yourself

1. What type of neurons carry information from the central nervous system to muscles?
2. What names are given to the four lobes of the cerebral cortex?
3. Which division of the peripheral nervous system innervates skeletal muscle and the sense organs?
4. Which neurotransmitter is used by: (a) sympathetic nerves; (b) parasympathetic nerves at their target organs?
5. What is the main parasympathetic nerve supplying the organs of the thorax and abdomen?

18 The cardiovascular system

Basic concepts:
Circulation of the blood around the body is vital to ensure the supply of oxygen and nutrients to the tissues and the removal of waste. Oxygen is carried by haemoglobin within red blood cells. The white blood cells are key components in the body's defence mechanisms against infection. Circulation of blood through the arteries and veins is accomplished by the pumping action of the heart. The activity of the heart is closely regulated to enable sufficient pressure to be maintained in the circulation to ensure an adequate blood supply to all tissues, especially the brain. An understanding of the factors which are involved in the regulation of the cardiovascular system is essential to an appreciation of its central role in the maintenance of life.

18.1 Introduction

The cardiovascular system is designed to ensure the circulation of oxygenated blood to all organs of the body. Oxygenated blood from the lungs (see Chapter 19) is returned to the left side of the heart in the pulmonary veins and is then pumped out through the aorta into the systemic circulation in which it is distributed to the organs of the body via blood vessels known as *arteries*. Blood flows through these organs in narrow diameter, highly permeable capillaries. These facilitate exchange of substances between blood and the interstitial fluid bathing the cells of the organ. Deoxygenated blood is then returned, in blood vessels known as *veins*, to the right side of the heart. The blood is then pumped back to the lungs via the pulmonary arteries. Blood in the systemic circulation conveys nutrients and oxygen to the tissues of the body and carries away waste products. Blood also acts to transport chemical signalling molecules (hormones) from their site of production (endocrine glands) to their sites of action.

18.2 Blood

Blood has three main constituents:

- plasma (50–60% of volume)

- red blood cells (40–50% of volume)
- white blood cells and platelets (1% of volume).

The white blood cells are part of the immune system and are considered in more detail in Chapter 23. Blood is circulated around the body within blood vessels and functions primarily to transport essential substances from one organ to another.

18.3 Plasma

Plasma comprises a protein-rich solution in which the cellular constituents of the blood are suspended. The ionic composition of plasma is similar to that of interstitial fluid and it has an osmolarity of 290 mosmol/l, of which Na^+ and Cl^- are the major ions. Plasma also contains K^+ ions at 4 mmol and Ca^{2+} ions. Much of the plasma calcium exists not as free ions but in a bound form attached to the major plasma protein albumin. Minor fluctuations of plasma potassium can have significant consequences for excitable tissues as the resting membrane potential is predominantly determined by the K^+ ion concentration gradient across the cell membrane (see Chapter 14) .

Apart from the immunoglobulins the majority of plasma proteins are synthesised in the liver. Plasma proteins have a variety of functions which are now outlined.

Transport

Many substances are transported in blood bound to plasma proteins. Some, such as steroid hormones, are transported attached to albumin, the most common plasma protein. Others have their own specific carrier proteins, such as iron which is transported bound to ferritin.

Defence against infection

See Chapter 23 for a full discussion of this function.

Blood clotting

It is important that blood vessels, if they become damaged, do not continuously leak and allow significant blood loss. Plasma contains a group of proteins and some microcellular components known as *platelets* which work cooperatively to produce a blood clot to plug damaged vessels. In general, clotting is initiated by exposure of specific proteins in the epithelial lining (endothelium) of damaged vessels. The final component of the protein-clotting cascade is a filamentous protein, *fibrin*, which is cross-linked by the action of the enzyme thrombin. The fibrin meshwork that is formed traps platelets which help to build up the plug. Platelets, in turn, release the signal molecule serotonin which acts to cause neighbouring blood vessels to constrict and further minimise blood loss.

Clinical example: Haemophilia

One of the proteins of the clotting cascade is Factor VIII. This is a vital component in the chain reaction leading from endothelial damage to blood clot formation. The gene coding for Factor VIII is present on the X chromosome and mothers with one copy of the faulty gene have a 50% chance of passing this to their sons. Since boys have only one X chromosome those that do inherit the faulty gene will not synthesise any Factor VIII and will not be able to form blood clots. Even small wounds are potentially fatal to sufferers of this condition. Haemophilia has been treated by infusion of Factor VIII purified from the blood of donors – it has led to problems when the donated blood also contained infections such as HIV or CJD.

Oncotic pressure

The presence in plasma of proteins, such as albumin, which carry a high negative charge helps to create what is known as plasma oncotic pressure. This is a combination of the osmotic forces due to the protein particles themselves, which cannot pass between capillary and interstitial fluid, and additional osmotic forces due to the cations which these proteins attract and hold. Plasma oncotic pressure is equivalent to 25 mmHg and is the key force drawing fluid back into the venous end of capillaries once it has circulated through the interstitial spaces. In situations where plasma oncotic pressure is reduced, such as in liver failure when plasma protein levels drop, then fluid stays in the interstitial space resulting in tissue swelling (known as *oedema*).

18.4 Red blood cells

Red blood cells (*erythrocytes*) are continuously produced in the bone marrow. Their formation is under the control of the hormone erythropoietin which is produced by the kidney in response to decreasing arterial oxygen levels. They are shaped as a biconcave disc which allows both a large surface area to volume ratio and flexibility as they pass through small capillaries. Red blood cells do not have a nucleus and the cytoplasm is packed with the protein *haemoglobin*, which serves to transport oxygen around the body. Erythrocytes stay in circulation for about 150 days before being broken down in the spleen.

Haemoglobin is a protein consisting of four subunits (α-globin chains) each of which has at its core a haem ring. This is a structure containing reduced iron which allows the binding of oxygen. There is an interaction between the four subunits such that when O_2 is bound to one unit it becomes easier for O_2 to bind to the remaining subunits. The binding of O_2 to haemoglobin is normally represented graphically (Fig. 18.1) by plotting the partial pressure of oxygen (pO_2) against the % saturation (amount bound/ total capacity × 100) of the haemoglobin. This sigmoidally-shaped graph shows that at a normal lung pO_2 of 100 mmHg the haemoglobin will be 97.5% saturated, whereas at a normal tissue pO_2 of 40 mmHg the saturation will drop to 70%. Thus when haemoglobin within erythrocytes arrives in tissues, approximately 25% of the carried oxygen will dissociate from the haemoglobin and diffuse into cells.

Figure 18.1 Haemoglobin dissociation curve

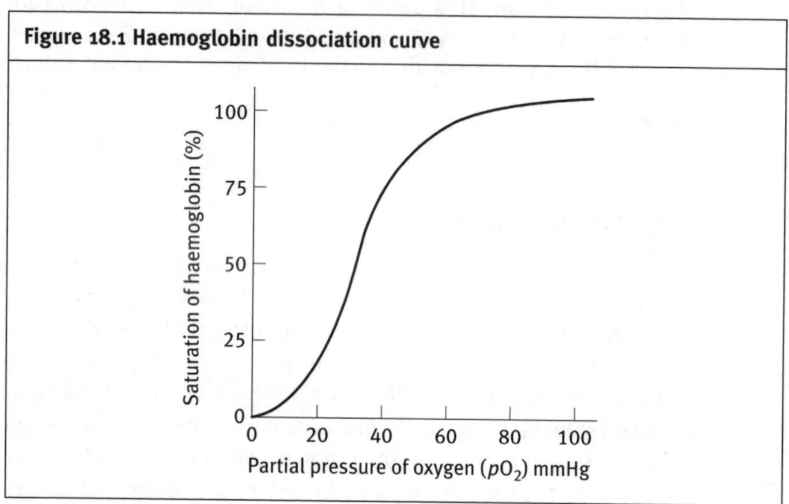

18.5 Platelets

Platelets are small subcellular fragments formed by being pinched off from their precursor cell, the *megakaryocyte*. They are important in the formation of blood clots (see Section 18.3).

18.6 The heart

Structure

The heart is a four-chambered pump which is responsible for maintaining blood flow within the systemic and pulmonary circulations –

Figure 18.2 The structure of the heart

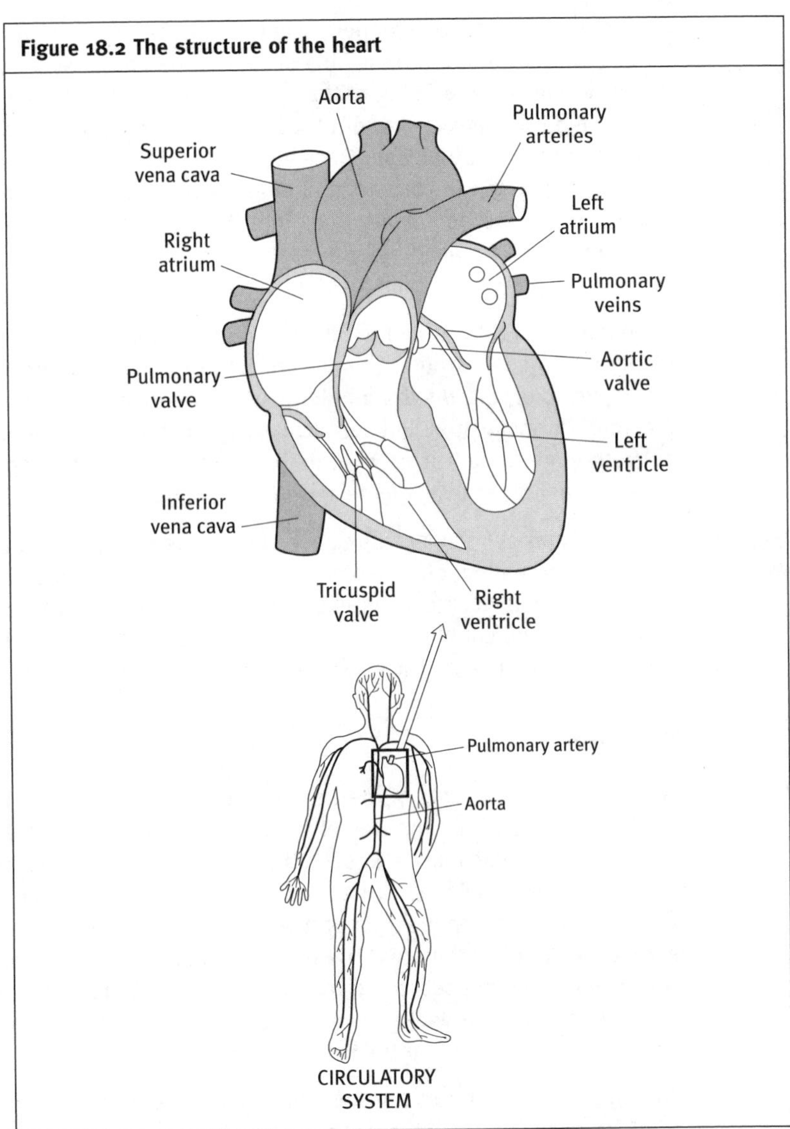

Fig. 18.2 shows a schematic diagram of its structure. The four chambers comprise two *ventricles* and two *atria* (singular *atrium*). The activity of the heart is divided into two phases – *systole*, during which the ventricles contract, and *diastole*, during which the ventricles relax and the atria contract.

The right atrium of the heart receives venous blood back from the systemic circulation via veins called the *superior and inferior venae cavae*. The right atrium is separated from the right ventricle by a valve (the *tricuspid valve*) which opens when the pressure in the ventricle is lower than that in the atrium. This occurs at the end of systole as the ventricle relaxes, and filling of the ventricle is achieved by a combination of passive flow of blood from the atrium and weak contraction of the atrial wall. Once ventricular filling is complete, the ventricle will begin to contract and blood is pushed out through a valve at the base of the *pulmonary trunk* into the *pulmonary arteries* and into the lungs. The increased blood pressure in the ventricle causes closure of the tricuspid valve which prevents backflow of blood into the atrium.

Venous blood return from the lungs is via the *pulmonary veins* which empty into the left atrium. This is connected to the left ventricle through the *bicuspid* or *mitral valve* and left ventricular filling also occurs during diastole. During systole left ventricular contraction forces blood into the aorta through the aortic valve at its base. The volume of blood ejected is known as the *stroke volume* and is approximately 50 ml.

The aorta and pulmonary trunk have elastic tissue in their walls and expand to accommodate the blood ejected from the heart. During diastole there is elastic recoil of the walls of these large vessels and this ensures the continued circulation of blood around the body or to the lungs while the ventricles are relaxing.

Function

The timing of the events of the cardiac cycle is achieved via specialised excitable tissue within the wall of the heart (see Chapter 14). The structure of cardiac muscle is such that the gap junctions between adjacent fibres provide electrical continuity, which means that depolarisation and contraction of any one cardiac muscle fibre will spread to all adjacent fibres. The muscle fibres of the atrial and ventricular walls are separated by a non-conducting fibrous *septum* but within either atria or ventricles a single point of depolarisation will produce contraction of all the muscle fibres.

The *sinu-atrial* (S-A) *node*, which is located in the wall separating the right from the left atrium, contains cells which are capable of generating spontaneous action potentials at a regular rate of about 70/min. The cells of the S-A node are connected via gap junctions to the muscle fibres of the atrial walls and so are able to initiate synchronised atrial contractions. The wave of depolarisation which spreads through the atrial wall is picked up by another specialised set of cells, the *atrio-ventricular* (A-V) *node*. This node transmits the action potential to the ventricles via a set of elongated muscle fibres known as the *A-V bundle*. This pierces the fibrous septum separating the atria from the ventricles and travels down the dividing wall between the two ventricles until it reaches the point of the ventricles furthest away from the aorta and pulmonary trunk. Contact is then made with the muscle fibres of the ventricular wall and contraction is initiated. There is a delay between the wave of excitation arriving at the A-V node and the initiation of ventricular contraction that allows time for filling of the ventricles to be completed.

The electrical events of the heart can be recorded by placing electrodes on the surface of the chest. The resultant trace is the electrocardiograph (ECG). An ECG (Fig. 18.3) has three main elements:

- a P wave which represents atrial depolarisation;
- this is followed after a delay by the QRS complex which represents ventricular depolarisation;
- finally a T wave which indicates ventricular repolarisation.

Figure 18.3 An electrocardiograph (ECG)

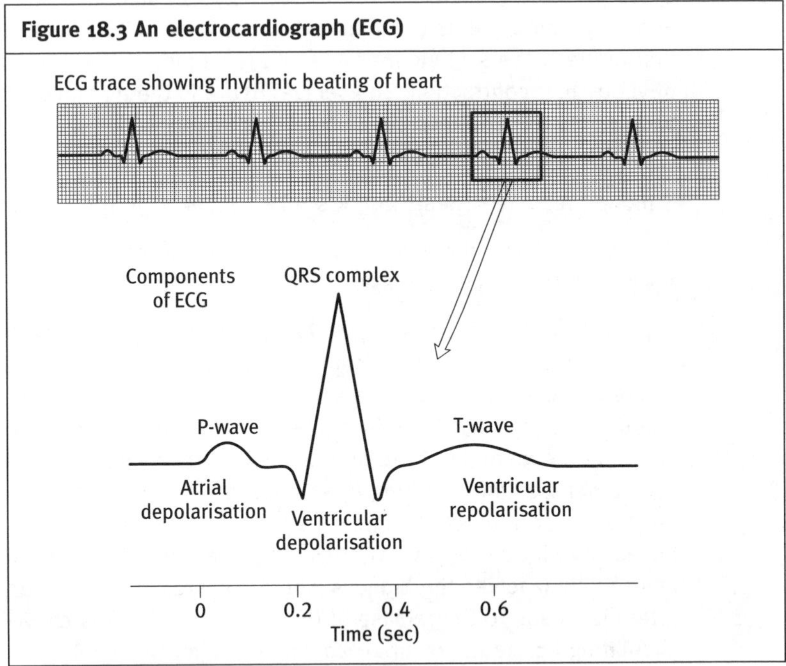

18.7 The circulatory system

The circulatory system consists primarily of *arteries* which take blood away from the heart and *veins* which return the blood to the heart. The two sides of the circulation are connected by *capillaries* which permit exchange between blood and tissues. Blood leaving the heart enters the *aorta* which contains many layers of elastic tissue in its wall. From here the blood passes to muscular arteries, such as the radial artery in the arm (from which the pulse rate is normally measured at the wrist). These arteries control the distribution of blood to the various organs of the body. Muscular arteries branch repeatedly to lead to narrow *arterioles* which connect to the capillary beds. The tone of the smooth muscle in the walls of the arteries and arterioles is controlled through the sympathetic nervous system.

Because of their narrow diameter and large total cross-sectional area, arterioles are the site of the major peripheral resistance to blood flow. From the capillaries blood passes back into the thin walled *venules* and *veins* for return to the heart.

The heart must work to force blood through the circulatory system against the resistance offered by narrow peripheral vessels. This generates a pressure in the circulatory system known as *blood pressure*. Blood pressure is measured during both systole and diastole. It is normally expressed as two figures – for example 120/80 mmHg. The first figure is the systolic pressure and is an indication of the strength of ventricular contraction. The second figure is the diastolic pressure and relates to peripheral resistance. The factors which influence blood pressure are described by the formula:

$$\text{blood pressure} \propto \text{heart rate} \times \text{stroke volume} \times \text{peripheral resistance.}$$

This means that if the amount of blood pumped by the heart or the resistance to its flow increase then blood pressure will rise.

Regulation of blood pressure can be achieved by three key mechanisms. One is a *neurogenic* mechanism which involves *baroreceptors* (pressure receptors) located in the aorta and arteries in the neck. Decreasing blood pressure leads to an increase in signals sent to the cardiovascular centre in the medulla of the brain. This in turn triggers an increase in sympathetic output resulting in an increase in heart rate and an increase in peripheral resistance as the result of vasoconstriction. Increasing blood pressure will reverse these effects. The second mechanism involves the kidneys where a decrease in blood pressure in arterioles leads to the release of the enzyme renin. This converts the circulating hormone precursor angiotensinogen to angiotensin I. Angiotensin I is then converted by an enzyme in the walls of blood vessels to angiotensin II, which acts to increase peripheral vasoconstriction and thus raises blood pressure. A third regulatory mechanism involves regulation of fluid volumes by the kidney and will be dealt with in Chapter 22.

Clinical example: Hypertension

A diastolic blood pressure consistently over 80 mmHg is classed as *hypertension.* A prolonged increase in blood pressure is linked with an increased risk of heart attack and stroke, as well as damage to the kidneys and to the eyes. The causes of hypertension are frequently unknown but risk factors may include obesity and stress. Lifestyle modifications, such as healthy eating and exercise, are commonly the first line of treatment. Pharmacological treatments may involve beta blockers to slow down the heart rate or inhibitors of the enzyme which converts angiotensin I to II (ACE inhibitors).

18.8 Test yourself

1. What filamentous protein is cross-linked by thrombin to form clots?
2. What name is used to describe the phases of the cardiac cycle in which: (a) contraction of the ventricles and (b) contraction of the atria occurs?
3. What valve separates: (a) the right atrium from the right ventricle; (b) the left atrium from the left ventricle?
4. Which chambers of the heart receive: (a) oxygenated blood returning from the lungs; (b) deoxygenated blood returning from the tissues?
5. Which three physiological factors affect blood pressure?

19 The respiratory system

Basic concepts:
The respiratory system ensures a constant supply of oxygen to the body and the removal of carbon dioxide. During respiration the lungs draw in air from outside the body and mix it with air held within small thin-walled sac-like structures known as alveoli. These are in close contact with pulmonary capillaries and this permits the rapid diffusion of oxygen from air to blood and of carbon dioxide in the reverse direction. Air is then returned to the atmosphere during expiration and waste carbon dioxide is thus removed. It is important to understand the factors which control the movement of air into and out of the lungs and the mechanisms of gaseous exchange.

19.1 Introduction

Cellular respiration (see Chapter 7) involves the utilisation of oxygen and the generation of carbon dioxide during the aerobic production of energy. Oxygen is carried to the tissues bound to the haemoglobin in red blood cells (see Chapter 18) and carbon dioxide is transported away in a variety of forms but mainly as bicarbonate ions formed as a result of the reversible reactions

$$H_2O + CO_2 \leftrightarrow H_2CO_3 \leftrightarrow H^+ + HCO_3^-$$

The respiratory system is designed to ensure that the blood remains oxygenated and that excess CO_2 is removed to the atmosphere.

19.2 Structure of the respiratory system

The respiratory organs in mammals are the *lungs*, which lie within the *thoracic cavity*. The surface of each lung is covered by a thin *pleural membrane* which is continuous with the pleural membrane lining the inside of the wall of the thoracic cavity. This arrangement is best visualised by imagining a partially inflated balloon (the pleura) contained within a box (the thorax). If you were to push your fist (representing the lung) into the balloon then there would be a layer of balloon coating your fist and another layer lining the inside of the box, and between them would be an air-filled space, the pleural cavity (Fig. 19.1). In reality this

Figure 19.1 The lungs

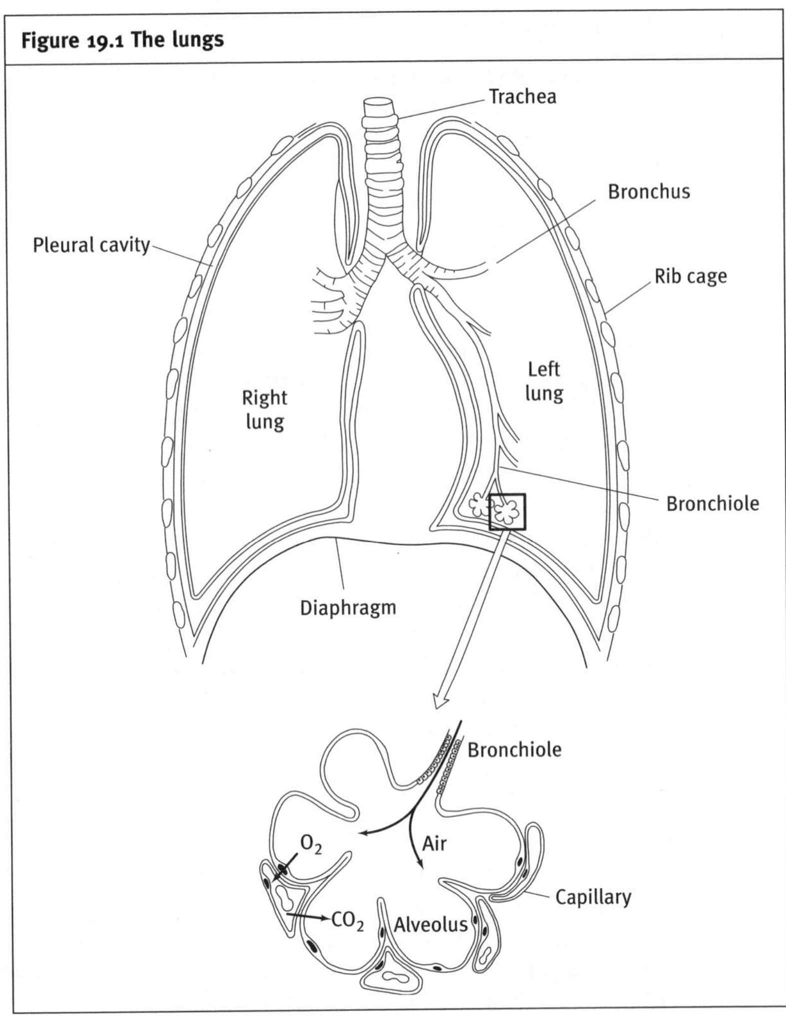

pleural cavity, which lies between the lung and the wall of the thorax, contains only a small amount of fluid and the two layers of pleural membrane are closely related to each other. Because the pleural cavity is a closed space any movement of the thoracic wall is coupled to movement of the outer surface of the lung within it. Thus when the volume of the thorax is increased during inspiration (see below) the volume of the lung is increased by an equivalent amount and air is drawn in.

The respiratory system is divided into two main parts – the *conducting portion* and the *respiratory portion.*

Conducting portion

The function of the conducting portion of the respiratory system is to moisten, warm and clean inspired air before it is delivered to the respiratory portion of the system where gaseous exchange between air and blood takes place.

Air is initially inspired through the nose and then passes through the *nasopharynx* at the back of the mouth, through the *larynx* (voicebox) and into the *trachea* (windpipe). From the trachea the airways branch repeatedly through *bronchi* and *bronchioles*, which become progressively narrower until they terminate at the small sac-like *alveoli* (singular *alveolus*) comprising the respiratory portion of the system. The wider parts of the airways are maintained open by rings of cartilage in their wall. The major resistance to airflow into the lungs is found in medium-sized airways and increasing the diameter of these airways by relaxation of smooth muscle in their walls is important when increased air flow into the lungs is required. The airways are lined by a specialised *respiratory epithelium* which contains a mixture of mucus-secreting *goblet cells* and *columnar cells* with hair-like protrusions called *cilia* (singular *cilium*). The mucus acts as a sticky layer to trap inhaled dust particles and bacteria. The cilia constantly beat to move the mucous layer up the airways towards the back of the throat where it is swallowed.

Respiratory portion

This is the part of the lung in which exchange of gases takes place between the inspired air and the blood entering the lung in the pulmonary artery. The basic structural unit is the alveolus which is a sac-like structure with walls made up of squamous cells (see Chapter 12). These cells share a basal lamina with the squamous endothelial cells of the pulmonary capillaries and together they form the diffusion barrier. This is normally less than 1.5 μm thick and permits rapid diffusion of both O_2 and CO_2 between blood and the air in the alveolus. There are approximately 300 million alveoli in the lungs with a total surface area of 80 m^2.

19.3 Respiration

With respect to the respiratory system, the term *respiration* is used to describe the act of breathing, through which air is moved into (*inspiration*) and out of (*expiration*) the lungs. Because the lungs are tightly linked to the walls of the thoracic cavity the volume of the lungs is increased by increasing the volume of the thoracic cavity. There are two ways in which this happens. The first involves movements of the *diaphragm*. This is a domed muscular sheet which separates the thorax from the abdomen and which, at rest, lies level with the lowest part of the *sternum* (breastbone). During inspiration the diaphragm moves downwards and so the height of the thoracic cavity is increased. The second way involves movements of the rib cage which act to increase the diameter of the thoracic cavity (Fig. 19.2). If, as you breathe in, you place your hand on the lowest part of your sternum you will feel it move upwards and outwards.

Air in the lungs is normally equilibrated with atmospheric pressure but as the lungs expand then their internal pressure drops and air is drawn in. The lungs have an inherent resistance to expansion which is contributed to by having to stretch elastic fibres in the walls of the alveoli and by the surface tension of the squamous alveolar cells. Thus inspiration requires active muscular effort in the rib cage and the diaphragm to overcome this resistance. The ease with which a lung can be expanded is known as its *compliance*. Surface tension in the lungs is lowered, and thus compliance increased, by the secretion onto the surface of the alveolar cells of a lipid substance known as *surfactant*. Once the inspiratory muscles relax, the lungs recoil and return to their resting state. During this recoil phase air is forced out of the lungs.

Fig 19.2 Movement of the diaphragm in respiration

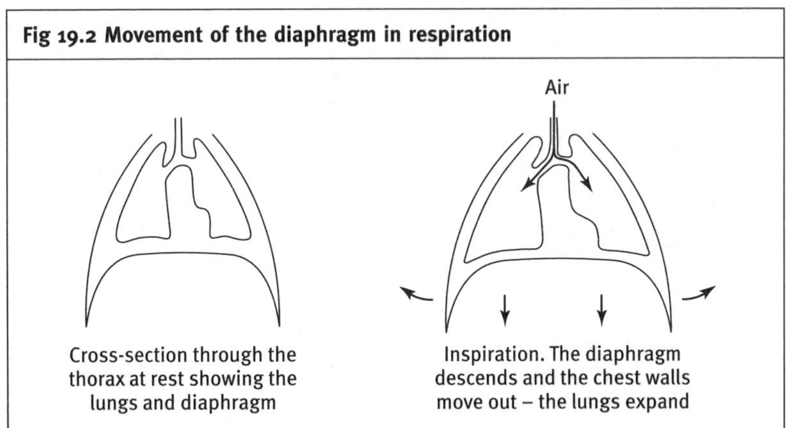

Cross-section through the thorax at rest showing the lungs and diaphragm

Inspiration. The diaphragm descends and the chest walls move out – the lungs expand

19.4 Gaseous exchange

The levels of O_2 and CO_2 in blood and air are described in terms of their partial pressures. The SI unit of pressure is the pascal (Pa) and partial pressures can be expressed in kilopascals (kPa). However, because blood pressure is always expressed clinically in millimetres of mercury (mmHg) other physiological pressure measurements also tend to be expressed in this older unit. The partial pressure of oxygen in atmospheric air is written as pO2 and normally has a value of 160 mmHg. In oxygenated blood the partial pressure is written p_aO_2 and has a range 75–100 mmHg. Blood heading for the lungs in the pulmonary arteries has a p_vO_2 of around 40 mmHg and a p_vCO_2 of around 45 mmHg. During inspiration a fixed volume of air is drawn into the lungs (the tidal volume) and mixes with the amount of air left in the lungs at the end of expiration (the *residual volume*). This residual air has been partially depleted of oxygen and contains higher levels of carbon dioxide than are found in the atmosphere. As a result of the mixing of fresh inspired air with this air left over from previous breaths, the p_AO_2 in the alveoli at the end of inspiration is approximately 100 mmHg and the p_ACO_2 is approximately 40 mmHg. Diffusion of a gas between blood and air will occur when the pressures are unequal on either side of the diffusion barrier. Thus, under normal circumstances there will be diffusion of oxygen from the alveolar air into the capillary and of carbon dioxide in the reverse direction (Fig. 19.3). The amount of gas exchange that can occur between the capillary and the alveolar air is normally limited only by the rate of flow of blood through the capillaries – the greater the rate of flow, the more gas that can be exchanged.

Fig 19.3 Gaseous exchange between air and blood in the lung

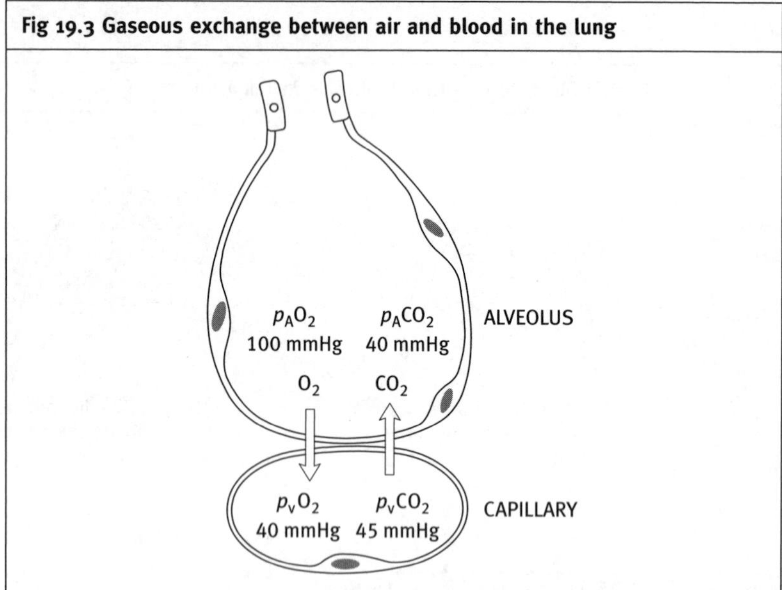

19.5 Control of respiration

Respiration is controlled by a group of neurons in the medulla of the brain known as the *respiratory centre* – this acts to control the activity of the muscles of respiration. The diameter of the airways is regulated via the autonomic nervous system. The respiratory centre neurons are directly activated by a rise in p_aCO_2 and this is the primary stimulus to increased respiratory effort. There are also sensors in the blood vessels of the head and neck which can detect low levels of oxygen.

Clinical example: Asthma

In asthma an external stimulus, such as pollen or animal hair, triggers an inflammatory reaction in the walls of the airways which causes them to constrict. This reduces the amount of air which can move in and out of the lungs and leads rapidly to an increase in p_aCO_2 since this cannot be removed from the system. Respiratory activity is increased and the affected individual will be seen to be gasping for air. However, because of the narrowed airways it is virtually impossible to increase the flow rate to the alveoli and eventually p_aO_2 levels will drop. In severe attacks the patient may lose consciousness. Treatment is by administration of inhaled drugs which relax the muscle of the airways and allow air to be drawn in.

19.6 Test yourself

1. What name is given to the membrane that surrounds the lungs and lines the thoracic cavity?
2. What are the two main types of cell in respiratory epithelium?
3. What muscular sheet separates the thorax from the abdomen?
4. Which cells make up the diffusion barrier between blood and air?
5. To changes in what are cells in the medullary respiratory centre particularly sensitive?

20 The digestive system

Basic concepts:
All energy is derived from the food we eat and in order for this energy to be used by the body these foodstuffs must be broken down into molecules that can be absorbed from the digestive tract into the bloodstream. The primary functions of the digestive tract are thus digestion and absorption. The digestive system consists of a hollow tube, the gastrointestinal tract, which extends from the mouth to the anus and which receives the secretions of various glands to aid the digestive processes. Some regions are specialised for digestion and others for absorption. Understanding the function of the digestive system is an important step in understanding how the energy requirements of the organism are met.

20.1 Introduction

Survival depends on adequate intake of food and water. Food provides the raw materials for energy production (see Chapter 7) as well as the building blocks for cells. In addition, food contains nutrients such as essential vitamins and minerals – for example vitamin B_1 (see Chapter 7), iron for the production of haemoglobin (see Chapter 18) and calcium for the formation of bones (see Chapter 24). The functions of the digestive system can be summarised as *digestion*, *absorption* and *excretion*. Food is broken down into molecules small enough to be absorbed through the wall of the gut into the bloodstream. Unwanted or undigested components of the diet are then excreted along with other waste products.

The digestive system comprises the *gastrointestinal* (GI) *tract* which runs from the *oral cavity* (mouth) to the *anus* and includes the *oesophagus*, *stomach*, *small* and *large intestines* and *rectum*. Associated with the GI tract are a number of accessory glands (*salivary glands*, *pancreas* and *liver*) whose secretions enter the GI tract and play a vital role in the digestive process (Fig. 20.1).

20.2 Oral cavity

The process of digestion is initiated in the oral cavity where food is broken down into small pieces by the actions of the teeth and tongue during chewing. In the oral cavity, food is mixed with the secretions of

Figure 20.1 The gastrointestinal tract

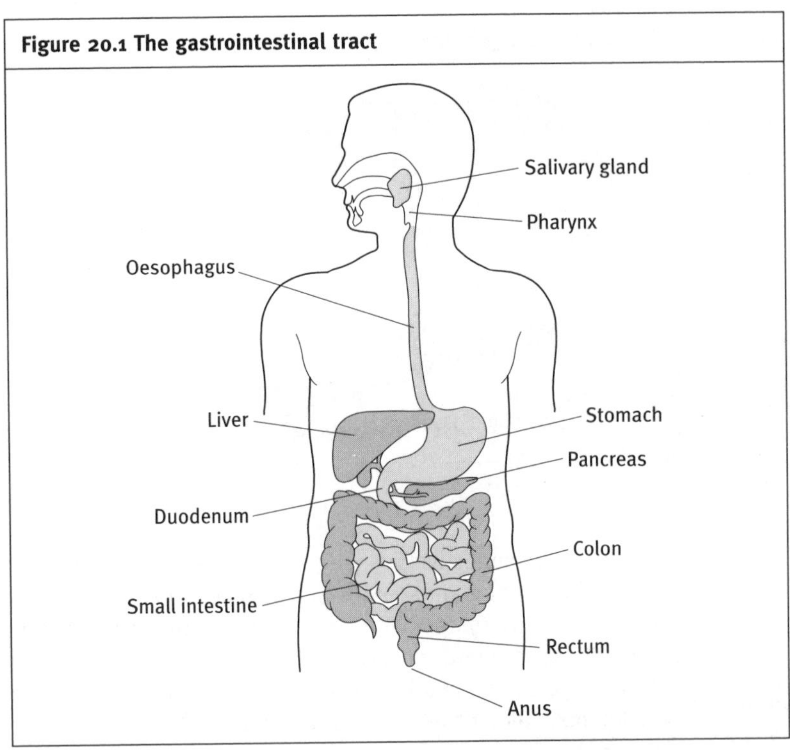

salivary glands, which produce a watery secretion containing the enzyme salivary α-amylase, which initiates the digestion of starch. The mixture of food and saliva forms a paste which is then passed into the oesophagus during the act of swallowing. The initial phase of swallowing is voluntary and depends on the actions of the tongue and of the muscles of the upper throat (*pharynx*).

20.3 Structure of the GI tube

The oesophagus connects the mouth to the stomach and is a hollow, muscular tube which forms the first part of the GI tract. The structure of the wall of the GI tube is relatively constant throughout its length and comprises three main layers which enclose the central space or *lumen* (Fig. 20.2). The innermost layer is the *mucosa* which lines the lumen with an epithelium. The structure of the epithelium varies depending on the functional demands of the various regions of the GI tract, but for the most part it is simple columnar. The mucosa is bounded internally by a layer of smooth muscle which allows the mucosa to be mobile and promotes local mixing of the contents of the lumen. Outside the mucosa

Figure 20.2 The gut tube

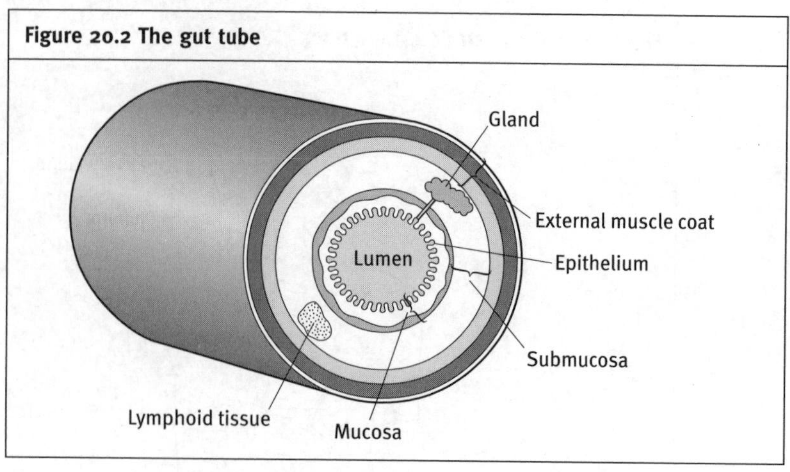

lies a layer of connective tissue known as the *submucosa*, which may contain additional glands and which also contains collections of cells from the immune system which make up the gut-associated lymphoid tissue (GALT). The outermost part of the gut wall is the *external muscle coat* which contains two layers of smooth muscle – one running longitudinally along the length of the gut and the other arranged in a circular or spiral fashion. This muscle coat is responsible for the wave-like movements of the gut wall (*peristalsis*) which help to push food along the length of the GI tract. The movements of the external muscle coat are regulated by the GI tract's own nervous system (the *enteric* nervous system). This contains up to 10 million neurons (as many as are found in the spinal cord!) which act locally in a reflex fashion to ensure coordinated movement of the gut wall. The activity of the enteric nervous system can be modified by the autonomic nervous system (see Chapter 17) and by locally produced peptide hormones.

20.4 The stomach

Contractions of the oesophagus cause the swallowed food to enter the stomach. This is a sac-like expansion of the GI tube which lies just below the diaphragm in the upper-left quadrant of the abdominal cavity. The stomach is stretchable to allow it to accommodate the food taken in during a meal. The external muscle coat of the stomach is particularly thick and vigorous contractions produce a churning movement. The entrance and the exit of the stomach are controlled by thickened circular bands of muscle. The upper of these is called the *oesophageal sphincter* and prevents the stomach contents being forced back up into the oesophagus. Failure of this sphincter can allow acid gastric juices to enter the oesophagus where they cause a painful burning sensation – heartburn. The *pyloric sphincter* prevents the passage of food from the stomach into the *duodenum* (the first part of the small intestine) before stomach functions are complete. On average food spends 4–6 hours in the stomach. During this time it is mixed with the secretions of the gastric glands. These are extensions of the epithelium lining the stomach wall and contain two key cell types – *parietal* and *chief* cells.

Parietal cells

These are responsible for the secretion of gastric acid. This is hydrochloric acid whose secretion maintains a low pH in the stomach. Acid secretion by parietal cells is regulated by three factors:

- the sight and smell of food triggers the vagus nerve to release acetylcholine (ACh) in the stomach wall where it acts on receptors on parietal cells;
- the presence of protein in the stomach triggers the release of a hormone, gastrin, which also acts directly on parietal cells to increase acid secretion;
- both gastrin and ACh trigger the local release of histamine which acts on parietal cells to enhance acid release.

In addition, parietal cells secrete intrinsic factor which is required for the intestinal absorption of vitamin B_{12}, an essential compound in haemopoiesis.

Chief cells

These secrete the digestive enzymes pepsinogen, amylase, lipase and prorennin. All of these enzymes are active at acidic pH. Both pepsinogen and prorennin are converted to their active forms, pepsin and rennin, by the stomach acid and all work optimally at low pH. Pepsin initiates the

Clinical example: Gastric ulcers

Gastric ulcers occur when the mucosal barrier in the stomach is overwhelmed by the amount of acid produced. Once the mucus is breached then the acid begins to erode the wall of the stomach. This initially causes discomfort, but if the erosion penetrates the wall of the stomach then nearby blood vessels can also be destroyed leading to massive bleeding. One factor that has been identified as causing increased acid production is the presence in the stomach of the bacterium *Helicobacter pylori*. Treatment with antibiotics to kill the bacteria can eliminate gastric ulcers in many patients.

digestion of proteins and rennin is particularly important in young animals where it breaks down milk proteins.

In addition to these cell types, the epithelium of the stomach has many cells which produce a neutral or alkaline mucus. This protects the surface of the stomach from the damaging effects of hydrochloric acid.

20.5 Small intestine and accessory glands

Food passes through the pyloric sphincter from the stomach into the duodenum. In total the small intestine is 6 m long and its coiled tubes fill most of the abdominal cavity. In the duodenum the secretions of the liver and the pancreas are added to the stomach contents.

The pancreas

This is a gland which produces large numbers of digestive enzymes in a bicarbonate-rich fluid. The secretion of these enzymes is stimulated by the emptying of gastric contents into the duodenum. Most of the pancreatic enzymes are secreted as inactive precursors which are activated on entry into the GI tract. For example, the proteolytic enzyme trypsin is produced from the inactive precursor trypsinogen and itself then activates a number of other enzymes.

The liver

This vitally important organ has two key functional roles. Firstly it is a gland which is connected to the duodenum and which secretes *bile* through the *bile duct* into the GI tract. Prior to its secretion bile is stored in the *gall bladder*. Bile is a solution of bile salts which act to emulsify dietary lipids and are essential for their further digestion and absorption. Bile also contains the pigment *bilirubin*, which is derived from the breakdown of the haem component of red blood cells. The oxidation of bilirubin in the GI tract into a brownish pigment is what gives faeces their characteristic colour. Secondly, the liver has key metabolic roles. It is structured so that the blood returning from the GI tract filters past the walls of the cells of the liver, the *hepatocytes*, and these cells act to modify blood composition in the following ways.

- *Glucose regulation* – the liver is the primary site of glucose storage in the body and, following a meal, glucose is taken up into the liver and stored as glycogen. As glucose levels in the blood drop then the liver releases glucose to maintain a constant level.
- *Plasma protein synthesis* – the liver manufactures the majority of plasma proteins including albumin and the elements of the clotting cascade.

- *Detoxification* – the liver is responsible for taking potentially toxic substances, such as drugs and alcohol, out of the circulation and breaking them down into metabolites that can be excreted.
- *Lipid metabolism* – the liver is responsible for the storage and metabolism of fatty acids and triglycerides.

Once the stomach contents have been mixed with the secretions of the pancreas and liver then the resultant fluid (*chyme*) moves along the small intestine. The cells of the epithelium of the small intestine possess a surface coated with small finger-like projections called *microvilli*. This is known as a *brush border* and it increases the surface area of the cells vastly. The brush border is coated with a layer of enzymes which have been secreted by the epithelial cells themselves and which are responsible for the terminal digestion of the food. Thus, on this surface, disaccharides are converted to monosaccharides and dipeptides are converted to individual amino acids. Absorption of these final products of digestion is then achieved mostly through Na^+-coupled transport mechanisms (see Chapter 8) in the membrane of the epithelial cells (Fig. 20.3). The absorbed substances pass into the capillaries which run in the wall of the intestine. From there the blood travels to the liver before re-entering the systemic circulation in the inferior vena cava.

Figure 20.3 Absorption of glucose in the small intestine

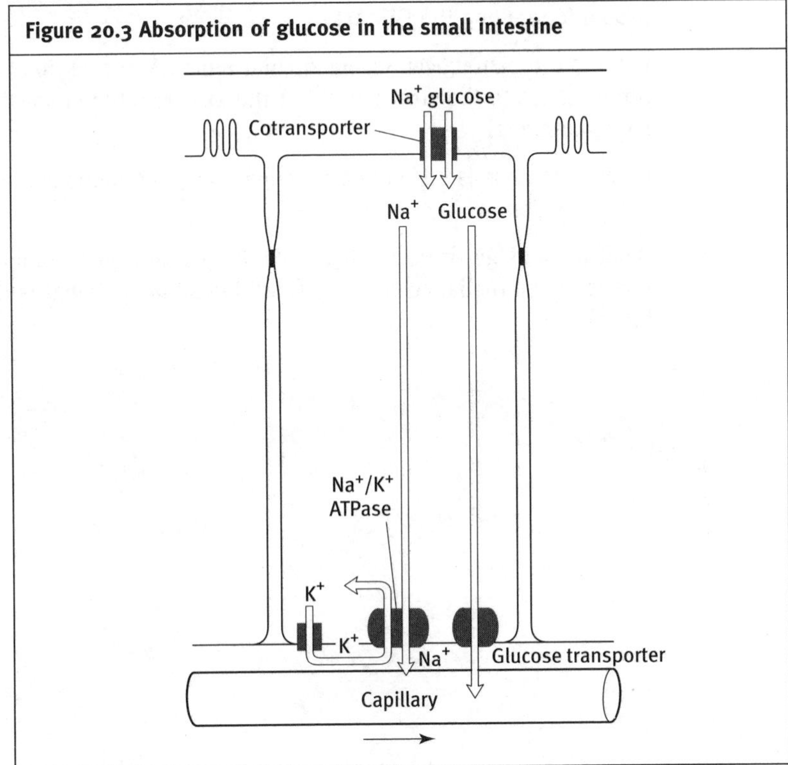

20.6 Large intestine

The primary function of the upper parts of the small intestine is to digest and absorb proteins, carbohydrates and nucleic acids. As the small intestine joins the large intestine, the functional focus shifts to the absorption of lipids, water and electrolytes. As the chyme passes along the large intestine through the *colon* and into the *rectum*, water and electrolytes are absorbed so that the faeces become compacted. As the rectum, which lies at the end of the large intestine nearest to the anus, fills with faeces the mechanical stretching triggers a desire to relax the sphincter around the anal canal. The opening of this sphincter is under voluntary control. Once the sphincter is open contractions of the rectum force its contents out through the anal canal in the act of *defaecation.*

20.7 Test yourself

1. What are the three main layers of the wall of the gastrointestinal tube?
2. What term describes the wave-like movements of the gut wall which propel food along the GI tract?
3. What is the name given to the circular band of muscle that lies: (a) at the entrance to the stomach; (b) at the exit from the stomach into the small intestine?
4. What substance is secreted by the liver into the duodenum and what is its function?
5. What name is given to the finger-like projections on the luminal surface of epithelial cells of the small intestine and what is their function?

21 The reproductive system

Basic concepts:
Reproduction ensures the perpetuation of the species. It provides the mechanism whereby the male and female gametes (sperm and eggs) can develop and be brought together so that fertilisation and the subsequent development of an embryo can occur. The regulation of all stages of reproduction from the formation of gametes to the act of birth itself is under complex hormonal control and it is important to understand this in order to be able to understand how sexual function and fertility may be regulated.

21.1 Introduction

The function of the male and female reproductive systems is to ensure the survival of the species by generating male and female *gametes* (*spermatozoa* and *ova* respectively) and providing a means of these being brought together so that *fertilisation* may occur (see Chapter 10). Following fertilisation, the female reproductive system is responsible for supporting the developing embryo and its placenta until the baby has grown sufficiently to be born.

21.2 Male reproductive system

The male reproductive system (Fig. 21.1) comprises paired gonads (*testes*) which are each connected via the *epididymis* and the *vas deferens* to the *urethra*, a muscular tube which opens at the tip of the *penis*. The testes and epididymis are the sites of sperm production and maturation. This occurs optimally at 34°C, which is why they are located outside the abdominal cavity in a muscular sac known as the *scrotum*. The vas deferens runs from each testis to join with the urethra at the base of the bladder. The urethra conveys both urine and sperm to the exterior. During sexual intercourse sperm are released from the penis via the urethra in a process known as ejaculation. Ejaculated sperm are contained in about 5 ml of seminal fluid, which is derived from the secretions of accessory glands such as the prostate and seminal vesicles.

Figure 21.1 The male reproductive system

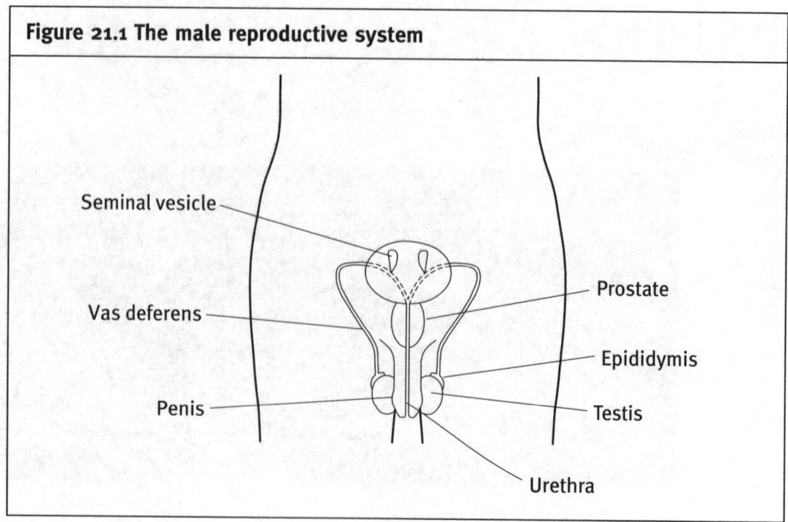

21.3 Spermatogenesis

Each testis contains an extensively coiled tube, the *seminiferous tubule*, within which sperm production takes place. Spermatozoa are derived initially from the mitotic division of primary germ cells known as *spermatogonia* which produces many *spermatocytes* which undergo meiotic division to produce haploid *spermatids*. The spermatids are embedded in *Sertoli cells* which support their differentiation into spermatozoa under the influence of FSH (see Chapter 16). Maturation of the spermatozoa from inactive to active cells takes place in the epididymis.

Within the connective tissue surrounding the seminiferous tubules are located *Leydig cells* which, under the influence of LH (see Chapter 16), are responsible for the production of the male sex hormone *testosterone*. Testosterone acts in the testis to promote spermatogenesis and is responsible for the development and maintenance of secondary sexual characteristics (e.g. beard growth) and sexual behaviour. Sperm are produced throughout the reproductive life of the male and, though sperm production drops gradually with age, sperm from a 90-year-old male are still capable of fertilising an ovum.

21.4 Female reproductive system

The female reproductive system (Fig. 21.2) comprises paired gonads (*ovaries*) which are linked via the *Fallopian tubes* to the *uterus*, and then through the *cervix* and *vagina* to the exterior. The ovaries lie within the

Figure 21.2 The female reproductive system

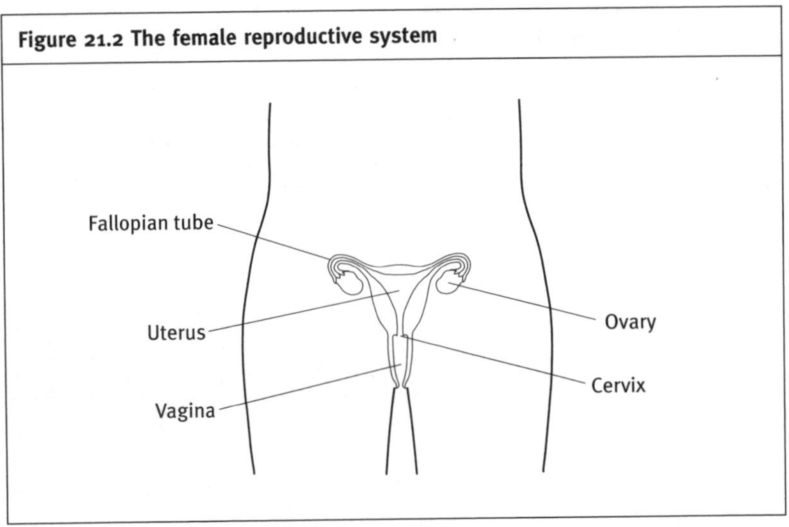

abdominal cavity and the ova which they release have to travel a short distance within the abdominal cavity before being drawn into the Fallopian tubes by the beating of ciliated cells. The Fallopian tubes enter the uterus, which is a pear-shaped muscular sac whose lining is designed to support the fertilised ovum during its development into a baby. The uterus opens through a narrow aperture, the cervix, into the vagina which is a muscular canal opening to the outside. The vagina acts as a receptacle for the erect penis during copulation (sexual intercourse) and as the birth canal during *parturition* (the act of giving birth).

21.5 Oogenesis and the menstrual cycle

Female germ cells undergo extensive mitotic division in the foetal ovary. At birth, the ovaries contain about 1 million *primary oocytes*. These are cells arrested in the first stage of their meiotic division. Only 200–400 of these will ultimately be released as mature ova and be available for fertilisation – the rest will degenerate. A small number of these primary oocytes become surrounded by a layer of *granulosa cells* to form a *primary follicle*. Between birth and puberty these follicles are not able to mature appropriately. In a female who has passed puberty there is a monthly *menstrual cycle* (Fig. 21.3) in which a single follicle matures and releases its oocyte in a process known as *ovulation*. The oocyte together with protective layers which surround it is known as an *ovum*. This cycle is under hormonal regulation and cycles only cease during pregnancy and following the menopause, which occurs after the age of approximately 50 when no more follicles are available to mature.

Figure 21.3 The menstrual cycle

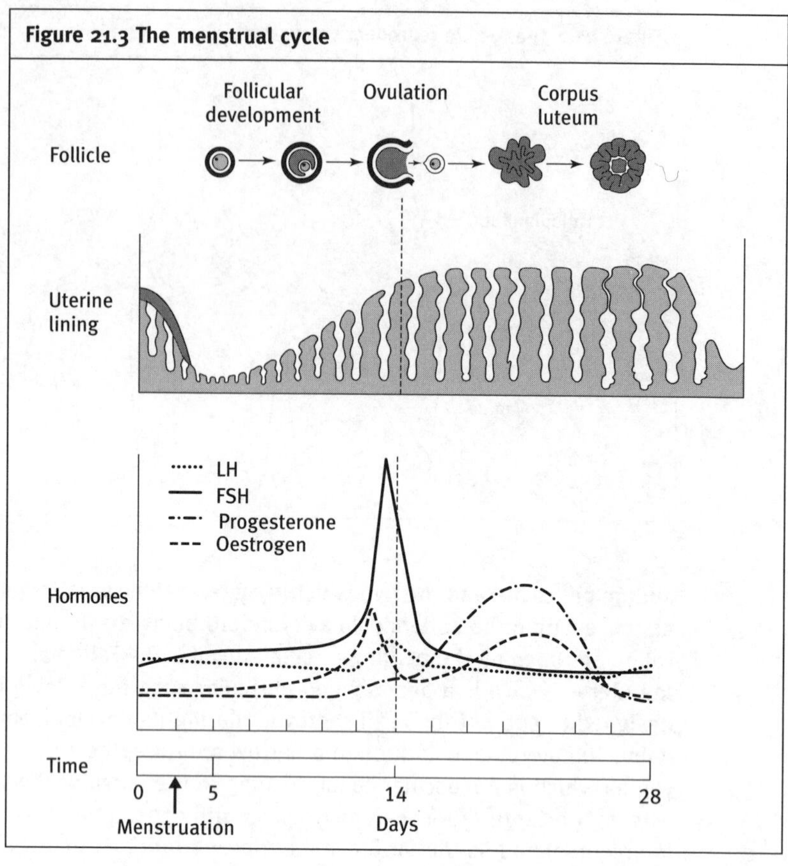

Menstruation marks the end of one menstrual cycle and the beginning of the next and involves the shedding of the lining of the uterus along with a small amount of blood. Each menstrual cycle lasts approximately 28 days. During the first phase of the cycle, which begins with the onset of menstruation, raised levels of FSH (see Chapter 16) recruit a cohort of primary follicles which begin to mature. As they do so they secrete increasing levels of oestrogen which stimulates the lining of the uterus to begin to proliferate. As the follicles mature, one becomes gradually dominant and the others regress. Just before the midpoint of the cycle, at about 14 days, rapidly rising oestrogen levels trigger a surge in LH production which causes ovulation. The follicle, in response to LH, then transforms into a structure known as the *corpus luteum* which secretes the hormone progesterone. This stimulates further growth of the lining of the uterus. Towards the end of the menstrual cycle one of two things may happen:

- a successful fertilisation of the released ovum will have occurred and it will implant into the uterine lining;

- in the absence of fertilisation, the corpus luteum will cease hormone production and the uterine lining will degenerate and be shed, thus signalling the onset of menstruation and the start of a new cycle.

Clinical example: The contraceptive pill

The contraceptive pill contains a mixture of synthetic oestrogens and progesterones. These act to provide negative feedback to the pituitary and hypothalamus where they suppress the production of FSH and LH. This prevents follicle development and ovulation. However, the synthetic hormones do allow growth of the lining of the uterus, and so stopping the drugs for 7 days out of every 28 allows menstruation to occur normally.

21.6 Copulation and fertilisation

Copulation involves the insertion of the erect penis into the vagina for the purposes of triggering ejaculation and the delivery of seminal fluid containing 150 to 600 million sperm into the female reproductive tract. The sperm then use their flagella to drive themselves through the cervix into the uterus and into the Fallopian tubes where fertilisation normally occurs. Only about 100 sperm complete this journey successfully and on the way they undergo a process known as *capacitation*, which renders them capable of penetrating and fertilising the ovum. When the sperm encounters an ovum it passes through the glycoprotein coat which surrounds the ovum and its head fuses with the cell membrane of the oocyte. This triggers the oocyte to alter the structure of its wall to prevent fertilisation by other sperm and to complete its meiotic division. The haploid sperm and oocyte nuclei then fuse to create a new diploid cell. *Conception* has occurred.

21.7 Implantation and pregnancy

Following fertilisation, the ovum begins to divide and soon creates a small ball of cells known as a *blastocyst*. This will eventually give rise to both the embryo and the placenta. In its early stages the blastocyst begins to secrete the hormone human chorionic gonadotrophin (hCG) which signals the corpus luteum not to regress and thus prevents the uterine lining from being shed. Detection of hCG is used in pregnancy-testing kits. The blastocyst lies in the uterine cavity for about 72 hours before attaching to and invading the uterine wall in the process of implantation. As the cells in the blastocyst continue to divide they form a hollow ball of cells. The cells on the outer wall of this ball will form the fetal part of the placenta and the inner cell mass will develop into the fetus. As the developing fetus burrows further it induces cells of the uterine wall to participate in forming the placenta. The placenta consists of blood vessels from the fetus in close proximity with blood from the mother. This allows exchange of oxygen, carbon dioxide, nutrients and waste products between the fetal and maternal circulations.

The fetus will normally spend about 9 months in the uterus. During the first three months (*trimester*) most of the vital organs are formed, and the remaining six months are largely devoted to growth.

21.8 Birth

As the result of a number of factors, including endocrine signals and mechanical stretching of the uterus, the woman will go into labour. This involves rhythmic contractions of the uterus stimulated by the release of bursts of oxytocin from the posterior pituitary which increase in frequency as labour progresses. These contractions are accompanied by hormonally regulated relaxation and dilation of the cervix and vagina to allow the passage of the baby to the outside world.

Clinical example: Ectopic pregnancy

An ectopic pregnancy is one in which implantation occurs in a site other than the uterus. Ectopic pregnancies commonly occur in the Fallopian tubes but can also occur on the outside of abdominal or pelvic organs if the ovum escapes into the abdominal cavity. It can be particularly dangerous because the fetus is developing in a site which is not adapted to receive it. As the fetus grows it can cause rupture of the wall of the organ into which it is embedded and lead to excessive bleeding. A woman suffering from a ruptured ectopic pregnancy will complain of sudden abdominal pain and show signs of excessive blood loss such as increased heart rate and a pale, clammy skin.

21.9 Test yourself

1. What name is given to: (a) the male gonads; (b) the muscular sac in which they are contained?
2. Down which structures must the ova pass after leaving the ovary in order to reach the uterus?

3. Which hormones stimulate the lining of the uterus to proliferate and are: (a) secreted by maturing primary follicles early in the cycle; (b) secreted by the corpus luteum later in the cycle?
4. What hormone is secreted by the blastocyst after fertilisation and can form the basis of a pregnancy test?
5. Which pituitary hormone stimulates the uterus to contract during labour?

22 The urinary system

Basic concepts:
The maintenance of body fluid and electrolyte levels is vital to ensure functioning of all cells, but particularly the excitable tissues. For example, small fluctuations in plasma potassium levels can lead to heart failure and paralysis. The kidney is responsible for this regulation. This is achieved by a process of filtering the fluids and electrolytes from plasma and then reabsorbing sufficient quantities of each into the blood to maintain appropriate plasma levels. Reabsorption in the kidneys is under hormonal control. The kidney is also involved in the excretion of the waste nitrogen formed from the breakdown of protein. Kidney failure is fatal and so it is vital to understand the functioning of this complex organ.

22.1 The kidney

The kidney is one of the main homeostatic organs in the body. Chapter 16 explains how hormones secreted by the kidney control red blood cell production and peripheral resistance. The kidney is primarily responsible for *osmoregulation* – the regulation of fluid and electrolyte balance within the body. In association with this function, the kidney is also responsible for *excretion* – the removal of waste products. These include *urea* (waste nitrogen) and other metabolites. The output of the kidney is *urine*, a concentrated solution of urea. This is passed to the bladder for storage before it is voided to the exterior via the urethra (Fig. 22.1).

Figure 22.1 The urinary system

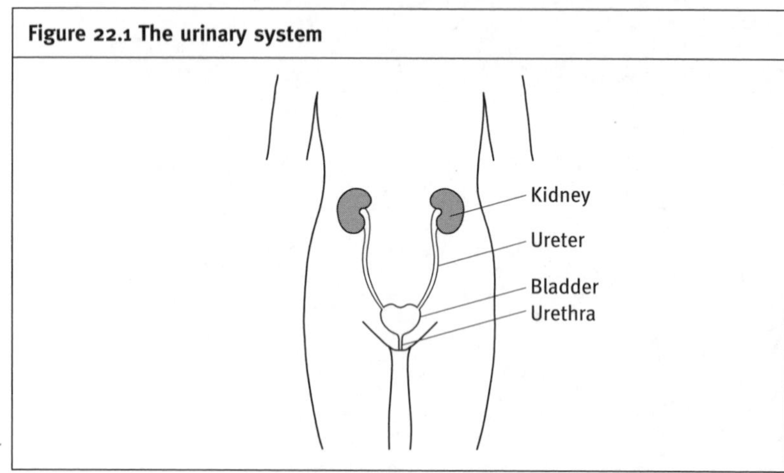

The kidneys are paired organs situated on the posterior wall of the abdomen, where they are protected from damage by the lower ribs. They receive a blood supply from the aorta through the *renal arteries* and return filtered blood to the circulation through *renal veins.* A cross-section through a kidney (Fig. 22.2) shows three main areas – an outer *cortex*, an inner *medulla* and the *renal pelvis* where urine is collected and drains into the *ureter* and thence to the *bladder.*

Figure 22.2 Cross-section of the kidney

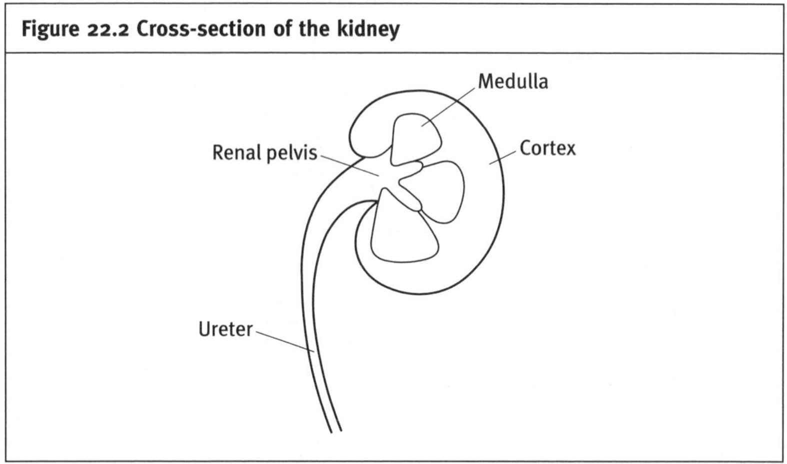

The basic functional unit of the kidney is the *nephron* (Fig. 22.3) which is a tubular structure closed at one end and open at the other. Each kidney contains between one and two million nephrons.

At the closed end of each nephron is *Bowman's capsule*. This is invaginated by a knot of capillaries, the *glomerulus*. Most glomeruli are located in the renal cortex. In this region, filtration of the plasma takes place. Plasma proteins do not pass through Bowman's capsule. Approximately 6 litres of plasma is filtered by the kidneys each hour. The filtrate is iso-osmolar with plasma and contains water, electrolytes, urea, glucose and amino acids. The function of the nephrons is to reclaim the required amounts of useful materials whilst reducing the 6 l/hr of filtrate to approximately 60 ml/hr of a concentrated urea solution.

This process begins in the proximal tubule with the active reabsorption of glucose, amino acids, NaCl and other electrolytes. The movement of these solutes generates an osmotic force which draws water along with them. About 80% of the filtered water is reabsorbed here, along with all the glucose and amino acids. Reabsorption of sodium in the proximal tubule is under the control of the hormone angiotensin II and this is the major site for regulation of blood volume and pressure (see Chapter 18).

Figure 22.3 The nephron

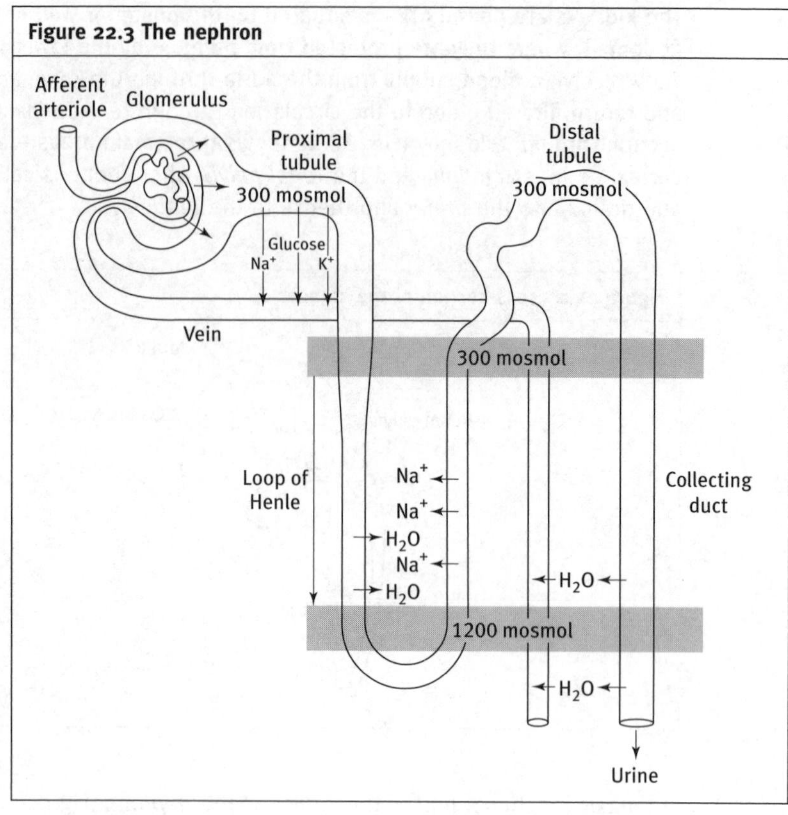

The filtrate then passes into the descending limb of the *loop of Henle*. This part of the nephron dips into the renal medulla. The descending and ascending limbs of the loop of Henle operate a countercurrent system which results in an osmolarity gradient being established between the top of the loop and the bottom. At the bottom of the loop of Henle, the tubular and interstitial fluids have an osmolarity of 1200 mosmol – about four times that of plasma. The gradient is established because the ascending limb is impermeable to water but actively pumps sodium into the descending limb, thus increasing its solute concentration.

Fluid from the ascending limb then enters the distal convoluted tubule where more sodium is actively pumped out along with water. The removal of sodium here is under the control of the hormone aldosterone which is produced by the adrenal gland in response to decreased plasma Na^+ ion concentration or in response to dehydration. Increased sodium reabsorption also increases water re-uptake.

The final removal of water takes place in the collecting ducts which dip back down through the renal medulla towards the renal pelvis. As they pass through the renal medulla they run through the area of high osmolarity created at the bottom of the loop of Henle. The presence of specific water channels (*aquaporins*) in the cells of the epithelium of the collecting duct allows rapid transfer of water out of the duct and into the interstitial fluid. The aquaporins are less permeable to the movement of urea than of water, with the consequence that urea becomes concentrated within the urine.

The number of aquaporin channels present in the collecting duct is regulated by antidiuretic hormone (ADH) which is produced in the hypothalamus in response to decreasing water content of blood (rising plasma osmolarity). This triggers the insertion of more aquaporin channels and thus allows more water to be reabsorbed from the filtrate. In the event of a rise in plasma volume (for instance after drinking large volumes of water) ADH secretion is reduced, aquaporin channels are withdrawn and more water flows through the collecting ducts into the ureter and down to the bladder.

22.2 The bladder

This is a distensible muscular sac with a wall containing a thick layer of smooth muscle. The bladder is capable of storing up to 1 litre of urine. The bladder is connected to the exterior via the *urethra*. Stretching the bladder wall generates the urge to pass urine, but relaxation of muscular sphincters around the urethra is needed before emptying can take place. The neural control of these sphincters is through a combination of voluntary and involuntary mechanisms.

Clinical example: Renal failure

There are many causes of renal failure but one of the most common occurs in response to prolonged high blood pressure. This initially causes damage to the filtration barrier in the kidney glomerulus and leads to the excretion of protein and glucose in the urine. The presence of these substances in urine can prevent water being reabsorbed and lead to the patient producing large quantities of dilute urine. Eventually the damaged nephrons die and then the patient will not have enough functional kidney tissue to enable them to clear all the toxic metabolites from their blood. In this situation an artificial kidney (a dialysis machine) is used to help clear the patient's blood of metabolites until they can receive a kidney transplant.

22.3 Test yourself

1. What is the basic functional unit of the kidney?
2. What is the primary function of Bowman's capsule and the glomerulus?
3. Where in the kidney does most reabsorption of glucose, sodium and other electrolytes take place?
4. Which hormone acts to regulate reabsorption of sodium in the distal convoluted tubule?
5. What effect does antidiuretic hormone (ADH) have on the kidney?

23 The immune system

Basic concepts:
The body is under constant threat from microorganisms in the external environment and these seek to use the body's resources to ensure their own survival. Such infections are potentially fatal and the survival of the individual depends on being able to destroy or neutralise these infective agents. The immune system functions to protect the body against infection by recognising invading microorganisms and destroying them. It does this by employing a wide range of molecules and cell types which interact to produce a coordinated response. Occasionally this response can be targeted against the body itself and can be severely debilitating. An understanding of the interactions between the cells of the immune system and their targets is important in appreciating how microorganisms interact with the host to cause disease. Knowledge of the immune system can also help to control both infectious and other non-infectious diseases.

23.1 Infection and infectious agents

The greatest threat to human health worldwide is infectious disease. Within the developed world enormous progress has been made in fighting infections, due to the development of good sanitation and drugs such as *antibiotics*. Nevertheless, emerging infections still pose a significant threat. The body has its own system of defence against infection, the *immune system*, which works effectively against all but the most problematic infectious agents. We can help the immune system in its fight against infection by immunisation, and through such programmes killer diseases of the past, such as smallpox, have been eradicated.

Clinical example: Emerging diseases

Microorganisms have the capacity to mutate very rapidly and so there is an ever present danger of new infections appearing. The HIV–AIDS syndrome was first identified in the early 1980s but this virus now accounts for millions of deaths a year. In the late 1980s BSE was identified as a new disease of cattle. Subsequently this was found to be transferable to humans where it causes a fatal degenerative neurological condition called Creutzfeldt–Jakob disease (vCJD for short).

> The 'flu' virus is able to mutate, giving rise to new strains which can cause pandemics. The 1918–19 'flu' pandemic killed between 20 and 40 million people; more than were killed in the First World War. Currently there is concern about a new strain of bird 'flu', H5N1, which, if it acquires the ability to spread from person to person, could have devastating effects.

Types of infectious agent

Infectious agents can take various forms. Some diseases are caused by *viruses.* Viruses cannot replicate themselves unless they infect a cell of an appropriate host, hence they are described as *obligate intracellular parasites.* Structurally a virus consists of a nucleic acid genome, which can be either DNA or RNA, enclosed in a protein *capsid.* In some viruses there is a lipid outer *envelope* containing proteins or glycoproteins. Diseases caused by viruses include HIV–AIDS, chickenpox and influenza ('flu' for short).

Another group of microorganisms that can cause disease is the *bacteria.* Bacteria are prokaryotic, single-celled organisms surrounded by a cell wall. They are larger than viruses and, unlike viruses, they do not need to infect cells in order to replicate. Diseases caused by bacteria include tuberculosis (TB), leprosy and food-poisoning due to Salmonella.

The largest infectious agents are the eukaryotic parasites. These can be single-celled (protozoa) or multicellular organisms including worms. Malaria is a protozoal disease and a major cause of death in some parts of the world.

It is important to note that most microorganisms we encounter are harmless, and it is only a very few species that cause infectious disease in man or other animals. Those organisms that can cause disease are referred to as *pathogens.* Pathogens can be transmitted in various ways. Some, such as colds and 'flu', infect the respiratory tract and can be transmitted from person to person via droplets produced by coughs and sneezes. Other infections, such as Salmonella and cholera, are acquired by eating or drinking contaminated food or water. Some infections are transmitted sexually, such as HIV and syphilis. In other cases microorganisms may enter the body through cuts or other damage to the skin – this is the case for tetanus, a bacterium that is found in soil. Some infectious agents, such as malaria, are spread by insects when they bite humans or other animals. In this case the insect is described as a *vector* for the disease. The following table shows some of the major causes of death due to infectious diseases (Source: *WHO World Health Report, 2004*)

Disease	Causative agent	Deaths in 2002
Total deaths due to infectious and parasitic diseases	Various	10.9 million
HIV–AIDS	RNA virus	2.8 million
Tuberculosis	Bacterium	1.6 million
Malaria	Single-celled eukaryotic organism, Plasmodium	1.2 million
Measles	RNA virus	0.6 million

23.2 Immune responses to infection

The body's system of defence against infection is the immune system. The immune system has a large number of different weapons in its armoury to enable it to deal with different types of infection arising in different parts of the body. These weapons consist of various cells and molecules that are able to attack infectious agents.

Barriers to infection

The first line of defence against infection is provided by the barriers that exist between our bodies and their surroundings. The skin is impervious to most infectious agents and is protected by acidic secretions that are hostile to most microorganisms. Mucosal surfaces, such as those lining the respiratory tract, are protected by a layer of mucus, a highly viscous secretion that effectively traps microorganisms and can be moved along the mucosal tract by cilia on epithelial cells. Secretions such as tears and digestive juices contain the enzyme lysozyme, which attacks bacterial cell walls.

Molecules and cells of the immune system

If an infectious agent manages to breach these barrier defences and gain access to the body it will then come into contact with further molecules and cells of the immune system. *Phagocytes* are cells that can ingest microorganisms and then destroy them. There are two main types of phagocyte – *neutrophilic granulocytes* and *macrophages.* Granulocytes are the predominant type of white blood cell. They circulate round the body and home in on sites of infection. Macrophages are found in certain tissues, such as the lungs, where they provide a resident population of immune cells. Both granulocytes and macrophages produce a range of molecules that can attack any ingested microorganisms. Microorganisms are internalised into a vesicle, which then fuses with lysosomes of the cell, exposing the microorganism to their various enzymes. A range of molecules found in blood and tissue fluids can also attack infectious agents. One of the most important of these is *complement*. Complement is a group of plasma proteins which, in the presence of infection, acts to promote phagocytosis. It does this by binding to the surface of microorganisms and marking them as foreign for ingestion by phagocytes (Fig. 23.1). Complement can also act to promote the process of inflammation.

Figure 23.1 The stages of phagocytosis

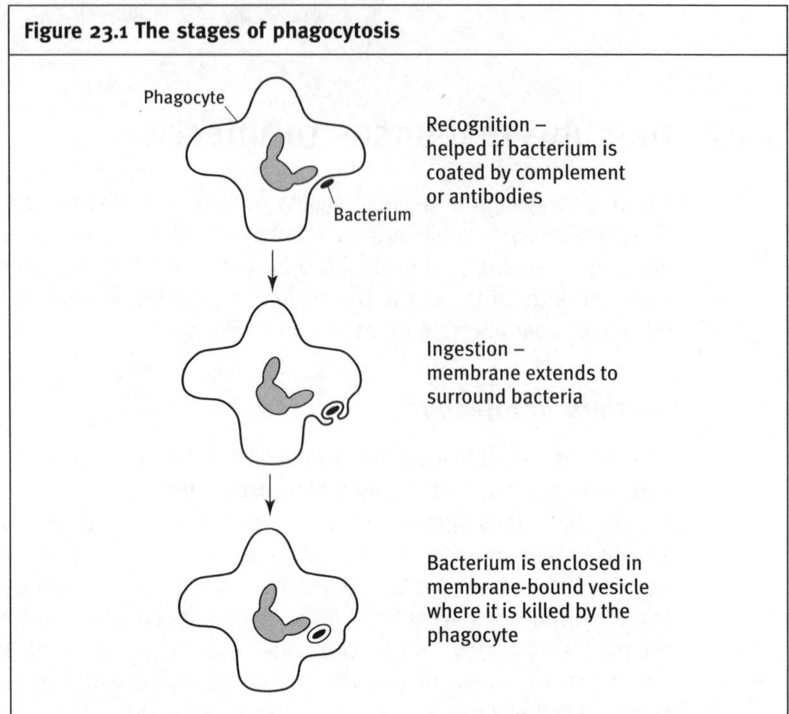

23.3 Inflammation

When an infectious agent enters a tissue of the body it is important that the cells and molecules of the immune system are brought into the site of infection so they can start to launch an attack. This is brought about by the process of *inflammation*. Inflammation involves changes in the blood vessels that supply a tissue making the vessel walls more permeable. This allows circulating cells and fluid to leave the bloodstream and enter the tissues. Interaction of microorganisms with tissue cells induces the production of molecules, which in turn act on the blood vessels to bring about the increased permeability. One of the principal cells found at site of inflammation is the granulocyte, which can help dispose of the infectious agent by phagocytosis.

23.4 Lymphocytes and the specific immune response

The defences against infection that have been described so far are present in the body, even before an infection comes along, and form part of the *innate* immune system.

The innate defences are fast-acting and can help to keep an infection at bay while the body develops its second line of defence – a *specific immune response*. Specific immune responses are produced by *lymphocytes* which circulate round the body, moving between the blood and tissues and seeking out infection. There are two main types of lymphocytes, *B cells* and *T cells*. Both T and B cells have receptors on their surface that are highly specific for individual molecular structures of different infectious agents. Each cell has only one type of receptor and can therefore only recognise and respond to one type of infection. The structures that are recognised by the T and B cell receptors are called *antigens*. Antigens can include proteins, polysaccharides and other molecules produced by microorganisms. The number of different antigens that can be recognised by the specific immune system is enormous, meaning that the body is able to defend itself against any microorganism it might encounter. This is because of the very large number of different T and B cell receptors, which has been estimated to be in the region of 10^9.

T and B cells have different mechanisms for fighting infection. When a B cell comes into contact with its specific antigen, it differentiates into plasma cells which secrete proteins called *antibodies*. Antibodies are found in the blood, tissue fluids and mucosal secretions. They are able to bind to antigens with a high degree of specificity. This can help fight infection in several ways – antibodies binding to antigens on the surface of a virus can block that virus from infecting a body cell; antibodies can also promote phagocytosis in a similar way to complement, by binding to microorganisms and marking them as foreign.

While antibodies can act against microbes that occur in extracellular locations within the body, T cells are particularly good at recognising intracellular infections such as viruses. *Cytotoxic T cells* can recognise cells of the body that contain a virus infection and kill these cells, so limiting the replication and spread of the virus. *Helper T cells* act by secreting proteins called cytokines that promote the activities of various other cells of the immune system. For example, helper T cells are important in activating B cells to allow them to secrete antibodies, and in interacting with macrophages to up-regulate their killing mechanisms. The functions of T and B lymphocytes are summarised in the table below.

Type of cell	Function
B lymphocyte	Gives rise to antibody-secreting plasma cells
Helper T cell	Secretes cytokines that help other cells – for example activate macrophages; help B cells become antibody-secreting cells
Cytotoxic T cell	Kills virus-infected body cells

23.5 Diseases of the immune system

HIV and immunodeficiency

The importance of the immune system in defending against infection is clearly seen in individuals whose immune function is compromised. Such people are described as being *immunodeficient*. Immunodeficiency can be due to a number of causes including malnutrition, certain types of cancer and some drug treatments. Infection can also cause immunodeficiency, and the best illustration of this is the HIV virus.

HIV is an RNA retrovirus. It infects cells of the immune system including helper T cells and macrophages. After infection the RNA genome is converted into DNA by the viral enzyme *reverse transcriptase*. This DNA can integrate itself into the DNA of the host cells and, in this form, the virus can remain latent for years, with the infected person showing no symptoms. After a period of time the virus becomes activated and this leads to the depletion of helper T cells. When numbers of helper T cells fall below a critical level the patient starts to succumb to a range of infections. Some of these infections involve microorganisms that pose no threat to healthy individuals and these are described as *opportunistic* infections. If untreated, the HIV-infected patient will eventually die, normally as a result of infection.

Treatment against HIV involves drugs that act against the virus. Two types of drug are commonly used – inhibitors of reverse transcriptase and inhibitors of the viral protease. The viral genes are transcribed and translated into a large protein which is then cleaved by a viral protease to give rise to the proteins of the virus. If this cleavage is blocked by protease inhibitors then virus production is prevented.

Allergy and autoimmunity

Although the immune system is crucial in protecting against infection, unfortunate consequences for the host can arise when immune responses are made inappropriately. Two examples of this are diseases arising from *allergy* and *autoimmunity*.

In allergy, an immune response occurs against a substance that would not normally cause problems for the host. For example, hay fever sufferers produce antibodies against grass pollen. These antibodies cause inflammation in the upper respiratory tract leading to the symptoms of the disease – runny nose, sneezing, etc. Some allergies are more serious. Asthma can be caused by allergic reactions lower down the respiratory tract leading to problems with breathing that can be life-threatening. The most serious type of allergic reaction is seen to antigens that enter the systemic circulation. This can occur with some food antigens, such as those derived from peanuts. In this instance the inflammation resulting from the allergy can affect blood vessels throughout the whole body and can very quickly lead to collapse and death if treatment is not administered rapidly.

In autoimmunity, the immune response starts to attack the body's own cells and tissues instead of being directed against an infectious agent. Under normal circumstances lymphocytes that have receptors for the body's own molecules (self-antigens) are eliminated or rendered non-responsive – a process known as immunological tolerance. Autoimmune disease arises when something goes wrong with this process. Autoimmune diseases can affect a range of tissues of the body and include conditions such as diabetes, rheumatoid arthritis and multiple sclerosis.

23.6 Test yourself

1. What two main constituents are present in *all* viruses?
2. What name is given to a molecular structure recognised by the receptors of T and B lymphocytes?
3. Which cells of the immune system recognise and kill virus-infected cells?
4. Which type of phagocytic cells circulate in the blood and enter sites of inflammation?
5. What molecules secreted by plasma cells bind to a specific antigen?

24 The musculoskeletal system

Basic concepts:
The action of muscles on the bony skeleton generates movement at joints. Control of movement is achieved through the action of motor neurons which are regulated by local reflexes in the spinal cord and by descending control from the brain. Bone is a connective tissue which acts as an important store of calcium, the plasma levels of which are under hormonal control. Adequate levels of calcium in the blood are essential for normal functioning of nerve and muscle and so it is important to understand these regulatory mechanisms.

24.1 Introduction

Locomotion is one of the primary attributes of animals. In most vertebrates, movement of the limbs is achieved by the action of muscles pulling on a jointed skeleton. The skeleton is composed predominantly of the connective tissue, *bone*. Bones articulate smoothly with each other at *synovial joints* which act as low-friction shock absorbers due to the presence of *cartilage* on their surfaces. Muscles acting on the skeleton are known as skeletal muscles and are controlled by somatic nerves which arise from the brainstem and spinal cord (see Chapter 17).

24.2 Bone

Bone is the major component of the skeleton. It is a strong, rigid connective tissue. Like all other connective tissues in the body, bone is a living tissue and depends on adequate supplies of nutrients and oxygen to stay alive. Connective tissues (see Chapter 13) are composed of three main elements – cells and fibres embedded in a protein, and proteoglycan matrix. Bone differs from other connective tissues only in the presence within this matrix of the crystalline mineral hydroxyapatite – $Ca_{10}(PO_4)_6.(OH)_2$ – which gives bone its rigidity.

Bone is constantly being remodelled, with old bone being eaten away by cells known as *osteoclasts*. These are multinucleated phagocytic cells derived from blood-borne precursors. Their activity can be stimulated by increased levels of thyroid and parathyroid hormone and inhibited by oestrogen. New bone is laid down in the spaces which osteoclasts create by cells known as *osteoblasts*. These cells secrete the molecules which make up the extracellular matrix of bone and can increase their activity in response to growth hormone and insulin-like growth factor. Following deposition of the matrix the process of mineralisation, during which hydroxyapatite crystals grow in the matrix, is largely dependent on plasma calcium levels.

In the adult skeleton the activity of osteoclasts and osteoblasts needs to be kept in balance in order to maintain bone density. This is achieved via a complex pattern of hormonal regulation which is partially related to plasma calcium levels since bone is the principal calcium store in the body. There is also direct coupling between osteoblast and osteoclast activity.

24.3 Regulation of calcium levels

Plasma calcium levels are tightly regulated. Variations in plasma calcium above or below the normal range can have profound effects on excitable tissues. Plasma calcium levels depend on three main processes:

- amount of dietary calcium absorbed from the gut;
- amount of calcium deposited in or released from bone;
- amount of calcium excreted by the kidney.

Alteration of any one of these processes will have consequences for the others. Thus a decrease in calcium uptake from the gut will lead to an increase in calcium release from bone and contribute to a demineralisation of the bone matrix.

Plasma calcium levels are detected by cells in the parathyroid gland, kidney, brain and other organs. The two main substances which control calcium levels are parathyroid hormone (see Chapter 16) and vitamin D.

Clinical example: Osteoporosis

Osteoporosis is the loss of bone tissue due to excessive activity of osteoclasts. It is a particular problem in post-menopausal women because of the sudden loss of the inhibitory effect of oestrogen on osteoclasts. In a sufferer from osteoporosis, bones become more brittle and susceptible to fracture. Typically this is seen in the fractured hips which occur as the result of seemingly innocuous falls in the elderly.

Vitamin D can be derived from the diet and the most important dietary sources are fish, liver and milk. However, the major source of vitamin D for most individuals is from synthesis by cells in the skin as a direct result of the action of sunlight. The actions of vitamin D are to:

- increase absorption of calcium from the gut;
- increase removal of bone by osteoclasts.

24.4 Cartilage

In addition to bone, the skeleton also contains cartilage. This is found on the surfaces of bones at joints and plays an important role during the growth of bones. Cartilage is a connective tissue which contains a dense extracellular matrix containing collagen and sulphated proteoglycans. Sulphate ions attract Na^+ ions into the matrix which in turn draw and hold in water by osmosis. The presence of water in the matrix makes cartilage resistant to compression and it is thus ideally structured to resist forces transmitted through joints during locomotion.

24.5 The skeleton

The bones in the skeleton can be classified as *long* bones or *flat* bones. A typical long bone is the femur which is located in the thigh. This has a long, straight hollow shaft containing a cavity filled with *bone marrow.* Bone marrow is the site of production of red and white blood cells. At both ends of the femur are heads which form a movable synovial joint with the adjacent bone at either the hip or the knee. A typical flat bone is the frontal bone (forehead), one of the bones which make up the skull. Flat bones tend to be protective in function and their joints with adjacent bones are often rigid. They also contain a marrow cavity.

24.6 Synovial joints

Synovial joints enable adjacent bones to move relative to each other (Fig. 24.1). For example, between the femur in the thigh and the tibia in the lower leg is the knee joint which allows the movements of flexion (bending) and extension (straightening). The joint is surrounded by a capsule of tough connective tissue which is lined by the synovial membrane. This secretes fluid into the joint cavity which acts both as a lubricant for the joint surfaces and to convey nutrients to the cartilage covering the ends of the bones. The cartilage on the bones provides a low-friction surface to facilitate movement and acts as a shock absorber to minimise the jarring effects of walking or running. The stability of

Figure 24.1 A synovial joint

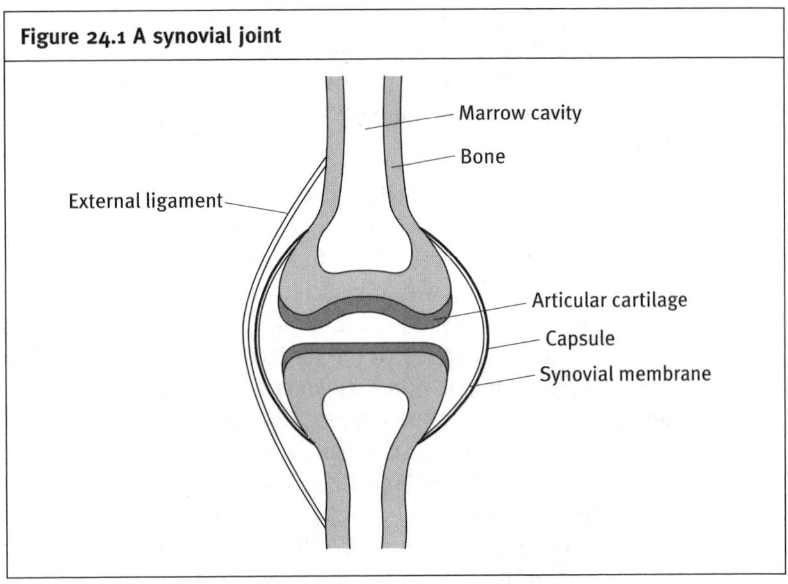

synovial joints is increased by the presence of strong ligaments which surround the joint and by the muscles which act on the joint. This is why physiotherapy for the muscles around a joint is important to athletes recovering from joint damage.

24.7 Muscles and locomotion

The structure and function of muscle fibres is covered in Chapter 14. Individual muscles, such as the quadriceps muscles in the front of the thigh, are made up of many muscle fibres embedded in a connective tissue matrix. Muscles which move a joint are often inserted directly into the bone above the joint and then give rise to a tendon which crosses the joint to be moved and is then inserted into the bone below the joint. During walking or running, the knee joint undergoes alternate flexion and extension movements. This is achieved by a controlled pattern of contraction and relaxation of the muscles in the front and back of the thigh. The *extensor* muscles in the front of the thigh straighten the leg as an individual pushes off from the ground and the *flexor* muscles bend the knee as the free leg swings forward. These alternate movements are controlled by sets of neurons in the spinal cord which act as *locomotor pattern generators* and are not subject to conscious regulation.

Individual motor neurons within the spinal cord send axons to innervate one or more muscle fibres. The number of muscle fibres controlled by a single neuron is known as the *motor unit* and varies depending on the level of control required. Thus, muscles which control coarse movements, such as those in the thigh, will have large motor units (up to 200 fibres) while in the fingers precision of movement is achieved by having a one-to-one relationship between a motor neuron and a muscle fibre.

In order for muscle contraction to be regulated effectively it is important that the central nervous system is constantly aware of the state of contraction of the muscle. This information is provided by specialised sensory fibres located within muscles and joints and is known as *proprioception*. An example of proprioception in action can be seen in the knee-jerk reflex, in which the tendon at the front of the knee joint is stretched by tapping it with a hammer and the thigh muscles immediately straighten the leg to counteract the stretch (Fig. 24.2). The neural basis for this is that sensory nerve endings in the muscle are stimulated by the tendon being stretched and their axons run back into the spinal cord where they directly excite the motor neurons which innervate the quadriceps muscle. They also indirectly cause relaxation of the muscles in the back of the thigh. Reflexes like this operate continuously during locomotion to counteract unexpected movements of the joint, such as those produced when individuals travel over rough ground.

Figure 24.2 The knee-jerk reflex

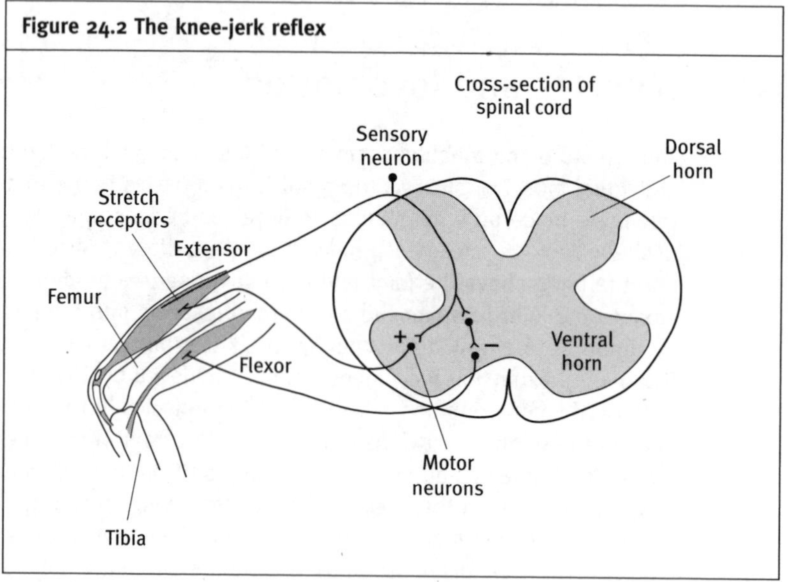

During locomotion, individual muscles are not under direct conscious control but deep nuclei in the brain send signals via the cerebral cortex to the spinal cord which allow the local pattern generators to produce movement. The smooth execution of movement is due to the actions of the cerebellum (see Chapter 17). The brain can also exert voluntary control over individual muscles by sending signals directly from neurons in the cerebral cortex to individual motor neurons in the spinal cord.

24.8 Test yourself

1. Which cells are responsible for the resorption of bone?
2. What are the two main actions of vitamin D?
3. What lies in the centre of the shaft of a long bone?
4. What is a motor unit?
5. In the knee-jerk reflex, which muscles contract in response to tapping the tendon at the front of the knee joint?

Answers to 'test yourself' questions

Chapter 1
1. On the inside of the molecule
2. Intracellular fluid, interstitial fluid, plasma
3. (a) Na^+; (b) K^+
4. Carbon
5. Protein

Chapter 2
1. Amino acids
2. Amino (NH_2) and carboxyl (COOH)
3. The sequence of amino acids
4. (a) On the outside of the helix; (b) every 3.6 residues; (c) hydrogen bonds
5. (a) Homodimer; (b) Heterodimer

Chapter 3
1. $C_6H_{12}O_6$
2. Glycosidic bond
3. Amylase
4. Glycosyl transferase
5. Glycogen

Chapter 4
1. Saturated
2. Glycerol
3. Two
4. Amphipathic
5. Adipose tissue

Chapter 5
1. Deoxyribonucleic acid
2. Adenine (A), thymine (T), guanine (G) and cytosine (C). A pairs with T; C pairs with G
3. A codon
4. Promoter
5. Transfer RNA (tRNA)

Chapter 6
1. Lysosomes
2. In mitochondria
3. Mitochondria
4. Actin
5. Ribosomes

Chapter 7
1. Pyruvate
2. Adenosine triphosphate
3. In mitochondria
4. Oxidative phosphorylation
5. Lactic acid

Chapter 8
1. 3 osmolar
2. True
3. Na^+
4. Sodium/potassium ATPase
5. No, because it can freely diffuse through the cell membrane

Chapter 9
1. G1, S and G2
2. The mitotic spindle
3. The centromere
4. Telomeres
5. Cyclin-dependent kinases

Chapter 10
1. Gametes (eggs and sperm)
2. Meiosis
3. Chiasmata
4. Zona pellucida
5. Embryonic stem cells

Chapter 11
1. Homozygous
2. Insertion
3. They are likely to be linked – i.e. on the same chromosome
4. (a) XX; (b) XY
5. Selective advantage

Chapter 12
1. One
2. Stratified squamous
3. Basal lamina
4. Tight junctions
5. Desmosomes

Chapter 13
1. Glycosaminoglycans
2. Collagen
3. Chondroblasts
4. Chondrocytes
5. To promote inflammation

Chapter 14
1. K^+
2. Actin and myosin
3. Acetylcholine
4. The cell body, the axon and dendrites
5. To increase the rate of conduction by providing electrical insulation for the axon

Chapter 15
1. Insulin
2. Glucagon
3. The hypothalamus
4. The blood flow to the skin is reduced due to narrowing of the peripheral blood vessels
5. Brown fat

Chapter 16
1. Steroids
2. Adrenalin
3. GTP
4. Cortisol
5. (a) Thyroid stimulating hormone (TSH); growth hormone (GH); adrenocorticotrophic hormone (ACTH); luteinising hormone (LH); follicle stimulating hormone (FSH); and prolactin (PRL)
   (b) Antidiuretic hormone (ADH); oxytocin

Chapter 17
1. Motor neurons
2. Frontal, parietal, occipital and temporal
3. The somatic division
4. (a) Noradrenalin; (b) acetylcholine
5. The vagus nerve

Chapter 18
1. Fibrin
2. (a) Systole; (b) diastole
3. (a) Tricuspid valve; (b) bicuspid or mitral valve
4. (a) Left atrium; (b) right atrium
5. Heart rate, stroke volume and peripheral resistance

Chapter 19
1. The pleural membrane
2. Ciliated columnar cells and goblet cells
3. The diaphragm
4. Squamous alveolar cells and capillary endothelial cells
5. The partial pressure of CO_2 in the blood

Chapter 20
1. The mucosa, the submucosa and the external muscle coat
2. Peristalsis
3. (a) Oesophageal sphincter; (b) pyloric sphincter
4. Bile – this contains bile salts which serve to emulsify lipids
5. Microvilli – these greatly increase the surface area for absorption

Chapter 21
1. (a) Testes; (b) scrotum
2. Fallopian tubes
3. (a) Oestrogen; (b) progesterone
4. Human chorionic gonadotrophin (hCG)
5. Oxytocin

Chapter 22
1. Nephron
2. Filtration
3. In the proximal convoluted tubule
4. Aldosterone
5. It causes the insertion of water channels into the collecting ducts to increase reabsorption of water

Chapter 23
1. Nucleic acid (DNA or RNA) and a protein capsid
2. An antigen
3. Cytotoxic T cells
4. Granulocytes – more specifically, neutrophilic granulocytes
5. Antibodies

Chapter 24
1. Osteoclasts
2. It increases absorption of calcium from the gut and increases osteoclast activity
3. The marrow cavity containing bone marrow
4. The number of muscle fibres innervated by one motor neuron
5. The extensor muscles of the thigh

Glossary

absolute refractory period: the period during which a section of the axonal membrane cannot be depolarised following the passage of an action potential

acetylation: the addition of an acetyl group $-COCH_3$ to another molecule

acetylcholine: an excitatory neurotransmitter molecule used at various synapses in the central nervous system, in the neuromuscular junction and by the parasympathetic nervous system

acetylcholinesterase: an enzyme located in the synaptic cleft of cholinergic synapses which acts to break down acetylcholine and thus to terminate its action

acrosome: a structure at the tip of a sperm head that is responsible for enabling the sperm to penetrate the layers surrounding the egg and initiating fertilisation

actin: a globular protein which can be assembled into a microfilament; it plays an important role in cell motility

action potential: a wave of depolarisation which travels away from the cell body of a neuron along the axon and which triggers transmitter release at the axon terminal

active transport: a mechanism by which substances are moved across a cell membrane against their concentration gradient using energy derived from ATP

adenine: a purine base found in DNA and RNA; it is paired with thymine in DNA and uracil in RNA

adenosine triphosphate (ATP): the major source of usable energy in cell metabolism; composed of adenine, ribose, and three phosphate groups

adenylate cyclase: an enzyme located in the cell membrane which can be activated by the α-subunit of G-proteins to convert ATP into cyclic AMP

adipocyte: the main cell type of fat; capable of storing triglycerides in a single large globule within its cytoplasm

adipose tissue: a connective tissue, made up largely of adipocytes, which acts as the major site for storage of triglycerides within the body

adrenaline: a hormone (also known as epinephrine) released by the medulla of the adrenal gland during the 'flight or fight' reaction; it produces a variety of effects around the body including increasing heart rate

adrenocorticotrophic hormone (ACTH): a hormone produced by corticotrophs in the anterior pituitary; it acts on the adrenal cortex to stimulate the production and release of corticosteroids

aggrecan: a large aggregate of proteoglycan which together with hyaluronic acid forms the basis of the extracellular matrix in many connective tissues

albumin: the most abundant and most negatively charged plasma protein; it is produced in the liver and acts as a carrier molecule for substances being transported in the bloodstream; it is a major contributor to plasma oncotic pressure

aldosterone: a hormone produced by the adrenal cortex which acts on the distal convoluted tubule in the kidney where it regulates sodium reabsorption

allele: one of the alternative forms of the same functional gene; alleles occupy the same position (locus) on homologous chromosomes

alpha helix: a secondary structure in proteins in which the polypeptide chain twists upon itself in a highly regular helical manner; each turn of the helix represents 3.6 amino acid residues

alveoli: in the lungs the alveoli are thin-walled air-filled sacs which lie at the end of the respiratory tree and in which gaseous exchange takes place; the term alveoli is also used to describe clusters of secretory cells in exocrine glands

amino acid: an organic acid containing the amino group (NH_2) and serving as a building block for proteins

amphipathic: a molecule which has both hydrophilic and hydrophobic moeities

amylase: a digestive enzyme which can hydrolyse starch

amylose: an unbranched form of starch in which glucose residues have an α-1,4 linkage

anaerobic respiration: the breakdown of organic molecules and the release of energy in the absence of oxygen

anaphase: a stage of mitosis during which the sister chromatids separate and move away from each other

angiotensin: a hormone which acts to raise blood pressure; produced by the action of renin on angiotensinogen; the product angiotensin I is converted to the active form angiotensin II by the action of angiotensin converting enzyme (ACE) found on the walls of capillaries

anion: a negatively charged atom or group

antibody: a protein produced by plasma cells in response to a foreign organism or molecule; by binding to the foreign organism or molecule it inactivates it or marks it for destruction

anticodon: a sequence of three nucleotides in a transfer RNA molecule which hydrogen bonds to the complementary codon in a messenger RNA

anti-diuretic hormone (ADH): a hormone produced in the hypothalamus and released from the posterior pituitary; it acts on the collecting duct in the kidney to promote water reabsorption; also known as vasopressin

antigen: a substance which can be recognised by antibodies and T cell receptors to provoke an immune response

aquaporin: one of a family of membrane channels which permit the rapid movement of water across cell membranes

artery: a vessel which carries blood away from the heart

asexual reproduction: reproduction which does not involve the fusion of gametes; it produces an organism identical to the parent

ATP synthase: an enzyme complex on the inner mitochondrial membrane which utilises the energy from proton flow to convert ADP to ATP

atrium: a chamber of the heart which receives venous blood and passes it on to a ventricle

atrio-ventricular (A-V) node: specialised tissue in the heart wall which initiates ventricular contraction by transmitting impulses along the A-V bundle

autoimmunity: a process in which the immune system recognises components of the body as foreign and attacks them

autonomic nervous system: a division of the peripheral nervous system which is divided into sympathetic and parasympathetic branches; it innervates smooth muscle and glandular tissue

autosome: chromosome other than the sex chromosomes; in human cells this refers to chromosomes 1–22

axon: a long thin process of a neuron; it transmits an action potential away from the cell body; it ends at a synapse or neuromuscular junction where neurotransmitter is released

B cell (B lymphocyte): a cell of the immune system which gives rise to an antibody-producing plasma cell

bacteria (*singular bacterium*): prokaryotic organisms surrounded by a rigid proteoglycan cell wall

baroreceptor: sensory structure located in major arteries capable of detecting changes in blood pressure

base: in the context of DNA or RNA; the purine or pyrimidine component of nucleotides

basal lamina: a thin molecular matrix secreted from the base of epithelial cells which provides linkage between them and underlying

connective tissue basal transcription factors: a group of proteins which must bind to the promotor region of a gene before transcription can occur

beta cell: a cell in the pancreatic islets which secretes insulin in response to rising glucose levels

beta sheet: a secondary structure found in proteins in which the amino acids associate with each other through hydrogen bonds to form a rigid, folded sheet-like structure

bilirubin: a pigment released during the breakdown of red blood cells; derived from the iron-containing haem group of haemoglobin

blastocyst: an early embryonic stage prior to implantation consisting of a small ball of cells

Bowman's capsule: the closed end of the nephron into which the knot of capillaries known as the glomerulus is invaginated; the site of plasma filtration

brown fat: a specialised adipose tissue which can generate heat and is important in thermoregulation in infants

brush border: the microvilli on the surface of absorptive cells in the gut and elsewhere

calmodulin: a protein activated by calcium binding which modifies the activity of target enzymes

capacitation: a process which the head of a sperm undergoes in the female genital tract in preparation for fertilisation

capillary: a thin-walled vessel which joins the arterial to the venous sides of the circulation; the site where substances move between plasma and interstitial fluid

capsid: the outer protein coat enclosing the nucleic acid genome of a virus

carbohydrate: molecules, such as sugar and starch, with the general formula $(CH_2O)_n$

cation: a positively charged atom or group

cell cycle: the stages through which a cell passes whilst undergoing mitotic division; can be divided into interphase, during which DNA synthesis occurs, and mitotic (M)-phase, during which the cell divides

cell membrane: a phospholipid bilayer surrounding all cells

cellulase: an enzyme which can break down cellulose; often found in the digestive tracts of herbivorous animals but not in humans

cellulose: an unbranched polymer of glucose linked by β-1,4 glycosidic bonds which forms an important component of plant cell walls

central dogma: the concept that information flows from DNA to RNA to protein

centromere: the structure which, in a replicated chromosome prior to mitosis, holds together the two sister chromatids

centrosome: an area in the cell cytoplasm responsible for stabilising the microtubular component of the cytoskeleton either in interphase or during spindle formation in mitosis

checkpoint: a point during the cell cycle at which a decision is made whether or not to proceed to the next stage; progress through a checkpoint is regulated by cyclin-dependent kinases

chiasmata (*singular chiasma*): the points of connection at which parts of homologous chromosomes are exchanged during meiosis

chief cell: enzyme-secreting cell in the glands of the stomach

cholesterol: a lipid molecule based on a structure containing four linked hydrocarbon rings and a short hydrocarbon tail; it plays an important role in determining the fluidity and permeability of plasma membranes and forms the start point for the synthesis of steroid hormones

chondroblast: a cell responsible for the deposition of the extracellular matrix in newly forming cartilage

chondrocyte: a mature cell located within the matrix of cartilage responsible for the maintenance of the extracellular matrix

chromatid: following DNA replication (in S-phase) the two identical copies of each of the 46 chromosomes are known as sister chromatids and are joined to each other at the centromere

chromosome: a strand of DNA containing a number of genes together with associated proteins; human cells contain 46 chromosomes in 23 pairs; in each pair one chromosome is derived from the mother and one from the father

cilium (*plural cilia*): an extension of the apical surface of epithelial cells containing a specialised array of microtubules which allows it to move backwards and forwards in a wave-like fashion

citric acid cycle: a metabolic pathway located within the mitochondrion which oxidises acetyl groups derived from the breakdown of organic molecules to generate NADH and CO_2

clone: a cell or organism genetically identical to the cell or organism from which it was derived

codon: a sequence of three nucleotides within a messenger RNA molecule that specifies a particular amino acid or the start or stop of translation

collagen: a triple helical protein made up from three α-chains which can assemble into fibrils and fibres; it provides the strength and flexibility of connective tissues

complement: a group of plasma proteins which when activated can protect against infection by binding to the surface of infecting microorganisms and promoting their phagocytosis and by promoting inflammation

compliance: in pulmonary physiology describes the ease with which a lung can be expanded

condensation reaction: a chemical reaction in which water is removed

corpus luteum: a structure in the ovary derived from a post-ovulatory follicle which maintains the uterine lining in readiness for implantation of the fertilised ovum

cortisol: a hormone produced by the cortex of the adrenal gland which has stimulatory effects on many metabolic processes and is chronically elevated in stress

co-transporter: a membrane protein which facilitates the transport of one substance up its concentration gradient by linking it with the transport of another substance (usually Na^+) down its concentration gradient; also called symport

covalent bond: a chemical bond formed between two atoms in which they share a pair of electrons

C-terminus: The end of a polypeptide chain at which the amino acid has a free carboxylic acid group

cyclic AMP: adenosine monophosphate in a form in which the single phosphate links back to the ribose sugar; it is an important second messenger in many signalling pathways

cyclin-dependent kinase: a protein kinase involved in the regulation of the cell cycle; cyclin-dependent kinases are active when combined with regulatory proteins known as cyclins

cytokine: a type of signal molecule which is secreted from one cell and acts on another

cytokinesis: the final stage of mitosis during which the two daughter cells separate

cytoplasm: that part of the internal fluid compartment of the cell which is bounded by the plasma membrane but not contained within the nucleus or other cell organelles

cytosine: a pyrimidine base found in DNA and RNA; it pairs with guanine

cytoskeleton: the internal protein framework of the cell made up of microfilaments, intermediate filaments and microtubules

deletion: a mutation in which one or more nucleotides are lost from the DNA strand

dendrite: a branched extension from the cell body of a neuron onto which synapses are made by axons of other neurons

dense body: a point of attachment for the sarcomeres of smooth muscle cells which is equivalent to the Z-line of striated muscle

depolarisation: a change in the membrane potential of a cell such that the inside becomes less negative with respect to the outside; usually caused by the entry of cations into the cell

desmosome: a cell junction at which protein bridges provide strong mechanical linkage between two adjacent cells

diastole: that phase of the cardiac cycle during which the ventricles relax and the atria contract

differentiation: the process by which a cell acquires specialised structure and function via the expression of a specific set of genes

diffusion: the movement of molecules and small particles from an area of higher concentration to an area of lower concentration by random, thermally driven movements

dimer: a molecule such as a protein comprising two associated sub-units

disulphide bridge: a covalent bond formed between two -SH groups; occurs between cysteine residues in proteins

DNA: deoxyribonucleic acid (DNA); a polymer of deoxyribonucleotides; this is the genetic material for all life forms and adopts a double helical structure; the sequence of bases in the nucleotides provides the code for the synthesis of proteins

domain: a part of a protein with a specific structure and function

dominant: a gene whose phenotype is expressed in both the homozygous and heterozygous state

electrical potential: a voltage difference between the inside and the outside of a cell membrane caused by an uneven distribution of charge

electrical syncytium: a group of cells linked by gap junctions through which depolarisation can spread extremely rapidly

endocytosis: a process whereby membrane-bound vesicles are formed at the cell surface and taken into the cytoplasm, leading to the internalisation of extracellular substances

endoplasmic reticulum: a series of membrane bound spaces within the cytoplasm of cells which act as the sites of protein synthesis and other metabolic processes

endothelium: the simple squamous epithelial lining of blood vessels and the chambers of the heart

enhancer: a region of DNA, often located some distance away from a gene, to which regulatory proteins may bind and increase transcription of that gene

enteric nervous system: a collection of neurons located in and regulating the activity of the gastro-intestinal tract

enzyme: a protein that catalyses a specific chemical reaction

epidermis: the outer epithelial layer of the skin

erythrocyte: a red blood cell

erythropoietin: a hormone produced in the kidney in response to low oxygen levels in the blood which stimulates the production of erythrocytes

eukaryote: an organism based on cells which possess a distinct nucleus and cytoplasm

exocytosis: the process by which membrane-bound vesicles fuse with the cell membrane and release their contents into the extracellular space

exon: a section of a eukaryotic gene that encodes the amino acid sequence for part of a protein

extracellular fluid: the fluid which lies outside cells; in the body it is divided into interstitial fluid which is in direct contact with cells and plasma which is contained within blood vessels

extracellular matrix: the extracellular component of connective tissues; it contains proteoglycans, glycoproteins and water; collagen and elastic fibres are embedded within it

facilitated diffusion: a process in which the rate of movement of a substance across a cell membrane and down its concentration gradient is enhanced by the presence in the membrane of transporter proteins or channels

fatty acid: a molecule that has a carboxylic acid group attached to a long hydrocarbon tail; used as a major energy source during metabolism; forms the hydrophobic tails of membrane phospholipids

fertilisation: the process which begins when a sperm and an egg come together; it leads to their fusion and the initiation of the development of the embryo

fibrin: a protein formed from fibrinogen by the action of thrombin; it is the major protein component of blood clots

fibroblast: a cell found in fibrous connective tissues such as tendon and responsible for the secretion of the extracellular matrix; fibroblasts proliferate in response to injury causing fibrosis or scarring

fibrocyte: a mature cell of fibrous connective tissues such as tendon; responsible for the maintenance of the extracellular matrix

flagellum: a long protrusion on a cell which propels the cell through a fluid medium by its beating; e.g. the sperm tail

follicle stimulating hormone (FSH): a hormone secreted by gonadotrophs in the anterior pituitary; in females FSH stimulates ovarian follicles to mature; in males FSH promotes spermatogenesis

gamete: a haploid cell (carrying only one set of chromosomes) formed as part of sexual reproduction by meiotic division; a sperm or an egg

gametogenesis: the process by which gametes are formed

gap junction: a cell-to-cell junction which permits the passage of ions and small molecules from one cell to another

gene: a region of DNA within a chromosome that codes for a specific protein

genetic code: the rules that determine which triplet of nucleotides (codons) in DNA code for which amino acid in protein synthesis

genetic locus: the position occupied by a gene on a chromosome

genotype: the full set of genes carried by an individual cell or organism

gestation period: the time between fertilisation and birth

glia: supporting cells in the nervous system; responsible for maintenance of a constant micro-environment, myelination and reaction to injury

glomerulus: the knot of capillaries which lies within Bowman's capsule in a nephron; may also be used to describe the capsule plus the capillaries

glucagon: a hormone produced by α-cells in the pancreas in response to falling glucose levels; it stimulates glycogen breakdown in liver and muscle

glucose: a six carbon sugar that has the formula $C_6H_{12}O_6$; a major source of metabolic energy which can be stored as glycogen (animals) or starch (plants)

glyceride: a lipid formed when the carboxylic acid group on a fatty acid reacts with glycerol

glycogen: a branched polysaccharide of glucose; a major energy store in liver and muscle cells

glycolipid: a membrane lipid in which the polar head group has been glycosylated

glycolysis: a metabolic pathway in which glucose is broken down into two three-carbon pyruvate molecules with the production of a small amount of ATP and NADH; takes place in the cytoplasm

glycosylation: the addition of sugars to another biological molecule

glycoprotein: a protein which has been glycosylated

glycosaminoglycan: a high molecular weight polysaccharide containing repeated subunits of amino and carboxylic acid sugars which is prevalent in the matrix of connective tissues

glycosidase: an enzyme which catalyses the breakdown of a glycosidic bond

glycosidic bond: a covalent bond formed between two monosaccharides by a condensation reaction

Golgi apparatus: an organelle consisting of stacked membrane sacs in which newly synthesised proteins are sorted and packaged into membrane-bound vesicles for transport to the cell surface or to lysosomes

G-protein: a trimeric intracellular GTP-binding protein associated with the cytoplasmic domain of a transmembrane receptor; following receptor-ligand interaction, the G-protein sub-units diffuse away from the receptor and initiate the production of second messengers

grey matter: that part of the central nervous system containing neuronal cell bodies, dendrites, synapses and glial cells

growth factors: cell signalling molecules which stimulate growth and differentiation of their target cells or organs

growth hormone: a hormone produced by the somatotrophs in the anterior pituitary; it has multiple trophic effects on its target tissues; necessary for normal development of an individual

guanine: a purine base found in DNA and RNA; it pairs with cytosine

haemoglobin: a tetrameric protein found in red blood cells and responsible for the transport of oxygen from the lungs to the tissues of the body

haploid: carrying only one half of the normal chromosomal complement of a cell; i.e. a single copy of each chromosome

helicase: an enzyme which unwinds the DNA double helix prior to DNA replication

hepatocyte: the basic cell type of the liver; responsible for production of plasma proteins, glucose uptake and storage, detoxification and bile production

heterodimer: a molecule comprising two non-identical sub-units

heterozygous: in diploid organisms, having two different alleles at the same locus of homologous chromosomes

hexose: a sugar, such as glucose, which has six carbon atoms per molecule

histone: a protein component of eukaryotic chromosomes around which the DNA strand is coiled

homeostasis: the maintenance of a constant internal environment

homodimer: a molecule containing two identical sub-units

homologous chromosome: one of two copies of a specific chromosome in a diploid cell; one copy is inherited from the father and one from the mother

homozygous: in diploid organisms, having identical alleles at the same locus of homologous chromosomes

hormone: a signal molecule released from cells at one site and transported through the blood to exert a specific effect on its target cells at a different site

human chorionic gonadotrophin: a hormone released by the placenta which prevents regression of the corpus luteum; detection of its presence in urine can be used in pregnancy tests

hydrogen-bond: a bond formed between a hydrogen atom in one molecule and an electronegative atom (e.g. O or N) in another molecule or another part of the same molecule

hydrophilic: a molecule or part of a molecule which can readily interact with or dissolve in water; 'water-loving'

hydrophobic: a molecule or part of a molecule which is unable to interact with or dissolve in water; 'water-hating'

hydrophobic interaction: a repulsive interaction between a molecule and water; the coming together of hydrophobic groups because of their exclusion by water

hydroxyapatite: crystalline calcium phosphate; the mineral component of bone

hyperosmolar: describes a solution or fluid compartment having a higher osmolarity than another

hyperpolarisation: movement of the resting membrane potential to a more negative value than normal

hypertonic: describes a solution which because of its high effective osmolarity would cause water to move out of cells placed in it

hypo-osmolar: describes a solution or fluid compartment having a lower osmolarity than another

hypotonic: describes a solution which because of its low effective osmolarity would cause water to move into cells placed in it

immunoglobulins: proteins produced by the immune system which protect against infection; antibodies

inflammation: a tissue response to insult or injury characterised by increased blood flow, localised oedema and infiltration of tissues by cells of the immune system

inspiration: the act of breathing in

insulin: a hormone produced by β-cells of the pancreas in response to increased plasma glucose; it stimulates glucose uptake and storage in cells of muscle and liver

intercalated disc: the junction between two adjacent cardiac muscle cells; it contains alternating desmosomes and gap junctions

intermediate filaments: strong intracellular filaments assembled from short rod-like proteins; the component of the cytoskeleton which provides mechanical stability e.g. keratin

interphase: the period in the cell cycle between cell divisions; comprises G1, S and G2 phases

interstitial fluid: the component of extracellular fluid which directly bathes the surface of cells

intracellular fluid: the 60% of body water which lies within cells and forms the solvent for cytoplasmic and nuclear contents

intron: a section of DNA within a gene which does not code directly for a polypeptide sequence

ionic bond: a chemical bond resulting from the attraction between oppositely charged ions

iso-osmolar: describes a solution or fluid compartment having the same osmolarity as another

isotonic: describes a solution which causes no net movement of water into or out of cells placed in it

keratin: a tough intermediate filament protein; found in high levels in skin, nails and hair

lectin: a protein which recognises and binds to specific carbohydrate side chains present on molecules such as glycoproteins and glycolipids

leptin: a hormone produced by cells in adipose tissue which acts to suppress appetite

Leydig cells: cells in the testis responsible for the production of testesterone

ligand: a molecule such as a hormone or neurotransmitter that binds to a specific site on a receptor protein

ligase: an enzyme which joins together DNA molecules

linkage: the tendency of certain genes to be inherited together as a result of their proximity on a chromosome

lipid: an organic molecule which is insoluble in water

luteinising hormone: a hormone produced by gonadotrophs in the anterior pituitary which in the female triggers ovulation and formation of the corpus luteum; in the male it stimulates Leydig cells to produce testosterone

lymphocyte: a cell of the immune system involved in the immune response

lysosome: a cytoplasmic organelle with a low internal pH containing hydrolytic enzymes involved in the breakdown and recycling of cellular macromolecules

lysozyme: an enzyme which attacks peptidoglycan in bacterial cell walls

macrophage: a cell specialised for the phagocytosis of particulate matter

mast cell: a cell found in connective tissues which is involved in local inflammatory responses

meiosis: cell division which leads to the production of four haploid gametes from a single diploid parent cell

messenger RNA (mRNA): a molecule produced as a copy of the DNA coding strand during transcription; each mRNA codes for a specific protein

metaphase: a stage of mitosis in which the chromosomes arrange themselves in the centre of the cell, equidistant from each pole of the spindle

microfilament: a thin intracellular protein filament such as actin or myosin; the part of the cytoskeleton which is primarily involved in cellular motility

microtubule: a tubular cytoskeletal component made from sub-units of the protein tubulin; it can be rapidly assembled and disassembled; important in intracellular transport and in the mitotic spindle

microvilli (*singular microvillus*): small finger-like membrane projections specialised for absorption which increase the surface area of epithelial cells

mitochondria (*singular mitochondrion*): cytoplasmic organelles which contain the enzymes necessary for the citric acid cycle and oxidative phosphorylation; they are responsible for the production of ATP during aerobic respiration

mitosis: the part of the cell cycle during which the cell divides and the sister chromatids that have been produced as a result of DNA replication are separated and distributed equally between the two daughter cells

mitotic spindle: a specialised array of microtubules formed during mitosis which is responsible for separation of the sister chromatids during anaphase and elongation of the cell during telophase

molecular chaperone: a molecule which assists in the folding or intracellular transport of a polypeptide

monosaccharide: a simple sugar molecule with the general formula $(CH_2O)_n$

motor neuron: a neuron whose axon innervates one or more muscle fibres

motor unit: the total number of muscle fibres innervated by the axonal branches of one neuron

mucin: a large glycoprotein containing a high proportion of O-linked oligosaccharides; a major constituent of mucus

mucus: a sticky glandular secretion containing high levels of mucin; it protects and lubricates epithelial surfaces

mutation: a random change in the nucleotide sequence of DNA within a chromosome

myocyte: a muscle cell

myelination: the process of forming a myelin sheath

myelin sheath: a layer of electrical insulation around an axon; formed from the cell membranes of glial cells which wind around the axon

myofibrils: highly organised bundles of the microfilaments actin and myosin found within the cytoplasm of muscle cells; myofibrils are made up of repeated units known as sarcomeres and shorten during muscle contraction

myosin: a protein which can break down ATP to drive its movement relative to actin microfilaments; one of the key components of muscle cells

Na^+/K^+ ATPase: a membrane pump found in many cells which uses energy derived from ATP to pump three sodium ions out of the cell and two potassium ions into the cell

negative feedback loop: a process in homeostatic or metabolic regulation by which the product of a reaction feeds back to inhibit that reaction and thus slows down or terminates its own production

nephron: the basic functional unit of the kidney; responsible for filtration of plasma and the production of urine; has a central role in osmoregulation

neuromuscular junction: a specialised point of contact between the terminal of the axon of a motor neuron and a muscle fibre at which the release of acetylcholine triggers muscle contraction

neuron: an excitable cell within the nervous system; comprises a cell body, axon and dendrites

neurotransmitter: a chemical released from an axon at a synapse or neuromuscular junction which binds to a receptor on the post-synaptic target cell and triggers a response

neutrophilic granulocyte: a phagocytic white blood cell; important in defence against infection

nicotine adenine dinucleotide (NAD^+): a carrier molecule which accepts a hydride ion (H^-) from a donor molecule to produce NADH; participates in energy production from the breakdown of glucose and other molecules

node of Ranvier: a small area of an axon between myelinated sections where the axonal membrane is in direct contact with interstitial fluid and where depolarisation events occur during saltatory conduction of an action potential

noradrenaline: a neurotransmitter (also known as norepinephrine) used at various synapses in the central nervous system and in the periphery by the sympathetic nervous system

N-terminus: the end of a polypeptide chain at which the amino acid has a free amino group

nucleic acid: a polymer assembled from nucleotides; the two principal nucleic acids are deoxyribonucleic acid (DNA) and ribonucleic acid (RNA)

nucleosome: the basic structural unit of the eukaryotic chromosome comprising DNA coiled around a core of histone proteins

nucleotide: a molecule containing a purine or pyrimidine base, a pentose sugar (ribose or deoxyribose) and a phosphate group

nucleus: a cellular organelle characteristic of eukaryotic cells; has a double membrane with pores surrounding the genetic material of the cell; contains the enzymes and proteins necessary for DNA transcription

oedema: the accumulation of fluid in the interstitial spaces between cells; swelling

oesophagus: a muscular tube used to convey food from the mouth to the stomach

oestrogens: a class of steroid hormones (e.g. oestradiol) produced in the ovary which initiate growth of the uterine lining during the menstrual cycle and are responsible for the development of secondary sexual characteristics in the female

oligodendroglia: glial cells which myelinate axons in the central nervous system

oncotic pressure: the osmotic force which acts to draw water back into capillaries from the interstitial fluid

oocyte: a female gamete

oogenesis: the process of production of primary oocytes within the ovary

operator: a region of an operon which binds a repressor or inducer to regulate the expression of the genes in the operon

operon: a sequence of DNA found in bacteria containing an operator, a promotor and a structural gene or genes transcribed as a single unit

opsonisation: the coating of an infectious agent with molecules such as antibodies to enable recognition by phagocytes

organelle: a structure or compartment within a eukaryotic cell which is specialised to carry out a specific function

osmolality: a measure of the number of particles of solute per mass of solvent; expressed as osmol/kg

osmolarity: a measure of the number of particles dissolved within a solution; determined by the molarity of the solute multiplied by the number of particles into which it dissociates; expressed as osmol/l.

osmoregulation: a homeostatic mechanism by which the osmolarity of body fluids is maintained

osmosis: the bulk transfer of water across a semi-permeable membrane from an area of low solute concentration to one of higher solute concentration; the membrane must be less permeable to the solute than to water

osteoblast: a cell responsible for secreting the extracellular matrix of developing bone and for initiating mineralisation

osteoclast: a multinucleated phagocytic cell which resorbs bone and releases calcium into the blood-stream

osteocyte: a mature cell of bone responsible for the maintenance of the extracellular matrix

ovum: an oocyte together with the protective layers that surround it

oxidative phosphorylation: a process located on the inner membrane of mitochondria in which energy derived from the breakdown of organic molecules is used to generate ATP

oxytocin: a hormone produced in the hypothalamus and released from the posterior pituitary which stimulates contraction of the uterus during birth and milk release during suckling

parasympathetic nervous system: a division of the autonomic nervous system which slows down the heart and promotes digestive processes

parathyroid hormone: a hormone secreted by the parathyroid glands in response to decreasing plasma calcium, which stimulates the release of calcium from bone

parietal cell: a cell found in glands in the stomach; secretes HCl and vitamin B_{12} intrinsic factor

partial pressure: a measure of the amount of dissolved gas in a solution; equivalent to the proportion of atmospheric pressure that would be provided by that gas if it were present in the atmosphere at a level sufficient to equilibrate with the solution

parturition: the act of giving birth

passive transport: the movement of a molecule across the cell membrane down its concentration gradient; occurs by diffusion or facilitated diffusion and does not require energy

pathogen: an organism which can cause disease

pentose: a sugar with the general formula $(CH_2O)_5$; e.g. ribose

peptide bond: a bond formed by a condensation reaction between the carboxylic acid group of one amino acid and the amino group of another

peptidyl transferase: an activity within ribosomes which catalyses the formation of peptide bonds between the NH2 group of an incoming amino acid on a tRNA molecule and the free COOH group of the preceding amino acid

peripheral resistance: the resistance to blood flow provided predominantly by small arterioles in the peripheral circulation

peristalsis: the wave-like contraction of a muscular tube such as the gut to move matter along its length

phagocyte: a cell specialised to perform phagocytosis; important in defence against infection e.g. a macrophage

phagocytosis: ingestion of particulate matter by a cell

phenotype: the observable characteristics of an organism 3′-5′

phosphodiester bond: a bond in nucleic acids in which the –OH on the 3rd carbon of one sugar is covalently bound to the phosphate group, which in turn forms a bond with the –OH on the 5th carbon of the next sugar

phospholipid: an amphipathic lipid found in cell membranes; many phospholipids are based on glycerol to which are linked two fatty acid chains, a phosphate group and a polar head molecule

phosphorylation: the addition of a phosphate group ($-PO_4$) to a molecule

placenta: the organ that develops during pregnancy to provide the interface between the mother and developing foetus

plasma: the fluid component of the blood

plasma cell: a cell that secretes antibody; derived from differentiation of a B lymphocyte

platelet: a cellular fragment found in large numbers in the blood; important for clotting

point mutation: a change in a gene involving a single base

polar: describes a molecule or group with an uneven distribution of charge

polygenic: describes a characteristic that is dependent upon multiple genes

polymer: a molecule, linear or branched, composed of multiple sub-units

polymerase: an enzyme that catalyses the addition of sub-units to form a polymer; e.g. DNA polymerase, RNA polymerase

polymorphic: in relation to a gene, refers to having multiple variants (alleles) within the population

polypeptide: a polymer of amino acids; polypeptides comprise a single chain whereas proteins may contain one or more polypeptide chains

polyunsaturated: a type of fatty acid containing multiple double bonds

postsynaptic membrane: the dendritic membrane adjacent to the synaptic cleft in which neurotransmitter receptors are localised

presynaptic membrane: the axonal membrane adjacent to the synaptic cleft from which neurotransmitter is released

prokaryote: an organism based on cells that lack a nucleus; e.g. bacteria

prolactin: a hormone produced by lactotrophs in the anterior pituitary; it initiates and helps maintain milk production

prometaphase: a stage of mitosis in which the nuclear membrane starts to break down so the chromosomes can become attached to the mitotic spindle

promoter: a sequence upstream of a gene to which RNA polymerase binds to initiate transcription

prophase: a stage of mitosis in which the chromosomes condense and the mitotic spindle starts to form

proprioception: the sense of the position of the body, or parts of the body, in space

protein: a polymer of amino acids comprising one or more polypeptide chains; one of a large group of molecules found in all living organisms

protein kinase: an enzyme that adds phosphate groups from ATP to specific amino acid residues in protein molecules

proteoglycan: a macromolecule consisting of a core protein with covalently-linked, negatively-charged sulphated glycosaminoglycans; e.g. aggrecan, a major component of the extracellular matrix

protozoa: a single-celled eukaryotic organism; some species can cause infectious disease, e.g. malaria

purine: a molecule based on a nitrogen-containing 6-membered ring fused to a second 5-membered ring found in the nucleotides of DNA and RNA; includes adenine and guanine

pyrimidine: a molecule based on a nitrogen-containing 6-membered ring found in the nucleotides of DNA and RNA; includes thymine, cytosine and uracil

pyruvate: a three-carbon breakdown product formed from glucose by glycolysis

receptor: a molecule that binds specifically to another molecule (a ligand) normally with some biological consequence

recessive: a gene whose phenotype is only expressed in the homozygous state

recombination: a process in which two strands of nucleic acid break and become rejoined in a new relationship; occurs during the pairing of homologous chromosomes in meiosis

relative refractory period: a period following the absolute refractory period during which the membrane is slightly hyperpolarised and more difficult to depolarise again

renin: an enzyme secreted by the kidney that catalyses the production of angiotensin from an inactive precursor, leading to an increase in blood pressure

repressor: a protein that binds to DNA and reduces the transcription of a gene or genes

residual volume: that volume of air which is left in the lungs after expiration

resting membrane potential: the difference in electrical potential between the inside and outside of a cell membrane when the cell is in an unstimulated state

retrovirus: a virus with an RNA genome which uses reverse transcriptase to synthesise DNA; e.g. human immunodeficiency virus (HIV)

reverse transcriptase: an enzyme that catalyses the synthesis of DNA on an RNA template

ribosomal RNA: RNA that is part of the ribosome

ribosome: a subcellular particle comprising RNA and protein that catalyses protein synthesis

ribozyme: RNA with catalytic activity

ribonucleic acid (RNA): a polymer of ribonucleotides; occurs as messenger (m)RNA, transfer (t)RNA and ribosomal (r)RNA

saltatory conduction: a process in which an action potential jumps from one node of Ranvier to the next, allowing an increased conduction velocity

sarcomeres: repeated units that make up the myofibrils of striated muscle

Schwann cell: a glial cell which myelinates axons in the peripheral nervous system

seminal fluid: a secretion in which spermatozoa are ejaculated

sensory neuron: a neuron that transmits impulses from the periphery to the central nervous system

Sertoli cell: a cell within the testis that supports the production of spermatozoa

sex chromosome: one of the pair of chromosomes that determine the sex of an individual; X and Y in humans

sexual reproduction: a type of reproduction involving the mixing of genetic material from two parents to produce offspring that are genetically different from the parents and each other

smooth endoplasmic reticulum: endoplasmic reticulum that is not associated with ribosomes; plays a role in lipid synthesis, intracellular Ca^{2+} storage and detoxification processes

solute: a substance dissolved in solution

spermatid: an immature form of the spermatozoon

spermatozoon: a male gamete

spinal cord: part of the central nervous system lying within the vertebral canal

spinal nerve: a nerve originating from the spinal cord

stem cell: an undifferentiated cell that can divide indefinitely and give rise to one or more types of differentiated cells

steroid hormone: one of a family of hormones structurally related to cholesterol that acts via intracellular receptors to regulate gene expression

stroke volume: the volume of blood expelled from a ventricle as it contracts

surfactant: a lipid substance secreted by type II cells in the alveolus which reduces the alveolar surface tension, making the lungs easier to inflate

sympathetic nervous system: a division of the autonomic nervous system which controls the ‘fight or flight’ reaction

synapse: a structure found between the axon of one neuron and dendrites of another, across which nerve impulses can be transmitted

systole: that phase during the cardiac cycle in which the ventricles contract and the atria relax

T cell (T lymphocyte): a lymphocyte that undergoes maturation in the thymus and recognises antigens present on cells of the body

telomere: the end of a chromosome

telophase: a stage of mitosis in which a nuclear envelope is formed around the two separated sets of chromosomes prior to cell division

testosterone: a steroid hormone produced by the testis in the male and involved in the development and maintenance of secondary sexual characteristics

tetramer: a molecule made up of four sub-units e.g. haemoglobin

thermoreceptors: receptors that detect changes in temperature

thrombin: an enzyme that converts fibrinogen to fibrin during the formation of blood clots

thymine: a pyrimidine base found in DNA; it pairs with adenine

thyroid hormone: a hormone produced by the thyroid gland that stimulates metabolic activity

thyroid stimulating hormone: a hormone produced by thyrotrophs in the anterior pituitary; it stimulates the thyroid gland to produce thyroid hormone

tidal volume: the volume of air that is drawn into the lungs during inspiration

tight junction: a junction between epithelial cells where the adjacent cell membranes are held together very closely so as to prevent the passage of most substances from one side of an epithelial sheet to the other

tonicity: describes the effective osmolarity of a solution; an indication of the effect a solution is likely to have on cells placed in it

transcellular fluid: extracellular fluid that is not plasma or interstitial fluid; e.g. fluid in the bladder or the gut

transcription: the process by which a complementary strand of RNA is synthesised on a DNA template

transcription factor: a protein that regulates the transcription of a gene or genes

translation: the process by which a protein is synthesised on a strand of mRNA

transfer RNA (tRNA): a type of RNA molecule that carries the anticodon for lining up amino acids with codons on messenger RNA; each amino acid has a specific tRNA molecule

triglyceride: a derivative of glycerol with three fatty acids attached

trisomy: an abnormality in which three rather than two copies of a chromosome are present

tropomyosin: a filamentous protein associated with actin filaments in striated muscle; it blocks the actin domains to which myosin may bind

troponin: a protein associated with actin filaments in striated muscle

trypsin: a proteolytic enzyme secreted by the pancreas

trypsinogen: the inactive precursor of trypsin

t-tubules: membrane invaginations that run into the centre of a striated muscle fibre; they play a role in triggering the opening of voltage dependent Ca^{2+} channels in the smooth endoplasmic reticulum

tubulin: a protein found in microtubules

tyrosine kinase: an enzyme that adds phosphate groups derived from ATP to tyrosine residues in protein molecules

uniport: a membrane protein that transports a molecule from one side of the membrane to the other down a concentration gradient

uracil: a pyrimidine base found in RNA; it is equivalent to thymine in DNA and pairs with adenine

urea: a molecule ($CO(NH_2)_2$) which is excreted in urine; it is used to remove waste nitrogen from the body

van der Waals forces: weak, short-range, inter- or intra-molecular attractions or repulsions between non-polar atoms

vein: a blood vessel which carries blood to the heart

ventricle: a large chamber of the heart responsible for pumping blood to the lungs (right ventricle) or the rest of the body (left ventricle)

virus: an infectious agent consisting of nucleic acid and protein that must infect a host cell in order to replicate

voltage sensitive channel: an ion channel that opens in response to a change in membrane potential

white matter: that part of the central nervous system containing myelinated axons

Z-line: a feature of striated muscle; a protein band into which actin filaments at either end of the sarcomere are inserted

zona pellucida: the coating surrounding the oocyte which the sperm must penetrate for fertilisation to occur

zygote: a diploid cell formed after fusion of an egg and sperm

Index

Bold entries represent definitions in the *Glossary*.

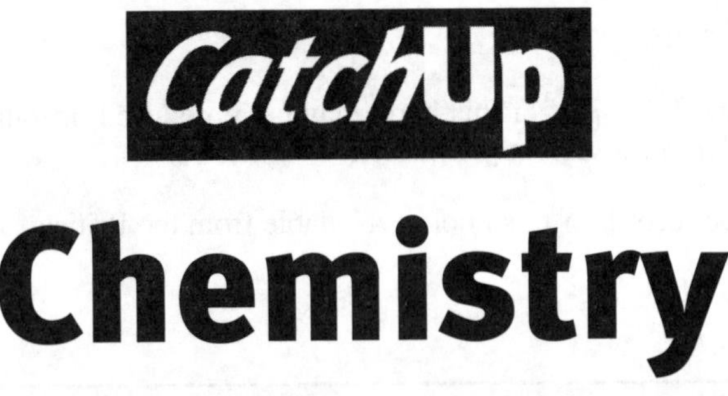

Chemistry

For the life and medical sciences

Mitch Fry

Faculty of Biological Sciences, University of Leeds, Leeds, UK

and

Elizabeth Page

School of Chemistry, University of Reading, Reading, UK

Scion

First published 2005

A CIP catalogue record for this book is available from the British Library.

ISBN 1 904842 10 0

Scion Publishing Limited
Bloxham Mill, Barford Road, Bloxham, Oxfordshire OX15 4FF
www.scionpublishing.com

Important Note from the Publisher

The information contained within this book was obtained by Scion Publishing Limited from sources believed by us to be reliable. However, while every effort has been made to ensure its accuracy, no responsibility for loss or injury whatsoever occasioned to any person acting or refraining from action as a result of information contained herein can be accepted by the authors or publishers.

Typeset by Phoenix Photosetting, Chatham, Kent, UK
Printed by Biddles Ltd, King's Lynn, UK, www.biddles.co.uk

Contents

Preface

Life is a collection of interacting chemical processes, designed to be self-sustaining. The carefully managed oxidation and release of energy from foodstuffs allows life-forms to maintain a precisely controlled cellular environment, and to build ever increasing levels of metabolic and structural complexity. All life on earth is built upon basic chemical principles; in particular it has taken the element carbon as a building block for producing a huge and diverse range of biological molecules, all designed to function within the universal solvent of life, water.

Nowadays, in our rush to achieve academically, there is a tendency to 'package' information. While this may 'fast-track' our progress, it actually dilutes our understanding. Seeing the whole picture is so much more rewarding. Studying the basic chemical principles of life is necessarily the first step in developing that picture.

You don't need a degree in chemistry to be aware of these principles. This short textbook explores those chemical principles which are readily extrapolated to the biological scenario, providing a basic foundation to facilitate an understanding of biological processes. We have developed these principles in an easy to understand way, without assuming any prior knowledge of chemistry beyond that covered in secondary school. You cannot divorce biology from chemistry; the former is a special extrapolation of the latter. Nor should you avoid taking the time to understand 'first principles'; your effort will be amply rewarded.

As you embark upon your chosen bioscience course, take the time to wonder at the marvel that is life; ask why? and how? Engage your enthusiasm; it's what being a biological scientist is all about!

Mitch Fry and Elizabeth Page
Leeds and Reading, May 2005

About the authors

Dr Mitch Fry BSc PGCE PhD is a biochemistry graduate who has worked as a senior research scientist in the pharmaceutical industry and as a science teacher in both secondary and higher education. His main role is in the teaching, support and supervision of life science undergraduate students at the University of Leeds, including 'pre-university' awareness activities and university admission procedures.

Dr Elizabeth Page BSc PhD PGCE is Director of Undergraduate Studies in the School of Chemistry at the University of Reading. She has had experience in teaching chemistry to biologists and other life sciences students for the past ten years. She has particular interest in supporting first year undergraduates as they make the transition to tertiary education.

1 Elements, atoms and electrons

Basic concepts:
We begin by considering basic atomic structure and the nature of isotopes. Isotopes play an important role in biology and are the subject of a 'Taking it Further' section. We look at electron distribution and configuration in atoms and explore the concept of atomic orbitals. An appreciation of atomic orbitals is essential to an understanding of the reactivity and bonding behaviour of atoms, and in particular to those elements that constitute the major building blocks of biological systems.

The ninety-two naturally occurring elements can combine in a variety of ways to form the matter that constitutes the world we live in. An **element** is a single substance which cannot be split by chemical means into anything simpler; for example, carbon is an element, oxygen is an element. Every element is represented by a symbol, an upper case letter, followed in some cases by a lower case letter; for example, carbon = **C**, calcium = **Ca**, nitrogen = **N**, sodium = **Na**. Elements are made up of many tiny, but identical particles, called **atoms**. An atom can be described as the smallest particle into which an element can be divided, while still retaining the properties of that element. Thus the element carbon is composed entirely of carbon atoms, the element oxygen is composed entirely of oxygen atoms, and so on. Atoms themselves consist of smaller units known as **sub-atomic particles.** There are three main types of sub-atomic particles, known as **protons, neutrons** and **electrons.**

The relatively massive centre of an atom is referred to as the **atomic nucleus.** The atomic nucleus is composed of protons and neutrons. These two sub-atomic particles are distinguished by their charge; each proton carries a single positive charge, whereas neutrons have no net charge (hence their name). The number of protons present in the nucleus is unique to each element. The **atomic number (*Z*)** of an element is equal to the number of protons in its nucleus. The sum of the number of protons and neutrons in the nucleus of an atom is equal to the **mass number (*A*)** of that element.

Reminder

Atomic number (Z) = number of protons

Mass number (A) = number of protons + number of neutrons

1.1 Isotopes

Sometimes the number of neutrons in an element's nucleus can vary, giving rise to different **isotopes**.

The nuclear composition of an atom is shown by $^{A}_{Z}\mathbf{X}$ where A = the mass number and Z = the atomic number.

Because the atomic number of an element is specific to that element this symbol for the atom is often reduced to ^{A}X.

Figure 1. Isotopes of hydrogen

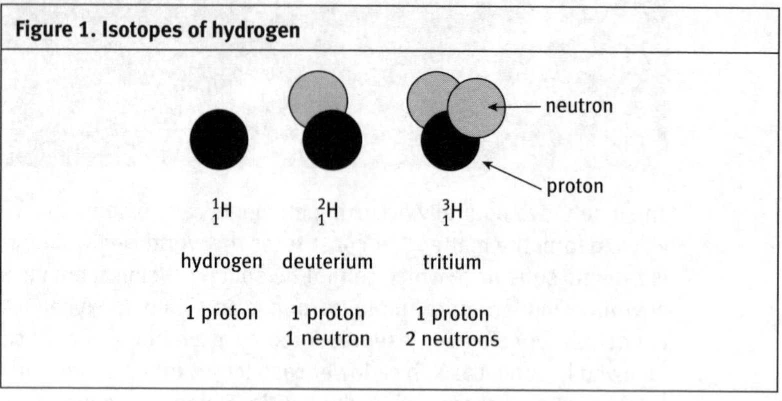

Hydrogen may exist as $^{1}_{1}H$ (also written hydrogen-1), $^{2}_{1}H$ (also written hydrogen-2) or $^{3}_{1}H$ (also written hydrogen-3). In other words, $^{1}_{1}H$ (hydrogen, the commonest form) has 1 proton only in its nucleus and has a mass number of 1. $^{2}_{1}H$ (referred to as deuterium) has 1 proton plus 1 neutron in its nucleus, giving it a mass number of 2. $^{3}_{1}H$ (referred to as tritium) has 1 proton plus 2 neutrons, giving it a mass number of 3. In each case there is only one proton present, and so the element is hydrogen; these are all isotopic forms of hydrogen (Fig. 1). Likewise, carbon can exist as three different isotopes, $^{12}_{6}C$, $^{13}_{6}C$, and $^{14}_{6}C$ of which $^{12}_{6}C$ and $^{14}_{6}C$ are the most common. Each of these is an isotope of carbon, each contains 6 protons, but the number of neutrons varies from 6 to 8.

Reminder

Isotopes of the same element differ from each other in their number of neutrons, not in their number of protons

Isotopes may be either **stable** or **radioactive**. In the examples we have used, $^{1}_{1}H$ and $^{2}_{1}H$ are stable isotopes of hydrogen, whereas $^{3}_{1}H$ is radioactive. Similarly, $^{12}_{6}C$ and $^{13}_{6}C$ are stable isotopes of carbon, but $^{14}_{6}C$ is radioactive.

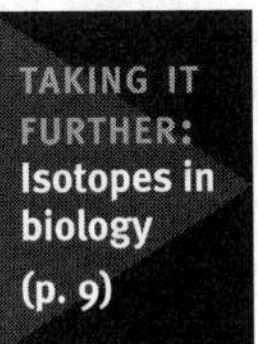

Isotopes are extremely useful to biologists and are used in a variety of applications.

Reminder

Radioisotopes are unstable and decay, releasing radioactive emissions

1.2 Electrons

The number of electrons in an atom equals the number of protons in the nucleus of that atom. It is the arrangement of electrons in an atom which determines its chemical reactivity.

Electrons can be thought of as sub-atomic particles having almost negligible mass. The single negative charge (–1) on an electron is equal and opposite to the positive charge on the proton. The number of electrons in an atom balances the number of protons, so an atom has no overall net charge.

Electrons move around the nucleus of the atom with speeds close to that of the speed of light. It is almost impossible to say exactly where an electron is at any specific point in time; this is the basis of the uncertainty principle and we talk about 'the probability of finding an electron in a particular space at any point in time'.

Experiments carried out at the beginning of the twentieth century demonstrated that electrons do not freely circulate around the nucleus but are restrained to specific **energy levels** or **shells**. Energy levels are given numbers, $\boldsymbol{n}$, starting with $n = 1$; under normal conditions electrons occupy the lowest energy levels first. (*The terms 'energy level' and 'shell' tend to be used interchangeably.*)

Reminder

While the atomic number of an atom defines the element, it is the arrangement of electrons that determines the chemical reactivity

Within each energy level are sub-levels, which are specific locations within an energy level where there is a high probability of finding an electron. These regions within an energy level are referred to as **atomic orbitals.** Atomic orbitals have particular shapes and can hold a maximum of two electrons. Closest to the nucleus is the lowest energy level with the value $n = 1$. The particular type of orbital within the $n = 1$ level is called an **s** orbital. Because it is in the $n = 1$ energy level it is called the **1s** orbital. The s orbital is spherical in shape.

Reminder

Atomic orbitals are regions in space where there is a high probability of finding electrons

We can define the **1s** orbital as a spherical region of space close to and surrounding the nucleus where there is a high probability of finding an electron (Fig. 2). The small black spot at the centre of each diagram represents the nucleus of the atom. The diagram is not to scale. Most of the space outside the nucleus of the atom is empty. In fact the radius of the nucleus is roughly one ten-thousandth of the size of the whole atom.

Figure 2. Arrangement of *s* and *p* atomic orbitals

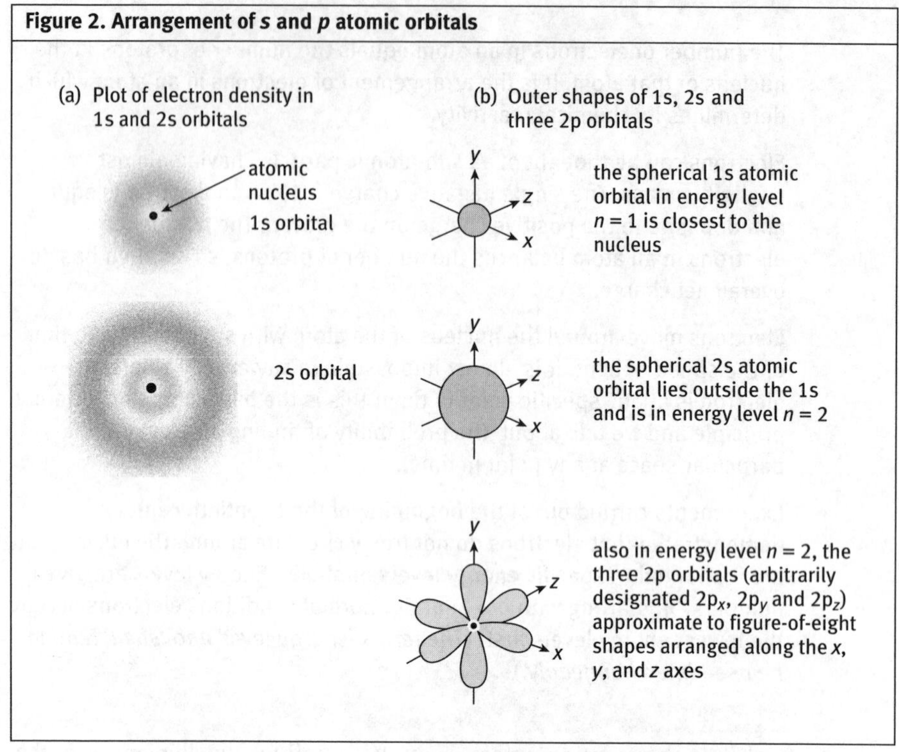

The second energy level ($n = 2$) is at a slightly higher energy than the first. Within the second energy level there are two types of regions in space where an electron can be found. One of these regions is a spherical orbital, as in the first energy level. However, the name of this orbital is the 2s orbital because it is in the second energy level.

The second type of atomic orbital at this energy level ($n = 2$) is called a **p** orbital. There are three p orbitals within an energy level, and each p orbital has a specific orientation along the *x*, *y* or *z* axis. p orbitals represent regions in space which are shaped like a figure of eight or a dumb-bell. The labels given to the three p orbitals within the second level are **$2p_x$**, **$2p_y$** and **$2p_z$**. As we move further from the nucleus the levels become increasingly higher in energy and the orbitals within them become more complex and numerous.

There are some important rules governing the way electrons fill energy levels and orbitals.

- **Electrons fill orbitals of the lowest energy first.** So the first level ($n = 1$) with its 1s orbital is filled before an electron can occupy the second level.
- **Within an energy level electrons fill the orbitals with the lowest energy first.** s orbitals represent a lower energy configuration than p orbitals and so electrons will fill the s orbital within an energy level before filling the p orbitals.
- **Individual orbitals can hold a maximum of two electrons.** Each s orbital (1s, 2s, 3s etc.) can only accommodate two electrons at the most. When two electrons occupy the same orbital they spin in opposite directions, otherwise their negative charges would force them apart.

It is rather difficult to draw electron orbitals but much easier to think of the way in which electrons fill orbitals as **'electrons in boxes'**. This is usually referred to as the **electron configuration** of the atom.

For example, hydrogen is the simplest atom with only one proton in the nucleus and therefore one electron outside the nucleus. We can write the electron configuration of hydrogen as:

$$1s^1$$

This means that the single electron of the hydrogen atom occupies the 1s orbital, which is in the lowest energy level, $n = 1$. The next element is made by adding one more proton to the nucleus and one more electron to the outer energy level. This element is helium, He, and it has two neutrons in its nucleus. The symbol for helium is ${}^{4}_{2}He$. The two electrons of the helium atom occupy the 1s orbital and so the electron configuration of helium is:

$$1s^2$$

The first energy level is now full and the next element lithium, Li with an atomic number of 3, has three electrons. Two of the electrons from lithium occupy the first level and the remaining electron must go into the s orbital of the second level, $n = 2$. The electron configuration of lithium is therefore:

$$1s^2 2s^1$$

When the 2s orbital is full the electrons start to fill the 2p orbitals, which are of slightly higher energy (but still in energy level $n = 2$). So, by the time we reach carbon, which has 6 electrons, the electron configuration can be written as:

$$1s^2 2s^2 2p^2$$

The three 2p orbitals ($2p_x$, $2p_y$, $2p_z$) are considered to have identical energies. Therefore, an electron will enter an empty p orbital before pairing with another electron in a half-full p orbital. As all the 2p orbitals are equivalent in energy it is not possible to say which of the 2p orbitals ($2p_x$, $2p_y$ or $2p_z$) is filled first. As there are three p orbitals in each level (except $n = 1$) and each can hold a maximum of two electrons, a total of six electrons can occupy the p orbitals in any energy level.

The information above can be summarised pictorially by using the 'electrons in boxes' representation. This is shown here for carbon, C (Fig. 3).

Figure 3. Electron filling in carbon

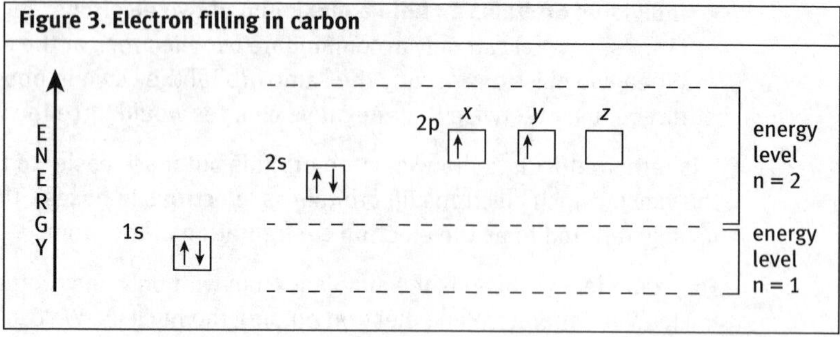

In energy level $n = 1$, the 1s orbital is full, containing two electrons of opposite spin direction, shown by the two arrows pointing in opposite directions. Likewise, at a higher energy level ($n = 2$), the 2s orbital is full, and at a slightly higher energy still, but within the second level, there is one electron in two of the three p orbitals and one p orbital is empty. Here the electrons have been arbitrarily assigned to the $2p_x$ and $2p_y$ orbitals.

Each element in the periodic table is formed by placing one more proton in the nucleus and one electron in the outer energy levels to balance the charge. After carbon, with atomic number of 6, is nitrogen (atomic number 7); the electron configuration of nitrogen is shown in Fig. 4.

Figure 4. Electron filling in nitrogen

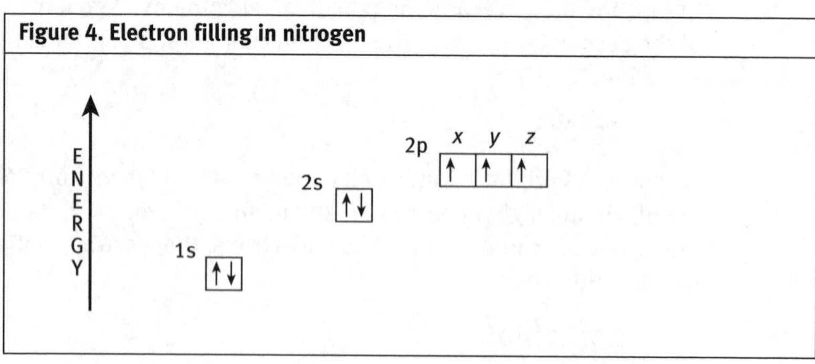

In the nitrogen atom each p orbital contains one electron and the 2p orbitals are half-filled.

Introducing one more proton in the nucleus and one electron in the outer energy level gives the element oxygen with atomic number 8. The extra electron of oxygen must now enter a half-filled orbital and so adopt an opposite spin. Thus one of the 2p orbitals is now full (Fig. 5).

Figure 5. Electron filling in oxygen

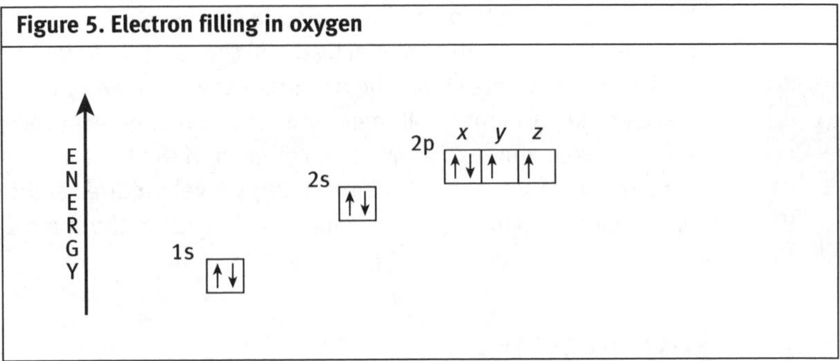

Fluorine (atomic number 9) has two electrons in two of the 2p orbitals and one electron in the third 2p orbital. By the time we reach neon (atomic number 10), at the end of the second row in the periodic table, all the orbitals in the second level are full.

Neon is an unreactive element and a member of the inert or noble gas group. Its lack of reactivity resides in the fact that all its outer electron orbitals are full. This means that neon is reluctant to either gain or lose electrons to other atoms and so does not easily take part in bonding. The electronic configuration of neon is shown in Fig. 6.

Figure 6. Electron filling in neon

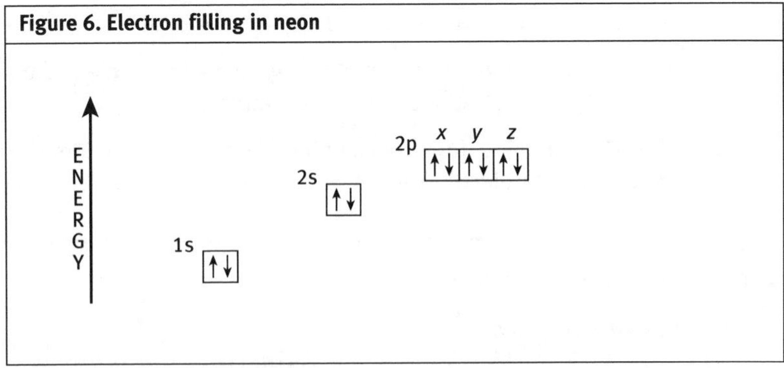

It is simplistic, but useful, to think of atoms reacting together in an attempt to fill their outer electron energy levels and in doing so to reach a stable state (the **octet rule**). This model can be used to describe the way elements in the second row of the periodic table undergo bonding. Elements in the group which is furthest to the right of the periodic table, such as helium, neon, argon and krypton, all have filled outer energy levels and are very stable and unreactive. Reactivity, which results in the formation of new bonds between atoms, is most often achieved through the 'sharing' of electrons in the formation of covalent bonds. By sharing electrons, atoms are able to effectively fill their outer energy levels and so achieve a more stable state. The octet rule states that atoms will react together in an attempt to fill their outer electron energy levels, and therefore reach a more stable state. For most of the lighter elements, a complete and stable outer (valence) energy level requires eight electrons (i.e. two electrons in 2s and two electrons in each of the three 2p atomic orbitals).

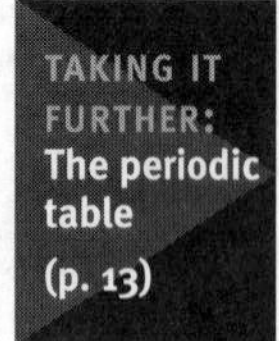

1.3 Summing up

1. The number of protons in an atom's nucleus is unique for each element, but the number of neutrons may vary, giving rise to isotopic forms of the same element.
2. Isotopes are very important in biology [see 'Taking it further: Isotopes in biology'].
3. Electrons are found in atomic orbitals, named 1s, 2s, 2p and so on, in increasing energy levels as they get further from the nucleus.
4. Atomic orbitals are spaces where there is a high probability of finding an electron(s); electron configurations and 'electrons in boxes' are simple ways of depicting the arrangement of electrons in an atom.
5. A maximum of two electrons occupy an s orbital or a single p orbital.
6. Elements that have full outer energy levels are stable and unreactive (see 'Taking it further: The periodic table').
7. Atoms combine (form covalent bonds) by sharing electrons or losing and gaining electrons in an attempt to fill their atomic orbitals.

1.4 Test yourself

The answers are given on p. 159.

Question 1.1
What are the mass numbers (A) of the isotopes hydrogen-1, hydrogen-2 and hydrogen-3, respectively?

Question 1.2
(a) How many types of atomic orbital are present in energy level $n = 1$?
(b) How many types of atomic orbital can be present in energy level $n = 2$?

Question 1.3
Define 'atomic orbital'.

Question 1.4
(a) How many electrons are present in an atom with full 1s, 2s and 2p orbitals?
(b) Would such an atom be likely to react?

Question 1.5
(a) Arrange the following atomic orbitals in order of their energy, lowest to highest; $2p_x$, $2p_y$, $2p_z$, 2s, 1s.
(b) What does this tell you about the energy of atomic orbitals with respect to their distance from the nucleus?

Taking it further

1.5 Isotopes in biology

Isotopes of a number of elements are unstable and will readily lose mass or energy in order to reach a more stable state; these are **radioisotopes**. Radioisotopes decay, and in doing so can emit one or more forms of radiation, namely alpha particles, beta particles and gamma radiation (or gamma rays). After this decay process an entirely different element remains.

The employment of radioactive isotopes in biology probably began in 1923 at the University of Freiburg with the work of Georg Hevesy who was measuring the uptake and distribution of radioactive lead in plants (for which he later received the Nobel Prize in 1943). By the mid-1930s, the cyclotron had been invented and with it came the capability of creating artificial radioactive isotopes such as carbon-14, iodine-131, nitrogen-15, oxygen-17, phosphorus-32, sulfur-35, tritium (hydrogen-3), iron-59 and sodium-24.

There are a number of properties of radioisotopes that particularly lend them to a variety of uses in biology.

1. Radioisotopes can be detected; with the use of sophisticated monitoring equipment, we can detect, measure and follow radioisotopes in organisms and in the environment.
2. Radioisotopes of a particular element are not distinguished from the naturally occurring stable element by organisms.
3. Radioisotopes have a characteristic half-life. The half-life of a radioisotope is the time taken for the radioactivity to drop by one half; different radioisotopes decay at different rates.
4. Radioisotopes can be damaging to biological molecules.

Ecology and the environment

We can study the distribution of nutrients in ecosystems, the impact of sewage discharges on sea and land fauna, the dispersion of invertebrate pests, and many more effects, by following the distribution of incorporated radioisotopes. The radioisotopes chosen need to have a suitable half-life; long enough that the experiment can be completed, but not too long that they remain in the environment and exert possible damaging effects.

Dating of sedimentary rock

Accurate dating of sedimentary rocks can be obtained using **radiometric dating**. This uses the phenomenon of radioactive decay of isotopes. When sedimentation occurs radioactive isotopes are incorporated in the rock, and these decay to form other atoms at a known rate. This rate is measured as the **half-life** of the isotope, defined as the time taken for half the parent atoms to decay to the daughter atoms. For example, potassium-40 (^{40}K) decays to form argon-40 (^{40}Ar), which is trapped in the rocks. The amount of argon can be measured. The half-life of ^{40}K is 1.3×10^6 years, so it is useful for dating very old rock (as old as the Earth), the minimum age being 100 000 years.

Carbon-14 dating

Carbon-14 is continuously formed in the upper atmosphere by the action of cosmic rays on nitrogen-14. Carbon-14 is a beta emitter that decays to nitrogen-14 with a half-life of 5730 years. This isotope of carbon is often used to date the remains of anything which contains carbon. There is a naturally occurring ratio of one carbon-14 atom to about one billion carbon-12 atoms. All life is based on carbon compounds and as an organism grows it incorporates carbon-14 continuously in this ratio. When the organism dies the uptake of carbon ceases and no additional carbon-14 will be added. The concentration of the carbon-14 will decrease steadily with time by decaying to nitrogen-14. If the carbon-14 to carbon-12 ratio in a sample is measured, the age of the sample can be estimated with reasonable accuracy. The results of this technique agree to within 10% of historical records.

Isotopes in medicine

It has long been known that radiation kills cancer cells, but unfortunately it also kills other healthy cells. **Radioimmunotherapy** provides an opportunity to deliver more specific radiation to tumour cells while sparing normal tissue. The principles of radioimmunotherapy involve first making an antibody which will specifically recognise an antigen associated with the cancer cells. The antibody is then radio-labelled; this involves chemically attaching a radioisotope to the antibody. The radio-labelled antibody is then injected into the body, where it eventually ‘seeks out’ the cancer cells and attaches to them. Being in such close proximity to the cancer cell, the radiation from the radioisotope will kill those cells, without causing too much damage to surrounding healthy tissues.

Iodine-131 has been available for many years for treating thyroid cancer. It is now being used to attach to an antibody to provide a way of performing radioimmunotherapy. Iodine-131 has a half-life of eight days; it is a beta and gamma emitter.

The decay of iodine-131, to xenon, is shown in the equation below.

$${}^{131}_{53}\text{I} \longrightarrow {}^{131}_{54}\text{Xe} + \underbrace{{}^{0}_{-1}\text{e}}_{\beta\ \text{particle}} + \text{gamma radiation}$$

The high energy electron emitted in this decay, a beta particle, originates from the atomic nucleus of iodine; a neutron changes to a proton, emitting a beta particle, increasing the atomic number to 54 (= xenon), but without any change to the mass number. The iodine-131 itself is chemically linked to the antibody. The beta particles emitted by iodine-131 have sufficient energy to travel about 5 mm in biological tissues; this helps to localise the damaging effects of the beta particles to the tumour itself but also enables a significant area of cancer cells to be targeted. An additional advantage of using iodine-131 is that this radioisotope is also a gamma emitter. Gamma radiation can be detected by specialist imaging equipment. This enables doctors to learn more about the diseased tissues, particularly their extent and localisation. Medical isotope diagnostic procedures often facilitate an earlier and more complete disease diagnosis and therefore a more rapid and effective treatment.

In clinical medicine another example of an isotope-based detection method is the carbon-13 breath tests used for the detection of *Helicobacter pylori*. (Note that carbon-13 is a stable isotope, not a radioisotope.) This bacterium is responsible for causing stomach ulcers. It uses an enzyme called urease, which breaks urea down to carbon dioxide (Fig. 7). If you give the patient a 'meal' of urea that contains carbon-13, then carbon-13-labelled CO_2 ($^{13}CO_2$) will be exhaled and can be detected in the patient's breath, if the bacterium is present.

Figure 7. Conversion of urea to ammonia and carbon dioxide, catalysed by urease

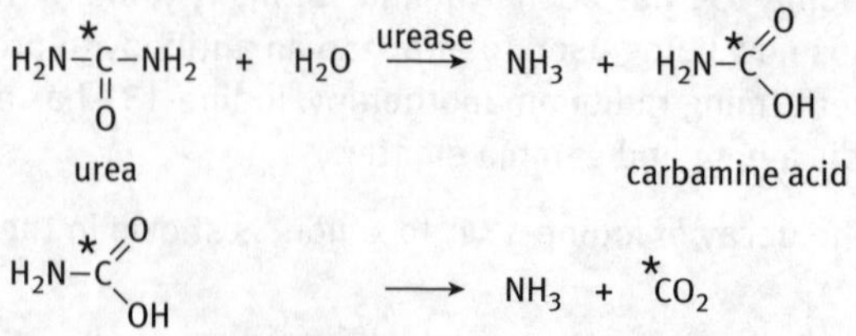

$$H_2N-\overset{*}{C}(=O)-NH_2 + H_2O \xrightarrow{\text{urease}} NH_3 + H_2N-\overset{*}{C}(=O)OH$$

urea — carbamine acid

$$H_2N-\overset{*}{C}(=O)OH \longrightarrow NH_3 + \overset{*}{C}O_2$$

The action of the enzyme urease results in the formation of ammonia and carbon dioxide; the carbon-13 (shown by the asterisk) in the urea is transferred to carbon dioxide; this can then be detected in exhaled breath.

Isotopes in research

A huge range of applications and tools are now available to the research scientist. Commonly available isotopes, such as carbon-14, iodine-131, nitrogen-15, oxygen-17, phosphorus-32, sulfur-35, tritium (hydrogen-3), iron-59 and sodium-24, can be attached to a variety of biomolecules (fats, carbohydrates, proteins, nucleic acids).

There is almost no limit to the number and kind of compounds which can be labelled and traced. We can use radio-labelled biomolecules in order to trace their metabolism and to elucidate metabolic pathways, or to deduce the mechanism of an enzyme-catalysed reaction. Isotopes used as labels have proven particularly useful in the analysis and detection of molecules. For example, consider that we might need to detect the presence of minute quantities of a specific protein. A good way to do this would be to produce and use a highly specific antibody to that protein. But how do we then specifically detect the antibody (or to be more accurate, the antibody–protein complex)? Answer, attach a radioisotope to the antibody first, then we can look for the antibody by detecting radioactivity. A relatively simple way of detecting radioactivity is through its effect on a photographic film, a technique called autoradiography.

So we can date rocks, animal and plant remains, employing naturally occurring isotopes. We can produce artificial isotopes to radio-label and follow the fate of chemicals in the environment, or in the human body. We can target radioisotopes in medicine to kill cancer cells and to localise diseased tissue. We can attach radioisotopes to biomolecules for purposes of analysis and detection.

Taking it further

1.6 The periodic table

Time line
1789 Antoine Lavoisier defines a chemical element and constructs a table of 33 of them.
1829 Johann Dobereiner announces his law of triads.
1843 Leopold Gmelin publishes his famous *Handbuch der Chemie*.
1858 Stanislao Cannizzaro assigns atomic weights to elements.
1862 Beguyer de Chancourtois makes use of atomic weights to reveal periodicity.
1865 John Newlands draws up a table of periodicity, his 'law of octaves'. Politically ridiculed, Newlands' work is not recognised until 1887.
1868 Julius Lothar Meyer constructs 'primitive' periodic table in 1864; a more sophisticated version is produced in 1868 but not published until his death in 1895.
1869 Mendeleev publishes his periodic table, based on atomic mass and chemical valency of the elements. Meyer publishes too late to claim priority over Mendeleev, but in time to confirm that the latter's discovery is based on sound chemical principles.

Historical perspectives

Since the earliest days of chemistry, attempts have been made to arrange the known elements in ways that revealed similarities between them. Ever since Antoine Lavoisier defined a chemical element and drew up a table of 33 of them for his book 'Traité élémentaire de chimie', published in 1789, there have been attempts to classify them. It took the genius of Dimitri Ivanovich Mendeleev to 'discover' the periodic table which today is so widely used and accepted. Today, the periodic table is firmly based on the properties of atomic number and the electron energy levels which surround the nucleus. Both of these concepts post-date Mendeleev by several decades; he, however, perceived them indirectly through the relative properties of atomic mass and chemical valency, and arrived at the periodic table in 1869. At this time, 65 elements were known; Mendeleev arranged these in his table, pointing out the many unoccupied positions in the overall scheme. He took the much bolder step of predicting the properties of these missing elements; moreover, the gap in atomic masses between cerium (140) and tantalum (181) suggested to him that a whole period of the table remained to be discovered. By the end of that century, most of the elements predicted by Mendeleev had been isolated, including the 'missing' period, the lanthanides.

The modern periodic table

The modern periodic table has probably reached its final form, firmly grounded on atomic theory. The periodic table shown in Fig. 8 is simple in the extent of the data it contains. Each box of the table contains the element's identity (agreed chemical symbol), the atomic mass, and the atomic number (number of protons). The elements are arranged in rows or **periods** (across), and columns or **groups** (1 to 18, down), increasing in atomic number from left to right, and in mass number. Elements in the same column have similar chemical properties. Mendeleev had noted that various properties of the elements seemed to go through cycles as the atomic number was increased. Boiling points did not simply increase with atomic number, but went through peaks and troughs; similarly for ionisation energy (the energy required to remove an electron). The number of chemical bonds that an element could form with another element varied with atomic number. Elements with atomic numbers 3, 11 or 19 could combine with only one atom of another element (Group 1 elements are said to be **monovalent**); those with atomic numbers 5 and 13 were able to form three bonds (Group 13 elements are **trivalent**). Elements with atomic numbers 2, 10, 18 etc. do not readily form bonds with any other elements; these are the stable **inert** or **noble** gases (Group 18). In other words, the properties of the elements exhibit a **periodicity**; the Periodic Law states that 'the properties of the chemical elements are a periodic function of atomic number'.

Figure 8. The modern periodic table

atomic number → 6
atomic symbol → C
relative atomic mass → 12.01

Metals
Transition Metals
Metalloids
Nonmetals

PERIODS / GROUPS	1	2	3	4	5	6	7	8	9	10	11	12	13	14	15	16	17	18
1	1 H 1.01																	2 He 4.00
2	3 Li 6.94	4 Be 9.01											5 B 10.81	6 C 12.01	7 N 14.01	8 O 16.00	9 F 19.00	10 Ne 20.18
3	11 Na 22.99	12 Mg 24.31											13 Al 26.98	14 Si 28.09	15 P 30.97	16 S 32.07	17 Cl 35.45	18 Ar 39.95
4	19 K 39.10	20 Ca 40.08	21 Sc 44.96	22 Ti 47.90	23 V 50.94	24 Cr 52.00	25 Mn 54.94	26 Fe 55.85	27 Co 58.93	28 Ni 58.70	29 Cu 63.55	30 Zn 65.41	31 Ga 69.72	32 Ge 72.64	33 As 74.92	34 Se 78.96	35 Br 79.90	36 Kr 83.80
5	37 Rb 85.47	38 Sr 87.62	39 Y 88.91	40 Zr 91.22	41 Nb 92.91	42 Mo 95.94	43 Tc 97.91	44 Ru 101.07	45 Rh 102.91	46 Pd 106.42	47 Ag 107.87	48 Cd 112.41	49 In 114.82	50 Sn 118.71	51 Sb 121.76	52 Te 127.60	53 I 126.90	54 Xe 131.29
6	55 Cs 132.91	56 Ba 137.33	57-71*	72 Hf 178.49	73 Ta 180.95	74 W 183.84	75 Re 186.21	76 Os 190.23	77 Ir 192.22	78 Pt 195.08	79 Au 196.97	80 Hg 200.59	81 Tl 204.38	82 Pb 207.2	83 Bi 208.98	84 Po 208.98	85 At 209.99	86 Rn 222.02
7	87 Fr 223.02	88 Ra 226.03	89-103†	104 Rf 261.11	105 Db 262.11	106 Sg 266.12	107 Bh 264.12	108 Hs 277	109 Mt 268.14	110 Ds 271	111 Rg 272	112 Uub 285	113 Uut 285					

* Lanthanide series:	57 La 138.91	58 Ce 140.12	59 Pr 140.91	60 Nd 144.24	61 Pm 144.91	62 Sm 150.36	63 Eu 151.96	64 Gd 157.25	65 Tb 158.93	66 Dy 162.50	67 Ho 164.93	68 Er 167.26	69 Tm 168.93	70 Yb 173.04	71 Lu 174.97
† Actinide series:	89 Ac 227.03	90 Th 232.04	91 Pa 231.04	92 U 238.03	93 Np 237.05	94 Pu 244.06	95 Am 243.06	96 Cm 247.07	97 Bk 247.07	98 Cf 251.08	99 Es 252.08	100 Fm 257.10	101 Md 258.10	102 No 259.10	103 Lr 262.11

Valence and the octet rule

Atoms whose outer (**valence**) electron energy levels hold a complete set of electrons are substantially more stable than those which do not. The electronic configuration of sodium (Na) is $1s^2\,2s^2\,2p^6\,3s^1$. If sodium were to lose one electron, and form the Na^+ ion, it would gain the electronic configuration of the stable element neon ($1s^2\,2s^2\,2p^6$). That such an ionisation can occur is supported by the relatively low ionisation energy for sodium (and other Group 1 elements). Similarly, chlorine has the electronic configuration $1s^2\,2s^2\,2p^6\,3s^2\,3p^5$; in gaining one electron chlorine would attain the electronic structure of the stable element argon ($1s^2\,2s^2\,2p^6\,3s^2\,3p^6$). Hence, the Cl^- ion is readily formed. For most of the lighter elements a complete outer or valence shell requires eight electrons.

Consequently, the observations noted above are said to obey the **octet rule.**

$$\underset{1s^22s^22p^63s^1}{Na} \longrightarrow \underset{1s^22s^22p^6}{Na^+} + e^-$$

$$\underset{1s^22s^22p^63s^23p^5}{Cl} + e^- \longrightarrow \underset{1s^22s^22p^63s^23p^6}{Cl^-}$$

The properties of the elements exhibit **trends**. These trends can be predicted using the periodic table and can be explained and understood by considering the electron configurations of the elements. Elements tend to gain or lose valence electrons to achieve a stable octet formation. Stable octets are seen in the inert gases, Group 18 of the periodic table. In each group the elements have the same outer energy level electron configuration, and therefore the same **valency**. For example, Group 1 elements require the input of a relatively small amount of energy to lose an electron, i.e. to ionise. In contrast it is extremely difficult to remove an electron from a Group 17 element; these are close to a stable octet and complete outer electron energy level and therefore have a strong tendency to acquire electrons. Electrons are added one at a time, moving from left to right across a period. As this happens, across a row the electrons in the outermost energy level experience increasingly strong nuclear attraction and become closer and more tightly bound. Furthermore, moving down a group, the outermost electrons become less tightly bound to the nucleus. This happens because the number of filled principal energy levels increases downward in each group, and so the outermost electrons are more shielded from the attraction of the nucleus. These trends explain the periodicity observed with respect to atomic radius, ionisation energy, electron affinity and electronegativity in moving from left to right across a period.

Atomic radius

The atomic radius (defined as half the distance between the centres of two neighbouring atoms) decreases across a period from left to right and increases down a given group. From left to right across a period, electrons are added one at a time to the outer energy shell. Electrons within the same energy level cannot shield each other from the nuclear attraction. Since the number of protons is also increasing, the effective nuclear charge increases across a period. Moving down a group, the number of electrons and filled electron energy levels increases, but the number of valence electrons stays the same. The outermost electrons are exposed to approximately the same effective nuclear charge, but are found further from the nucleus as the number of filled energy levels increases; therefore, the atomic radii increase.

Ionisation energy

The ionisation energy, or ionisation potential, is the energy required to completely remove an electron from a gaseous atom or ion. The closer and more tightly bound an electron is to the nucleus, the more difficult it will be to remove. Ionisation energies increase moving from left to right across a period (decreasing atomic radius), and decrease moving down a group (increasing atomic radius). Group 1 elements have low ionisation energies, with the loss of an electron forming a stable octet.

Electron affinity

Electron affinity reflects the ability of an atom to accept an electron. Atoms with stronger effective nuclear charge have greater electron affinity. For example, Group 17 elements, the halogens, have high electron affinities; the addition of an electron to such elements results in a completely filled outer electron energy level.

Electronegativity

Electronegativity is a measure of the attraction of an atom for the electrons in a chemical bond. The higher the electronegativity of an atom, the greater its attraction for bonding electrons. Electronegativity is related to ionisation energy. Atoms with low ionisation energies have low electronegativities, because their nuclei do not exert a strong attractive force on the electrons, and *vice versa*. In a period, moving from left to right, electronegativity increases (as ionisation energy increases). Going down a group in the periodic table, electronegativity decreases as both atomic number and atomic radius increase, and ionisation energy decreases.

Biological life

About 25 of the 92 natural elements are known to be essential to life. Carbon, oxygen, hydrogen and nitrogen make up 96% of living matter. Phosphorus, sulfur, calcium, potassium and a few other elements account for most of the remaining 4% of an organism's weight. Biomolecules are constructed using carbon to provide the framework, and hydrogen, nitrogen and oxygen to provide functionality. Sulfur and phosphorus are important constituents of some biomolecules. Trace elements are required by organisms in only minute quantities. Trace metals, such as iron, magnesium and zinc, are important catalytic components present in enzymes.

2 Bonding, electrons and molecules

Basic concepts:
Understanding the nature of covalent bonding is an essential prerequisite to predicting the behaviour of biological molecules. Here we explore the nature of covalent bonding and types of covalent bonds.

Atoms react together to form molecules. Biological life forms are able to produce complex and extremely large molecules. Inherent in the stability of such molecules are the **intramolecular forces** between atoms which hold the molecules together, producing bonds which are strong and relatively stable. Such bonds may be referred to as covalent, dative covalent, polar covalent or ionic.

2.1 What is a covalent bond?

A covalent bond is formed between two atoms, commonly by the sharing of two electrons, one electron being donated by each of the two atoms. Consider the hydrogen atom. Remember hydrogen has one proton in the nucleus and one electron in a 1s orbital.

Using our diagrams of atomic orbitals, we can consider the hydrogen molecule, H_2, as being formed from the overlap of two 1s orbitals, one from each hydrogen atom (Fig. 9). In this way, each hydrogen atom effectively has two electrons at any time. In other words, each hydrogen atom has a full 1s orbital. Furthermore, since both hydrogen atoms in this bond are identical, we can assume that the electrons are shared equally between the two atoms, producing a symmetrical covalent bond. This 'head-to-head' merging of atomic orbitals forms **sigma molecular orbitals**. It is the formation of bonding molecular orbitals which results in covalent bonding.

Figure 9. Formation of a covalent bond in the hydrogen molecule

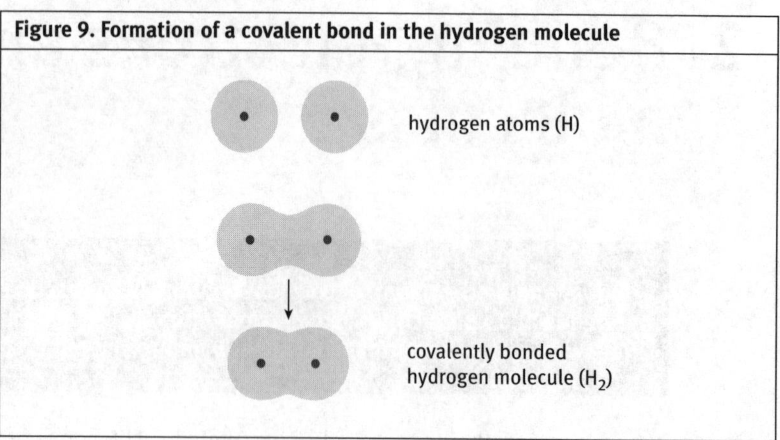

Question

How can two electrons fill the 1s orbital of both H atoms? Remember, orbitals represent spaces where there is a high probability of finding an electron, and electrons move close to the speed of light. So, statistically there is a high probability that at any one time both electrons will be associated with one or other hydrogen atom!

Sigma molecular orbitals are just as easily formed between p orbitals, again in a 'head-to-head' merging (Fig. 10).

Figure 10. Sigma molecular orbital formation between p orbitals

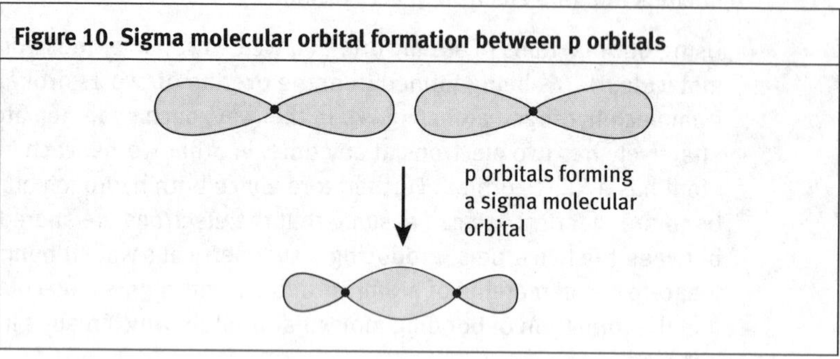

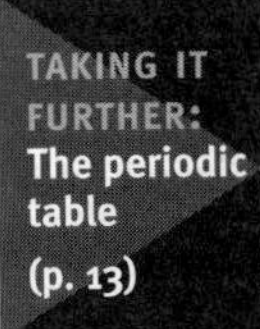

Question

Why do atoms form covalent bonds?

In the hydrogen molecule, both H atoms effectively gain a full 1s orbital, and therefore attain a stable state. For hydrogen this simply means filling its 1s orbital. In the case of nitrogen, this means filling its three 2p orbitals (see Fig. 4). Atoms whose outer (valence) electron energy level holds a complete set of electrons are substantially more stable than those which do not. By sharing electrons through covalent bonding, atoms are able to effectively fill their outer electron energy levels and so gain greater stability. For most of the lighter elements a complete outer or valence shell requires eight electrons (two electrons in the 2s orbital plus six electrons in the three 2p orbitals).

2.2 Non-bonding electrons – lone pairs

In bonding, atoms attempt to attain a complete set of electrons in their valence shells as this arrangement generally leads to stability. In many molecules this means that atoms have electrons in their valence shells which are not involved in bonding to other atoms. These non-bonded electrons are known as **lone pairs**.

Reminder

Atoms share electrons in covalent bonds in order that each attains a full outer electron energy level, and hence greater stability

Consider what happens when two fluorine atoms react together to form a molecule (Fig. 11). The electron configuration of fluorine is $1s^2 2s^2 2p^5$ and so each atom of fluorine has seven outer electrons. When the half-filled p orbitals of the fluorine atoms, each containing a single electron, overlap, a sigma bond is formed. By sharing electrons in this way each fluorine atom now has a filled outer valence shell of eight electrons. However, six of the electrons on each fluorine atom are not involved in bonding. They are paired together in non-bonding orbitals and are called lone pairs. Lone pairs of electrons have an effect upon the reactivity of a molecule and also upon its shape.

In the water molecule the two lone pairs and two bonding pairs of electrons on the oxygen atom are arranged tetrahedrally. This is the arrangement in which the negative centres experience the minimum repulsion. The lone pairs exert a larger repulsive effect than the bonding pairs and so the hydrogen atoms are pushed together slightly and the bond angle becomes slightly less than the tetrahedral angle, 109°. Overall the molecule is therefore V-shaped (Fig. 11).

Figure 11. Lone pairs of electrons

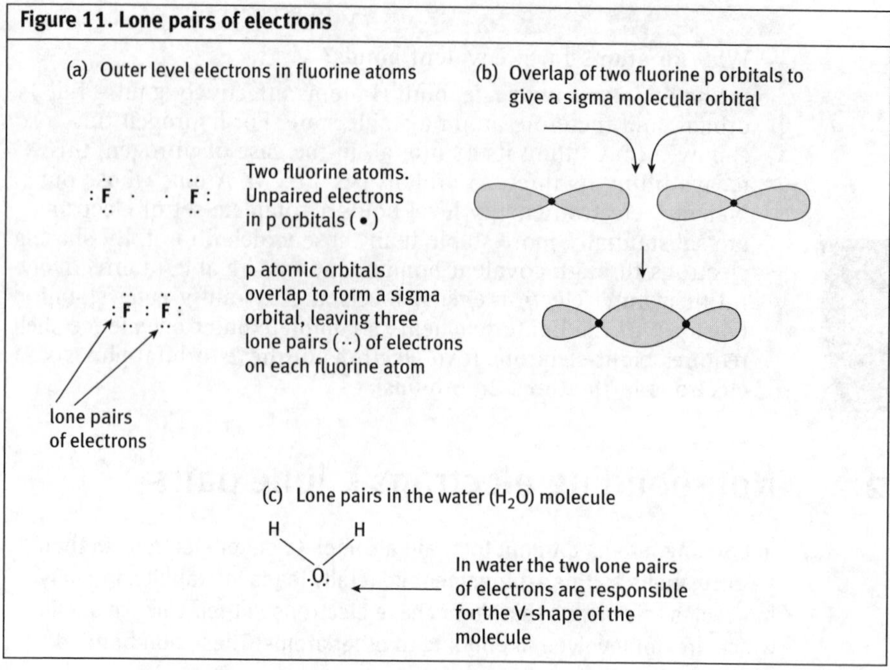

2.3 Pi molecular orbitals

In addition to forming sigma molecular orbitals, electrons in p orbitals can also overlap in a 'side-to-side' fashion, forming pi molecular orbitals (Fig. 12).

Figure 12. Formation of pi molecular orbitals

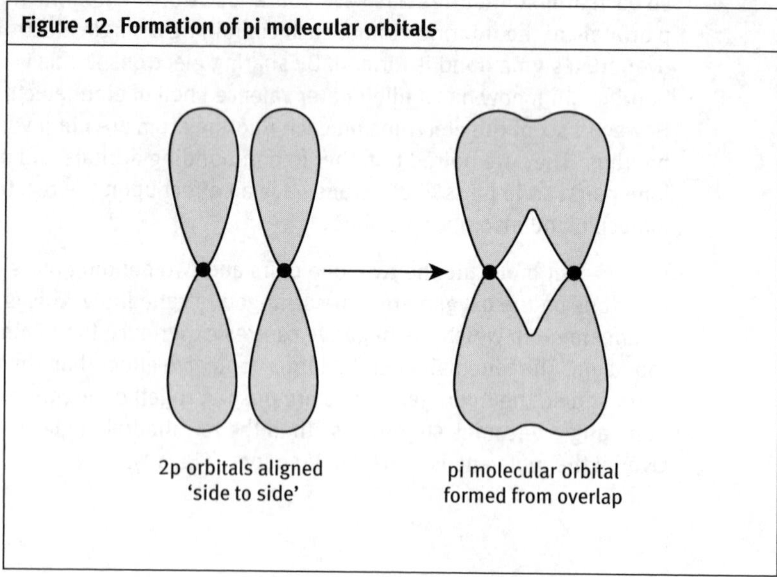

Pi bonding between atoms occurs in addition to, rather than instead of, sigma bonding. When both a sigma molecular orbital and a pi molecular orbital are formed between two atoms, a 'double bond' results. Whereas sigma bonding, shown by shorthand as a single line connecting two atoms, allows complete rotation about the bond, pi bonding, shown in shorthand as a double line between atoms (one sigma bond plus one pi bond) restricts rotation about that bond.

Yes – free rotation about bond

single bond

No – rotation about bond is restricted

double bond

Pi bonding has important consequences in determining the shapes (conformations) of biological molecules, and particularly protein molecules.

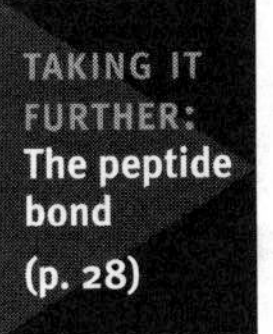

Reminder

Covalently bonded atoms can rotate freely about a sigma molecular orbital, but not about a pi molecular orbital

2.4 Coordinate bonds

A covalent bond is formed by two atoms sharing a pair of electrons. The atoms are held together because the electron pair is attracted by both of the nuclei. In the formation of a simple covalent bond, each atom supplies one electron to the bond, but that doesn't have to be the case. A **coordinate** bond (also called a **dative** covalent bond) is a covalent bond (a shared pair of electrons) in which both electrons come from the same atom. In simple diagrams, a coordinate bond is shown by an arrow. The arrow points from the atom donating the lone pair to the atom accepting it. In Fig. 13, the nitrogen atom in the ammonia molecule is donating its pair of electrons to the empty 1s orbital of the positive hydrogen ion, a proton. A new dative covalent bond is formed in the ammonium ion, NH_4^+. Once formed, each of the N-H bonds is equivalent, irrespective of the source of the electrons.

Figure 13. Coordinate covalent bond formation in the ammonium ion

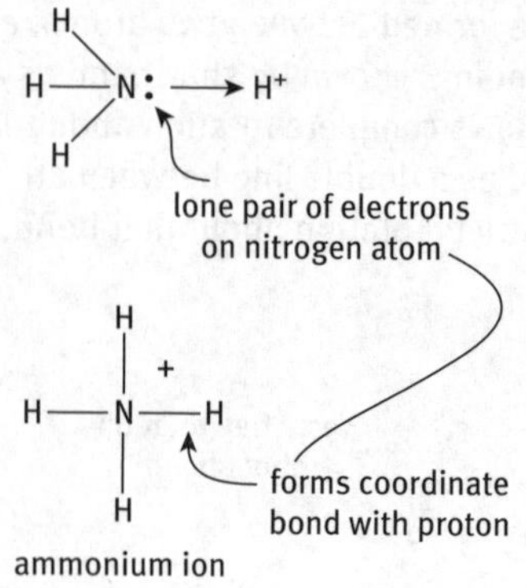

H
H
N:
H
H+
lone pair of electrons
on nitrogen atom
H
+
H
N
H
H
forms coordinate
bond with proton
ammonium ion

2.5 Electronegativity and polar covalent bonds

TAKING IT FURTHER: The periodic table (p. 13)

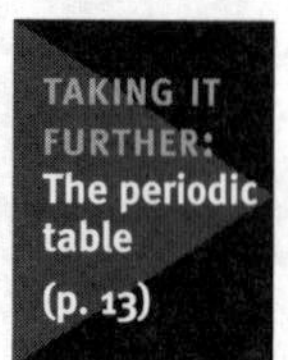

In covalent bonds between like atoms, electrons are shared equally between the two atoms in the bond. However, certain types of atoms are able to exert a greater pull on electrons than others. This ability to attract electrons within a bond is called the **electronegativity** of that element. The electronegativity of an atom is a property dependent on the size of the atom, and the degree of 'shielding' of the positively charged nucleus by the negatively charged electrons. The section of the periodic table below (Fig. 14) shows the relative electronegativities of a number of atoms; the larger the value the greater their electronegativity.

Figure 14. Electronegativity values of some elements of the periodic table

H 2.2						
Li 1.0	Be 1.5	B 2.0	C 2.5	N 3.1	O 3.5	F 4.1
Na 0.9	Mg 1.2	Al 1.5	Si 1.7	P 2.1	S 2.4	Cl 2.8
K 0.8	Ca 1.0	Ga 1.8	Ge 2.0	As 2.2	Se 2.5	Br 2.7

In general, elements on the right-hand side of the periodic table are more electronegative than those on the left-hand side. In addition, as atoms get bigger (i.e. going down a group in the periodic table), electronegativity decreases [see 'Taking it further: The periodic table']. In biological molecules, oxygen and nitrogen are particularly important; their electronegativity has important consequences in terms of the reactivity and associations of molecules.

Reminder

Electronegativity is a measure of the degree to which an atom 'draws' electrons towards itself in a bond

2.6 What effect does electronegativity have on covalent bonds?

In a sigma molecular orbital between carbon and oxygen, both atoms contribute one electron to the bond. Oxygen draws electrons towards itself because it is relatively more electronegative than carbon. Consequently, the two electrons in the bond are more likely to be found closer to the oxygen atom than the carbon atom (Fig. 15). This unequal electron distribution results in the formation of a **polar covalent bond**.

Figure 15. Unequal electron distribution in a polar covalent bond

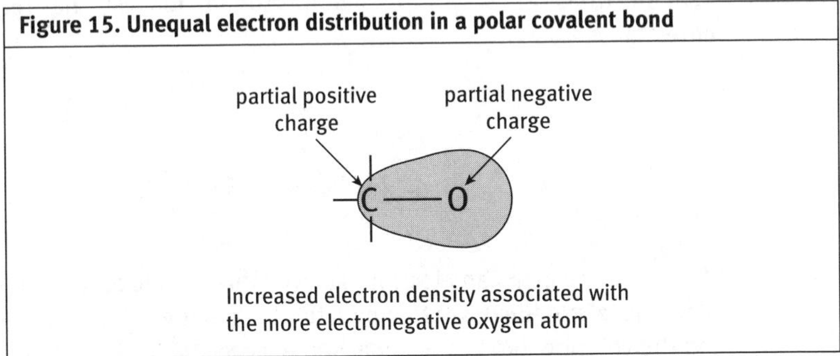

Reminder

A **polar covalent bond** is a distorted covalent bond in which there is an unequal distribution of electrons

This unequal electron distribution results in one end of the bond being slightly positive (electron deficient) and the other end being slightly negative, and thus a **dipole** is produced.

Reminder

A **dipole** is produced when a pair of electric charges, of equal magnitude but opposite polarity, are separated by some (usually small) distance

Polar covalent bonds play an important role in biological molecules. They invariably form the basis of **functional (reactive) groups** on biological molecules. Such groups are responsible for the reactivity between molecules, their solubility in water and contribute to the **intermolecular forces** between molecules.

2.7 Ionic bonds

Ionic bonds occur when there is a complete transfer of electron(s) from one atom to another resulting in two ions, one positively charged and the other negatively charged. Electrons are not 'shared' as in a covalent bond, but are 'lost' or 'gained'. For example, when a sodium atom (Na) donates its one electron in its outer 3s electron energy level to a chlorine (Cl) atom, which needs one electron to fill its outer 3p electron shell, sodium chloride results. The bond between the two new ions (Na^+Cl^-) is an ionic bond.

When sodium loses an electron from the 3s atomic orbital, all available atomic orbitals in the $n = 2$ energy level are filled, containing a total of eight electrons ($2s^2 + 2p^6 = 8$), and the sodium ion, Na^+, has reached a more stable state.

$$\begin{array}{ccc} Na & \rightarrow & Na^+ + e^- \\ 1s^2 2s^2 2p^6 3s^1 & & 1s^2 2s^2 2p^6 \end{array}$$

When chlorine gains an electron, all available atomic orbitals in the $n = 3$ energy level are filled, containing a total of eight electrons ($3s^2 + 3p^6 = 8$), and the chloride ion, Cl^-, has reached a more stable state.

$$\begin{array}{ccc} Cl + e^- & \rightarrow & Cl^- \\ 1s^2 2s^2 2p^6 3s^2 3p^5 & & 1s^2 2s^2 2p^6 3s^2 3p^6 \end{array}$$

As with covalent bond formation, the 'driving force' in forming ionic bonds is to achieve a full, and therefore stable, outer electron energy level. In ionic bond formation it is the ion, rather than the atom, that reaches a more stable state and electrons are lost or gained, rather than shared, to achieve this state.

2.8 The concept of the chemical bond

We have described the covalent bond, the coordinate bond, the polar covalent bond and the ionic bond. In reality, the intramolecular forces between atoms lie within a spectrum of bonding, with covalent and ionic representing extreme ends of this spectrum. The type of bonding which prevails depends upon the electronegativity difference between the atoms involved. Pure covalent bonds exist where the electronegativity difference lies between 0 and 0.4, polar covalent bonds where the electronegativity difference is between 0.4 and 2.1, and ionic bonds where the electronegativity difference is greater than 2.1 (Fig. 16).

Figure 16. A spectrum of bonding between atoms

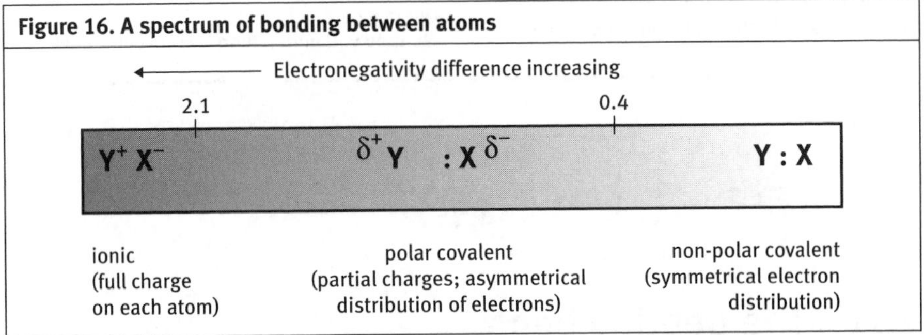

The intramolecular forces between atoms provide the inherent stability for molecular construction. However, central to biological life are the rapid and transient interactions that must occur between different molecules. Such intermolecular forces provide the basis for interaction and recognition at the molecular level of life, and are dealt with in Chapter 3.

2.9 Summing up

1. A covalent bond is formed by the sharing of two electrons between two atoms, normally one electron being provided by each atom.
2. Coordinate (dative covalent) bonds are formed where both electrons in the bond are provided by just one of the atoms.
3. Atoms form covalent bonds in an attempt to fill their outer atomic orbitals and so reach a more stable state.
4. Both sigma ('head-to-head'), and pi ('side-to-side') molecular orbitals are possible through the merging of atomic orbitals.
5. Rotation about a double covalent bond is restricted (see 'Taking it further: The peptide bond').
6. Some atoms are more electronegative than others, which can result in the formation of polar covalent bonds.

7. Polar covalent bonds form the basis of functional groups on molecules.
8. Ionic bonding involves the gain or loss of electrons, the ions so formed achieving stability through attaining full outer atomic orbitals.

2.10 Test yourself

The answers are given on p. 159.

Question 2.1
Describe the relationship between atomic orbitals, molecular orbitals and covalent bonds.

Question 2.2
Define the difference between a sigma molecular orbital and a pi molecular orbital.

Question 2.3
What do you understand by a 'polar covalent bond'?

Question 2.4
What is unusual about a dative covalent bond?

Question 2.5
What do you understand by the 'octet rule'?

Taking it further

2.11 The peptide bond

Proteins are polymers of amino acids. Amino acids are linked together by the formation of peptide bonds, between the carboxyl group of one amino acid and the amine group of another amino acid, in a condensation reaction (Fig. 17). A **condensation reaction** is a chemical reaction in which a molecule of water is lost.

Figure 17. Formation of a peptide bond between two amino acids in a condensation reaction

$$^{+}NH_3-C(R1)(H)-COO^{-} + {}^{+}NH_3-C(R2)(H)-COO^{-} \longrightarrow {}^{+}NH_3-C(R1)(H)-C(=O)-N(H)-C(R2)(H)-COO^{-} + H_2O$$

alpha carbon

alpha carbon

N-terminus

Peptide bond

C-terminus

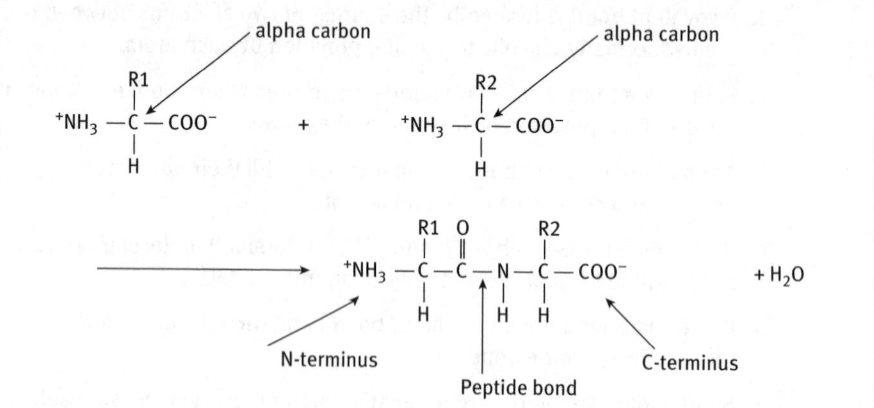

The sequence of amino acids in a protein determines the primary structure of that protein. The flexibility and folding of the polypeptide chain is responsible for the specific three-dimensional shape of the protein, which is the main determinant of the structure–activity characteristics of that protein. Proteins derive their name from the ancient Greek sea-god **Proteus** who could change shape; the name acknowledges the many different properties and functions of proteins.

We have seen that there is free rotation about a single sigma covalent bond, but that rotation is restricted about a double (sigma + pi) covalent bond. The peptide bond is special in the sense that it is a partial double bond, and so rotation about the bond is restricted. This property of the peptide bond has a profound effect in determining the conformation of the polypeptide chain.

The partial double bond character of the peptide bond

The partial double bond character of the peptide bond (Fig. 18) is a consequence of the electronic configuration of the nitrogen atom, and of the pi bonding in the carbonyl group.

Figure 18. The peptide bond

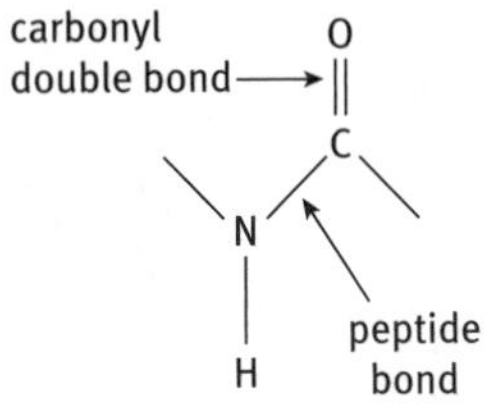

Nitrogen (N), atomic number 7, has the electronic configuration

$$1s^2\ 2s^2\ 2p^3$$

Prior to covalent bonding, the atomic orbitals in the $n = 2$ energy level of nitrogen are hybridised. (Hybridisation is covered further in Section 5.2.)

Hybridisation of atomic orbitals in the same energy level occurs in order to maximise the number of covalent bonds which can be formed. The more covalent bonds which can be formed, the more stable is the atom. Nitrogen is capable of a number of types of hybridisation; in the case of peptide formation the hybridisation is referred to as sp^2 hybridisation; the s refers to the 2s orbital, and p^2 to the fact that two of the three p orbitals are hybridised. We can see this using 'electrons in boxes' diagrams (Fig. 19).

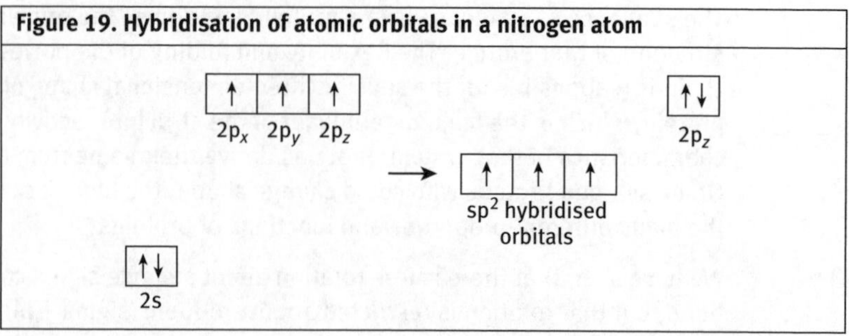

Figure 19. Hybridisation of atomic orbitals in a nitrogen atom

Electrons in the 2s orbital are raised to the slightly higher energy level of the 2p atomic orbitals; three of the resulting four orbitals are hybridised (sp^2), leaving one p orbital unchanged.

Drawing the resulting atomic orbitals (Fig. 20) we can see that the three sp^2 hybridised orbitals are available to form sigma covalent bonds, leaving the p orbital above and below the plane of the page (the three sp^2 hybrid orbitals are arranged in a planar orientation).

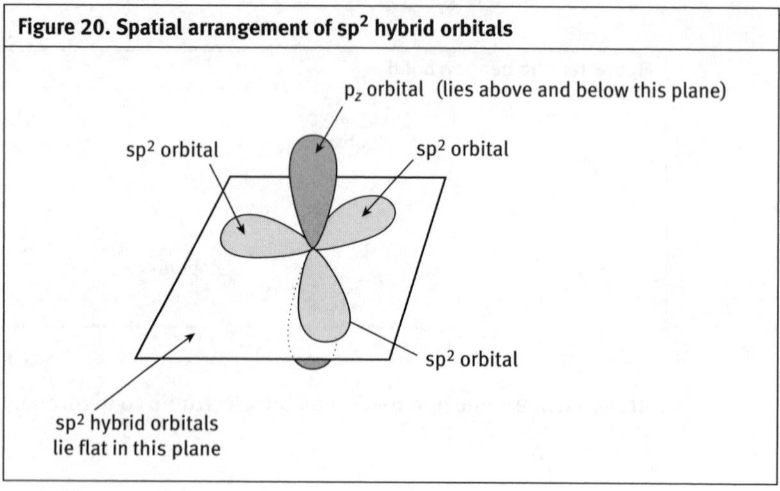

Figure 20. Spatial arrangement of sp^2 hybrid orbitals

In Fig. 21 we have shown just the p orbitals in the peptide bond. The p orbital on the nitrogen atom is close enough to interact with the p orbitals on the carbon and oxygen atoms of the carbonyl group. The 'side-to-side' merging of p orbitals forms pi covalent bonds.

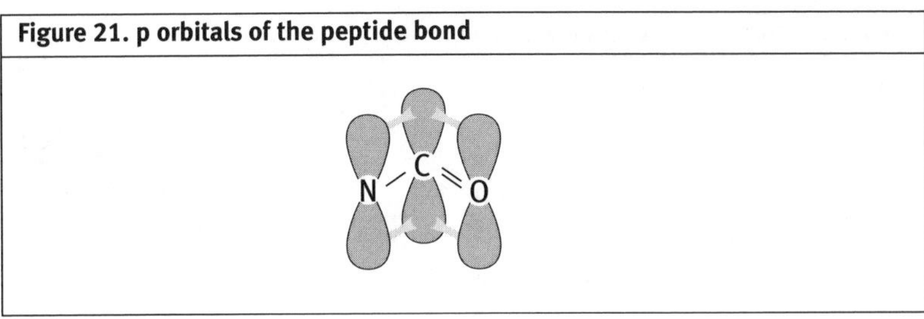

Figure 21. p orbitals of the peptide bond

Electrons can move between the p orbitals (electrons are said to be delocalised) in the formation of pi covalent bonds. In fact, the peptide bond is a resonance structure, shown in Fig. 22.

Figure 22. The peptide bond is a resonance structure

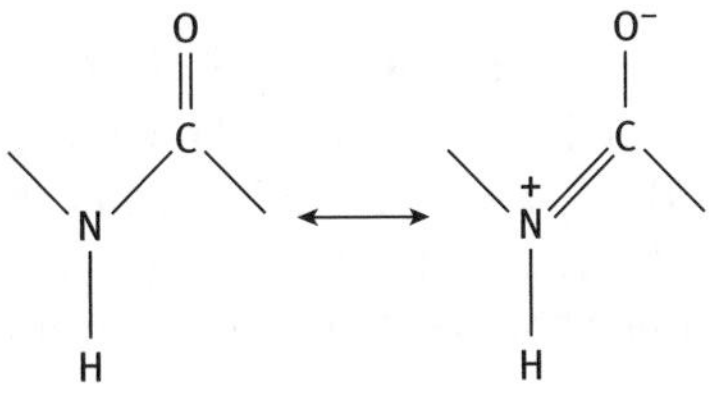

So we see that the peptide bond has a partial double bond character; there is a variable amount (about 40%) of pi bonding between the nitrogen and carbon, sufficient to restrict bond rotation.

The important properties and outcomes of the peptide bond can be summarised as follows.

- The rigidity of the peptide bond reduces the degrees of freedom of the polypeptide during folding.
- Due to the double bond character, the six atoms involved in the peptide bond group are always planar (Fig. 23).

Figure 23. Atoms around a peptide bond lie in a planar conformation

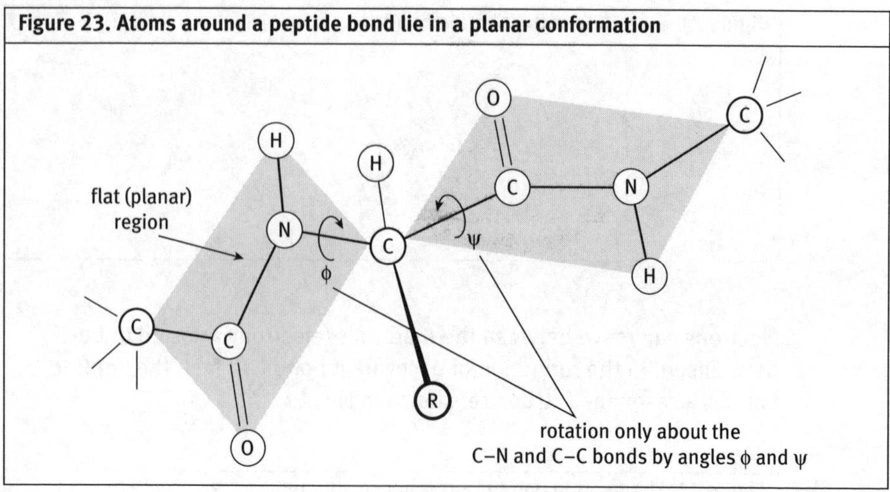

- Therefore, rotation about the C–N and C–C bonds by angles of φ and ψ respectively defines the shape of the polypeptide. If all psi and phi angles are the same the peptide assumes a repeating structure. For certain combinations of angles this can take the form of a helical structure (the alpha-helix) or a beta-sheet structure (Fig. 24).

Figure 24. Alpha-helices and beta-sheets are common protein conformations

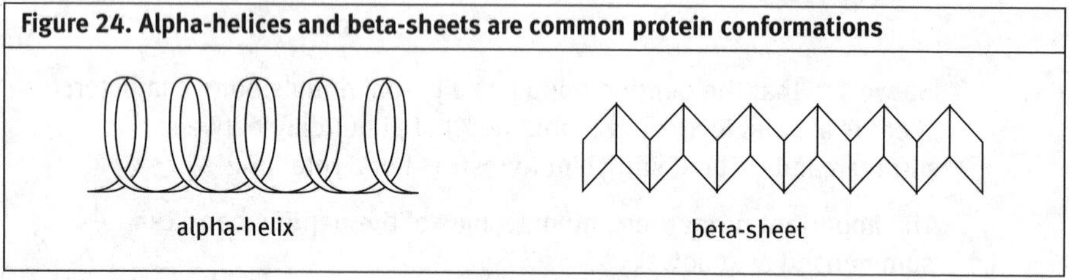

Furthermore:

- The peptide bond is invariably found in the ***trans*** conformation, i.e. alpha-carbon atoms are on **opposite** sides of the C-N peptide bond (Fig. 25). This avoids steric hindrance between groups attached to the alpha-carbon atoms (this is a form of isomerism; see Section 6.3).

Figure 25. *cis* and *trans* conformations of the peptide bond

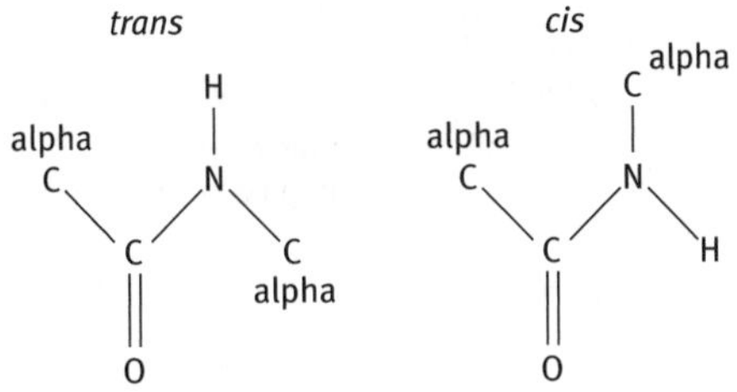

- The resonance donation of electrons by the nitrogen atom makes the carbonyl less electrophilic (electron-seeking). As a result the amide is comparatively unreactive. This is a good thing otherwise proteins would be too reactive to be of much use in biological systems.

The 'special' amino acid proline, a secondary amine, as opposed to other naturally occurring amino acids that are all primary amines, is unusual in that its amino group forms part of a rigid and planar ring structure (Fig. 26).

Figure 26. Rotation is possible about a proline peptide bond

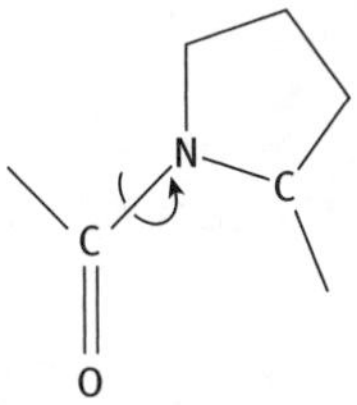

Proline forms peptide bonds with other amino acids, but because of the ring structure, the peptide bond so formed lacks a partial double bond character. Thus, rotation about this peptide bond is possible. Therefore, when proline is incorporated into a polypeptide chain (Fig. 27), rotation about its peptide bond, plus lack of rotation about the phi -N-C- bond (now part of the ring structure of proline), causes a 'kink' in the polypeptide chain. The presence of proline inhibits alpha- and beta-chain conformations in proteins; proline is also found at the 'bends' in polypeptide chains.

Figure 27. Peptide bonds in a polypeptide chain

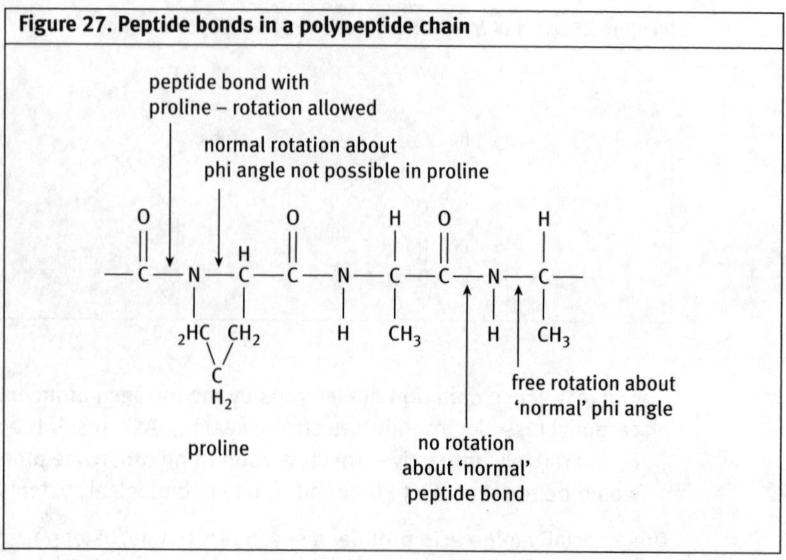

Thus we can see, at the level of the primary structure of proteins, how the special characteristics of the peptide bond are so important in constraining the three-dimensional conformations of polypeptide chains.

3 Interactions between molecules

Basic concepts:
Covalent bonds can be thought of as the 'glue' which holds molecules together; they are strong, stable and require significant energy to break. On the other hand, intermolecular interactions are relatively weak, and in water form and break rapidly in a reversible manner. Weak though they are, intermolecular interactions are collectively strong and vitally important in understanding the interactions of biological molecules.

Every biological event involves molecular interactions. Molecules must interact (bind) to initiate an action, and then separate. Whether it is an enzyme binding to its substrate, a hormone binding to its receptor on a membrane, or RNA being transcribed on a DNA template, all require to be bound in a precise orientation for a short time. Such 'recognition' events are made possible through intermolecular interactions.

Reminder
Intermolecular interactions are weak, but collectively strong

Intermolecular interactions may be broadly classified as **electrostatic** or **hydrophobic** in nature. **Electrostatic interactions** include a number of types that we can classify as hydrogen bonding, charge–charge interactions, and short-range electrostatic interactions (van der Waals forces).

3.1 Hydrogen bonding

This is an important, and relatively strong, form of intermolecular interaction. A 'classic' example of hydrogen bonding occurs between water molecules (Fig. 28).

Figure 28. Hydrogen bonding in water and ice

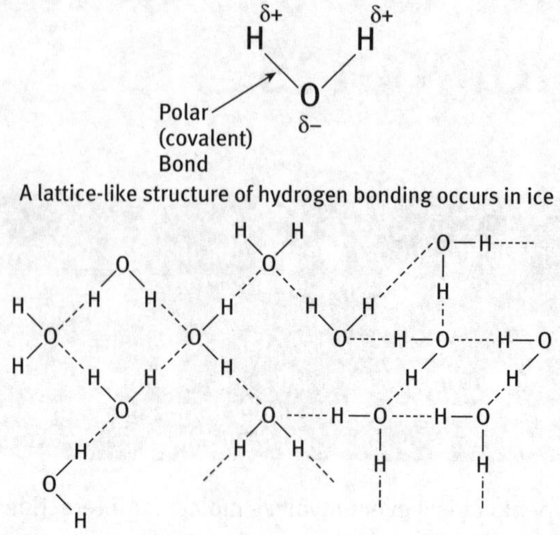

δ+ H — O — H δ+
δ–
Polar (covalent) Bond

A lattice-like structure of hydrogen bonding occurs in ice

In water, the electronegative oxygen draws electrons towards itself, causing the two hydrogen atoms to carry a slight positive charge. There is a large dipole within the O – H bonds. The oxygen atom of the water molecule possesses two **lone pairs** of electrons whereas the hydrogen atoms have lost much of their share of the electrons to the electronegative oxygen atom. Hydrogen bonding involves the oxygen atom of one molecule of water sharing a lone pair of electrons with the slightly positively charged hydrogen atom of another water molecule. Water molecules therefore interact with each other through hydrogen bonding. In liquid water, such interactions rapidly form and break, but in ice a lattice-like arrangement is formed. This property of water is paramount in determining its unique characteristics as the 'solvent of life'.

Hydrogen bonds can form whenever a strongly electronegative atom (e.g. oxygen, nitrogen and fluorine) approaches a hydrogen atom which is covalently attached to a second strongly electronegative atom.

For example, hydrogen bonding can occur between a carbonyl and amino group.

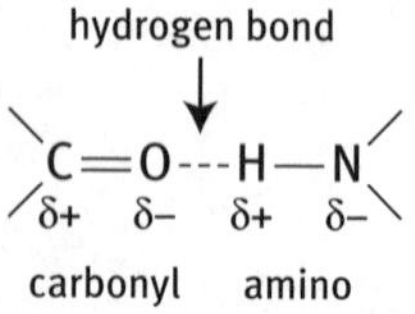

hydrogen bond

C=O---H—N

δ+ δ– δ+ δ–

carbonyl amino

A variety of such groups, containing an electronegative atom, occur in biological molecules.

hydroxyl		**carbonyl**		**amino**	
in alcohols, carbohydrates, proteins and nucleotides	$—\overset{\delta-}{O}—\overset{\delta+}{H}$	in organic acids, carbohydrates, proteins and nucleotides	$O^{\delta-}$ double-bonded to $—C^{\delta+}—$	in proteins and nucleotides	$H^{\delta+}$ bonded to $—N^{\delta-}—$

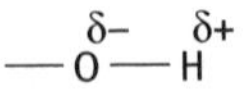

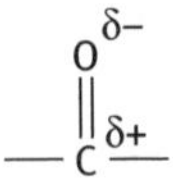

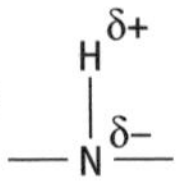

In every hydrogen bond there is the acceptor, a hydrogen atom attached to an electronegative atom, and the donor, an electronegative atom (O, N or F) with one or more lone pairs of electrons.

Reminder

Hydrogen bonds can form whenever a strongly electronegative atom (O, N or F) approaches a hydrogen atom that is covalently attached to a second strongly electronegative atom

The hydrogen bond is of particular importance in biological systems. It may look like a simple charge–charge interaction, but the hydrogen bond has special characteristics. With this type of intermolecular interaction the optimum hydrogen bond has a distance of about 2.8 Å and is linear (an ångström, Å, is 1×10^{-10} metre, of the order of the size of an atom). For example, in the hydrogen bond between a carbonyl and amino group shown above, we could draw a straight line between the oxygen and the nitrogen, along which the bond would be aligned. In other words, the direction of the bond is relatively constrained. Compared to other intermolecular interactions, the hydrogen bond is also relatively strong. These properties of the hydrogen bond have important consequences in biology, constraining as they do the distances and orientations between molecules. Hydrogen bonds are significant in determining the structures of proteins, and in holding together the double-stranded DNA molecule.

Reminder

Hydrogen bonds have both an optimum length and directionality

Hydrogen bonds occur between the purine and pyrimidine bases of opposite DNA strands, constraining the bases to lie flat (planar) and at a defined distance from one another. Furthermore, and because of hydrogen bonding, only thymine will bond with adenine, and only cytosine with guanine (Fig. 29).

So, our genetic code is determined by hydrogen bonding!

Figure 29. Hydrogen bonding between DNA bases

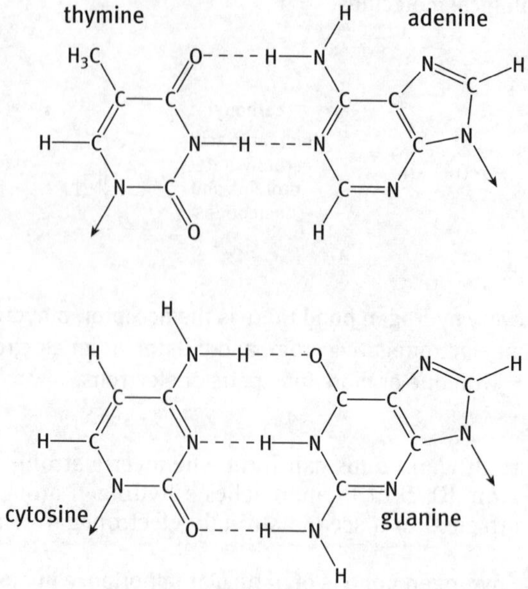

thymine adenine

cytosine guanine

3.2 Charge–charge interactions

Charge–charge interactions are electrostatic in nature (as indeed are hydrogen bonds) but, unlike hydrogen bonds, the distance between interacting groups is not 'fixed' and the alignment of the interaction is not so important. Any group or atom in a molecule which carries a charge, as a result of either the electronegativity difference between atoms, the presence of an acidic or basic group, or an atom with a lone pair of electrons, can engage in charge–charge interactions. That interaction may be attractive (with an oppositely charged group) or repulsive (with a similarly charged group). Both are important in the interaction of biological molecules.

Groups which commonly carry a charge at physiological pH, and which occur particularly in proteins, are the carbonyl (negatively charged) and the amino (positively charged) groups. The charge on groups which have acidic and basic characters are necessarily very dependent on the pH of the solvent. Low pH (acid, with a higher H^+ concentration) leads to protonation; for example the amino group NH_2 exists as NH_3^+. A higher pH (more alkaline) favours dissociation of groups, for example the carboxyl group COOH exists as COO^-.

Proteins consist of amino acids, linked together through peptide bonds to produce long polypeptide chains (Fig. 30). The R groups (side groups,

Figure 30. Side groups in polypeptide chain

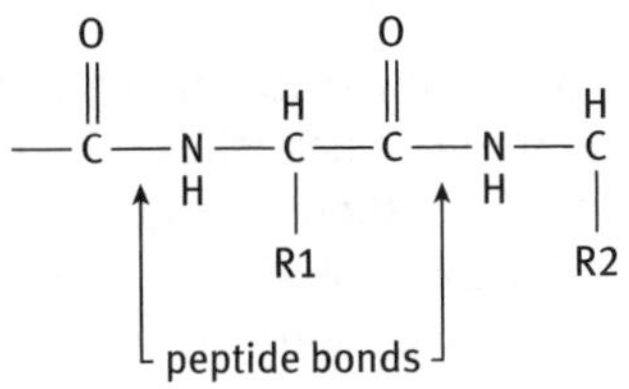

e.g. R1 and R2 in Fig. 30) of amino acids in the chain frequently contain carboxyl (-COOH) or amino (-NH_2) groups. At physiological pH, the carboxyl group dissociates, losing a proton (H^+) and becoming negatively charged, -COO^-. At physiological pH, the amino group associates with a proton to become positively charged (-NH_3^+). On cellular proteins, therefore, there are likely to be numerous charged groups, where charge–charge interactions become possible between molecules, or indeed within molecules.

Reminder

Interaction between charged groups is likely to be very pH-dependent

3.3 Short range charge–charge interactions

At very short distances between molecules, another type of charge–charge interaction becomes possible, referred to as **van der Waals forces**. These forces are very weak attractions (or repulsions) that occur between atoms or molecules at close range. In a covalent bond between atoms, the two electrons in that bond will, for most of the time, be found equally distributed between the two atoms. But electrons are not fixed. Statistically, there will be times when both electrons are next to one or other of the atoms; this will momentarily make that atom negatively charged, leaving the other atom in the bond positively charged, and so we will have a **temporary dipole**. (The distribution of charges in a water molecule is an example of a **permanent dipole**.) Alternatively, if we approach a covalent bond with, say, a positively charged amino group, then we could **induce** a dipole in that bond, by drawing electrons towards one end of the bond. This constant movement and redistribution of electrons in molecules produces complicated and fluctuating changes in attraction and repulsion.

Reminder

Dipoles are constantly being produced in molecules in solution

3.4 Hydrophobic interactions

Hydrophobic (meaning literally 'water hating') interactions differ from those intermolecular interactions already described in that they are not electrostatic in nature. Nor is there any attraction, or repulsion, between hydrophobic groups. The basis of hydrophobic interactions resides in the behaviour of water. Hydrophobic, or **apolar**, groups are so called because they are insoluble in water and do not interact with it. **Hydrophilic** ('water loving'), or **polar**, groups are soluble in water, and do interact with it. The **functional** groups that we have so far come across, such as carboxyl (-COOH), amine (-NH_2), hydroxyl (-OH), are all polar groups. They all contain an electronegative atom that will induce a dipole and an unequal distribution of charge. Water is a polar molecule; it will interact with other polar groups through electrostatic interactions. Molecules which contain polar groups are generally **soluble** in water.

On the other hand, apolar (hydrophobic) groups do not contain any particularly electronegative atoms. For example, the apolar methyl group (-CH_3) consists of carbon covalently bonded to three hydrogen atoms. The electronegativities of carbon and hydrogen are similar and there is therefore little tendency for electrons to assume an unequal distribution (no dipole is realised). The methyl group is therefore 'neutral'. Neutral groups have no way of interacting with water (no charge–charge interaction is possible). The water solvent will respond to such apolar groups by 'excluding' them. Apolar groups do not attract or repel each other, but rather coalesce (come together) through the action of the water solvent.

Reminder

Apolar (hydrophobic) groups do not interact with water; they are 'neutral' groups in which the electronegativity of the constituent atoms is approximately equal

The 'hydrophobic effect' is of major importance in biology. Biological membranes form a hydrophobic barrier which defines a cell, or cellular organelle. Protein molecules fold and adopt specific three-dimensional structures, driven by the 'expulsion' of hydrophobic groups from water. The hydrophobic groups in proteins are found buried within the protein's interior, where there is no water.

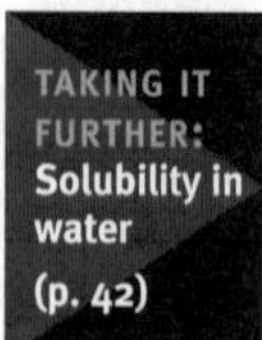
TAKING IT FURTHER: Solubility in water (p. 42)

Reminder

All biological molecules must interact (bind) in a short-lived, reversible manner to invoke a response. Such phenomena are mediated by intermolecular interactions

3.5 Summing up

1. Relative to covalent bonds, intermolecular interactions are weak, but collectively strong. They form the basis of recognition and binding between biological molecules.
2. Hydrogen bonds are a relatively strong form of intermolecular electrostatic interaction; they can form whenever a strongly electronegative atom (e.g. oxygen, nitrogen or fluorine) approaches a hydrogen atom which is covalently attached to a second strongly electronegative atom. Distance and direction are important in hydrogen bonds.
3. Charge–charge interactions (attraction or repulsion) occur between groups that carry a net electrical charge. Such interactions also occur at very small distances between molecules as a result of temporary, induced or permanent dipoles.
4. Hydrophobic, or apolar, interactions occur with groups that show no polarity (no unequal distribution of charge), and are therefore unable to interact with water. They are insoluble in water and are excluded from this solvent.

3.6 Test yourself

The answers are given on p. 159.

Question 3.1
(a) What type of intermolecular interaction is likely to occur when a strongly electronegative atom approaches a hydrogen atom that is covalently attached to a second electronegative atom?
(b) Give three examples of common biological functional groups which may participate in this type of intermolecular interaction.

Question 3.2
Give three characteristics of the hydrogen bond which distinguish it from other types of charge–charge interaction.

Question 3.3
What are the differences which distinguish intermolecular interactions from intramolecular (covalent) interactions?

Question 3.4
Induced or temporary dipoles in molecules are important in what type of intermolecular interaction?

Question 3.5
Which of the following groups are considered 'hydrophobic'?
(a) $-CH_2OH$
(b) $-SH$
(c) $-CH_2CH_3$
(d) $-CH_2NH_2$
(e) 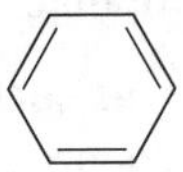
(f) 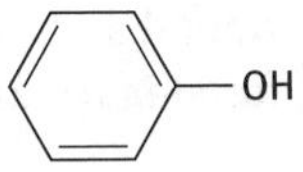 OH

Taking it further

3.7 Solubility in water

We have discussed the fact that atoms and molecules are held together by various types of intermolecular interaction, including hydrogen bonds, charge–charge interactions and van der Waals forces. These forces are intricately involved in solubility because it is the solvent–solvent, solute–solute, and solvent–solute interactions that govern solubility.

Water is a polar molecule. As we have seen, its strongly electronegative oxygen atom induces a charge distribution in the molecule (a dipole). This gives rise to hydrogen bonding between water molecules; thus water is a very ordered structure!

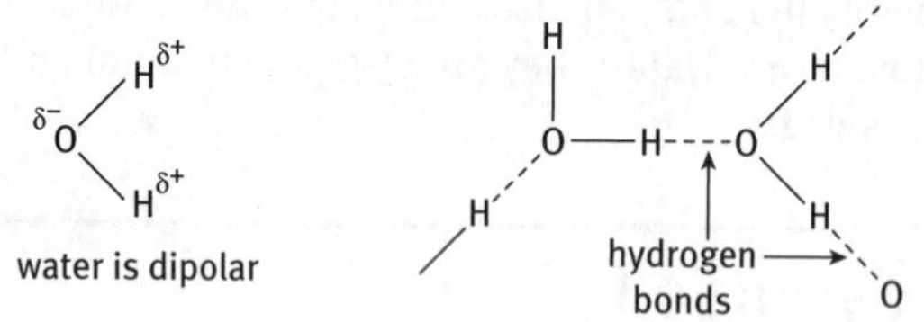

water is dipolar

hydrogen bonds

In order for substances to dissolve in water, room has to be made within this ordered structure to accommodate the substance.

If such a substance can interact with water molecules, through hydrogen bonding or charge–charge interactions, then it will be accommodated (it is a polar molecule) and consequently it will dissolve. By definition, a polar substance is one which interacts with water. Polar molecules are often referred to as hydrophilic, 'water loving'.

For simple substances such as salt, the small and charged Na^+ and Cl^- ions are readily accommodated within the water structure. Water molecules surround and interact with ions through charge–charge interactions, forming a 'cage' or hydration sphere to surround the ion (Fig. 31).

Figure 31. Water molecules surround ions in solution

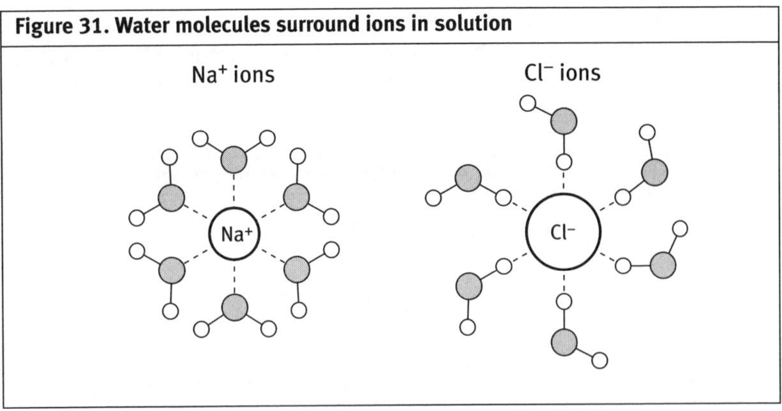

For larger molecules which contain polar groups, the situation is somewhat more complex but the principle is the same. Biological molecules can contain a variety of polar functional groups, any of which will confer solubility in water. Hydroxyl groups (-OH), carboxyl groups (-COOH), carbonyl groups (-C=O), aldehydes (-CHO), sulphydryl groups (-SH) and amino groups (-NH_2) all contain an electronegative atom and all carry a charge under physiological conditions. As a general rule, the more polar groups on a molecule, the more soluble that molecule is.

Large, polar molecules, will naturally be surrounded by a relatively large hydration sphere (Fig. 32).

Figure 32. Hydration spheres surround larger polar molecules in solution

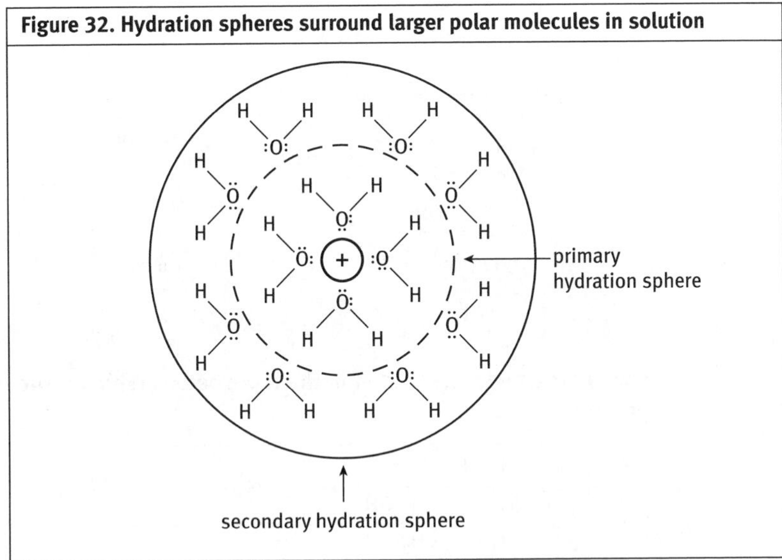

The hydroxyl group (-OH) on biomolecules is particularly important in conferring solubility through its ability to form hydrogen bonds (shown by the dashed lines in Fig. 33).

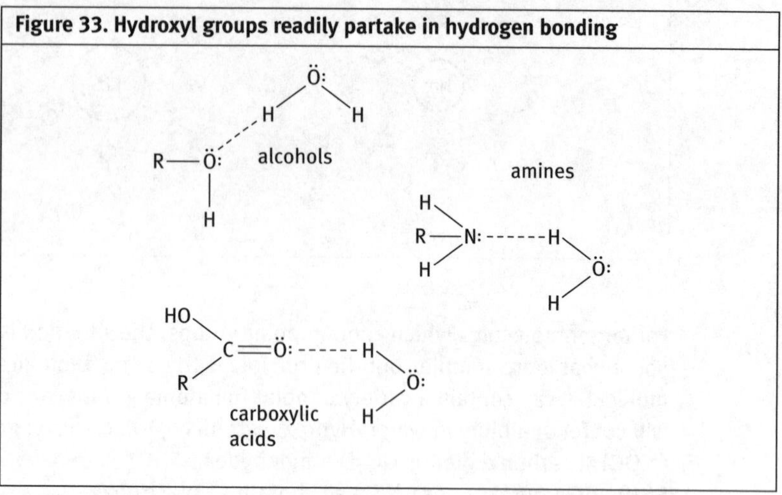

Figure 33. Hydroxyl groups readily partake in hydrogen bonding

As a general rule, the more polar groups a molecule has (and the smaller the molecule) the more soluble it is in water. Simple sugars such as glucose and fructose, with an abundance of hydroxyl groups, are very water soluble.

glucose

fructose

Question

Look at the molecules listed in the table below. Which is the most polar?

A	CH_3-CH_2-OH
B	CH_3-CH_2-CH_2-CH_2-OH
C	CH_3-CH_2-CH_2-CH_2-CH_2-CH_2-OH
D	CH_3-CH_2-CH_2-CH_2-CH_2-CH_2-CH_2-CH_2-CH_2-CH_2-OH

Answer: **A** is the most polar; it has a hydroxyl group and is a small molecule. The solubility of these molecules in water decreases from A to D. The methyl (-CH_3) and methylene (-CH_2) groups are both apolar; they are hydrophobic 'water hating' groups. There is no 'electronegative effect' in these groups, no opportunity for hydrogen bonding with water, no dipole and therefore no charge–charge interaction.

A major group of biological molecules is the lipids, which includes fats, oils and waxes. A major constituent of lipids are fatty acids. Fatty acids consist of a long hydrocarbon chain that is apolar (hydrophobic), which terminates in a polar (carboxyl) group. The molecule is said to be amphipathic; it has both a polar end and an apolar end (Fig. 34).

Figure 34. An amphipathic fatty acid molecule

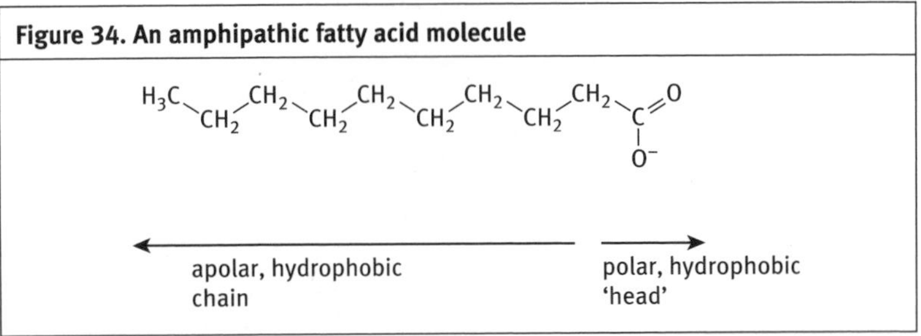

Fatty acids are linked together with glycerol and phosphate to form phospholipids, the major constituent of biological membranes. The highly polar phosphate group makes this a very amphipathic molecule (Fig. 35).

Figure 35. Amphipathic phospholipids are major constituents of biological membranes

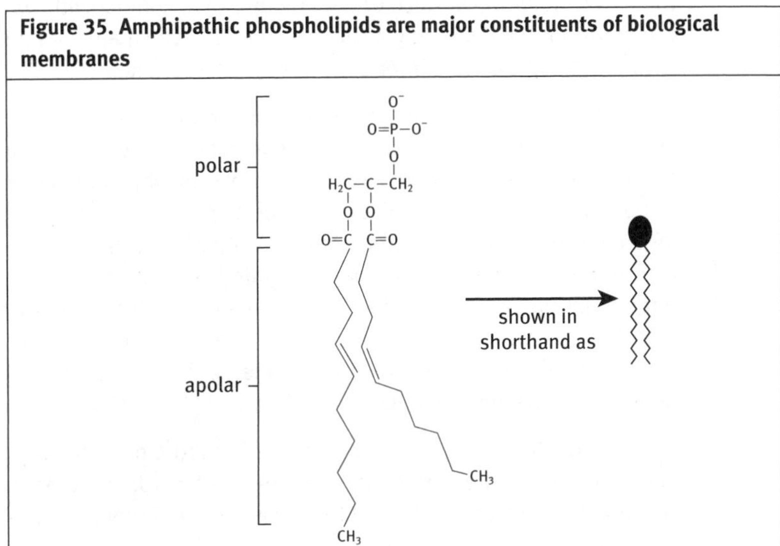

In water these phospholipid molecules align themselves to form a 'bilayer', the basic structure of biological membranes (Fig. 36). Polar 'head groups' face and interact with the water environment, whereas the apolar hydrophobic fatty acid 'tails' form a hydrophobic environment from which water (and other polar molecules) are excluded.

Figure 36. The basic structure of a biological membrane is a phospholipid bilayer

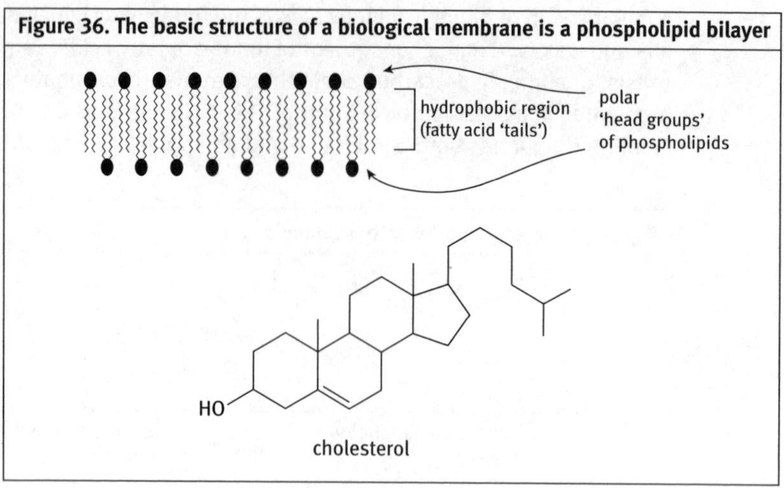

The steroid molecule cholesterol (Fig. 36) is a common constituent of the outer membranes of cells. The amphipathic character of this molecule can be appreciated through the hydroxyl group, which orientates this end of the molecule to the surface of the membrane in contact with water, and the hydrophobic fused ring structure and 'tail' which anchors the molecule within the hydrophobic interior of the cell membrane.

Although biological organisms are composed of greater than 60% by weight of water, they also contain an extensive network of biological membranes. In essence we have two completely opposite environments; the polar water environment and the apolar membrane environment. For molecules to move around the body they must traverse these two environments. Molecules will partition into these two environments dependent upon their polarity. This has enormous consequences for the distribution and pharmacological action of drugs. The science of pharmacology studies the absorption, metabolism, distribution and activity of drugs on the body.

As early as 1847, observations suggested that the more hydrophobic a substance, the more permeant it was; in other words, how more readily it was absorbed by the body, and these early observations remain valid today. It appears, however, that all hydrophilic molecules, and most hydrophobic molecules, require specific carriers (transporters) for their movements across cell membranes. Mammalian genomes probably contain in excess of 1000 genes encoding such transporters!

4 Counting molecules

Basic concepts:
Atoms join together to make molecules, molecules join together to make larger molecules or react to form new molecules. Since molecules come in a variety of shapes and sizes, it is essential that we have some simple way of comparing and counting them. For example, we might need to make up a solution with a specific ratio of compounds at a certain concentration, or express the activity of an enzyme in terms of how many molecules of substrate it converts in a given time. We can compare the concentration of molecules in solution through the concept of the mole. This is a 'must know' concept in almost every branch of biology.

4.1 Moles

As biologists, we are interested in the interaction between molecules, so we need a system whereby we can compare the number of molecules of substances in solution. That system uses the **mole**. A mole is a certain number of molecules. The system works like this:

- Firstly we need a standard. The standard we use is the carbon-12 atom, $^{12}_{6}C$, the stable isotope of carbon containing 6 protons and 6 neutrons. A convention was adopted for expressing the atomic mass of an element in terms of atomic mass units (amu). This was defined such that the carbon-12 atom is exactly equal to 12.0 amu. Thus 1 amu is equal to one-twelfth the mass of an atom of carbon-12.
  1 amu = 1.661×10^{-24} g.
- Every element in the periodic table can be assigned an atomic mass, which is its mass relative to that of one-twelfth the mass of an atom of carbon-12. This is known as the relative atomic mass (RAM) scale.
- Because the amu is very small it is more convenient to express the relative atomic mass in grams. We therefore define the quantity which is equivalent to 12g of carbon-12 to be equal to 1 mole.
- 1 mole, or 12 g of carbon-12, contains 6.022×10^{23} atoms of carbon-12.
- 1 mole, or the amount equal to the atomic mass in grams of any substance contains 6.022×10^{23} atoms of the substance.
- The number 6.022×10^{23} is known as Avogadro's number or Avogadro's constant.

One mole of a substance is the amount of substance equivalent to the formula mass in grams of the substance and contains Avogadro's number, 6.022×10^{23} atoms, molecules or ions of the substance.

From our definition, one mole of a compound has the identical number of mass components as there are atoms in 12 g of carbon-12. So, 12 grams of carbon is equivalent to 1 mole and contains 6.022×10^{23} atoms of carbon.

The mass of a water molecule (H_2O) is 18 amu (one oxygen =16 amu + two hydrogens = 2 amu). Therefore 18 grams of water is 1 mole and contains 6.022×10^{23} molecules of water.

Glucose ($C_6H_{12}O_6$) has a molecular mass of 180 amu (by adding the atomic mass units of all the atoms in a glucose molecule), so 180 grams of glucose is 1 mole (containing 6.022×10^{23} molecules of glucose).

One formula unit of sodium chloride, NaCl, has a mass of 58.45 amu. Therefore 1 mole of sodium chloride has a mass of 58.45 g. This mass of sodium chloride contains 6.022×10^{23} sodium (Na^+) ions and 6.022×10^{23} chloride (Cl^-) ions. This information is summarised in the table.

Formula	Formula mass of 1 unit	Formula mass of 1 mole	Number of single units in 1 mole	Number of atoms or ions in 1 mole
$^{12}_{6}C$	12 amu	12 g	6.022×10^{23} atoms C	6.022×10^{23} atoms C
H_2O	18 amu	18 g	6.022×10^{23} molecules H_2O	6.022×10^{23} atoms of O
				$2 \times 6.022 \times 10^{23}$ atoms of H
$C_6H_{12}O_6$ glucose	180 amu	180 g	6.022×10^{23} molecules glucose	$6 \times 6.022 \times 10^{23}$ atoms C
				$12 \times 6.022 \times 10^{23}$ atoms of H
				$6 \times 6.022 \times 10^{23}$ atoms of O
NaCl	58.45 amu	58.45 g	6.022×10^{23} units NaCl	6.022×10^{23} Na^+ ions
				6.022×10^{23} Cl^- ions

Because 12.0 g of carbon is an easily manageable amount of substance to use, the mole became the standard unit for counting numbers of atoms or molecules.

The MOLE is a number just like a dozen or a century! There are exactly the same number of molecules in a mole of glucose as there are in a mole of insulin. We can therefore compare the numbers of molecules of different substances in solution directly, irrespective of whether they are small or large molecules. For example, most enzymes are very large molecules compared to their substrates. Nevertheless, one enzyme molecule reacts with one substrate molecule at a time, so it is important to be able to work out how many molecules of each you might have in solution.

Question

How many moles are there in 57.5 g of sodium?

Sodium has 23 atomic mass units, so 23 g of sodium would be equal to 1 mole. So, in 57.5 g sodium, there must be 57.5/23 = 2.5 moles

4.2 Molecular mass

We have seen that the mass of one mole of any compound can be obtained by summing the relative atomic masses of all the elements which make up the compound and converting to grams. The mass of one mole of a compound in grams is called its **molar mass, *M*.**

So we have the relationship:

$$\text{Number of moles} = \frac{\text{mass in grams}}{\text{molar mass}} = \frac{m}{M}$$

A variety of symbols are used to describe the molecular mass of compounds.

- The molar mass (symbol M) of glucose is 180 g mol^{-1}. This is the mass of one mole.
- The relative molecular mass (symbol M_r) of glucose is 180. This is the relative mass of one molecule compared to one-twelfth the mass of one atom of carbon-12.
- The name 'dalton' (symbol Da), is used by biologists as an alternative to the cumbersome name of 'atomic mass unit' for one-twelfth of the mass of the atom of carbon-12. Thus, the molecular mass of glucose can be expressed as 180 Da.

These are all correct ways of saying the same thing by present recommendations. Note that M_r does not have units; this is because relative molecular mass is the ratio between the mass of a molecule and the mass of one-twelfth of a carbon atom, and as a ratio it has no units.

4.3 Moles and molarity

A **mole** refers to the amount of a substance whereas the **molarity** refers to its concentration.

Moles and molarities are a common source of confusion amongst students, partly because of the similarities in their names, so it's important to make sure that you know precisely what these terms mean.

> **Reminder**
>
> A mole is a number, an amount, whereas molarity is a concentration

A **mole** (abbreviation mol) is a measure of the amount of a substance.

A mole of a compound is the amount which contains a number of molecules equal to Avogadro's number (6.022×10^{23}). Or a mole of compound is the amount of a substance equal to the relative molecular mass expressed in grams.

Molarity is a measure of concentration. The molarity of a solution measures the number of moles of a substance in a certain volume.

> **Reminder**
>
> Avogadro's number is equal to the number of molecules in a mole ($= 6.022 \times 10^{23}$)

Now, in terms of numbers of molecules, it does not matter if you have 1 mole of a substance in 1 litre, or in 1 ml of solution, you still have the same number of molecules (because 1 mole $= 6.022 \times 10^{23}$ molecules). However, the concentration must change; 1 mole of substance in 1 ml is clearly a more concentrated solution than 1 mole in 1 litre. To define concentration, we refer to molarity.

The molarity of a solution is the number of moles per litre of solution (or number of moles per dm^3, if you prefer). The units of molarity are mol/l or mol l^{-1} or M.

A molar solution (abbreviation 1 M) contains one mole of a material in a volume of one litre.

So, given the M_r of glucose as 180, then 180 grams of glucose is 1 mole, and 180 grams of glucose dissolved in 1 litre of water would give a 1 molar (1 M) solution.

If we take just 1 ml of this solution, then the concentration of glucose in that 1 ml is still 1 molar (we have not changed the ratio of the mass of glucose to the volume of water). However, 1 ml of the solution contains 1/1000 the number of moles (1 mmol). This is because 1 ml of solution is 1/1000 of a litre.

So a 3.2 M solution has a concentration of 3.2 mol/l or 3.2 mol l^{-1}. The mass or amount of the compound in the solution depends on the volume that you use. 1 ml of this solution contains only 3.2 mmol, or .0032 mol, (i.e. one thousandth of the amount contained in a litre) but the concentration remains the same.

Reminder

$$\text{Concentration} = \frac{\text{number of moles}}{\text{volume (l)}} = \frac{n}{V} = c\ (\text{mol l}^{-1})$$

Question

What is the molarity of water?

The M_r of water (H_2O) is 18, so 18 grams of water is 1 mole. One litre of water has a mass of 1000 grams, so it must contain

$$\frac{1000\ \text{g}}{18\ \text{g mol}^{-1}} = 55.5 \text{ moles of water, thus}$$

the molarity of water is 55.5 mol l^{-1} (55.5 M)

4.4 A note on units

Biologists frequently deal with solutions that are much less than 1 molar concentration. The terms milli (one thousandth), micro (one millionth) and nano (one thousand millionth) are commonly used to express concentrations, amounts or volumes of substances. These are abbreviated to 'm' (milli), 'μ' (micro) and 'n' (nano).

milli = 10^{-3} = m
micro = 10^{-6} = μ
nano = 10^{-9} = n

For example, if we took 1 ml (1 millilitre = one thousandth of a litre) of a 1 molar solution of glucose (containing 1 mole of glucose per litre), then that would contain only 1 mmol of glucose (one thousandth of a mole of glucose).

When tackling calculations involving moles or molarity, work from first principles. More examples are given in 'Taking it further: Confidence with moles'.

Question

If we took 0.0105 g (10.5 mg) of a 10500 M_r protein, and dissolved this in 1 ml, then that 1 ml would contain 1 micromole (1 μmol) of the protein.

Why? because

$$\text{Number of moles} = \frac{\text{mass in grams}}{\text{molar mass}}$$

$$= \frac{0.0105\ \text{g}}{10\,500\ \text{g mol}^{-1}} = 0.000001\ \text{mol} = 1 \times 10^{-6}\ \text{mol}$$

or 1 micromole (1 μmol)

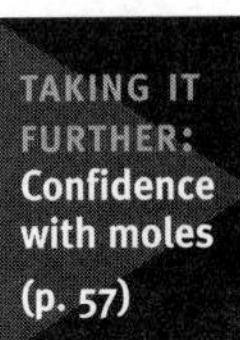

4.5 Dilutions

Equally important as understanding moles and molarity, is understanding dilutions, since solutions are often diluted to achieve the required molarity! There are basically two ways of diluting solutions; simple dilutions and serial dilutions.

1. **Simple dilution:** A simple dilution is one in which a unit volume of a liquid material of interest is combined with an appropriate volume of a solvent liquid to achieve the desired concentration. For example, one litre of a solution of glucose in water is combined with a further litre of water to give a glucose solution of half the original concentration. The dilution factor is the total number of unit volumes in which your material will be dissolved. In the case above, the dilution factor would be 1:2 (verbalise as '1 to 2' dilution). In a further example, a 1:5 dilution entails combining 1 unit volume of diluent (the material to be diluted) + 4 unit volumes of the solvent medium (hence, $1 + 4 = 5 =$ dilution factor). If you have a known volume and concentration of a substance, and you need to dilute this to give a different concentration, then the expression $V1C1 = V2C2$ is useful. V1 and C1 refer respectively to the original volume and concentration, and V2 and C2 to the desired volume and concentration.
2. **Serial dilution**: A serial dilution is simply a series of simple dilutions, which amplifies the dilution factor quickly, beginning with a small initial quantity of material. The source of dilution material for each step comes from the diluted material of the previous step. In a serial dilution the total dilution factor at any point is the product of the individual dilution factors in each step up to it. For example, say we start with 1 ml of a protein solution. If we take 0.1 ml (100 μl) of this, and add it to 0.9 ml of solvent, then we have made a 0.1:1 dilution (a 10-fold dilution). A further 0.1 ml from this, added to 0.9 ml solvent, gives another 10-fold dilution. The total dilution of the original protein solution is now $10 \times 10 = 100$. This is often the preferred way of making dilutions since it involves weighing out your substance only once, from which you can make a series of different concentrations. But remember, errors from dilutions can occur at every step and so the more dilutions that are made, the more errors that are possible.

More examples are given in 'Taking it further: Confidence with moles'.

4.6 Percent composition solutions

Sometimes, and usually out of convenience, solutions are made up as percent solutions.

Reminder

Percentage (%) means number of parts per hundred, i.e. 1% = 1 part per 100 total

1. **w/w percentage composition**: When both the solute (the substance being dissolved) and the solvent (the liquid doing the dissolving) are given in mass units (for example in grams), the percentage composition is written as % w/w. For example 20 g glucose dissolved in 480 g water has a % w/w of 2%. This means that by mass the glucose forms 2% of the total solution.

   Why? Because

   Mass solute (glucose) = 20 g
   Mass solvent (water) = 480 g
   Mass solution = 20 g + 480 g = 500 g
   % w/w = 20/500 × 100 = 2%

2. **w/v percentage composition**: This is the most common way of making up a solution in the laboratory. A certain mass of dry solute is weighed and placed in a container calibrated to contain a fixed volume. Solvent is added until the known, calibrated volume is reached. The concentration is then expressed in terms of percentage weight per volume of solution (% w/v). For example 10 g sodium chloride is dissolved in water to give 100 ml solution. The % w/v is therefore 10% w/v.

   Why? Because

   Mass solute (sodium chloride) = 10 g
   Volume of solution (sodium chloride and water) = 100 ml
   % w/v = (10/100) × 100 = 10%

Reminder

A liquid changes volume only slightly when a solute is dissolved in it, whereas its mass changes significantly

3. **v/v percentage composition**: When using liquid reagents, the percent concentration is based upon volume per volume, i.e. volume of liquid solute per volume of solution. So 10 ml of liquid substance added to 90 ml of buffer would give a 10% v/v solution. If you wanted to make 70% ethanol you would mix 70 ml of 100% ethanol with 30 ml water.

   Why? Because

   70% v/v ethanol = 70 ml ethanol in 100 ml total solution
   The volume of water solvent is therefore 100 ml – 70 ml = 30 ml.

4.7 Summing up

1. **Molecular mass**
   The mass of a molecule is the sum of the masses of the atoms from which it is made. The unit of mass – the atomic mass unit (amu or u), the dalton (symbol Da) – all mean the same thing and are defined as one-twelfth of the mass of an atom of carbon-12.

2. **Relative molecular mass**
   This is the relative mass of one mole (or one molecule) compared to one-twelfth of the mass of one mole (or one atom) of carbon-12.
3. **Mole**
   A mole is the quantity of a substance whose mass in grams is equal to the molecular mass of the substance. One mole of substance contains 6.022×10^{23} particles of the substance.
4. **Molarity**
   Molarity is a measure of the concentration of a solution. If you dissolve 1 mole of a substance in 1 litre of solution, you have made a 1 molar (1 M) solution.
5. **Dilutions**
   For simple dilutions, the dilution factor is the total number of unit volumes in which your material will be dissolved; serial dilutions are a combination of simple dilutions that produce a final dilution factor equal to the product of the individual dilution factors.
6. **Percent solutions**

   For a solid, $\frac{\text{mass of solute}}{\text{volume of solution}} \times 100 = \text{percent concentration (w/v)}$.

   For a liquid, $\frac{\text{volume of solute}}{\text{total volume of solution}} \times 100 = \text{percent concentration (v/v)}$.

4.8 Test yourself

The answers are given on p. 159.

Question 4.1
If a solution contains 0.01 grams of insulin, and the M_r of insulin is 6000, how many moles of insulin are present?

Question 4.2
How many grams of ethanoic acid ($C_2H_4O_2$) would you need to make 10 litres of a 0.1 M ethanoic acid solution? (Take the relative atomic masses of carbon as 12, oxygen as 16 and hydrogen as 1)

Question 4.3
50 ml of a glucose solution was prepared by dissolving 10 grams of glucose in 50 ml of water. Given that the relative molecular mass (M_r) of glucose is 180, how many moles of glucose are present in this solution, and what is the molarity of the solution?

Question 4.4
A stock solution of the amino acid glycine was 0.02 M. 1 ml of this solution was used in an enzyme assay, in a total volume of 3 ml. How many moles of glycine are present in the enzyme assay, and what is the molarity of glycine in the assay?

Question 4.5
1.2 grams of glycine ($M_r = 79$) was dissolved in 100 ml of water. In 1 ml of this solution, (a) how many moles of glycine are present, (b) what is the molarity of the glycine solution, (c) how many molecules of glycine are present?

Taking it further

4.9 Confidence with moles

Here are some examples in working out molarities, concentrations and dilutions.

Moles and molarity

A. To prepare a litre of a simple molar solution from a dry reagent.

Multiply the molecular mass by the desired molarity to determine how many grams of reagent to use:

If $M_r = 194.3$, and you need to make a 0.15 M solution, then use

$194.3 \times 0.15 = 29.145$ g/l

If you only need 30 ml of the above solution, then use $194.3 \times 0.15 \times 30/1000 = 0.87$ g

B. You take 50 μl of a 1 mM stock sugar solution and add it to the enzyme reaction mixture to give a final volume of 3 ml; what is the concentration of sugar in the reaction mixture?

50 μl (= 0.05 ml) into 3 ml is a dilution factor of $3/0.05 = 60$, so the final concentration of the sugar is $1/60 = 0.017$ mM (17 μM).

C. You take 50 μl of a stock sugar solution which contains 100 μmol of sugar per ml, and add this to 3 ml of enzyme reaction mixture. How many micromoles of sugar are there in the reaction mixture?

$0.05 \times 100 = 5$ μmol. This is what you have actually taken from the stock solution.

Adding this to the reaction mixture means you have 5 μmol in 3 ml, or 5/3 (= 1.66) μmol/ml.

D. You are given 10 ml of a stock solution of cytochrome *c* of 1 mg/ml and need to make 5 ml of a cytochrome *c* solution of 5 μg/ml.

(a) In other words, what volume of the stock solution do you need to add to a total volume of 5 ml to give 5 μg/ml?

In terms of V1C1 = V2C2, V1 is what we wish to know, C1 is 1 (mg/ml), V2 will be 5 (ml) and C2 needs to be 0.005 mg (=5 μg).

So, $V1 \times 1 = 5 \times 0.005$

From which V1 = 0.025 ml (or 25 μl). Therefore you would add 0.025 ml of the stock cytochrome *c* solution to 4.975 ml of water (or buffer) to give a final volume of 5 ml.

(b) Given that M_r for cytochrome *c* is 12 000, how many moles of cytochrome *c* are present in 1 ml of your new solution?

12 000 grams of cytochrome *c* would be equivalent to 1 mole. In 1 ml of solution you have 5 μg. Obviously the answer is going to be quite a small number; expressing in grams, 0.000005/12000 = 0.00000000042 moles, or 0.00042 μmol, or 0.42 nmol (nanomole, 10^{-9} of a mole).

(c) That's how many moles in 1 ml, so what is the molar concentration of our new cytochrome *c* solution?

Remember, molarity is expressed per litre. If we have 0.42 nmol in 1 ml, then we would have 0.42 × 1000 nmol in a litre (= 420 nmol). So this solution is 420 nM (nanomolar), or 0.420 μM, or 4.2×10^{-7} M

Dilutions

A. To convert from % w/v solution to molarity, multiply the percent solution value by 10 to get grams/l, then divide by M_r, the molecular mass.

$$\text{Molarity} = \frac{\text{M(\% solution)} \times 10}{M_r}$$

e.g. Convert a 6.5% solution of a substance with $M_r = 325.6$ to molarity, [(6.5 g/100 ml) × 10]/325.6 g/l = 0.1996 M

B. To convert from molarity to percent solution, multiply the molarity by the M_r and divide by 10:

$$\text{\% solution} = \frac{\text{molarity} \times M_r}{10}$$

e.g. Convert a 0.0045 M solution of a substance having M_r 178.7 to a percent solution:

[0.0045 moles/l × 178.7 g/mole]/10 = 0.08% solution

C. You are given 5 ml of a 10% solution of glucose ($M_r = 180$). What is the molar concentration of your solution?

A 10% solution of glucose is equivalent to 10 g/100 ml; you have only 5 ml which must contain 10 × 5/100 = 0.5 g

If 5 ml contains 0.5 g glucose, a litre would contain 0.5 × 1000/5 = 100 g

180 g of glucose in 1 litre would be 1 M, so 100 g is equivalent to $1 \times 100/180 = 0.55$ M

And how many moles of glucose are in 5 ml?

$0.55 \times 5/1000 = 0.00275$ moles, or 2.75 mmol (millimoles, 10^{-3} of a mole).

D. Look at the table below. You are given a stock protein concentration of 1.5 mg/ml and need to make a set of simple dilutions to give a range of protein concentrations, in a fixed final volume of 3 ml. The figures have been included for you.

Stock (ml)	Water (ml)	Total volume (ml)	Concentration of protein (mg/ml)	Amount of protein (mg)
3	0	3	1.5	4.5
2.5	0.5	3	1.25	3.75
2	1	3	1.00	3
1.5	1.5	3	0.75	2.25
1	2	3	0.50	1.5
0.5	2.5	3	0.25	0.75
0	3	3	0	0

Firstly working out a protein concentration in mg/ml (second column from the right of the table), each is calculated by taking a dilution factor, which depends upon the final volume (always 3 ml). So, for example, 2 ml of stock into a final volume of 3 ml is a dilution of 2/3, so $2/3 \times 1.5 = 1.00$ mg/ml, and 0.5 ml stock into a final volume of 3 ml is a dilution of 0.5/3, so $0.5/3 \times 1.5 = 0.25$ mg/ml, and so on. The final figure is a concentration, amount per volume.

Very often we wish to know only the total amount of protein present, rather than its concentration. This is given in the last column of the table. This is calculated simply by multiplying the volume of stock used (in ml) by the amount of protein present in 1 ml (= 1.5 mg). The final figure for the amount of protein is independent of the final volume of liquid; for example, in this case it is 3 ml but if it were 30 ml it would not change the amount of protein present in each case. **It would, however, change the concentration (mg/ml) of protein present.** For example, in the third row of the table, 2 ml stock into a final volume of 30 ml would be a dilution of 2/30, so $2/30 \times 1.5 = 0.10$ mg/ml, 10 times less than the original, which is not surprising considering the volume has been increased 10 times!

Always be sure you know what you are dealing with, or what you are trying to calculate; i.e. is it an amount (grams, micrograms, moles) or is it a concentration (g/ml, μg/ml, molar)?

E. In a microbiology lab students perform a three-step 1:100 serial dilution of a bacterial culture (see diagram below). The initial step combines 1 unit volume culture (10 μl) with 99 unit volumes of broth (990 μl) = 1:100 dilution. In the next step, one unit volume of the 1:100 dilution is combined with 99 unit volumes of broth, now yielding a total dilution of $1{:}100 \times 100 = 1{:}10\,000$ dilution. Repeated again (the third step) the total dilution would be $1{:}100 \times 10\,000 = 1{:}1\,000\,000$ total dilution. The concentration of bacteria is now one million times less than in the original sample.

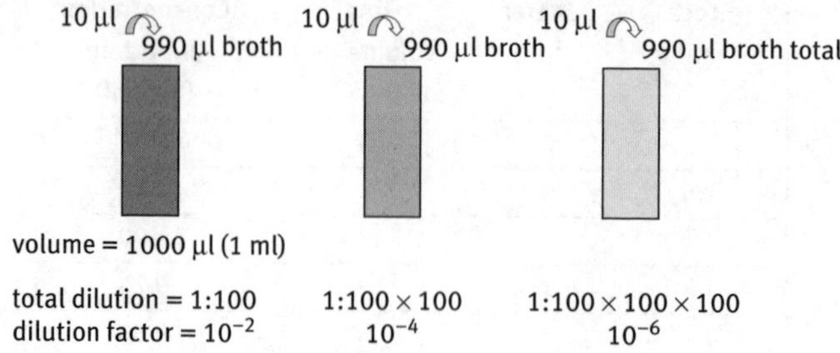

5 Carbon – the basis of biological life

Basic concepts:
The element carbon is a central building block in life forms. Here we relate the electronic structure of carbon to its special properties, which are evident in the types of covalent compounds that it forms. Such properties determine the types and shapes of biological molecules. Carbon forms the skeleton and backbone of biomolecules.

All biological molecules contain carbon; indeed, biological life is based upon carbon. To understand why and how carbon fulfils this central role, we need to look at its bonding behaviour.

5.1 The electronic structure of carbon

Carbon atoms have six electrons. Two of them will be found in the 1s atomic orbital close to the nucleus. The next two will go into the 2s orbital. The remainder will be in two separate 2p orbitals. This is because the p orbitals all have the same energy and the electrons prefer to be on their own if possible (Fig. 37). The electronic structure of carbon is normally written $1s^22s^22p^2$.

This configuration suggests that carbon has **two unpaired** electrons in its outer p orbitals that could participate in the formation of two covalent bonds. However, we know from observation that this is not so. The molecule CH_2 does not exist. The simplest compound of carbon is methane, CH_4, which has four equivalent covalent carbon–hydrogen bonds.

Figure 37. Outer electronic configuration of carbon

2s	$2p_x$	$2p_y$	$2p_z$
↑↓	↑	↑	

5.2 Hybridisation

To explain how it is possible for carbon to form four equivalent covalent bonds we use a modified theory of bonding called **orbital hybridisation**. **Orbital hybridisation** proposes that, prior to the formation of covalent bonds, carbon undergoes a change to its electron configuration (Fig. 38).

An electron from the 2s atomic orbital is elevated within energy level 2 to the empty 2p atomic orbitals (this requires a small input of energy). The four orbitals mix or hybridise to produce four equivalent hybrid orbitals. The electrons redistribute so that each of the now equivalent four hybrid atomic orbitals contains one electron. The new hybrid orbitals are called sp^3 orbitals. This then provides carbon with a maximum of four unpaired electrons that could form covalent bonds. In methane each half-filled sp^3 orbital overlaps with a hydrogen 1s orbital (containing one electron) to form a single covalent C-H bond.

Thus in methane, the carbon is said to be **sp^3 hybridised**; in other words, all four of the resulting hybrid orbitals are used in bonding. These have resulted from a 2s orbital and three 2p orbitals, hence sp^3.

Figure 38. Hybridisation in carbon

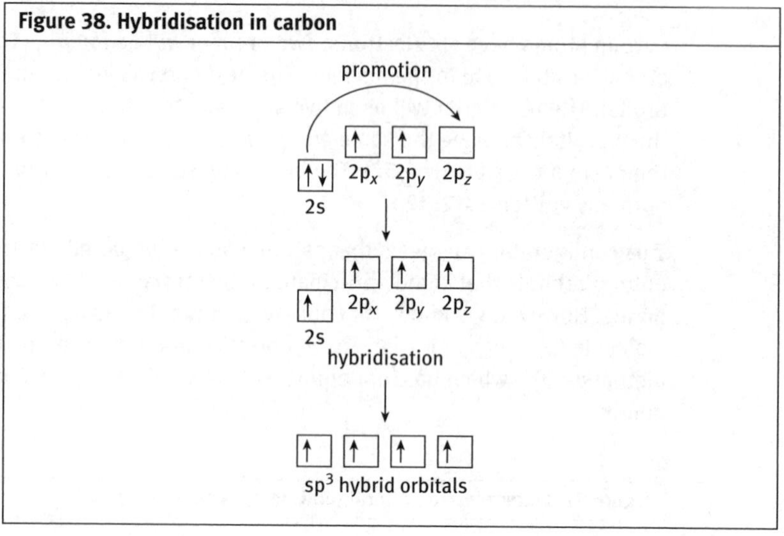

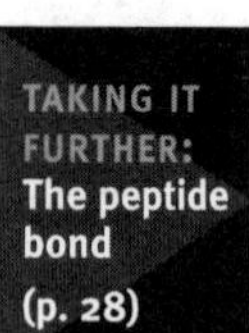
TAKING IT FURTHER: The peptide bond (p. 28)

Reminder

Orbital hybridisation occurs when atomic orbitals in an energy level mix, or hybridise, to form hybrid orbitals of the same energy

5.3 The tetravalency of carbon

When covalent bonds are formed, energy is released. The lower the energy content of a molecule, the more stable it is. Atoms will always attempt to make as many covalent bonds as possible to attain greater stability. Carbon is particularly good at this since it can form four covalent bonds. Carbon is said to be **tetravalent**. The tetravalency of carbon is central to its versatility as a biological building block. Carbon–carbon covalent bonds form readily, giving rise to both ring structures and chain compounds.

Every carbon atom in these rings has made four covalent bonds.

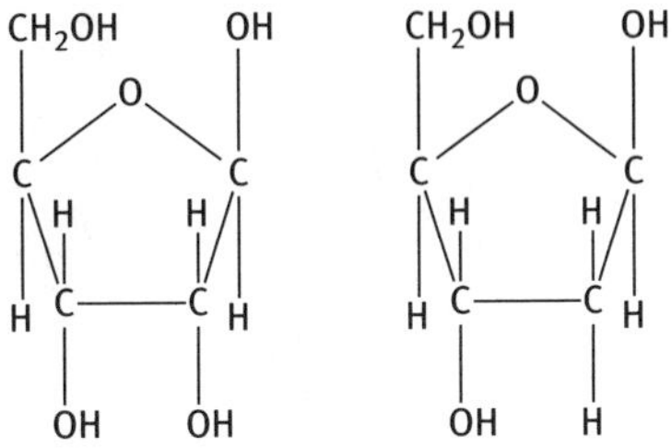

ribose

deoxyribose

The presence of carbon in ring structures is generally taken for granted (as indeed is the presence of hydrogen atoms attached to the carbon atoms), and so we would show deoxyribose, for example, in shorthand as

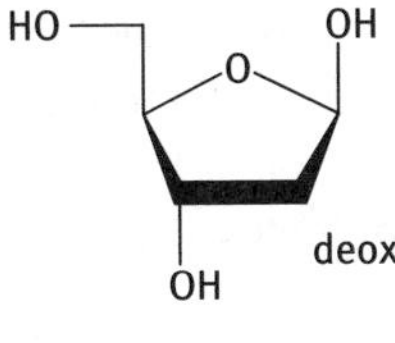

deoxyribose

Only the functional groups are shown, in this case the hydroxyl groups.

Long-chain structures are possible, such as those in the fatty acids. Again, each carbon in this palmitic acid chain is tetravalent.

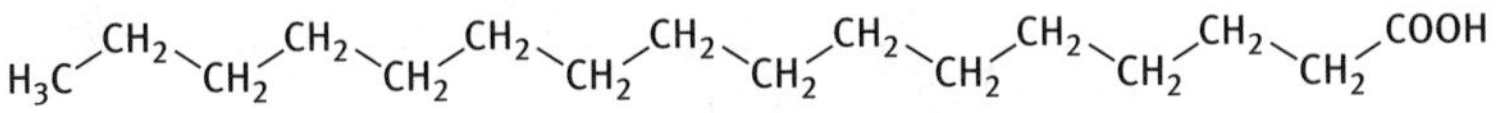

$H_3C-CH_2-CH_2-CH_2-CH_2-CH_2-CH_2-CH_2-CH_2-CH_2-CH_2-CH_2-CH_2-CH_2-CH_2-COOH$

palmitic acid

5.4 Shapes of molecules

When we draw structures such as those above, we are confined to a two-dimensional medium. Of course, many molecules are not flat, but rather three-dimensional. The shape of biological molecules is heavily dependent on the tetravalency of carbon. In an sp^3 hybridised carbon atom, the four hybrid atomic orbitals arrange themselves to be as far apart from each other as possible, each pointing to the corners of a regular tetrahedron. The shape of methane (Fig. 39) is therefore a tetrahedral structure.

Figure 39. Methane has a tetrahedral structure

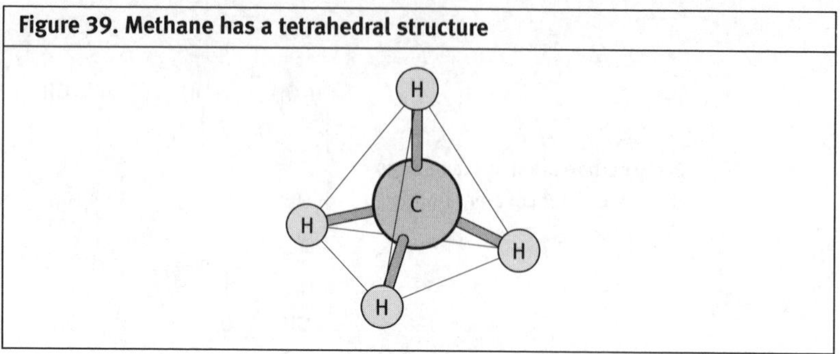

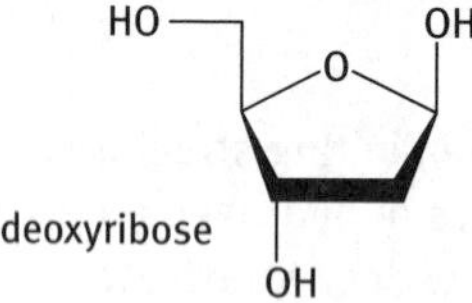

In this shorthand depiction of deoxyribose, the lower half of the ring is shown as a bold line; we use this to indicate that this part of the ring is coming out of the page. In other words, the ring is not flat; it has a three-dimensional shape. That shape is a consequence of the tetravalency of carbon.

In a similar way, we use different depictions to show the directions of covalently bonded groups in a molecule.

A normal bond is in the plane of the page; a wedge bond is coming out of the plane towards us, a hatched bond going into the plane away from us.

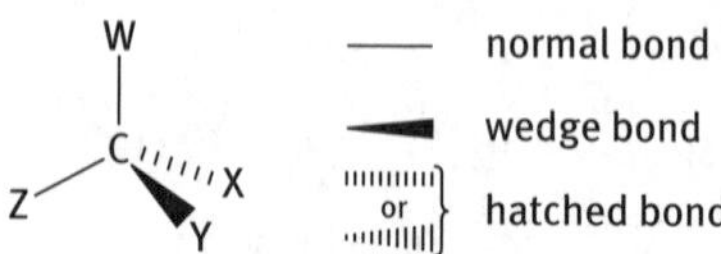

Ring structures in which carbon is bonded to four other atoms are not flat, but rather adopt 'chair' or 'boat' conformations. This is because the geometry at each sp^3 hybridised carbon atom is tetrahedral.

Reminder

sp^3 hybridisation means that carbon can form four covalent bonds

For glucose the chair structure is the most stable; in the boat form there is steric crowding, with the hydroxyl groups being pushed together.

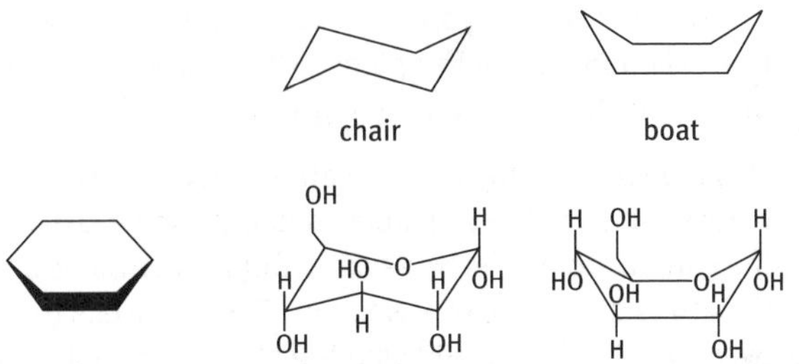

chair boat

glucose ring structure

With larger molecules, many shapes or conformations are possible. Whereas carbon makes this possible, the ultimate three-dimensional shape of such large molecules is determined by the variety of **intermolecular interactions** between **functional groups** on the molecule, which act to constrain and hold the molecule in a particular three-dimensional conformation. The structural and functional properties of large biological macromolecules are determined by their three-dimensional shapes.

Molecular shape is the determining factor in binding and recognition events in biology. The active site of an enzyme must be able to accommodate the shape of its substrate, which in turn imparts a high degree of specificity for the interaction. Similarly, membrane receptors will recognise a particular hormone, or membrane transporters will move only a particular molecule. With shape comes specificity, and with specificity comes recognition, control and order; the hallmarks of a biological system.

5.5 Carbon in chains and rings – delocalisation of electrons

In those carbon ring structures shown above, each carbon atom has adopted an sp^3 hybridisation and forms four covalent bonds. However, carbon is capable of other types of hybridisation.

In compounds of carbon which contain double bonds, sp^2 hybridisation takes place. The sp^2 orbitals are formed in the same way as sp^3 orbitals but only two of the p orbitals are hybridised, leaving one unchanged p orbital.

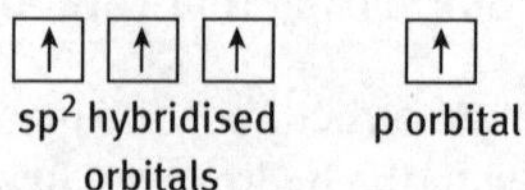

The three new sp^2 hybrid orbitals arrange themselves as far apart as possible and lie in a flat plane, 120° apart. The unchanged p orbital is perpendicular to the plane of the sp^2 orbitals.

This type of bonding occurs in ethene, C_2H_4, which possesses a double bond between the two carbon atoms and four single C-H bonds (Fig. 40). Each of the carbon atoms has three sp^2 hybridised orbitals and one p orbital. Each of the orbitals contains one electron. One of the electrons in the sp^2 orbitals overlaps ('head-to-head') with the electron from the sp^2 orbital on the second carbon atom to form a carbon–carbon sigma bond. The other two sp^2 orbitals on each carbon atom overlap with hydrogen 1s orbitals to give single C-H bonds. All six atoms are in the same plane; ethene is a planar (flat) molecule. The unchanged 2p orbitals on each C atom overlap sideways on to give a pi bond between the atoms. The pi bond is weaker than the sigma bond because only a small degree of overlap is possible.

Figure 40. sp^2 hybridisation and bonding in ethene

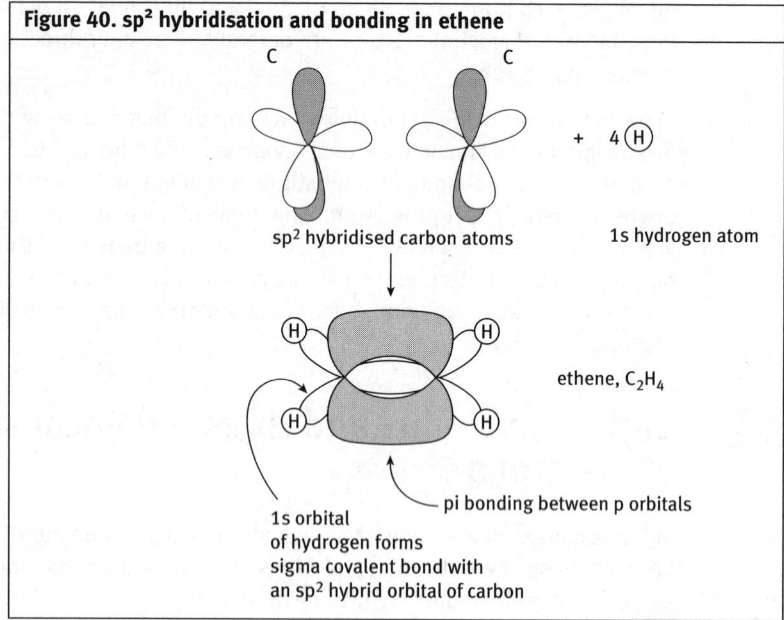

The classic example of sp^2 hybridisation in carbon is shown in the molecule benzene, C_6H_6 (Fig. 41).

Figure 41. sp^2 hybridisation and delocalisation of electrons in benzene

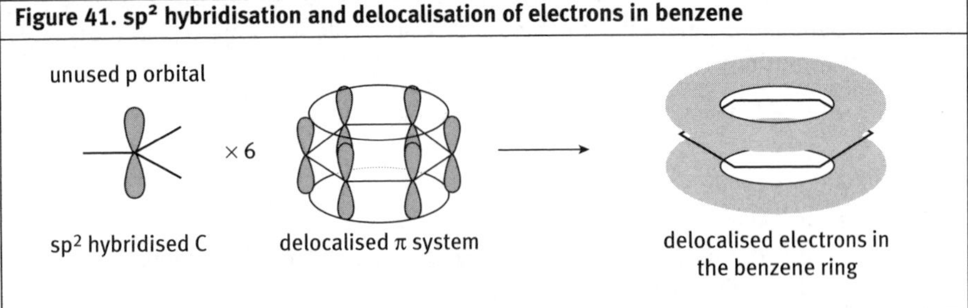

The benzene molecule consists of a ring of six carbon atoms, each attached to one hydrogen atom. Each carbon atom in benzene is sp^2 hybridised. The carbon uses its sp^2 orbitals to form three sigma bonds, one to each of the neighbouring carbon atoms and one to a hydrogen atom. The six carbon atoms and six hydrogen atoms are in the same plane. The six p orbitals, which are above and below the plane, are close enough to each other to overlap and form three pi bonds. This arrangement results in a continuous circular overlap of the six p orbitals. The electrons from each p orbital are free to move within this circular overlap region and are said to be **delocalised**.

Reminder

In sp^2 hybridisation, carbon forms three equivalent hybrid orbitals, but one p orbital remains unchanged. This unchanged p orbital is frequently able to participate in pi bonding

The shorthand structures shown below are all correct ways of depicting the benzene molecule.

5.6 Aromaticity

With electron delocalisation comes increased stability; benzene is a relatively unreactive compound, and the ring conformation is planar. Benzene is an aromatic compound. A molecule is aromatic, i.e. it displays **aromaticity**, if

- it is fully **conjugated** (i.e. there is a 2p orbital on every atom in the ring)
- it is cyclic
- it is planar
- it contains 6, 10, 14 (i.e. $4n + 2$, where n is any integer) etc. delocalised electrons (Huckel's rule).

Ring structures (and chain structures) may be conjugated without being aromatic. A conjugated system may be represented as a system of alternating single and double bonds.

$$-C=C-C=C-C=C-$$

In such systems conjugation is the interaction of one p orbital with another across an intervening sigma bond. The ring structure (a porphyrin ring) in the molecule haem (Fig. 42), is a conjugated ring structure; there is a system of alternating single and double carbon bonds around the full extent of the ring.

Figure 42. A conjugated porphyrin ring

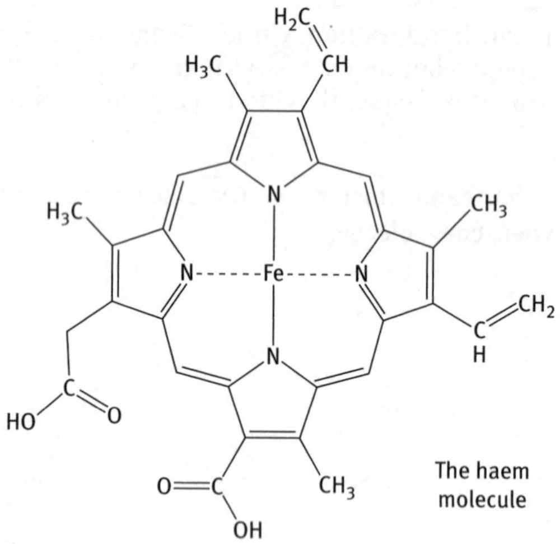

5.7 Summing up

1. Prior to forming covalent bonds, carbon undergoes hybridisation, in which 2s atomic orbital electrons are promoted to the same energy level as the 2p atomic orbital electrons.
2. sp^3 hybridisation involves the formation of four hybrid atomic orbitals, each containing one unpaired electron; carbon is therefore capable of forming four covalent bonds.
3. Carbon readily forms covalent bonds with itself, leading to a variety of possible structures, including ring and chain structures.
4. The tetravalency of carbon is important in determining the shapes of biological molecules.

5.8 Test yourself

The answers are given on p. 160.

Question 5.1
Explain the difference between a carbon atom which is sp^3 hybridised and one which is sp^2 hybridised.

Question 5.2
How would you explain the three-dimensional shape of the methane (CH_4) molecule?

Question 5.3
Which of the following hydrocarbons has a carbon–carbon double bond in its structure?
(a) C_3H_8
(b) C_2H_6
(c) C_2H_4
(d) CH_4

Question 5.4
In which of these ring structures would you find sp^2 hybridised carbon atoms?

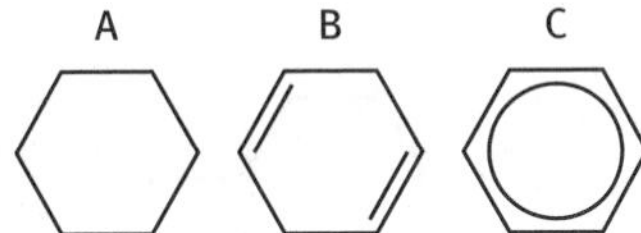

Taking it further

5.9 Carbon structures

The tetravalency of carbon and its different hybridisation states make a variety of shapes of molecules possible. Carbon readily bonds to other elements but also has a propensity to bond to itself, forming ring structures and chain structures. There are three known pure carbon structures, each of which has some remarkable properties. **Graphite** and **diamond** occur naturally; **fullerene** structures were only discovered in 1985. The different forms of the same element are called **allotropes**.

Structure of graphite

Naturally occurring graphite is called beta-graphite and it comes in a hexagonal form (Fig. 43). The hexagonal form of graphite has carbon atoms arranged in a hexagon; the hexagons form a plane. Each carbon atom in the hexagon is attached to three other carbon atoms. Carbon in graphite shows sp^2 hybridisation. The unhybridised p orbitals all lie parallel to each other and perpendicular to the graphite plane, generating a sea of delocalised pi electrons across this planar ring structure. The sheet-like structure with weak interlayer forces makes graphite a very soft substance; the layers can slide easily across each other and graphite is used as a lubricant.

Figure 43. Hexagonal graphite structure

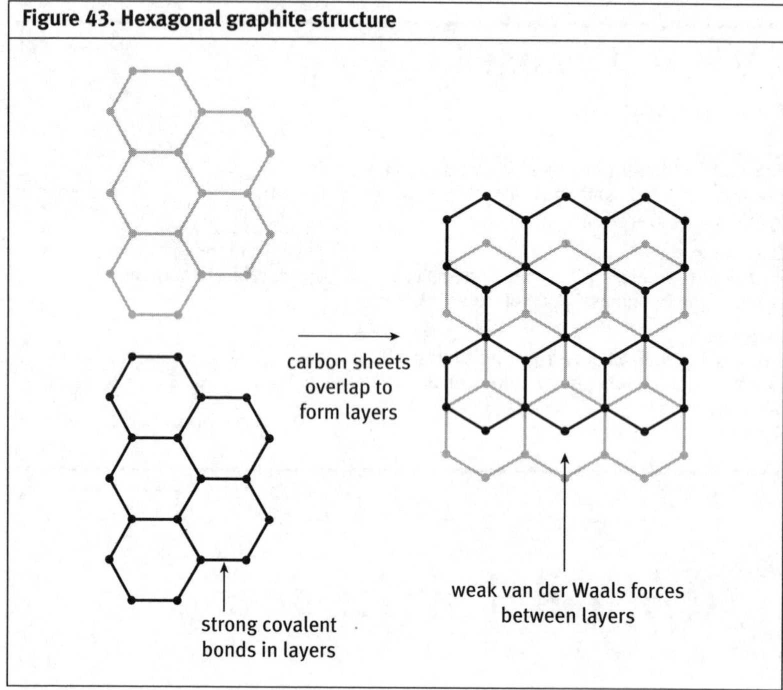

Structure of diamond

In diamond, each carbon atom adopts a sp^3 hybridisation; each carbon in the diamond structure is therefore at the centre of a tetrahedron, and diamond itself is a crystalline lattice (Fig. 44).

Figure 44. Diamond structure

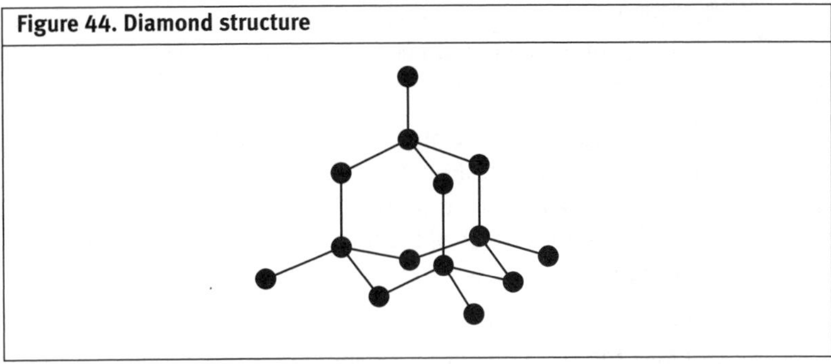

Each carbon atom is covalently bonded to four other carbon atoms. The sigma bonds formed have maximum electronic overlap with each other. The structure formed is therefore very rigid. This makes diamond one of the hardest naturally occurring substances.

Structure of fullerenes

The simplest type of fullerene is made up of 60 carbon atoms arranged in a series of interlocking hexagons and pentagons, forming a structure that looks like a soccer ball, and often referred to as 'buckyballs' by the 1996 Nobel Laureates Curl, Kroto and Smalley, who first discovered them (Fig. 45).

Figure 45. Simple fullerene structures

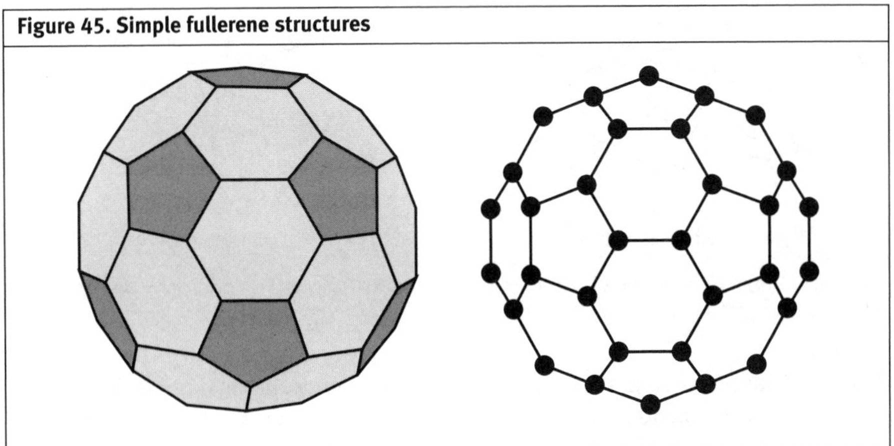

These are the only known molecules, composed of a single element, to form a hollow spheroid structure.

The pentagonal ring structures in the fullerenes, which are absent in graphite and diamond, provide the means for generating these curved structures.

Fullerene structures are providing some exciting ideas in biology. The hollow nature of this structure offers the potential for filling it, and perhaps using it as a novel drug-delivery system. Drugs could also be attached to the surface of the 'buckyball'; the spheroid structure might be expected to gain ease of entry to active sites of enzymes!

A 60-carbon fullerene is about one nanometre (10^{-9} m) in diameter, roughly the size of many small pharmaceutical molecules (in comparison, a human hair is as wide as 50 000 buckyballs). The fullerenes' unique structures might also be used as a scaffolding for building drug molecules.

A significant spin-off product of fullerene research are **nanotubes**, based on carbon or other elements (Fig. 46). These systems consist of graphitic layers seamlessly wrapped to form cylinders. They are only a few nanometres in diameter, but up to a millimetre long. Among the highlights of nanotube research to date is the demonstration that tubes can be opened and filled with a variety of materials, including biological molecules.

Figure 46. Carbon nanotubes.

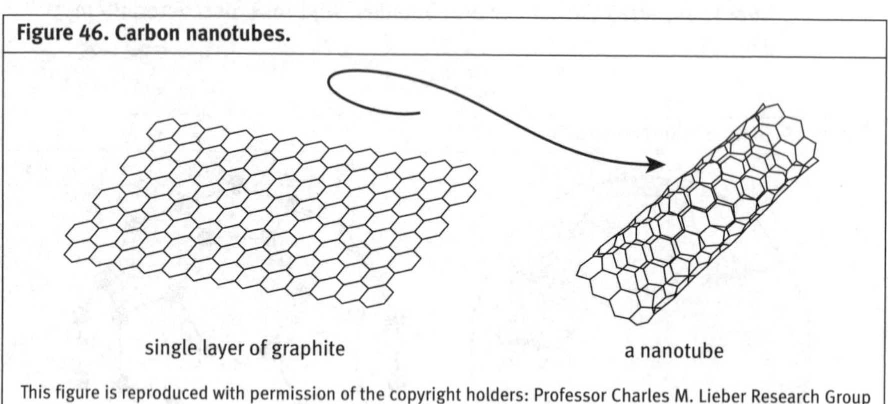

Nanoscience is a relatively new area of science, promising exciting new developments. Nanoscience is developing new power devices, sensors, new means of data storage, molecular electronics and nanotube gears to drive nanomachines! Just a taste of things to come!

6 The same molecule but a different shape

Basic concepts:
Molecules with the same chemical formula, but different structure and characteristics, are isomers. The tetravalency of carbon predicts the existence of mirror image isomers, so-called stereoisomers. Organisms in general impose a 'chiral environment'; for example we only use amino acids in their L-form, or sugars in their D-form. Enzymes will usually only recognise one type of isomer. The introduction of a 'wrong' isomer into the body can have disastrous consequences.

6.1 Isomers

In chemistry we define isomers as ' two or more different compounds with the same chemical formula but different structure and characteristics'.

Isomers can take many different forms, but there are two main forms of isomerism, namely **structural isomerism** and **stereoisomerism**.

In **structural isomers**, the atoms and functional groups are joined together in different ways. For example, the molecular formula C_2H_6O is that of ethanol, but it is also that of dimethylether (CH_3OCH_3). Both have the same elemental composition but are not the same compound; they are structural isomers.

In **stereoisomers** the bond structure is the same, but the geometrical positioning of atoms and functional groups in space differs. This class includes **optical isomerism** where different isomers are mirror images of each other, and **geometric isomerism** where functional groups at the end of a chain can be twisted in different ways.

6.2 Optical isomerism

As we have previously seen, carbon is a tetravalent atom capable of forming four covalent bonds. Once formed these bonds point towards the centre of a tetrahedron.

The tetrahedral carbon predicted the existence of **mirror image isomers.** When carbon makes four single bonds with four different groups, non-superimposable mirror image molecules (**enantiomers**) exist (Fig. 47).

Figure 47. Non-superimposable mirror image structures

H H
Br C F F C Br
Cl Cl
mirror

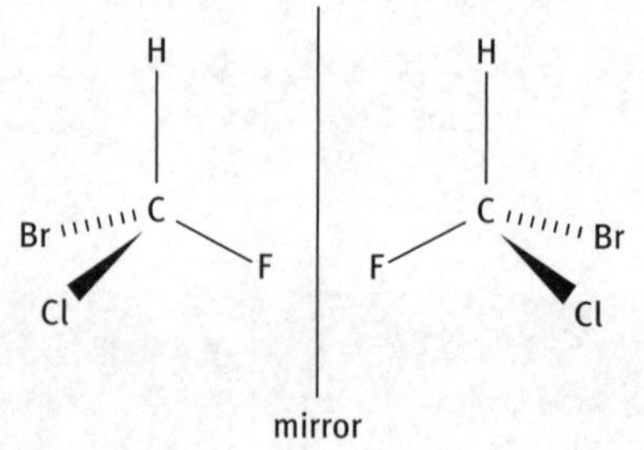

Derived from the Greek word *enantio* meaning opposite, enantiomers are non-superimposable mirror image structures.

Enantiomers are particularly important in biology. A molecule containing an **asymmetric**, or **chiral**, carbon can adopt two structures. An asymmetric carbon has four different atoms or groups attached to it. The amino acid structures shown in Fig. 48 are mirror images of each other. They have essentially identical chemical and physical properties. Physically they can be distinguished by their ability to rotate plane polarised light by equal amounts but in opposite directions (hence the D or L nomenclature). A solution of a D-isomer will rotate plane polarised light to the right (dextrorotatory = right-handed = clockwise); a solution of the L-isomer will rotate plane polarised light to the left (levorotatory = left-handed = counterclockwise). A solution of equal amounts of D- and L-isomers will not rotate plane polarised light; such a mixture is called a **racemic** mixture.

Figure 48. D- and L-alanine are enantiomers

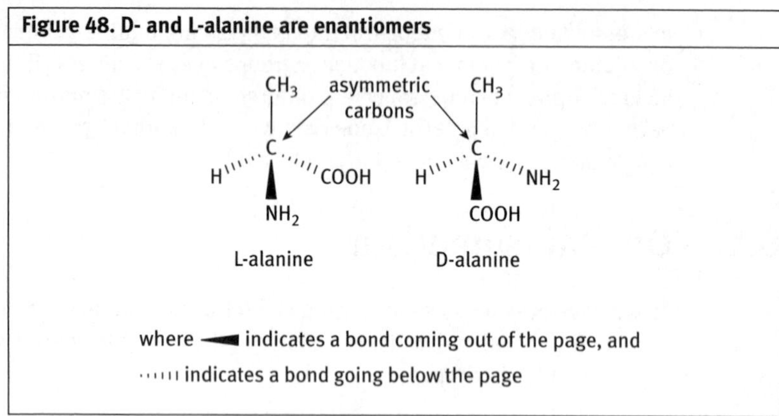

Sugars can also exist in D- and L-forms. The simplest three-carbon sugar is glyceraldehyde, shown in Fig. 49 in its D- and L-forms.

There is only one asymmetric (chiral) carbon in this structure. Looking at this molecule with the aldehyde group (-CHO) at the top and 'furthest away', if the OH group to the asymmetric carbon is on the right, then it is the D-form. If it is on the left it is the L-form.

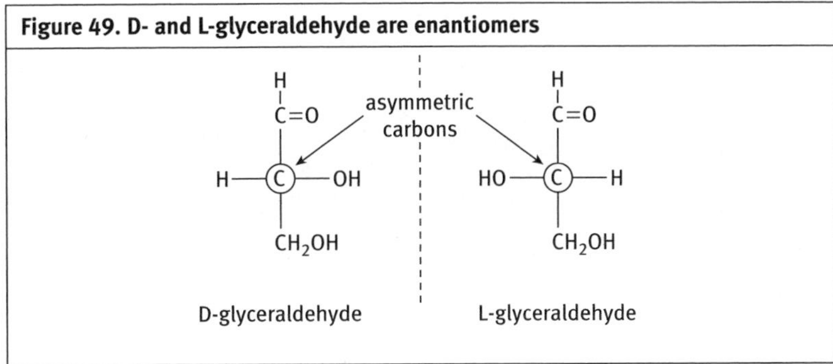

Figure 49. D- and L-glyceraldehyde are enantiomers

In more complex sugars the number of chiral carbons increases and consequently the number of possible isomers also increases. There are in fact 2^n stereoisomers possible for each chiral centre (n) in the molecule.

D-glucose and L-glucose are enantiomers, they are mirror images of each other. In glucose there are four chiral centres (carbons 2, 3, 4 and 5 in Fig. 50). In determining whether the structure is the D- or L-form, only the chiral carbon 'furthest' from the aldehyde group is considered (carbon 5 in Fig. 50). In D-glucose, the position of each group at each of the chiral carbons is 'reversed' relative to L-glucose; D- and L-glucose are therefore mirror images of each other.

Figure 50. D- and L-glucose are enantiomers

Enantiomers

stereoisomers that are mirror images of each other

D-glucose	L-glucose
^{1}CHO	^{1}CHO
H–2–OH	HO–2–H
HO–3–H	H–3–OH
H–4–OH	HO–4–H
H–5–OH	HO–5–H
$^6CH_2OH$	$^6CH_2OH$

On the other hand, D-glucose and D-galactose are referred to as **diastereomers**; they are stereoisomers that are not mirror images of each other. In Fig. 51, the determination of D- or L-forms is still made by considering the position of the OH group at the chiral carbon furthest from carbon atom 1, thus both the glucose and galactose structures shown are D-forms (the OH group is on the right at carbon atom 5). However, D-glucose and D-galactose are clearly not mirror images of each other.

Figure 51. D-glucose and D-galactose are diastereomers

Diastereomers stereoisomers that are not mirror images of each other

D-glucose	D-galactose
^{1}CHO	^{1}CHO
H–2–OH	H–2–OH
HO–3–H	HO–3–H
H–4–OH	HO–4–H
H–5–OH	H–5–OH
$^{6}CH_2OH$	$^{6}CH_2OH$

Considering the ring structure of glucose, another type of stereoisomer becomes apparent, namely **anomers**. Here the difference resides in the configuration of groups around one carbon atom, in this case carbon atom 1. In solution, glucose undergoes a process known as **mutarotation**. The open 'chair' form of glucose is in equilibrium with the ring structure, through the formation of a hemiacetal bond (Fig. 52).

Figure 52. Mutarotation in glucose

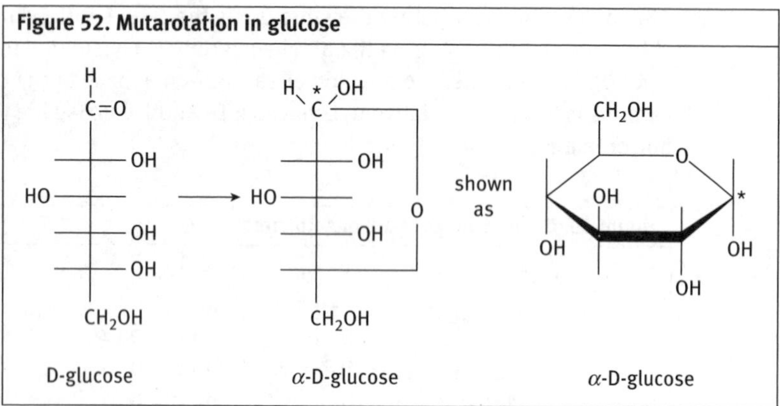

In Fig. 52, D-glucose cyclises to form α-D-glucose, in which the hydroxyl group at the anomeric carbon (shown by *) is shown pointing down. If the hydroxyl group were shown pointing up, then this would be β-D-glucose. Thus the ring form of glucose has one additional chiral centre than does the open chain form.

Furthermore, the alpha and beta ring forms of glucose will equilibrate in solution through the open chain form (Fig. 53).

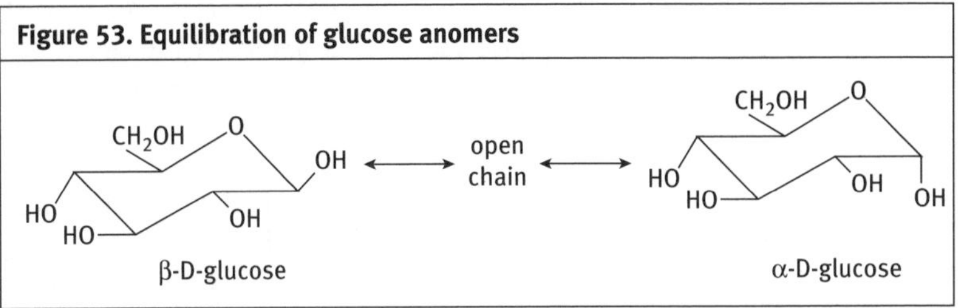

Figure 53. Equilibration of glucose anomers

In solution and at equilibrium, the beta ring form predominates (the hydroxyl groups are spaced further apart in the beta form, thus avoiding steric crowding).

6.3 Geometric isomerism

There are different types of geometric isomerism but a common one is that referred to as ***cis-trans* isomerism**. If a pair of stereoisomers contains a carbon–carbon double bond then it is possible to get a *cis* or *trans* arrangement of the substituents at each end of the double bond. These are referred to as ***cis-trans* isomers**. The simple example given here is that of butene.

CH_3 CH_3 / H H — *cis*-butene

CH_3 H / H CH_3 — *trans*-butene

With the methyl (CH_3) groups on the same side of the double bond, the *cis*-isomer is realised. The *trans*-isomer is realised when the methyl groups are on opposite sides of the double bond.

The carbon–carbon double bond effectively fixes the atoms or groups attached to it and prevents their rotation about the bond (remember, there is no rotation about a double bond).

In unsaturated fatty acids (i.e. those which contain a carbon–carbon double bond), orientation of groups about this bond may be either *cis* or *trans*.

cis-configuration *trans*-configuration

A *cis* configuration causes a 'kink' or bend in the carbon chain, whereas a *trans* configuration forms a straight chain (Fig. 54).

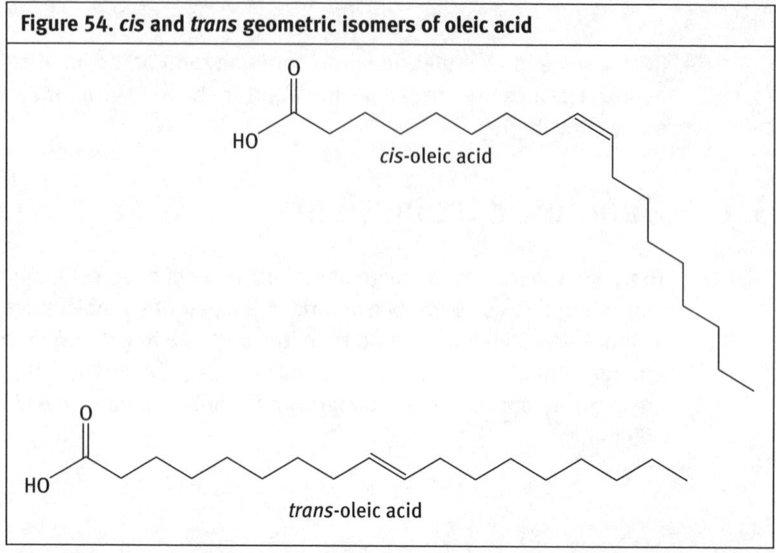

Figure 54. *cis* and *trans* geometric isomers of oleic acid

Naturally occurring unsaturated vegetable oils have almost all *cis* bonds, but using oil for frying causes some of the *cis* bonds to convert to *trans* bonds. If oil is constantly reused, more and more of the *cis* bonds are changed to *trans* until significant numbers of fatty acids with *trans* bonds build up. This is a health concern, since fatty acids with *trans* bonds have been shown to raise total blood cholesterol levels, thus increasing the risk of heart disease; they have also been shown to be carcinogenic, or cancer-causing.

There are, however, some striking examples in biology where the natural conversion between isomers is of central importance. In the mammalian eye, 11-*cis*-retinal (a form of vitamin A) forms part of the photoreceptor apparatus ('11' refers to the eleventh carbon atom in the chain). Exposure to light isomerises 11-*cis*-retinal to all-*trans*-retinal, in doing so triggering a series of reactions involved in the biochemical pathways of vision (Fig. 55).

Figure 55. The structure of 11-*cis*-retinal and all-*trans*-retinal

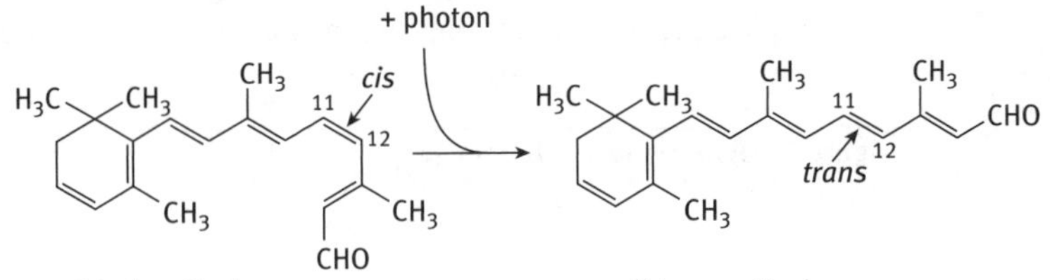

+ photon

11-*cis*-retinal

all-*trans*-retinal

6.4 Isomers as a problem

Life forms can distinguish isomers based on their different shapes. Normally, one isomer is biologically active and others are inactive. Our cells have imposed a **chiral environment**; in other words, only certain isomers are recognised and incorporated. We use only L-amino acids to construct our proteins, or D-sugars in carbohydrate metabolism. We favour *cis*-isomers in fatty acids and *trans*-peptide bonds in proteins. This molecular recognition resides at the level of the enzymes which orchestrate these processes. However, problems can arise when an inappropriate isomer is introduced into this environment.

In the 1960s, many pregnant women who had taken **thalidomide** gave birth to deformed babies. One of the enantiomers, isomer type 1, acted as a sedative as intended, but the other isomer, type 2, caused birth defects. Fig. 56 shows the two isomers of thalidomide which differ in the rotation of the ring group about the carbon marked *.

Figure 56. Two isomers of thalidomide

isomer type 2

isomer type 1

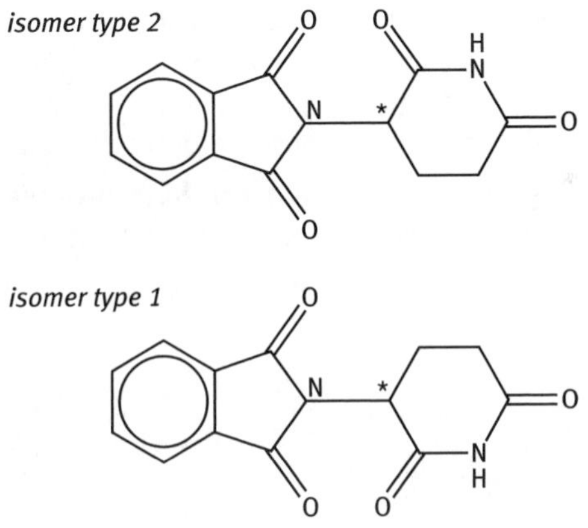

Sold over the counter in a number of pain remedies, **ibuprofen** is a mixture of two non-superimposable mirror image enantiomers (Fig. 57). Therapeutic activity is shown only when isomer type 2 is used.

Figure 57. Two isomers of ibuprofen

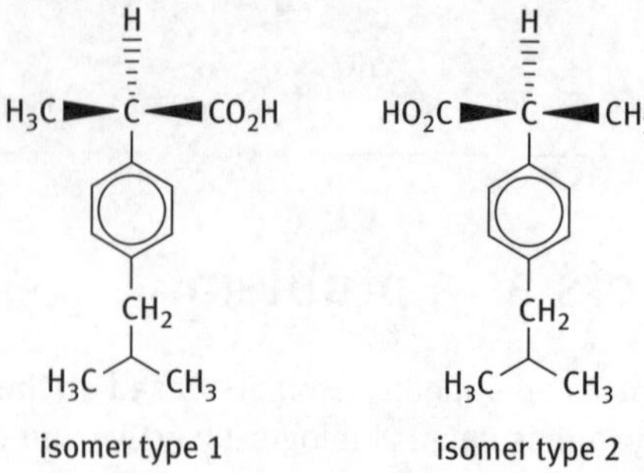

H
H_3C—C—CO_2H
CH_2
H_3C CH_3
isomer type 1

H
HO_2C—C—CH_3
CH_2
H_3C CH_3
isomer type 2

In today's pharmaceutical industry, legislation dictates that the production of new drugs must result in just one isomeric form to be marketed. Investment in new methods of chiral synthesis and methods of isomer separation have resulted in more potent and safer drugs.

6.5 Summing up

1. Isomers are defined as 'two or more different compounds with the same chemical formula but different structure and characteristics'.
2. By virtue of the tetravalency of carbon, stereoisomers (mirror images) exist for many biological molecules, including amino acids and sugars.
3. Molecules may contain one or more asymmetric (chiral) carbon atoms; the arrangement of groups around such atoms gives rise to different isomers, including enantiomers (non-superimposable mirror images), diastereomers (which are not mirror images), and anomers (which vary in their configuration of one group about one asymmetric carbon).
4. Biological life forms impose a chiral environment, in which only certain isomers are tolerated, e.g. L-amino acids and D-sugars.

6.6 Test yourself

The answers are given on p. 160.

Question 6.1
Choose the phrase which correctly describes the relationship between the molecules A and B:

(i) they are structural isomers
(ii) they are geometric isomers
(iii) they are enantiomers
(iv) they are isotopes

```
A     H               B  H   O
      |                   \ //
  H—C—OH                   C
      |                    |
     C=O               H—C—OH
      |                    |
  H—C—OH               H—C—OH
      |                    |
      H                    H
```

Question 6.2
What is the difference between pairs of molecules which are either enantiomers or diastereomers?

Question 6.3
What do you understand by a chiral centre?

Question 6.4
In a molecule with six chiral carbons, how many possible stereoisomers of that molecule could exist?

Question 6.5
How would you distinguish separate solutions of D- and L-glucose?

7 Water – the solvent of life

Basic concepts:
Water is the major component of living cells, comprising around 70% of a cell's total weight. The abundance of water in biological systems inevitably dictates the behaviour of the biological molecules it interacts with. Water may behave as both an acid and a base; the autoionisation of water is fundamental to an understanding of the acid–base behaviour of biological molecules and to pH control in biological systems.

In Chapter 3 we briefly introduced some very special properties of water, namely the ability of water molecules to undergo hydrogen bonding to one another. Hydrogen bonding between water molecules is a result of the charge distribution in the water molecule, imparted by the relatively high electronegativity of the oxygen atom compared to the two hydrogen atoms. The distribution of charge and the molecule's bent geometry give rise to a **dipole moment** in the water molecule (Fig. 58).

Figure 58. Dipoles and hydrogen bonding in water

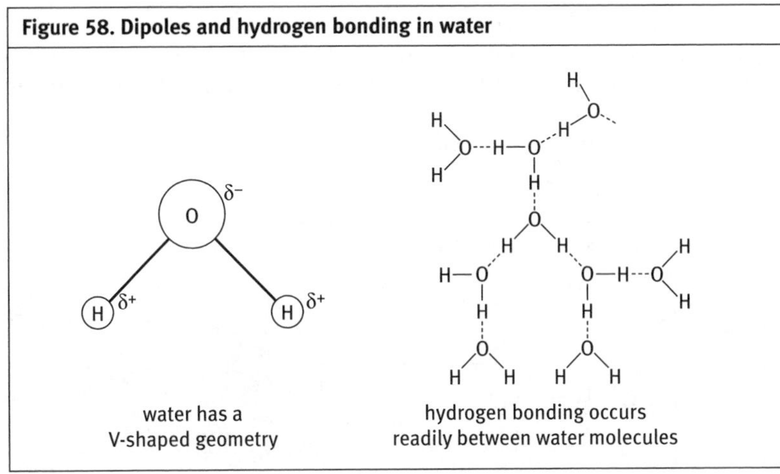

7.1 Bonding in the water molecule

Oxygen has an electronic configuration of $1s^2 2s^2 2p^4$. In forming covalent bonds with hydrogen, the oxygen atom undergoes sp^3 hybridisation. Ignoring the 1s electrons, the electron configuration of oxygen can be shown using the 'electrons in boxes' notation.

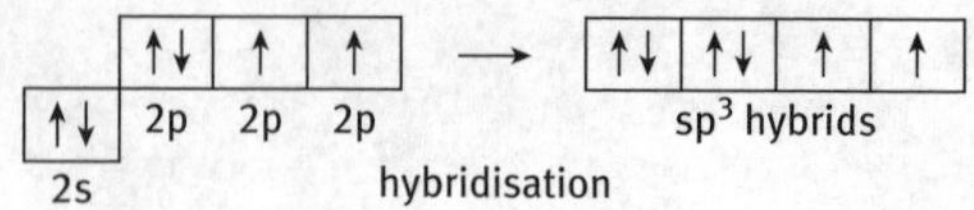

After hybridisation of the 2s and 2p orbitals, two unpaired electrons are available for bonding to hydrogen and two lone pairs of electrons remain in sp^3 hybridised orbitals on the oxygen atom (Fig. 59). The oxygen atom therefore forms two covalent bonds with hydrogen atoms by overlap of its half-filled sp^3 orbitals with the 1s orbitals of the hydrogen atoms. In fact, each water molecule is capable of forming a maximum of four hydrogen bonds to other water molecules. The two hydrogen atoms can each undergo hydrogen bonding with an oxygen atom of a neighbouring water molecule, and the oxygen atom (with two lone pairs of electrons) can hydrogen bond to hydrogen atoms on neighbouring water molecules.

Reminder

Hydrogen bonds can form whenever a strongly electronegative atom (O, N or F) approaches a hydrogen atom in a nearby molecule that is covalently attached to a second strongly electronegative atom

Figure 59. Electron distribution in the water molecule

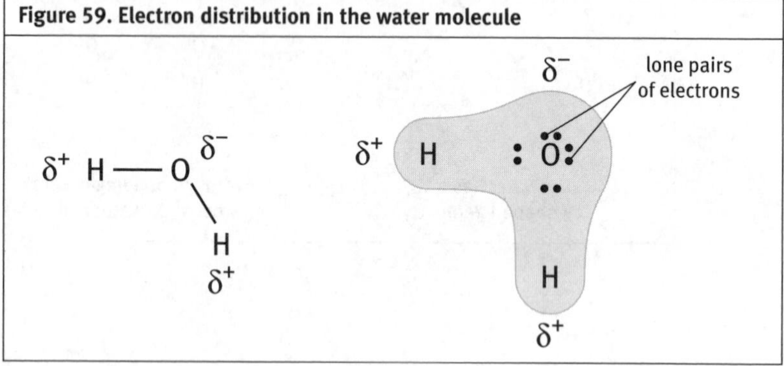

7.2 The dissociation (auto-ionisation) of water

Atoms within molecules are in constant motion, and one form of motion that atoms undergo is called vibration. Vibrational motion is the stretching and shortening of a bond as the atoms move closer together and then apart. Within liquid water, at any instant in time, the hydrogen atom of a hydrogen bond may find itself closer to the oxygen atom of a neighbouring water molecule than to the oxygen atom of its own molecule (Fig. 60a). The covalent O-H bond in the H_2O molecule stretches and the hydrogen bond between the molecules shortens (Fig. 60b). Eventually the covalent O-H bond stretches so far that it breaks and a new O-H bond is formed with the neighbouring molecule (Fig. 60c). The electrons of the hydrogen atom are left behind and the lone pair of electrons, formerly in the sp^3 orbital on the oxygen atom, is used in bonding to the hydrogen atom in a dative covalent bond. This process creates two new species, each of which has a full electric charge; a **hydroxonium ion**, H_3O^+, and a **hydroxide ion**, OH^-.

The process is known as the **dissociation** (or auto-ionisation) of water and scheme 1 represents the reaction (Fig. 61). The process is sometimes abbreviated, as in scheme 2. In pure water at room temperature only about one in a billion water molecules reacts in this way. The reactions in schemes 1 and 2 use the symbol $\rightleftharpoons$ to indicate that not all the molecules dissociate. The symbol $\rightleftharpoons$ is used for chemical reactions which have not gone to completion but reach an equilibrium. In reactions which reach equilibrium, only a certain number of reactant molecules (on the left-hand side of the equation) react and form products. However, when equilibrium is reached the forward and backward reactions occur at the same rate.

Figure 60. The dissociation of water

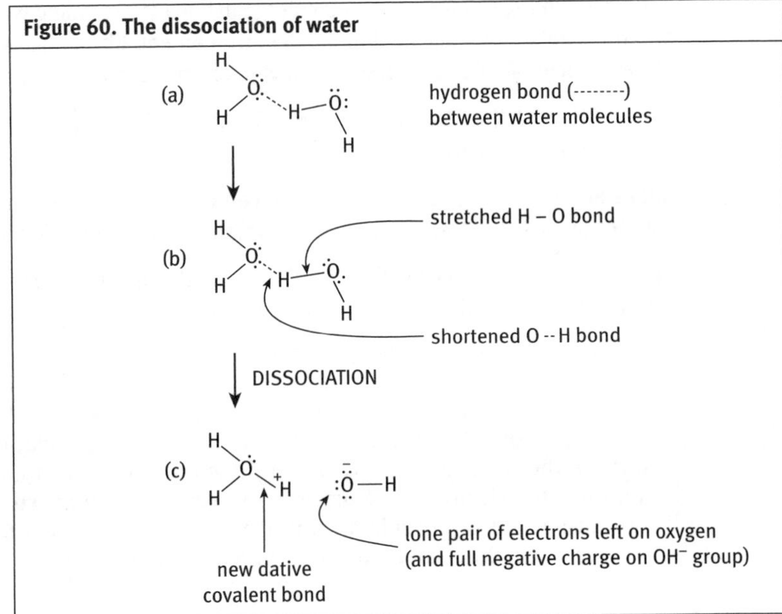

Figure 61. Equations which depict the auto-ionisaton of water	
Scheme 1	*Scheme 2*
$H_2O + H_2O \rightleftharpoons H_3O^+ + OH^-$	$H_2O \rightleftharpoons H^+ + OH^-$

7.3 Acids and bases

The behaviour of water provides the basis for understanding the concept of acids and bases.

An **acid** is defined as a substance which produces hydrogen ions (H^+) by dissociation.

For example, when hydrogen chloride, HCl(g), is added to water no HCl molecules are found, only H_3O^+ ions and Cl^- ions. Each HCl molecule reacts with water to form a hydroxonium ion and a chloride ion. The process can be represented by:

$$HCl(g) + H_2O(l) \longrightarrow H_3O^+(aq) + Cl^-(aq)$$

HCl is a strong acid because it dissociates completely in water. It is the acid which is secreted by the parietal glands in the stomach.

Reminder

The physical states of substances are denoted by (g) = gas, (aq) = aqueous, in water solution, (l) = liquid and (s) = solid

Not all acids dissociate completely in water. Most organic acids react only partially with water to produce low concentrations of hydroxonium ions, and leave undissociated acid molecules. Such acids are called **weak acids** and the symbol $\rightleftharpoons$ is used to indicate that an equilibrium is established. Ethanoic acid (acetic acid), the acid in vinegar, behaves in this way.

$$CH_3COOH(l) + H_2O(l) \rightleftharpoons CH_3COO^-(aq) + H_3O^+(aq)$$
(ethanoic acid)

Bases are defined as substances which can extract a proton, H^+. Frequently bases extract protons from water to leave hydroxide ions, OH^-.

Bases may produce hydroxide ions directly by dissociation, for example when potassium hydroxide is added to water:

$$KOH(s) + aq \longrightarrow K^+(aq) + OH^-(aq)$$

Reminder

Hydrogen ion concentrations are often discussed in relation to pH, despite the fact that free hydrogen ions do not exist in aqueous solution for all practical purposes. However, the hydroxonium ion will dissociate to provide a proton in situations where a proton is said to occur in theory, so the net effect is the same!

Bases which are completely dissociated in water, such as KOH, are called **strong bases.**

Other bases react with water to extract a proton and leave a hydroxide ion. For example ammonia, NH_3, is a base which reacts in this way. Because ammonia does not completely react with water and only produces a relatively small amount of hydroxide ions, ammonia is said to be a **weak base.**

$$NH_3(g) + H_2O(l) \rightleftharpoons NH_4^+(aq) + OH^-(aq)$$

Reminder

Acids produce H^+ ions by dissociation, bases extract a proton from water to generate OH^-, or produce OH^- directly by dissociation

7.4 Using pH as a measure of acidity

The acidity of a solution is measured by the molar concentration of hydroxonium ions. These concentrations can be very small, as in bases, or very large, as in strong acids. The range of hydroxonium ion concentrations is typically from 1 M to 1×10^{-14} M. In order to convert these very small numbers to numbers which are more convenient to use, the pH scale was devised. This is a logarithmic scale in which the pH of a solution is defined by:

$$\mathbf{pH = -\log_{10}[H^+]}$$

where $[H^+]$ is the hydrogen ion concentration in mol dm^{-3} or mol l^{-1}. The unit M can also be used to indicate mol dm^{-3} or mol l^{-1}.

Reminder

Square brackets [ ] denote 'concentration of'

$Log_{10}[H^+]$ means the logarithm to the base 10 of the hydrogen ion concentration. It is obtained on most calculators by using the 'log' function. From here on, log X will be used to indicate $\log_{10}X$.

Calculating the pH of a strong acid

0.1 M hydrochloric acid has a H^+ concentration of 0.1 M. The pH of this solution is given by: pH = –log[0.1]. The value of log [0.1] = –1.

Thus, pH = –(–1) = +1. The pH of this hydrochloric acid is therefore equal to 1

The pH scale can range from about 0 (strongly acidic) to +14 (very basic); a neutral solution has pH 7.0 (Fig. 62).

Figure 62. The pH scale

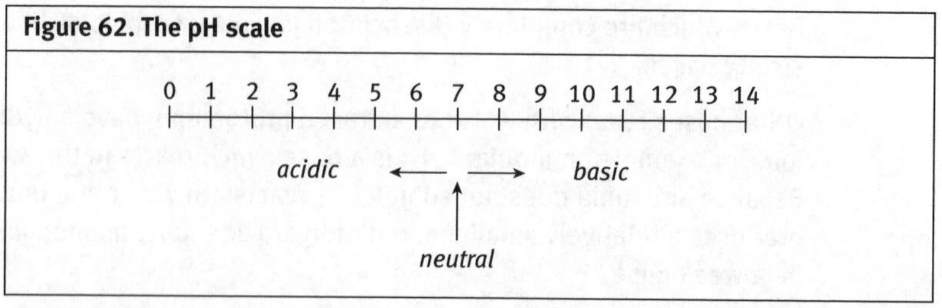

7.5 Calculating the pH of water

Water dissociates to a tiny extent. Hydrogen ions (H^+) associate with water molecules to form H_3O^+ (the hydroxonium ion).

$$H_2O + H_2O \rightleftharpoons H_3O^+ + OH^-$$

The **degree of dissociation** is measured by the equilibrium constant, which is the ratio of the concentration of the dissociated ions to the undissociated molecules. The equilibrium constant (K_{eq}) for this dissociation is defined as:

$$K_{eq} = \frac{[H^+]\,[OH^-]}{[H_2O]}$$

Because the concentration of water, $[H_2O]$, is a constant and simply equivalent to the density of water, this equation becomes:

$$K_w = [H^+]\,[OH^-]$$

where K_w represents the dissociation constant of water.

Reminder

One H_2O molecule dissociates to give one H^+ ion and one OH^- ion

The concentration of hydrogen ions (hydroxonium ions) in pure water must be equal to the concentration of hydroxide ions and is found experimentally to be equal to 1.0×10^{-7} M at 25°C.

Therefore we can write $[H^+] = [OH^-] = 1.0 \times 10^{-7}$ mol

Substituting these values in the expression for K_w we find:

$$K_w = [H^+]\,[OH^-] = 1.0 \times 10^{-7}\text{M} \times 1.0 \times 10^{-7}\text{M} = 1.0 \times 10^{-14}\ \text{M}^2$$

This tells us that very little of pure water exists in the form of its ions. It also tells us that because the value for K_w is a constant at a specific temperature, when the concentration of H_3O^+ decreases, the concentration of OH^- must increase by a similar amount, and *vice versa*.

Knowing the concentration of H_3O^+ in pure water we can calculate its pH from:

$$pH = -\log [H^+]$$

If $[H^+]$ is equal to 1.0×10^{-7} M

$$pH = -\log(1.0 \times 10^{-7}\ M) = -(-7) = 7$$

Thus, the pH of pure water is 7.

7.6 The dissociation of weak acids and weak bases in water

Weak acids are substances which dissociate only partially with water to give hydroxonium ions in solution. A weak base reacts only partially with H_2O to give hydroxide ions in solution.

As for water, the extent of dissociation of weak acids and bases can be expressed by a **dissociation constant.** The dissociation constant is the equilibrium constant for the decomposition of a species into its components.

The **acid dissociation constant, K_a,** is the equilibrium constant for the reaction in which a weak acid is in equilibrium with its conjugate base and the hydroxonium ion in aqueous solution. Similarly the **base dissociation constant, K_b,** is the equilibrium constant which represents the dissociation of a base into its conjugate acid and the hydroxide ion in aqueous solution. (Conjugate means 'joined together' especially in pairs. In chemistry, a conjugate species is related to its acid or base by the difference of a proton.)

The example below shows the dissociation of ethanoic (acetic) acid. Notice that in the equilibrium expression the concentration of water is not included. This is because water is vastly in excess and the value of $[H_2O]$ is a constant.

$$\underset{\text{(weak acid)}}{CH_3COOH(aq)} + H_2O(l) \rightleftharpoons \underset{\text{(conjugate base)}}{CH_3COO^-(aq)} + H_3O^+(aq)$$

The equilibrium constant for this dissociation is equivalent to the acid dissociation constant for ethanoic acid, K_a and can be expressed by:

$$K_a = \frac{[CH_3COO^-_{(aq)}]\,[H_3O^+_{(aq)}]}{[CH_3COOH_{(aq)}]}$$

For ethanoic acid, K_a is found to equal 1.8×10^{-5} M at 25°C, indicating this to be a weak acid (i.e. it does not dissociate appreciably in solution).

Therefore, the larger the value of K_a the stronger the acid. K_a values, like $[H^+]$ values, are often very small, and so the value is sometimes expressed as the logarithm of its reciprocal, or the negative logarithm, of the value. This is called the **pK_a**. Therefore,

$$\mathbf{p}K_a = -\log K_a$$

The smaller the value of pK_a the stronger the acid. For ethanoic acid, pK_a = $-\log(1.8 \times 10^{-5}\ \text{M}) = -(-4.74) = 4.74$.

Similarly the dissociation of a weak base, which results in the production of hydroxide ions, is represented by the base dissociation constant, K_b. Consider the reaction of ammonia with water:

$$\underset{\text{(weak base)}}{NH_3(aq)} + H_2O(l) \rightleftharpoons \underset{\text{(conjugate acid)}}{NH_4^+(aq)} + OH^-(aq)$$

The dissociation constant is:

$$K_b = \frac{[NH_4^+{}_{(aq)}]\,[OH^-{}_{(aq)}]}{[NH_{3(aq)}]}$$

The value of K_b for ammonia is 1.75×10^{-5} M.

Again **pK_b = –logK_b**.

So pK_b for ammonia = $-\log(1.75 \times 10^{-5}\ \text{M}) = -(-4.76) = 4.76$.

Reminder

The general case of a weak acid, HA, dissociating in water can be represented by:

$$HA(aq) \rightleftharpoons H^+(aq) + A^-(aq)$$

The acid dissociation constant, K_a, is given as:

$$K_a = \frac{[A^-]\,[H^+]}{[HA]}$$

7·7 Buffers and buffered solutions

The metabolic activity of cells results in the production of acids. For example, lactic acid is produced in muscle tissue. The function of proteins and enzymes requires conditions of constant pH. In particular, the pH of blood must be maintained within narrow limits and even a small variation can be fatal. Fortunately there are mechanisms in the body called **buffer systems** that act to minimise changes in pH.

A buffer solution is a solution which can resist changes in pH after the addition of hydroxonium or hydroxide ions. Buffer solutions consist of mixtures of weak acids or bases and their salts. For example carbonic acid, H_2CO_3, and sodium hydrogen carbonate, $NaHCO_3$; or ammonia, NH_3, and ammonium chloride, NH_4Cl. Such combinations of acids and bases and their salts are called **conjugate acid–base** pairs. Conjugate acid–base pairs are linked by the gain and loss of a proton (or hydoxonium ion).

For example, carbonic acid and the hydrogen carbonate ion, H_2CO_3/HCO_3^-

$$\underset{\text{(conjugate acid)}}{H_2CO_3(aq)} + \underset{\text{(base)}}{H_2O(l)} \rightleftharpoons \underset{\text{(conjugate base)}}{HCO_3^-(aq)} + \underset{\text{(acid)}}{H_3O^+(aq)}$$

ammonia and the ammonium ion, NH_3/NH_4^+

$$\underset{\text{(conjugate base)}}{NH_3(aq)} + \underset{\text{(acid)}}{H_2O(l)} \rightleftharpoons \underset{\text{(conjugate acid)}}{NH_4^+(aq)} + \underset{\text{(base)}}{OH^-(aq)}$$

Note that in each of these dissociations the water molecule and the hydroxonium ion formed are also a conjugate acid–base pair.

A buffer system is one that contains almost equal concentrations of conjugate acid and base.

An acidic buffer is one that will maintain a pH of less than 7. If a small amount of base in the form of the hydroxide ion (OH^-) is added to such a buffer, the conjugate acid will react with the hydroxide ions, thus preventing the rise in pH (to more basic conditions). If a small amount of acid in the form of hydroxonium ion (H_3O^+) is added to the buffer, the conjugate base will react with the H_3O^+ ions and prevent a fall in pH (to more acidic conditions). Such a buffer system can maintain an almost constant pH until either the conjugate acid or base is used up. Thus the higher the concentrations of conjugate acid and base, the greater the buffering capacity of the buffer.

TAKING IT FURTHER: Biological buffers (p. 97)

7.8 Calculating the pH of buffer systems using the Henderson–Hasselbalch equation

Every buffer system that is composed of a specific conjugate acid–base pair has a specific pH which it is able to maintain through its buffering action. Thus, different buffers will be used in different biological and chemical systems. The pH of a particular buffer system can be calculated using the **Henderson–Hasselbalch equation**.

This is derived in the following manner for a general conjugate acid–base pair, HA/A^-.

Consider the dissociation of the weak acid HA in water to its conjugate base, A^-:

$$HA(aq) + H_2O(l) \rightleftharpoons A^-(aq) + H_3O^+(aq)$$

The acid dissociation constant for the weak acid is defined as:

$$K_a = \frac{[A^-][H_3O^+]}{[HA]}$$

This can be written in the following way to show the ratio of conjugate base to acid at equilibrium:

$$K_a = [H_3O^+] \times \frac{[A^-]}{[HA]}$$

If we take logs to the base 10 of all the components of this equation in order to enable us to use the 'pK_a' notation we get:

$$\log K_a = \log [H_3O^+] + \log [A^-] - \log [HA]$$

and multiplying through by minus one (–1):

$$-\log K_a = -\log [H_3O^+] - \log [A^-] + \log [HA]$$

Substituting pK_a = –log K_a and pH = –log [H_3O^+] gives:

$$pK_a = pH - \log [A^-] + \log [HA]$$

which can be rearranged to:

$$pH = pK_a + \log \frac{[A^-]}{[HA]}$$

A more convenient way of writing this equation is:

$$\mathbf{pH = p}\boldsymbol{K}_\mathbf{a} \mathbf{+ \log \frac{[base]}{[acid]}} \text{ or } \mathbf{pH = p}\boldsymbol{K}_\mathbf{a} \mathbf{+ \log \frac{[proton\ acceptor]}{[proton\ donor]}}$$

This is known as the **Henderson-Hasselbalch equation.**

By using pK_a values, we are able to express the strength of an acid (i.e. its tendency to dissociate) with reference to the pH scale.

7.9 Life in water

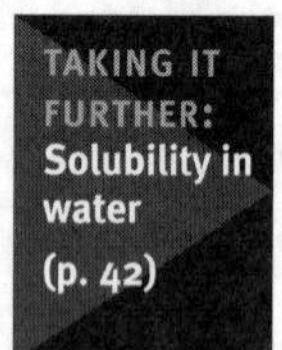
TAKING IT FURTHER: Solubility in water (p. 42)

The solubility of substances in water is determined by their structure. Solids which easily dissolve in water to form ions do so because the charged ions readily interact with the polar water molecules. Similarly, any organic molecule that possesses polar covalent bonds (e.g. the O-H bond in alcohols and carboxylic acids) is likely to be soluble in water.

Common functional groups on biological molecules include the hydroxyl group, the carboxyl group and the amino group.

The **hydroxyl group (OH)**, consists simply of a hydrogen atom bonded to an oxygen atom, which in turn is bonded to a carbon atom. The bond between the hydrogen and the oxygen is highly polar and thus this functional group strongly attracts water molecules, forming hydrogen bonds. The high solubility of sugars, for example, is due to the presence of hydroxyl groups. Organic molecules containing hydroxyl groups as the only functional group are called **alcohols**. Under normal biological conditions, the hydroxyl group does not dissociate.

The **carboxyl group (COOH)** consists of a carbon atom which is doubly bonded to an oxygen atom (as in the carbonyl group) and also bound to a hydroxyl group. Molecules containing a carboxyl group as the only functional group are called **carboxylic acids**, and the ethanoic acid found in vinegar is an example. The reason that the carboxyl group has acidic properties (is a proton donor) is that the bond between the oxygen and the hydrogen is so highly polar (more so due to the proximity of the carbon–oxygen double bond) that the hydrogen tends to dissociate from the molecules as a free H^+ ion (combining with water to form the hydroxonium ion) and produces a carboxylate ion (Fig. 63).

Figure 63. The dissociation of a carboxyl group

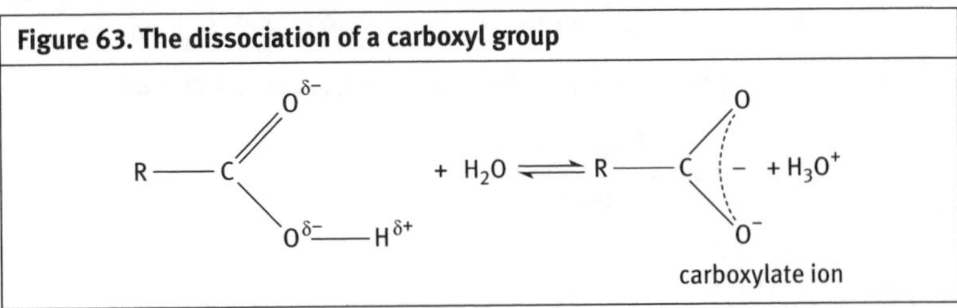

The dissociation constant K_a, for the carboxyl group, can be shown as:

$$K_a = \frac{[RCOO^-][H_3O^+]}{[RCOOH]}$$

The RCOOH acts as a weak acid and water as a base. The carboxylate ion formed can act as a base (accepting H^+) with water acting as an acid.

In forming the **amino group ($-NH_2$)**, nitrogen has atomic number 7 and therefore the electron configuration $1s^2 2s^2 2p^3$. When nitrogen hybridises it has three valence electrons (three hybrid sp^3 orbitals) and one lone pair of electrons.

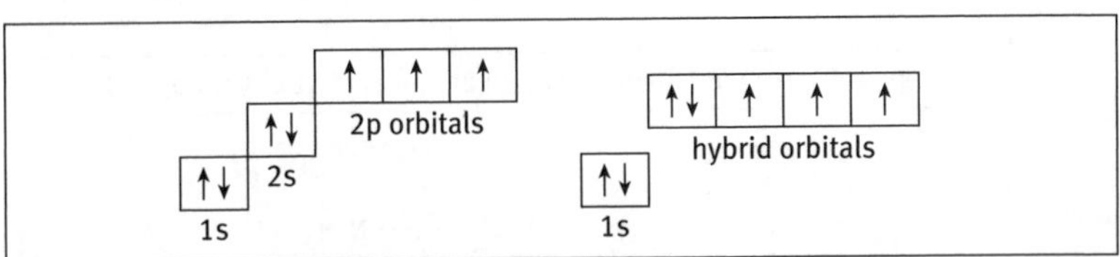

Thus, in the amino group the nitrogen atom possesses a lone pair of electrons. This can readily associate with a proton from water to form

$-NH_3^+$, which in turn can also dissociate (Fig. 64). The NH_2 group acts as a base and water as an acid. The $-NH_3^+$ group formed also acts as an acid by dissociating to lose a proton to water, which acts as a base in accepting the proton.

Figure 64. Protonation of the amino group

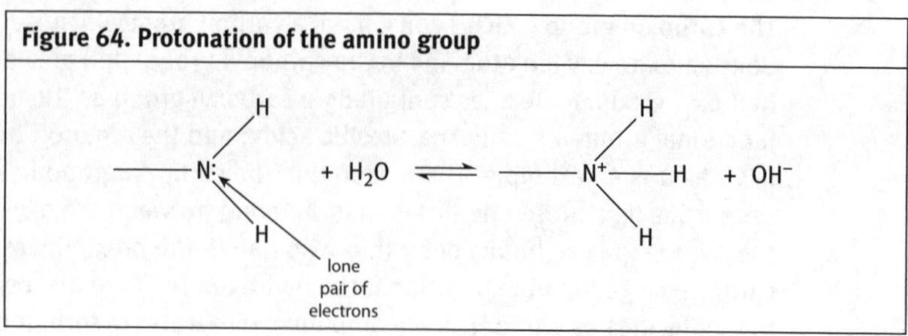

The dissociation of a proton from the protonated form of the amino group of an amino acid can be described in terms of its K_a value.

$$RNH_3^+ (aq) + H_2O (l) \rightleftharpoons RNH_2 (aq) + H_3O^+ (aq)$$

$$K_a = \frac{[RNH_2] [H_3O^+]}{[RNH_3^+]}$$

7.10 Amino acids

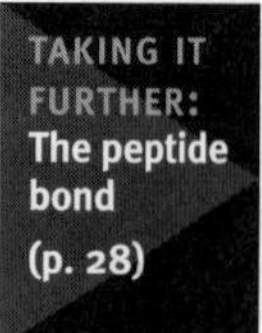

The general structure of an amino acid is shown in Fig. 65. All amino acids possess an amino group and a carboxyl group; these are both charged at physiological pH (pH 7). However, in proteins the amino acids are linked together through the formation of peptide bonds and so these charged groups are effectively 'lost'. However, the 'R' group is specific for each individual amino acid and includes groups which are polar, acidic, basic and apolar (hydrophobic). The R groups of amino acids, and therefore of proteins, are of central importance in determining the nature of intermolecular interactions within protein molecules (which determine their three-dimensional shapes) and between protein molecules and other molecules in the cell (such as between an enzyme and its substrate).

Figure 65. General structure of an amino acid at physiological pH

```
       COO⁻
        |
   H —— C —— N⁺H₃
        |
        O
```

Carboxyl and amino groups are commonly found in the side chains of amino acids. The pK_a values of these groups are such that at physiological pH 7, proteins will contain carboxylate groups (COO^-) and protonated amino groups (NH_3^+).

For example, glutamic acid possesses a carboxyl side group (R group). This has a pK_a of 4.07 and so at physiological pH will be dissociated and carry a negative charge.

glutamic acid

$$^{-}O{-}\overset{\overset{\large O}{\|}}{C}{-}CH_2CH_2{-}\overset{\overset{\large \overset{+}{N}H_3}{|}}{C}H{-}COO^-$$

R-group

Lysine possesses an amino side group (R group) with a pK_a of 10.53; this will be protonated at physiological pH and carry a positive charge.

lysine

$$H_3\overset{+}{N}CH_2CH_2CH_2CH_2{-}\overset{\overset{\large \overset{+}{N}H_3}{|}}{C}H{-}COO^-$$

R group

Reminder

How to 'read' pK_a. At a pH below the pK_a of a dissociable group, that group will be predominantly protonated. Thus, at physiological pH 7, the carboxyl group will be dissociated (COO^-), and the amino group will be protonated (NH_3^+)

7.11 Controlling cellular pH

The physiological pH is around neutrality, about pH 7.4. Changes in pH can have radical effects on biological molecules, affecting their structures and activities. Clearly it is important that the organism can control its internal pH; it needs to buffer against pH changes.

A note on logarithms

The pH scale is a logarithmic scale representing the concentration of H^+ ions in a solution. Remember, from algebra, that we can write a fraction as a negative exponent, thus 1/10 becomes 10^{-1}. Likewise 1/100 becomes 10^{-2}, 1/1000 becomes 10^{-3}, etc. Logarithms are exponents to which a number (usually 10) has been raised. For example log 10 (pronounced 'the log of 10') = 1 (since 10 may be written as 10^1). The log 1/10 (or 10^{-1}) = −1. pH, a measure of the concentration of H^+ ions, is the negative log of the H^+ ion concentration. If the pH of water is 7, then the concentration of H^+ ions is 10^{-7} M, or 1/10 000 000 M. In the case of strong acids, such as hydrochloric acid (HCl), an acid secreted by the lining of the stomach, $[H^+]$ is 10^{-1} M; therefore the pH is 1.

7.12 Summing up

1. Water is a dipolar molecule which readily undergoes hydrogen bonding with other water molecules.
2. The dissociation, or auto-ionisation, of water forms the hydroxonium ion (H_3O^+) and hydroxide ion (OH^-).
3. The degree of dissociation of water is measured by the equilibrium constant which is the ratio of the product of dissociated ions to the undissociated molecule.

$$K_w = [H^+]\,[OH^-]$$

4. An acid is defined as a substance which produces hydrogen ions (H^+) by dissociation; bases are defined as substances which can extract a proton, H^+, from the solvent water to leave a hydroxide ion, OH^-.
5. The pH scale is a logarithmic scale in which the pH of a solution is defined by:

$$\text{pH} = -\log\,[H^+]$$

6. Weak acids are substances which react only partially with water to give hydroxonium ions in solution. A weak base reacts only partially with H_2O to give OH^- ions in solution.
7. The value of the dissociation constant, K_a, for a weak acid is a small number, and therefore is frequently expressed as the pK_a (−log of the dissociation constant).

$$pK_a = -\log K_a$$

8. $$\text{pH} = pK_a + \log\frac{[\text{base}]}{[\text{acid}]}$$

This expression is referred to as the Henderson–Hasselbalch equation. It allows us to work out the pH of solutions of buffers, or the pK_a values of conjugate acid and base combinations at a given pH.

9. Carboxyl and amino groups are common functional groups on biological molecules which are dissociated and protonated respectively at physiological pH.

7.13 Test yourself

The answers are given on p. 161.

Question 7.1
Why is it unlikely that two neighbouring water molecules would be arranged like this?

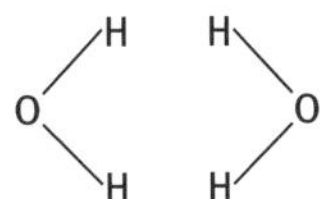

Question 7.2
(a) Give a simple definition of an acid.
(b) What is the difference between a strong acid and a weak acid?
(c) What type of scale is the pH scale?
(d) Compared to a basic solution at pH 9, the same volume of an acidic solution at pH 4 would have how many more hydrogen ions?

Question 7.3
(a) What is the pH of a 0.05 M solution of the strong acid HCl?
(b) What is the $[H^+]$ of a solution of pH 6.2?

Question 7.4
The equilibrium constant (K_a) for the dissociation of ethanoic acid was found to equal 1.8×10^{-5} M. What is the pK_a for ethanoic acid?

Question 7.5
An ethanoic acid and sodium ethanoate buffer system contained 0.10 M of the acid, and 0.05 M of the base. The pK_a of ethanoic acid is 4.75. What is the pH of this system?

Question 7.6
Tris is a weak base that is frequently used to prepare biological buffers. It has a pK_a of 8.08. The pH of the tris base solution is adjusted by addition of HCl. What is the pH of a tris buffer which contains 0.186 M of tris base, and 0.14 M of HCl?

Question 7.7
Molecules X, Y and Z have pK_a values of 4.2, 6.8 and 8.2 respectively. Which is the strongest acid, and what would be the $[H^+]$ in each case? Assume molecules X, Y and Z are present in solution at 1 M concentration.

Taking it further

7.14 Biological buffers

Many chemical reactions are affected by the acidity of the solution in which they occur. In order for a particular reaction to occur, or to occur at an appropriate rate, the pH of the reaction medium must be controlled. Biochemical reactions are especially sensitive to pH. Most biological molecules contain groups of atoms which may be charged or neutral, depending on the pH, and whether these groups are charged or neutral, has a significant effect on the biological activity of the molecule.

In all multicellular organisms, the fluid within the cell and the fluids surrounding the cells have a characteristic and nearly constant pH. For example, the pH of the blood in a healthy individual remains remarkably constant at 7.35 to 7.45. This is because the blood contains a number of buffers which protect against pH change caused by acidic or basic metabolites. From a physiological viewpoint, a change of +0.3 or –0.3 pH units is extreme.

This pH is maintained in a number of ways, and one of the most important is through buffer systems. In the laboratory, buffers are typically prepared by combining a solution of a weak acid and a solution of its salt. Take, for example, the ethanoic acid and sodium ethanoate buffer system. In solution, sodium ethanoate ($CH_3COO^-Na^+$) ionises to produce the conjugate base of ethanoic acid, CH_3COO^-.

The equation for the equilibrium that exists in the solution is:

$$CH_3COO^- + H_3O^+ \rightleftharpoons CH_3COOH + H_2O$$

Buffers work by removing H_3O^+ (H^+) or OH^- ions from solution as they are added.

A buffer system obeys Le Chatelier's Principle, which states that:

'When a stress is applied to a system at equilibrium, the system will adjust to relieve the stress'.

Therefore, if hydrogen ions are added they combine with the base to form the conjugate weak acid:

$$\mathbf{H_3O^+} + CH_3COO^- \rightleftharpoons CH_3CO\mathbf{OH} + H_2O$$

(the equilibrium above is pushed to the right)

If hydroxide ions are added, the weak acid dissociates to provide H^+ ions, which combine with the OH^- to form H_2O:

$$CH_3CO\mathbf{OH} + \mathbf{OH^-} \rightleftharpoons CH_3COO^- + \mathbf{H_2O}$$

(the equilibrium above is pushed to the left)

In either case, the pH does not change dramatically since neither the concentration of H^+ ions nor OH^- ions changes appreciably. Thus, a buffer is able to resist substantial changes of pH.

There are two important biological buffer systems, namely the dihydrogen phosphate system and the carbonic acid system.

1. **The dihydrogen phosphate/hydrogen phosphate system** operates in the internal fluid of all cells. This buffer system consists of dihydrogen phosphate ions ($H_2PO_4^-$) as a hydrogen ion donor (acid), and hydrogen phosphate ions (HPO_4^{2-}) as a hydrogen ion acceptor (base). These two ions are in equilibrium with each other as indicated by the chemical equation below.

$$H_2PO_4^- + H_2O \rightleftharpoons H_3O^+ + HPO_4^{2-}$$

If additional hydrogen ions enter the cellular fluid, they are consumed in the reaction with HPO_4^{2-} and the equilibrium shifts to the left. If additional hydroxide ions enter the cellular fluid, they react with $H_2PO_4^-$, producing HPO_4^{2-} and shifting the equilibrium to the right.

We can see how this system acts as a buffer against pH change, but why does it work so well at neutral pH?

The ability of a compound to act as a buffer at a given pH is determined by how readily it will accept and donate protons (H^+) at that pH. For this reason, any compound which will both accept and donate a lot of protons at a given pH will be an excellent buffer at that pH.

The equilibrium constant expression for this equilibrium is given by:

$$K_a = \frac{[H_3O^+]\,[HPO_4^{2-}]}{[H_2PO_4^-]}$$

and the pK_a (for the dissociation of $H_2PO_4^-$) is found to equal 7.21. In other words, this system is most effective at accepting and/or donating protons at pH 7.21. Buffer solutions are most effective at maintaining a pH near the value of their pK_a. In mammals, cellular fluid has a pH in the range 6.9 to 7.4, and so the dihydrogen phosphate buffer is effective in maintaining this pH range.

2. **The carbonic acid/hydrogen carbonate system:** Another biological fluid in which a buffer plays an important role in maintaining pH is blood plasma. In blood plasma, the carbonic acid/hydrogen carbonate ion equilibrium buffers the pH. In this buffer, carbonic acid (H_2CO_3) is the hydrogen ion donor (acid) and hydrogen carbonate ion (HCO_3^-) is the hydrogen ion acceptor (base).

$$H_2CO_3 + H_2O \rightleftharpoons H_3O^+ + HCO_3^-$$

This buffer functions in the same way as the phosphate buffer. Additional H^+ is consumed by HCO_3^- and additional OH^- is consumed by H_2CO_3. The value of K_a for this equilibrium is 7.9×10^{-7} mol l^{-1}, and the pK_a is 6.1 at body temperature. In blood plasma, the concentration of the hydrogen carbonate ion is about twenty times the concentration of carbonic acid, and therefore provides sufficient buffering capacity to resist excessive changes in the plasma pH [the pH of blood plasma ranges between 7.32 and 7.45].

As we saw earlier, biological molecules contain a number of functional groups that have the capacity to dissociate, notably carboxyl groups and amino groups. So, we might expect proteins (which are polymers of amino acids) to act as buffers, and indeed they do. In fact, albumin (in the plasma) and haemoglobin (in the red blood cell), constitute the largest 'pools' of buffers in the body. As we have stated, compounds act as good buffers at a pH close to their pK_a values. Proteins would be expected to act as buffers at pH values close to the pK_a values of their dissociable amino acid side chains, which for the most part are carboxyl groups and amino groups. Under normal physiological conditions and a pH close to 7, the buffering capacity of such groups is, however, negligible. A look at the table below shows the pK_a of side-chain carboxyl groups to be around 4.0 (for aspartic and glutamic acid), and for amino groups to be around 10 to 13 (for lysine, arginine and asparagine). However, there is one amino acid whose side chain does have a pK_a value near neutrality, namely histidine.

Amino acid	**pK_a of R group**
Arginine	Amino – 13.2
Asparagine	Amino – 13.2
Aspartic acid	Carboxyl – 3.65
Glutamic acid	Carboxyl – 4.25
Histidine	5.97
Lysine	Amino – 10.28

Figure 66. Dissociation of the histidine pyrrole ring nitrogen

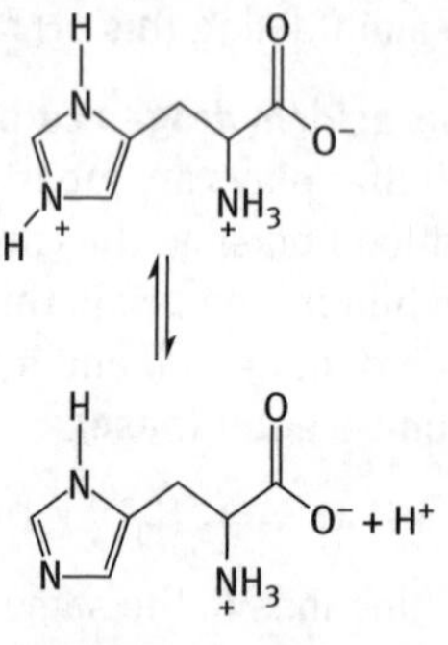

At a neutral pH, it is the pyrrole ring nitrogen in the histidine side chain which is in equilibrium,

$$-N^{+}H \rightleftharpoons -N + H^{+}$$

In other words, at neutral pH, it is the histidine in proteins such as albumin and haemoglobin which is responsible for their buffering action.

8 Reacting molecules and energy

Basic concepts:
Energy is a central concept in biological systems. Cellular processes are geared to obtaining energy from foodstuffs, and using that energy to do work. Here we consider how the cell obtains energy, how it must overcome 'energy barriers' in order for reactions to occur and the concept of 'free energy' and the laws of thermodynamics.

Energy is required by all organisms for a range of biochemical processes, such as the transport of molecules, the biosynthesis of molecules, the maintenance of pH or osmotic pressure, cellular motility and many others. Animals are chemotrophs; they obtain their energy from breaking down molecules (catabolism). They use much of this energy in the synthesis of new molecules (anabolism). The latter depends on the former, and managing these two processes is the key to 'successful metabolism'.

Reminder

Both the **calorie** and the **joule** are units of energy. A **calorie** is the amount of energy or heat needed to increase the temperature of one gram of water by 1° Celsius. One calorie has the same energy value as 4.186 joules (J). The nutritional calorie, Cal = 1000 cal = 4.186 kJ

8.1 Energy from molecules

When atoms form molecules through the formation of covalent bonds, energy is released. The energy of the product molecules is therefore less than that of the reactant molecules. It is for this reason that atoms form covalent bonds; to reach a lower energy and therefore a more stable state. For any particular chemical bond, say the covalent bond between hydrogen and oxygen, the amount of energy it takes to break that bond is exactly the same as the amount of energy released when the bond is formed. This value is called the **bond energy** (or bond dissociation energy). Bond energy is the energy required to break a covalent bond homolytically (into neutral fragments). Bond energies are commonly given in units of kcal mol^{-1} or kJ mol^{-1}, and are generally called bond dissociation energies when given for specific bonds, or mean bond energies when summarised for a given type of bond over many kinds of compounds.

Reminder

The mean **bond energy, ΔH_B** is the average energy change required to break one mole of a given type of chemical bond. e.g. ΔH_B (O-H) = 463 kJ mol^{-1}

Consider the following reaction, the oxidation of methane:

$$CH_4 + 2O_2 \longrightarrow CO_2 + 2H_2O + \textbf{HEAT}$$

Overall this reaction generates heat energy, it is an **exothermic** reaction. Looking at the two sides of the reaction, on the left-hand side of this equation covalent bonds are being broken and energy must necessarily be supplied for this to happen (initially in the form of a spark or flame), therefore this part of the reaction is **endothermic** (energy is absorbed). On the right-hand side of the equation covalent bonds are being formed and therefore energy is released.

Reminder

Covalent bonds are formed between atoms by sharing electrons; by sharing electrons and filling their outer energy levels, atoms reach a more stable state, i.e. one with a lower energy. It must follow therefore that energy is released when covalent bonds are formed, and absorbed when bonds are broken

Using published tables of bond energies, we can construct an energy balance for this reaction (ΔH_B = bond energy).

ENERGY IN			**ENERGY OUT**		
Energy to break four C-H bonds	4 × ΔH_B (C-H)	+1648 kJ	Energy to form two C=O bonds	2 × ΔH_B (C=O)	−1486 kJ
Energy to break two O=O bonds	2 × ΔH_B (O=O)	+992 kJ	Energy to form four O–H bonds	4 × ΔH_B (O–H)	−1852 kJ
		= +2640 kJ			= −3338 kJ
Energy balance =					**-698 kJ**

Thus, overall the reaction is exothermic and releases 698 kJ of energy per mole of CH_4. Note that a plus sign denotes energy that is put in to a reaction, a minus sign energy that is released from a reaction.

8.2 Getting molecules to react

Getting one molecule to react with another first requires the molecules to collide. The reacting molecules must have enough energy and collide in the correct direction such that one or more covalent bonds are broken. The energy needed to initiate this reaction (i.e. to break one or more covalent bonds) is referred to as the **activation energy, E_a**; it is the minimum amount of energy required for reaction to occur.

> **Reminder**
>
> The **activation energy**, E_a, of a reaction is the minimum amount of energy required for reaction to occur. It is specific for a specific reaction

The reacting molecules must be raised in energy to a **'transition state'**. In chemistry providing heat to the reaction is one way to raise molecules to the transition state; this increases the number of successful collisions and therefore increases the rate of the reaction. Another way to achieve this would be to provide an alternative way for the reaction to occur, one which has a lower activation energy. The alternative route is provided by a **catalyst**.

> **Reminder**
>
> A **catalyst** changes the rate of a reaction without being consumed in the reaction. It achieves this by providing an alternative mechanism for the reaction with a different value of E_a

We can summarise this in an 'energy profile' diagram (Fig. 67).

Figure 67. Energy profile diagram

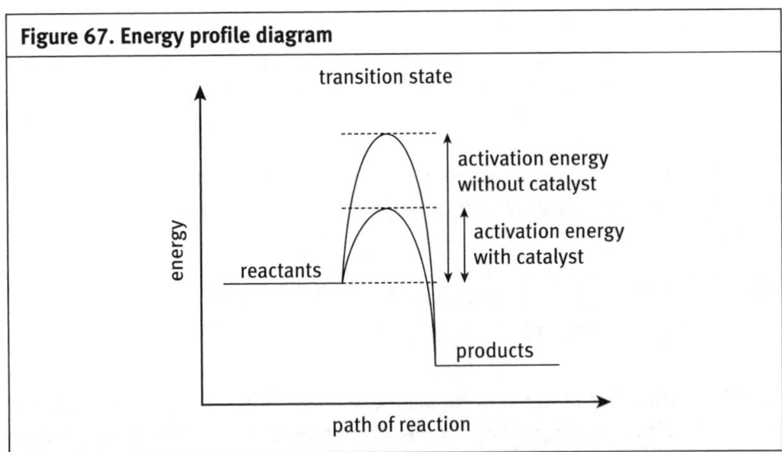

The reactants must be raised by a minimum amount of energy, the activation energy, to reach the transition state in order to initiate the reaction. In the presence of a catalyst the activation energy is smaller and so less energy needs to be supplied to initiate the reaction.

All reactions require an activation energy, whether the overall reaction is exothermic or endothermic. The two energy profiles in Fig. 68 show an exothermic reaction (A) and an endothermic reaction (B).

Figure 68. Energy profiles of an exothermic and an endothermic reaction

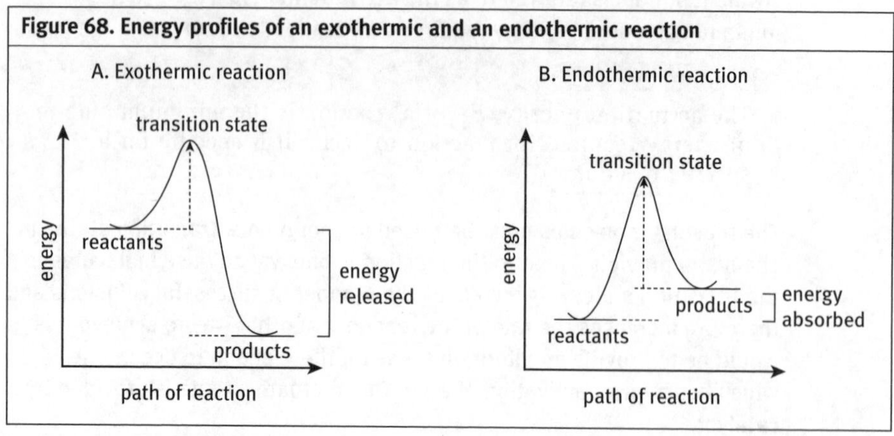

In reaction A, the energy content of the products is less than that of the reactants; energy must have been released in the reaction, so the reaction is exothermic. In reaction B, the energy content of the products is higher than the reactants so energy must have been absorbed in the reaction; the reaction is endothermic.

Reminder

An **exothermic** reaction is one in which heat energy is lost to the surroundings. The reaction mixture usually gets warm. An **endothermic** reaction is one in which heat energy is taken in from the surroundings. The reaction mixture usually gets cool

Biological enzymes, like catalysts, lower the activation energy of a reaction by providing an alternative reaction route.

8.3 Energy, heat and work: some basic terms of thermodynamics

All living things require a continuous throughput of energy. Their metabolism ultimately transforms the energy to heat, which is dissipated to the environment. A large portion of the cell's biochemical apparatus is therefore devoted to the acquisition and utilisation of energy. **Thermodynamics** (Greek: *therme* = heat and *dynamis* = power) is the science which describes the relationships between the different forms of energy.

Although complicated, biological systems obey the fundamental laws of thermodynamics.

In thermodynamics we refer to the 'system' (the object being studied) and its 'surroundings' (everything else in the universe). The system is often a cell, but may also be a biological polymer or molecule (the basic thermodynamic laws apply whether the system is 'living' or not).

The first two laws of thermodynamics may be stated as follows:

- The first law states that the total energy of the universe is always conserved. Energy may be converted into different forms but energy is neither created nor destroyed. Energy lost by the system, must be gained by the surroundings and *vice versa.*
- The second law of thermodynamics expresses the phenomenon that the universe tends towards maximum disorder, or in other words, the direction of all spontaneous processes is such as to increase the **entropy** of a system plus its surroundings. This simply reflects our commonsense understanding that things, if left alone, will not get more orderly!

Reminder

Entropy, S, is a measure of the disorder of a system

Such laws may appear fairly abstract but they are obeyed by all physical, chemical and biological systems without exception. How useful they are in a biological context depends on how we use them! In biology we seek to use thermodynamics to tell us how feasible a particular process is, in which direction it is likely to proceed and what the energy requirements are. We measure such changes under *standard conditions* (see Appendix 4).

8.4 Enthalpy

The heat change in a constant-pressure reaction (as is the case in biological processes) is given by the term delta H (ΔH), the **enthalpy change.** As we have seen above, reactions during which a system releases heat are known as exothermic processes. They have a negative value of ΔH (a negative value is used to show that the reaction is 'losing' heat). ΔH can help us keep track of energy changes and is useful since it depends only on the initial and final states of the system, but it does not indicate the favoured direction of a reaction. Many spontaneous reactions are exothermic, but some are endothermic.

Reminder

The **enthalpy change** in a reaction is equivalent to the heat change at constant pressure. It is measured in joules per mol (J mol^{-1})

Reminder

The Greek letter 'delta' (Δ) is used to denote 'a change in'. We cannot measure absolute values of enthalpy or entropy, but we can measure changes in them

8.5 Entropy

The **entropy, *S***, of a system is a measure of a system's degree of disorder; an increase in entropy is described by a positive ΔS. This is a statement of the second law of thermodynamics. All spontaneous reactions result in an increase in entropy. Entropy is a powerful indicator of the direction of a process, but in complex biological systems is extremely difficult to measure.

Because living systems exchange energy with their surroundings, both energy and entropy changes will take place, and both are important in determining the direction of thermodynamically favourable processes. All living things are open systems and, as such, exchange energy with their surroundings in two ways:

- by heat transfer
- by doing work on the surroundings (or having work done on them).

8.6 Gibbs free energy and work

Work can take many forms; expansion (e.g. lungs expanding), electrical (e.g. ion movement, nerve impulse), movement of a flagellum, contraction of muscle, etc. We need a thermodynamic function which includes both energy and entropy, and which will tell us how much work can be done. There are a number, but of prime importance in biology is the **Gibbs free energy**. In 1878 J.W. Gibbs devised a formula which combines both the first and second laws of thermodynamics. It introduces the term free energy, abbreviated to ***G*** in honour of Gibbs.

Reminder

The Gibbs free energy, G, is the amount of energy in a system which is free to do work at constant temperature and pressure

It is defined by the relationship:

$$\Delta G = \Delta H - T\Delta S$$ where T is the absolute temperature.

Biologists use the symbol $\Delta G^{o\prime}$ to denote the standard free energy change under biological conditions (see Appendix 4).

Reminder

Absolute temperature is measured on the Kelvin scale.
273 K = 0° C; −273° C = 0 K

Processes in which there is a negative enthalpy change (ΔH is negative – i.e. an exothermic reaction) and/or an increase in entropy (ΔS is positive) are typical of favourable reactions, producing a negative ΔG value.

A process with a negative ΔG is an **exergonic** process; free energy is released and may be available to do work.

A process with a positive ΔG is an **endergonic** process; free energy must be provided for the reaction to proceed.

ΔG is of fundamental importance in biology:

- it indicates whether a process will or will not occur
- it indicates in which direction a process will proceed
- it indicates how far from equilibrium a process is
- it indicates how much useful work may be available from a process.

The following three examples show how enthalpy and/or entropy contribute to the direction of chemical reactions. In each case the calculated value for ΔG is negative (the reaction is exergonic). We can conclude therefore that in each case the reaction is thermodynamically feasible, it will occur in the direction indicated and that the free energy released might be available to do work.

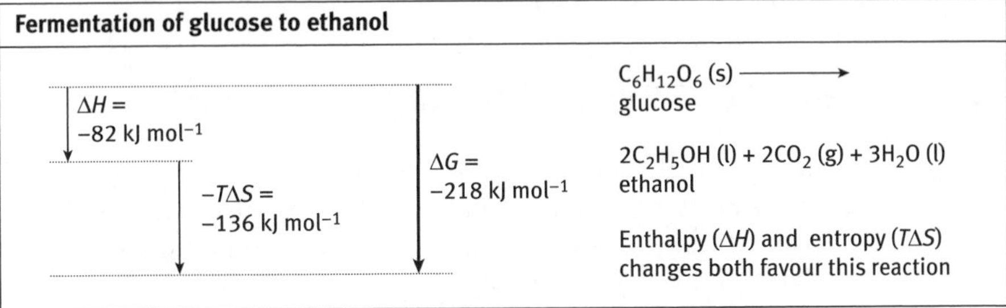

$$\Delta G = \Delta H - T\Delta S = (-82 - 136)\ \text{kJ mol}^{-1} = -218\ \text{kJ mol}^{-1}$$

Oxidation of ethanol

$\Delta H = -1367$ kJ mol^{-1}

$\Delta G = -1326$ kJ mol^{-1}

$-T\Delta S = +41$ kJ mol^{-1}

$C_2H_5OH\ (l) + 3O_2\ (g) \longrightarrow 2CO_2\ (g) + 3H_2O\ (l)$

The enthalpy change (ΔH) favours this reaction (it is 'enthalpy-driven'). The entropy change is slightly positive. If H_2O (g), i.e. water vapour, were the product, then the reaction would also be entropy-driven.

$$\Delta G = \Delta H - T\Delta S = (-1367 - (-41))\ \text{kJ mol}^{-1} = -1326\ \text{kJ mol}^{-1}$$

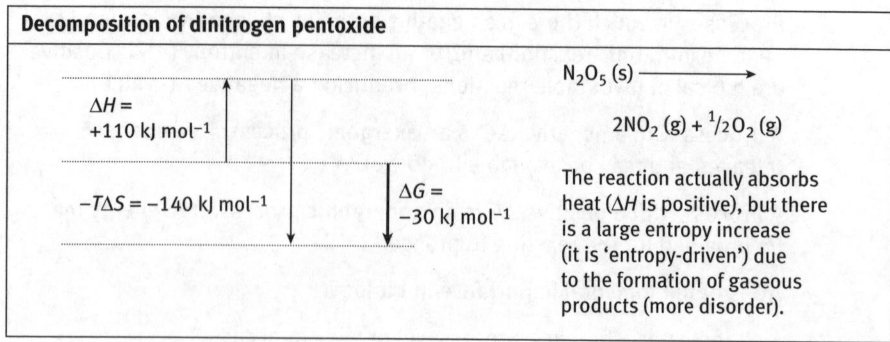

$$\Delta G = \Delta H - T\Delta S = (110 - 140)\ \text{kJ mol}^{-1} = -30\ \text{kJ mol}^{-1}$$

Biological processes for which ΔG is positive are 'energy-consuming' or endergonic. They take place only when **coupled** to a strongly exergonic process. The cell has to do work to render the process exergonic overall. Such coupling mechanisms are the key to understanding the many processes of metabolism. Energy released from the breakdown of biomolecules is used to provide the energy for the biosynthesis of others. As we stated at the beginning of this chapter, anabolism depends on catabolism, and the management of metabolism depends on the successful integration of the two.

8.7 Energy changes in biological reactions

Consider one further exothermic reaction, the oxidation of hydrogen. If you ignite a mixture of hydrogen and oxygen (the spark provides the necessary activation energy) the result is a dramatic explosion. The equation for this chemical reaction is:

$$2H_2\,(g) + O_2(g) \rightarrow 2H_2O(l)$$

And, as the explosion suggests, a release of energy occurs. In fact, the energy change is –447 kJ mol^{-1}.

Reminder

The cell has to do work, which it achieves by coupling an exergonic process with an endergonic process

Where does this energy go? In this case it is lost as heat and sound. But biological systems have learnt to control such processes such that the energy released can be usefully used. This chemical reaction may not seem very 'biological', but in fact it is a good model for the reaction at the very heart of life itself.

The subcellular organelles, mitochondria, exploit a similar reaction to the above to secure ‘free energy’. Mitochondria synthesise water using the hydrogen atoms removed from organic molecules like glucose, and the oxygen atoms they take in as they respire. The process is called **cellular respiration**. Cellular respiration results in a large negative ΔG.

The overall equation is:

$$C_6H_{12}O_6 + 6O_2 \rightarrow 6CO_2 + 6H_2O \quad \text{with the release of 2875 kj mol}^{-1}$$

This is energy which can be harnessed to do work. Mitochondria release this ΔG in small steps which they use to ‘drive’ an endergonic reaction, namely the synthesis of ATP (adenosine triphosphate). ATP is the ‘energy currency’ of the cell; its subsequent breakdown releases free energy which can in turn be used to drive other endergonic reactions. This is a major strategy by which the cell uses free energy derived from catabolism, to drive often unfavourable anabolic reactions. The success of the mitochondrion resides in the fact that released free energy is used to do work, rather than simply being lost as heat, as is the case in the equivalent chemical reaction. This strategy is further developed in ‘Taking it Further: Free energy and metabolic pathways’ and in Chapter 10.

We have seen that energy is required (bond energy) to break covalent bonds, that some reactions can result in a net energy release (exothermic and exergonic reactions), and that a portion of this energy released (free energy) might be used (coupled) to do work and to drive endergonic reactions. The energy changes which occur during the various cellular processes indicate how likely (how thermodynamically feasible) such processes are, in which direction they are likely to proceed and whether they are likely to provide or require energy. This understanding does not allow us, however, to quantify how fast a reaction will proceed, nor to know whether it will go to completion. Indeed it may tell us very little about the molecular mechanism of the reaction. Such information may be gained through the study of **kinetics**, which we consider in Chapter 9.

8.8 Summing up

1. Bond dissociation energy is the amount of energy required when a covalent bond is broken. This amount of energy must be supplied in order to subsequently break that bond. When the same covalent bond is formed, the same amount of energy is released.
2. Chemical reactions may be exothermic (energy is released, ΔH is negative) or endothermic (energy is absorbed, ΔH is positive).
3. Energy must always be supplied to reacting molecules in order to ‘activate’ the reactants; this is referred to as the activation energy, E_a.

4. The activation energy raises the reacting molecules to one or more transition states.
5. A catalyst provides an alternative reaction route with a different, usually lower, activation energy.
6. All biological systems obey the fundamental laws of thermodynamics; 'energy may be converted but is neither created nor destroyed', and 'spontaneous processes result in an increased disorder'.
7. Enthalpy change (ΔH), a measure of heat change, and entropy change (ΔS), a measure of disorder, are related to the Gibbs free energy (ΔG) by the expression $\Delta G = \Delta H - T\Delta S$
8. Reactions may be enthalpy- and/or entropy-driven.
9. ΔG is of fundamental importance in biology; it will indicate whether a process is feasible, its directionality, the extent of the process and whether that process can be used to do work.
10. A negative ΔG is indicative of an exergonic reaction and one which might provide free energy to do work. A positive ΔG is indicative of an endergonic reaction in which energy must be provided in order for the reaction to proceed.
11. Biological processes couple endergonic processes with exergonic processes, free energy from the latter being used to drive the former.

8.9 Test yourself

The answers are given on p. 161.

Question 8.1
Which of the following responses is correct?
Cells cannot use heat to perform work because:
(a) heat is not a form of energy
(b) cells do not have much heat
(c) temperature is usually uniform throughout a cell
(d) heat cannot be used to do work
(e) heat denatures enzymes

Question 8.2
Which of the following processes can occur without a net influx of energy from some other process?
(a) $ADP + P_i \rightarrow ATP + H_2O$
(b) $C_6H_{12}O_6 + 6O_2 \rightarrow 6CO_2 + 6H_2O$
(c) $6CO_2 + 6H_2O \rightarrow C_6H_{12}O_6 + 6O_2$
(d) amino acids $\rightarrow$ proteins
(e) glucose + fructose $\rightarrow$ sucrose

Question 8.3
Which of the following responses is correct?
An enzyme accelerates a metabolic reaction by:
(a) altering the overall free energy change for the reaction
(b) making an endergonic reaction occur spontaneously
(c) lowering the activation energy
(d) pushing the reaction away from equilibrium
(e) making the substrate molecule less stable

Question 8.4
Explain what is meant by exergonic and endergonic reactions.

Question 8.5
In the oxidation of glucose to carbon dioxide and water (glucose + oxygen $\rightarrow$ carbon dioxide + water) the enthalpy change ΔH of the reaction was found to be -2807.8 kJ mol^{-1}, and the free energy change ΔG was equal to -3089.0 kJ mol^{-1}.
(a) Calculate the entropy change ΔS per mole of glucose at 37°C, and
(b) Comment on the thermodynamic feasibility of this reaction.

Taking it further

8.10 Free energy and metabolic pathways

An organism's metabolism is geared to extracting energy from foodstuffs (catabolism) and using that energy to do work and drive otherwise unfavourable reactions. Central to this strategy is the molecule **ATP, adenosine triphosphate**, the 'energy currency' of the cell. The structure of ATP is given below, along with that of ADP for comparison.

ATP

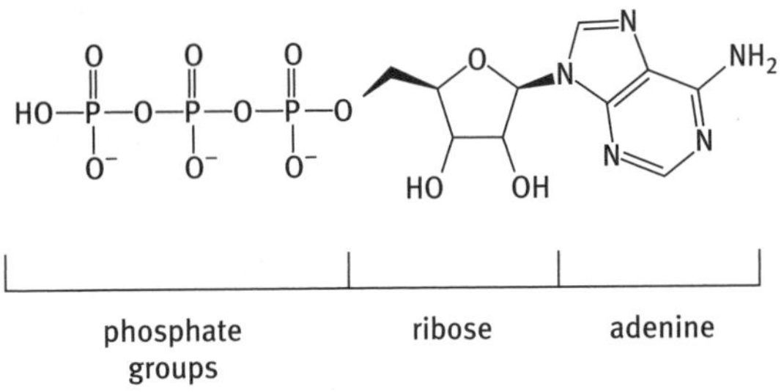

phosphate groups | ribose | adenine

ADP

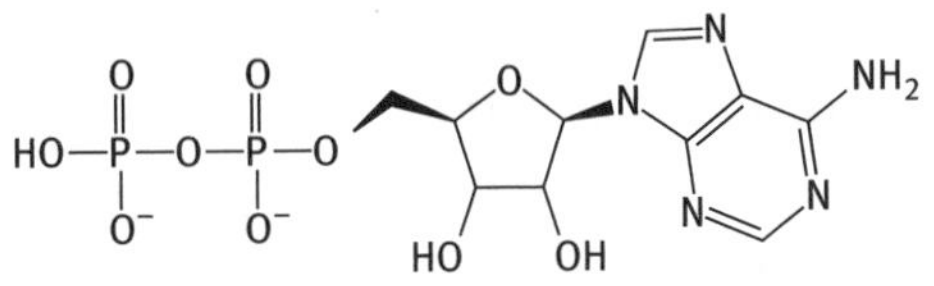

The ATP molecule is constructed from a base, adenine (which we also find in DNA and RNA), linked to a sugar, ribose (which we also find in RNA), which in turn is linked to a series of three phosphate groups. It is the presence of these phosphate groups which is of central importance to the role of ATP. Thermodynamically speaking, ATP is a rather unstable molecule. The three phosphate groups each carry a negative charge at physiological pH, and so there is a lot of repulsion between like charges. Removal of one (or two) of these phosphates alleviates this repulsion and produces a more stable molecule. Enzyme-catalysed hydrolysis of ATP (hydrolysis is the addition of water) yields ADP (adenosine diphosphate) by the removal of one phosphate group.

$$\text{ATP} + H_2O \rightarrow \text{ADP} + P_i \text{ (inorganic phosphate)}$$

This reaction is spontaneous and highly exergonic; it is associated with a large negative free energy change. The free energy change for this reaction is about 30.7 kJ mol^{-1}. Organisms **couple** the hydrolysis of ATP to drive the energy-consuming activities of the cell. ATP is not the only molecule in the cell which can provide such energy, but it is the main one; evolution has favoured those enzymes which can bind ATP and use its hydrolysis to drive endergonic reactions.

ATP powers most energy-consuming activities of the cell	
Activity	Examples
Anabolic reactions	Synthesis of proteins Synthesis of nucleic acids Synthesis of polysaccharides Synthesis of fats
Active transport of molecules	Transport of molecules and ions across biological membranes
Nerve impulses	
Maintenance of cell volume	Maintenance of osmotic gradients
Addition of phosphate groups to molecules	Phosphorylation of proteins alters their activity
Muscle contraction	
Cell motility	Movement of cilia, flagella and sperm
Bioluminescence	

Metabolic pathways

All metabolic pathways comprise a sequence of serial conversions, from initial reactant(s) through intermediates to final product(s). Intermediates may themselves participate in other metabolic pathways. There are some important concepts which become clear when we consider the many different metabolic pathways in an organism.

- All metabolic pathways are **unidirectional.** There may be individual reactions in the pathway which are clearly reversible, but as a whole the pathway proceeds in one direction only and results in a **net formation** of product.
- The net free energy change of any metabolic pathway is negative. In other words, all metabolic pathways are thermodynamically feasible under physiological conditions and proceed in a particular direction.

This latter point requires some further thought! Catabolic pathways are designed to extract and make available free energy from foodstuffs; they are necessarily associated with large negative free energy changes and by definition are feasible and unidirectional. But what of anabolic pathways? Anabolic processes are unfavourable and energy-consuming, so how can an anabolic pathway be thermodynamically feasible and unidirectional? The answer to this, which we will investigate further, is explained by the coupling of highly exergonic reactions (e.g. the hydrolysis of ATP) to unfavourable endergonic reactions. Of course, not all the reaction sequences of an anabolic pathway are endergonic, and with the input of free energy the overall free energy 'sum' of the whole metabolic pathway is made negative and therefore feasible.

There is one further point to consider.

- Although some metabolic pathways are catalysed by the same enzymes facilitating both the 'forward' (degradation) and 'backward' (synthesis) reactions, organisms always use two separate, non-reversible pathways, one for degradation and one for biosynthesis. This strategy allows the cell to exert metabolic control and balance its energy requirements.

We can consider the above points with a detailed look at two cellular metabolic pathways, namely **glycolysis** and **gluconeogenesis**.

Glycolysis and gluconeogenesis

Glycolysis is that catabolic sequence of reactions which begins the oxidative degradation of glucose. The glycolytic pathway is unidirectional and strongly exergonic, and results in the production of two molecules of ATP for every one molecule of glucose degraded. It is an essential component of the organism's overall energy maintenance; for example, skeletal muscles derive most of their energy (in the form of ATP) from glycolysis. However, organisms also need to synthesise glucose, which can then be stored (as glycogen) in order to provide a constant energy supply for when the body is not ingesting foodstuffs (the brain, for example, needs a constant supply of glucose). The synthesis of glucose is an anabolic process which occurs through the gluconeogenesis pathway ('gluconeogenesis' literally means 'synthesis of new glucose').

Glycolysis

The metabolic pathway for glycolysis is shown in Fig. 69.

Figure 69. Free energy changes in the glycolytic pathway

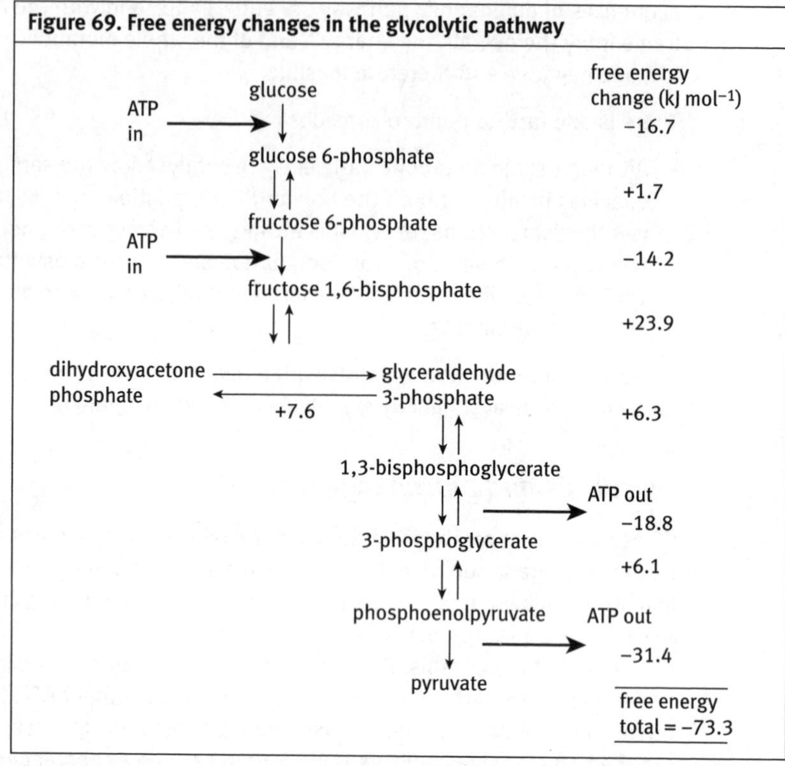

Firstly, and in common with most metabolic pathways whether catabolic or anabolic, the pathway has to be 'activated'. In glycolysis this involves the phosphorylation of glucose to glucose 6-phosphate. This reaction requires the input of energy and so is coupled to the hydrolysis of ATP; free energy from ATP hydrolysis drives this reaction forward (ATP_{in}). The reaction becomes unidirectional and is associated with a large negative free energy change. Such irreversible reactions are often referred to as a **'committed step'**. A second input of energy is required to convert fructose 6-phosphate to fructose 1,6-bisphosphate; again the reaction is unidirectional and a large negative free energy change is associated with hydrolysis of ATP. For catabolic pathways such as glycolysis, we can consider an **energy investment phase** and an **energy payoff phase**. In other words energy has to be supplied, to 'kick-start' the pathway, in order that energy can subsequently be derived. In glycolysis, two moles of ATP per mole of glucose are required to move the pathway forward, while four moles of ATP are recovered in two later exergonic reactions (one molecule of glucose provides two molecules of glyceraldehyde 3-phosphate). Thus, the net budget is an energy gain.

Although the more exergonic sequences in this pathway tend to be very unidirectional, there are clearly a number of reactions which are essentially reversible. One might ask the question 'why doesn't the pathway stop or even reverse?' The pathway cannot reverse, at least not completely, because of the exergonic unidirectional component reactions, and the reversible reactions are continuously displaced from equilibrium by the enzymes which rapidly remove the products and so 'pull' those reactions forward.

Gluconeogenesis

The metabolic pathway for gluconeogenesis is shown in Fig. 70. The pathway is shown side by side with glycolysis. Comparing gluconeogenesis with glycolysis, many of the reactions are common to the two pathways (shown by a → symbol) , whereas certain reactions are unique to gluconeogenesis (shown by the ●— symbol).

Figure 70. Gluconeogenesis and glycolysis compared

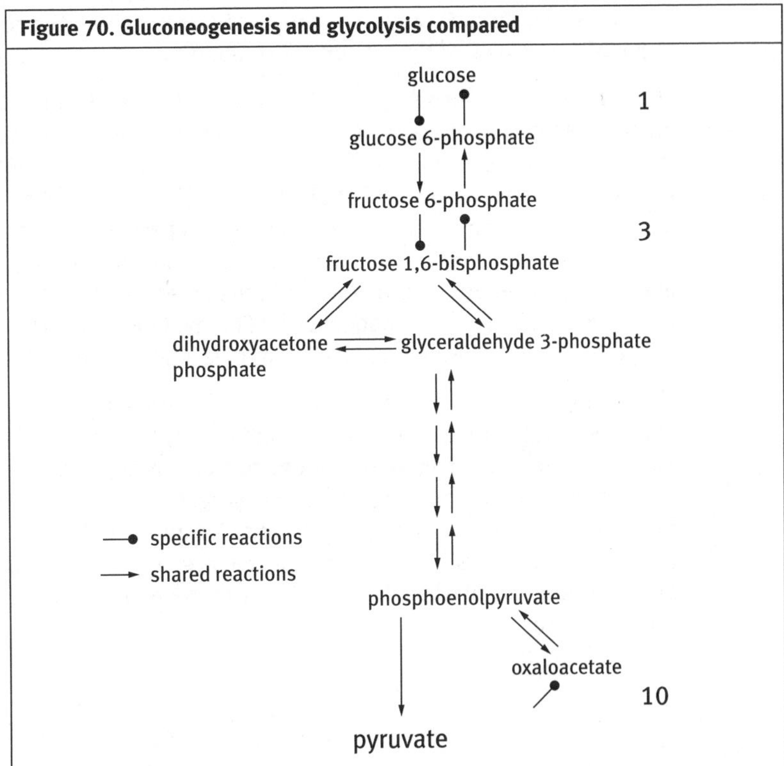

The three unique gluconeogenic reactions, **10, 3** and **1** in Fig. 70, and emphasised by the 'loops' in Fig. 71, correspond to those highly exergonic steps in glycolysis; these three glycolytic steps cannot simply be reversed and so must be 'by-passed' by other reactions.

Figure 71. Reaction sequences compared in glycolysis and gluconeogenesis

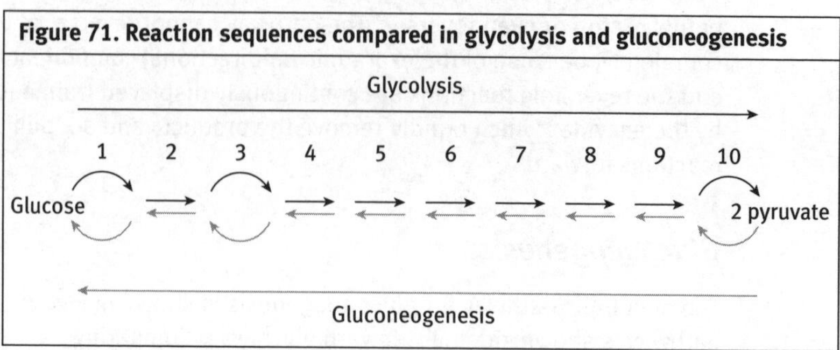

In gluconeogenesis these 'by-passes' are achieved by unique reactions, each requiring an input of energy in order to make them exergonic and unidirectional. The other reversible reactions are shared, although in fact the two metabolic pathways are actually separated within the cell by virtue of occurring in different cellular compartments.

The overall ΔG for gluconeogenesis is negative (–47.6 kJ mol^{-1}) thanks to an input of energy. Glycolysis too has an overall negative ΔG (–73.3 kJ mol^{-1}). The difference is that glycolysis accomplishes a negative ΔG whilst also yielding a net production of ATP, whereas the biosynthetic gluconeogenesis pathway requires an input of energy (partly through ATP hydrolysis) to achieve its overall negative ΔG. One pathway is catabolic and an energy producer (glycolysis) and the other pathway is anabolic and an energy consumer (gluconeogenesis). Energy, provided in the formation of ATP by catabolic pathways, is used to drive anabolic pathways.

The strategies seen in the consideration of glycolysis and gluconeogenesis are applicable to all metabolic pathways within the cell. An input of energy is generally required to 'commit' the pathway, usually through the free energy released by the hydrolysis of ATP. This ensures that the overall ΔG for the pathway is negative and that the pathway is unidirectional; reversible reactions are displaced from equilibrium, either by the rapid removal of products, or indeed through high concentrations achieved in the reactants.

9 Reacting molecules and kinetics

Basic concepts:
The successful management and integration of the cell's vast array of different metabolic reactions depends, to a large part, on the control and rates of the different reactions. The study of kinetics is a powerful tool for understanding the mechanisms and control of enzyme-catalysed reactions. Here we consider factors which affect the rates of reaction, rate-limiting steps in reaction schemes, and equilibrium states and their relationship to free energy changes.

Chemical kinetics is the study of the rate at which chemical reactions occur and the factors which affect rate. For living organisms the rate of reaction is extremely important; a 'successful' metabolic pathway requires that each individual reaction can occur at an optimal rate. Three factors are important in determining the rate of a reaction.

1. **Temperature**. The higher the temperature, the greater the kinetic motion of the molecules, the more energetic their collisions, and therefore the more likely that the activation energy will be reached or exceeded and a reaction will occur. Higher organisms control their temperature and so the direct effect of temperature on biochemical reactions is of lesser consideration.
2. **Catalysts**. Catalysts increase (or decrease) reaction rates while they themselves remain unchanged. Enzymes are biological catalysts. The physiological temperature of higher organisms is a balance; on the one hand it speeds up the rate and increases the feasibility of the enzyme-catalysed reaction, while on the other it is not so high as to cause undue damage to the delicate protein structure of the enzyme molecule.
3. **Concentration**. The more molecules there are present (the reactants), the more collisions which will occur in a given time, and therefore the greater the rate of the reaction (and the more products which will be formed). In biochemical pathways the concentration of reactants and products is especially important since the catalytic rate of many enzymes can be activated or inhibited by the levels of certain metabolites in those pathways (feedback inhibition).

Rates of reaction are generally expressed in terms of a change in concentration with time, in either the reactants or the products. For example, we might express a rate as moles of product used per second, mol s^{-1} or as the change in concentration of moles of product per second, mol dm^{-3} s^{-1}.

An expression for evaluating the rate of a reaction is

$$\text{rate} = \frac{-\Delta[\text{R}]}{\Delta t}$$

where Δ means a 'change in' and [R] means the molar concentration of the reactants. Of course the concentration of the reactants can only decrease with time and, since the rate can only be positive, a minus sign is included in the equation.

If a product in this reaction were a molecule, P, then we could also write a rate for the reaction in terms of the change in concentration of product per second. In this case the rate of reaction would be expressed as:

$$\text{rate} = \frac{\Delta[\text{P}]}{\Delta t}$$

In this case because the product concentration is increasing with time, the sign is positive.

Reminder

Rates of reaction are generally expressed as a change in concentration with time

9.1 Rate equations

Consider the reaction A + B → C + D. We can write a rate equation, for the rate of decrease in concentration of the reactants, as:

$$\text{rate} = -k[\text{A}]^x[\text{B}]^y$$

or indeed as the rate of increase in the concentration of the products as:

$$\text{rate} = k[\text{C}]^w[\text{D}]^z$$

k is the **rate constant** for the reaction. The rate constant isn't actually a true constant because it varies if you change the temperature of the reaction or add a catalyst. The rate constant is a constant for a given reaction only if all you are changing is the concentration of the reactants. For these equations, the powers of x and y are called the **orders of reaction** with respect to reactants A and B respectively; the orders of reaction are usually small whole numbers 0, 1 or 2. The **overall** order of the reaction would be given by $x + y$.

Reminder

A rate constant is only a 'true' constant if all that is changing is the concentration of the reactants

In a **zero order reaction** the reaction rate does not depend on the concentration of the reactants and so the order of reaction is zero; any number raised to the power of zero is one, and so the rate of such a reaction would be simply given as:

$$\text{rate} = -k$$

Zero order reactions are actually very common in biology. Where an enzyme is effectively saturated with substrate and is working at its maximal rate, it makes little difference to the rate of the reaction whether the concentration of the reactant (substrate) is slightly raised or lowered; the rate is determined only by the efficiency of the enzyme.

For a **first order reaction** the rate of reaction would depend on the concentration of just one of the reactants. We would write:

Rate $= -k[\text{A}]$, if the reaction was first order with respect to A only.

So, if we doubled the concentration of A, then we would double the rate of the reaction.

For a **second order reaction** there are two possibilities to consider. In the first case the rate of reaction may depend only on the concentration of one reactant squared (to the power 2). In this case if the concentration of the reactant is doubled, then the rate will increase 4-fold. The rate equation for a reaction which is second order with respect to A can be written as:

$$\text{rate}' = -k[\text{A}]^2 \qquad \text{(a)}$$

Therefore if [A] is doubled to [2A] the new rate, rate″ becomes:

$$\text{rate}'' = -k[2\text{A}]^2 = -k4[\text{A}]^2$$

If we compare the new rate, rate″, to the old rate, rate′, by dividing one by the other then:

$$\frac{\text{rate}''}{\text{rate}'} = \frac{-4k[\text{A}]^2}{-k[\text{A}]^2} = 4$$

Thus it can be seen that the new rate is 4 times the old rate if the concentration of reactant A is doubled.

In the second case (b), the rate of reaction depends on the concentrations of two reactants. In the rate equation:

$$\text{rate} = -k[\text{A}][\text{B}] \qquad \text{(b)}$$

doubling the concentration of either A or B will double the rate, and doubling both concentrations will increase the rate 4-fold.

In these rate equations:

- rate $= k[\text{A}]^1$ indicates the reaction is first order with respect to A (although we don't normally show the superscript '1')
- rate $= k[\text{A}]^2$ indicates the reaction is second order with respect to A

- rate = k[A][B] indicates the reaction is first order with respect to both A and B, and second order overall.

Orders of reaction can be found only by experiment. The order of a chemical reaction gives you information about which concentrations affect the rate of the reaction. You cannot look at an equation for a reaction and deduce what the order of the reaction is going to be – you have to do some practical work! Having found the order of the reaction experimentally, you *may* be able to make suggestions about the route or mechanism for the reaction, at least in simple cases.

Reminder

Orders of reaction give information on how the rate is affected by the concentration of the reactants

9.2 Reaction routes or mechanisms

In any chemical change, some bonds are broken and new ones are made. Very often, such changes do not happen in a single step. Instead, the reaction may involve a series of small steps one after the other; this is particularly true of enzyme-catalysed reactions. A reaction mechanism describes the one or more steps involved in the reaction in a way that makes it clear exactly how the various bonds are broken and made.

9.3 The rate-limiting step

Since many reactions occur through several steps, the rate for each step needs to be measured. There will always be one step which is the slowest and that step is called the **rate-limiting step**. The overall rate of a reaction is controlled by the rate of the slowest step. When you measure the rate of an overall reaction, you are actually measuring the rate of the rate-limiting step.

9.4 Considering the activation energy

In Section 8.2 we saw that the reactants must be raised in energy (the activation energy) to a transition state before covalent bonds could be broken and the reaction could proceed. In the diagram for reaction A (Fig. 72), only one transition state is evident, and mechanistically we would assume the reaction to be a relatively simple one.

Figure 72. Single and multiple transition states in a reaction

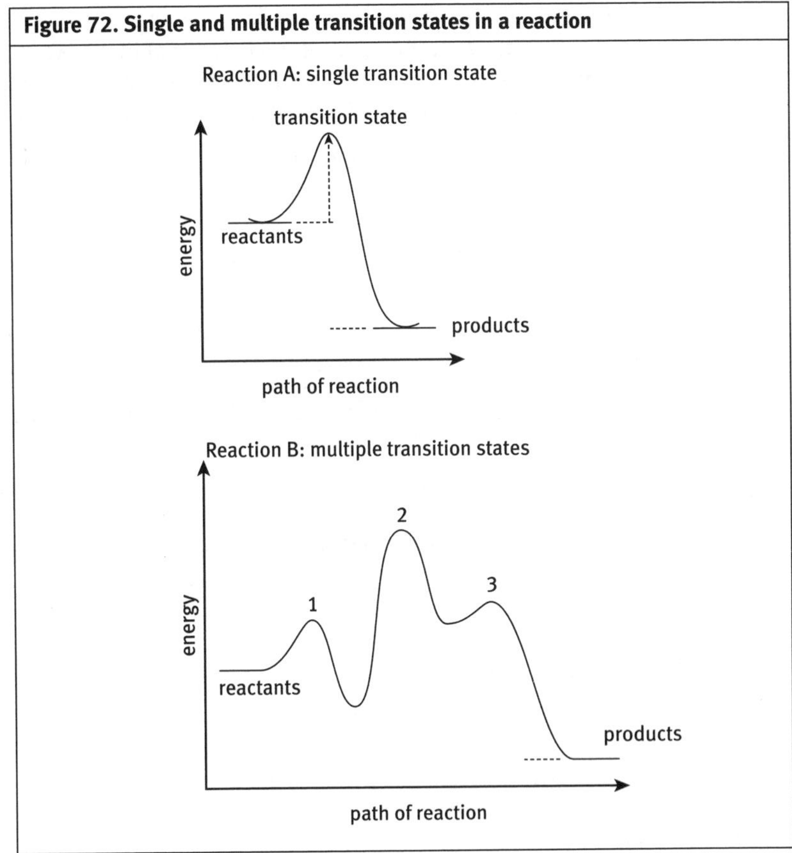

However, generally it is more likely that the reactant molecules would undergo a number of different rearrangements, and involve the breaking and formation of a number of covalent bonds. Each of these rearrangements (transitions) would have a characteristic activation energy. The diagram in reaction B (Fig. 72) shows three transition states. Transition state 2 has the highest activation energy and would therefore represent the rate-limiting step in this reaction.

Reminder

The rate-limiting step in an enzyme-catalysed reaction will involve the formation of the transition state with the highest activation energy

9.5 Equilibrium

The majority of chemical reactions, and certainly most biochemical reactions, do not proceed to completion. Instead a point is reached at which the products react together to re-form the reactants! The reaction is said to be **reversible.** At the point where the rate of the forward reaction (formation of product) equals the rate of the backward reaction (formation of reactant), the reaction is said to have reached **equilibrium.**

In a simple reaction, say the conversion of A to B, we can show the reaction as:

$$\mathrm{A} \underset{k^{-1}}{\overset{k^{1}}{\rightleftharpoons}} \mathrm{B}$$

where k^1 is the forward rate (formation of product) and k^{-1} is the backward rate (formation of reactant).

The rate of the forward reaction (formation of products) can be expressed as:

$$\text{rate f} = \frac{\Delta[\mathrm{B}]}{\Delta t} = k^1[\mathrm{A}]$$

and the rate of the backward reaction (formation of reactants) can be expressed as:

$$\text{rate b} = \frac{\Delta[\mathrm{A}]}{\Delta t} = k^{-1}[\mathrm{B}]$$

So, because rate f must equal rate b

$$\frac{\Delta[\mathrm{B}]}{\Delta t} = \frac{\Delta[\mathrm{A}]}{\Delta t}$$

therefore: $k^1[\mathrm{A}] = k^{-1}[\mathrm{B}]$

so: $\dfrac{k^1}{k^{-1}} = \dfrac{[\mathrm{B}]}{[\mathrm{A}]} = K_c$

where K_c is the equilibrium constant.

Reminder

At equilibrium, the forward rate (k^1) is equal to the backward rate (k^{-1})

Equilibrium constants, K, may have many different notations depending on the type of equilibrium being investigated (for example, K_w is the equilibrium constant for the ionisation of water). The symbol K_c specifies only that the equilibrium constant shall have some nominal unit. The most common concentration unit is mol per litre or mol l^{-1}. Biologists tend to use K_{eq} for the equilibrium constant of enzyme-catalysed reactions.

For the reduction of pyruvate, which occurs in skeletal muscle, to produce lactate,

$$\text{pyruvate} + \text{NADH} + \text{H}^+ \rightleftharpoons \text{lactate} + \text{NAD}^+$$

the equilibrium constant K_{eq} can be expressed as:

$$K_{eq} = \frac{[\text{lactate}][\text{NAD}^+]}{[\text{pyruvate}][\text{NADH}][\text{H}^+]}$$

If K_{eq} is very small, then the reaction would lie to the left with little product having been made.

To determine the amount of each compound which will be present at equilibrium you must know the equilibrium constant. To determine the equilibrium constant you first need to 'balance' the equation. Consider the generic equation:

$$\text{aA} + \text{bB} \rightleftharpoons \text{cC} + \text{dD}$$

The upper case letters are the reactants and products. The lower case letters are the coefficients which balance the equation.

The equilibrium constant, K_{eq}, for this general reaction would be expressed as:

$$\frac{[\text{A}]^a\,[\text{B}]^b}{[\text{C}]^c\,[\text{D}]^d}$$

For example,

$$N_2(g) + H_2(g) \rightleftharpoons NH_3(g)$$

This equation as written cannot obey the conservation of mass principle; it does not 'balance'. This is because there are different numbers of each type of atom on each side of the equation (there are 2 nitrogen atoms on the left, but 1 on the right; there are 2 hydrogen atoms on the left, but 3 on the right). The way we balance equations is with 'stoichiometric' coefficients, i.e. numbers in front of the formulae which denote how many 'items' or molecules of that chemical are involved.

Balancing this equation gives:

$$N_2(g) + 3H_2(g) \rightleftharpoons 2NH_3(g)$$

This equation now follows the law of conservation of mass. There are exactly the same numbers of atoms of each element on both sides of the equation. **Stoichiometry** refers to all quantitative aspects of chemical composition and reactions. The equilibrium expression for this reaction would therefore be written as:

$$K_c = \frac{[NH_3]^2}{[N_2][H_2]^3}$$

For example, if one litre of this reaction mixture at equilibrium was found to contain 1.60 moles NH_3, 1.20 moles H_2 and 0.80 moles N_2, then the equilibrium constant would be calculated as:

$$K_c = \frac{(1.60\ \text{M})^2}{(0.8\ \text{M})(1.20\ \text{M})^3} = 1.67\ \text{M}^{-2}$$

The unit for K depends on the units used for concentration. In the above example, concentrations are given as M (mol l^{-1}), and so K has a unit of $M^{2-(4)} = M^{-2}$.

Qualitatively we may look at how far an equilibrium lies towards the right (towards products) or left (towards reactants). The magnitude of the equilibrium constant gives us a general idea of whether the equilibrium favours products or reactants. If the reactants are favoured, then the denominator term of the equilibrium expression will be larger than the numerator term, and the equilibrium constant will be less than unity. If the products are favoured over the reactants, then the numerator term will be larger than the denominator term, and the equilibrium constant will be greater than unity.

9.6 The equilibrium position can change

An equilibrium is able to shift in either direction, either towards reactants or towards products, when a stress is applied. This stress may involve increasing or decreasing the amount of a reactant or product, changing the volume or pressure of a gas phase, or changing the temperature of the equilibrium system. The direction in which the equilibrium shifts may be predicted by using **Le Chatelier's Principle**. This states that if a stress is placed on an equilibrium, then the equilibrium will shift in the direction which relieves the stress. For example, in the following reaction, which represents the muscle conversion of pyruvate to lactate, the equilibrium lies very far to the right favouring the production of lactate.

$$\text{Pyruvate} + \text{NADH} + H^+ \rightleftharpoons \text{Lactate} + \text{NAD}^+$$

We can 'force' this reaction to the left (pyruvate production) by adding high concentrations of lactate, or indeed NAD^+. In muscle, the enzyme which catalyses this reaction is called lactate dehydrogenase; it is structurally designed to favour a high concentration of lactate (the enzyme has a low affinity for lactate). Lactate, from muscle, is distributed to the liver where it is converted back to pyruvate; here a structurally distinct lactate dehydrogenase favours the production of pyruvate (it has a higher affinity for lactate). This strategy is common in biology; structurally distinct forms of an enzyme (**isoenzymes**) promote different equilibrium positions. It is important to note, however, that whereas there may be an infinite number of equilibrium positions for a reaction, there is only one value of the equilibrium constant (at a defined temperature). The specific equilibrium position adopted by a system depends on the initial concentrations. To underline this fact consider the data below.

The table shows the results of three experiments for the reaction

$$N_2(g) + 3H_2(g) \rightleftharpoons 2NH_3(g)$$

Experiment	Initial concentration	Equilibrium concentration	K_{eq}
1	N_2 = 1.00 M	0.921 M	
	H_2 = 1.00 M	0.763 M	
	NH_3 = 0	0.157 M	$6.02 \times 10^{-2}\ M^{-2}$
2	N_2 = 0	0.399 M	
	H_2 = 0	1.197 M	
	NH_3 = 1.00 M	0.203 M	$6.02 \times 10^{-2}\ M^{-2}$
3	N_2 = 2.00 M	2.59 M	
	H_2 = 1.00 M	2.77 M	
	NH_3 = 3.00 M	1.82 M	$6.02 \times 10^{-2}\ M^{-2}$

Equilibrium concentrations were determined for N_2, H_2 and NH_3, starting with three sets of different initial concentrations. Calculating K_{eq} for each, using the equation

$$k_{eq} = \frac{[NH_3]^2}{[N_2][H_2]^3}$$

gave the same value in each case.

9.7 Free energy and equilibrium

In Chapter 8 we introduced the concept of free energy. Such concepts may appear somewhat abstract. However, by a mathematical combination of the equations for equilibrium constants and those for free energy, we can derive the equation:

$$\Delta G^{\circ\prime} = -RT \ln K_{eq}$$

(*R* is a constant called the Universal Gas Constant and *T* the absolute temperature)

Reminder

The equilibrium position can change, but the equilibrium constant is fixed at a constant temperature

In other words, the free energy change of a reaction is related to the equilibrium constant for that reaction. If we can measure the equilibrium constant of a biochemical reaction (which is a relatively straightforward procedure) then we can also calculate the free energy change (and *vice versa*). This can provide some very useful information, i.e. whether the process is feasible (likely to occur), in which direction it will proceed, and to what extent it will proceed. These are powerful data in trying to decipher the intricacies and interrelationships of cellular processes.

9.8 Free energy change is zero at equilibrium

Only when a reaction is moving towards equilibrium can there be a net free energy change. For example, for the reaction, A $\rightleftharpoons$ B, there may be a negative ΔG as the reaction moves towards B (the reaction is exergonic), or alternatively ΔG may be positive if the reaction is endergonic (and energy is being supplied to 'drive' the reaction). However, as soon as the reaction reaches equilibrium,

$$\mathrm{A} \underset{k^{-1}}{\overset{k^{1}}{\rightleftharpoons}} \mathrm{B}$$

TAKING IT FURTHER: Free energy and metabolic pathways (p. 111)

and the rate of production of B by the forward reaction equals the rate of production of A by the backward reaction, then by definition the net free energy change must be zero. Cellular reactions which are exergonic and generate a negative free energy change generally proceed in one direction only and are displaced from equilibrium by the rapid removal of products.

Reminder

At equilibrium, rate k^1 = rate k^{-1}, and ΔG is zero

If a cell's metabolic processes reach equilibrium, then that cell is effectively 'dead'! No free energy is produced to do work and there can be no **net** production of products. This is analogous to a battery which has reached equilibrium, i.e. a 'flat' battery; a battery at equilibrium cannot deliver an electromotive force and no current will flow.

9.9 Summing up

1. Rates of chemical reactions are affected by temperature (of lesser consideration for biological reactions), catalysts (enzymes are biological catalysts) and concentration of reactants. In enzyme-catalysed reactions, the concentration of both reactants and products is important because enzymes are often subject to feedback inhibition by their products.
2. Rates of reaction are generally expressed in terms of a change in concentration with time, in either the reactants or the products, and are expressed using rate equations.
3. Orders of reaction indicate the dependency of reaction rate on the concentration of the reactants; a zero order reaction is independent of reactant concentration, a first order reaction is dependent on the concentration of just one of the reactants.

4. The rate of a reaction depends upon the rate-limiting step, which corresponds to the formation of the transition state with the highest activation energy.
5. A reaction may be **reversible**. At the point where the rate of the forward reaction (formation of products) equals the rate of the backward reaction (formation of reactants), the reaction is said to have reached **equilibrium**.
6. The equilibrium constant of a reaction, K_c or K_{eq}, is obtained from the product of the product concentrations (raised to the power of their stoichiometric coefficients) divided by the product of the reactant concentrations (also raised to their stoichiometric coefficients). At a stated temperature, the equilibrium constant for a reaction is independent of initial concentrations; there is only one equilibrium constant for a reaction at a specific temperature, but there may be an infinite number of equilibrium positions.
7. A high value for the equilibrium constant indicates that the reaction lies in favour of the formation of products.
8. Le Chatelier's Principle states that if a stress is placed on an equilibrium then the equilibrium will shift in the direction which relieves the stress.
9. Changes in free energy of a reaction are related to the equilibrium constant for that reaction by the expression $\Delta G^{\circ\prime} = -RT \ln K_{eq}$
10. At equilibrium, a reaction has a zero free energy change.

9.10 Test yourself

The answers are given on p. 161.

Question 9.1
Write the equilibrium expression (K_{eq}) for the following reaction:
isocitrate + NAD^+ $\rightleftharpoons$ α-ketoglutarate + CO_2 + NADH

Question 9.2
If a reaction was found to have an equilibrium constant of 3.18×10^{520}, what would this tell you about the reaction?

Question 9.3
In the following reaction:
fumarate + water $\rightleftharpoons$ malate
the ΔG was found to equal −3.7 kJ mol^{-1}. Calculate the K_{eq} at 37°C for this reaction (take R = 8.314 J $K^{-1} mol^{-1}$), and comment on the value obtained.

Question 9.4
ATP is hydrolysed to produce ADP and inorganic phosphate, P_i; the ΔG for the reaction was found to equal −13 kJ mol^{-1}. Calculate the equilibrium constant for this reaction at 37°C (Take R as 8.314 J K^{-1} mol^{-1}).

10 Energy and life

Basic concepts:
The loss or gain of electrons by molecules defines the processes of oxidation and reduction. Life forms use the controlled oxidation of foodstuffs to release free energy, which can be chemically stored or coupled to less favourable reactions to do work. Every oxidation reaction is coupled to a reduction reaction; such redox reactions can be measured by their redox potentials, and in turn be related to the free energy change. Herein lies the very 'driving force' of life and the means by which organisms can build and maintain complex structures in an increasingly disordered environment.

Within the chloroplasts of green plants, light energy is harvested and converted into chemical energy which is stored in ATP and NADPH; these molecules are then used to 'fix' carbon dioxide and produce firstly sugars (carbohydrates), and subsequently amino acids (proteins) and fatty acids (fats). It is estimated that the process of photosynthesis makes some 165 billion tons of carbohydrate a year! Photosynthesis is an endergonic process, an unfavourable process with a positive free energy change and one that is only possible through the supply of energy from the sun (electromagnetic energy). Animals as well as plants then gear their metabolism to the breakdown (catabolism) of organic molecules, releasing free energy (which was originally derived from light) to support the myriad and often unfavourable anabolic processes in the cell. Life would seem to defy the second law of thermodynamics, that the Universe tends towards maximum disorder; our cells are highly ordered and exactly controlled, we are islands of low entropy in an increasingly disordered world! But the laws of thermodynamics are universal. In fact we spend all our time breaking organic molecules down to release energy in order to build others up, but our transfer of energy is far from 100% efficient and we lose much of it as heat, adding to the increasing entropy of the universe. In biological systems:

Energy release = ATP + heat!

Central to life is the ability of an organism to extract and transfer energy from 'food'. A number of strategies have evolved to enable this, the most successful involving the chemical reactions of oxidation and reduction.

10.1 Oxidation and reduction

The original definition of oxidation was 'reaction or combination with oxygen' and reduction was defined as 'reaction or combination with hydrogen'. Our cells oxidise sugars and fats to generate the free energy which we require. The complete oxidation of glucose is shown by the equation:

$$C_6H_{12}O_6 + 6O_2 \rightarrow 6CO_2 + 6H_2O + \text{heat}$$

Both the carbon and hydrogen atoms in glucose have been **oxidised** by bonding to oxygen. In our oxygen-rich atmosphere there is a natural tendency for things to become oxidised (fats go rancid, iron rusts), but a closer look at this equation also shows that the oxygen atoms have been **reduced**. Oxidation of one molecule is always linked to the reduction of another molecule. We speak of oxidation–reduction reactions, which we further abbreviate to **redox** reactions, to emphasise this link.

In biology, redox reactions are defined as those which involve transfer of electrons. All oxidation reactions involve the loss of electrons, and all reduction reactions involve the gain of electrons. The loss or gain of electrons is a better definition of oxidation and reduction since not all redox reactions necessarily involve oxygen or hydrogen. It can, however, be rather more difficult to decide whether electrons are being lost or gained in a reaction. In a simple case, the reaction between sodium and chlorine to form sodium chloride can be written as:

$$\mathbf{Na + \tfrac{1}{2}Cl_2 \rightarrow Na^+\,Cl^-}$$

Reminder

Oxidation Is Loss of electrons. Reduction Is Gain of electrons. The mnemonic 'OIL RIG' can be helpful in remembering this

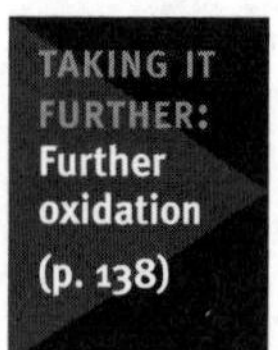
TAKING IT FURTHER: Further oxidation (p. 138)

The strongly ionic compound sodium chloride is formed as a result of sodium losing an electron (being oxidised) and chlorine gaining an electron (being reduced). Sodium, the electron donor, is the **reducing agent** because it reduces the chlorine. Chlorine, the electron acceptor, is the **oxidising agent** because it oxidises the sodium. Because an electron transfer involves both a donor and an acceptor, oxidation and reduction must always be linked. Not all redox reactions involve the complete transfer of electrons from one molecule to another, but rather can involve a change in the degree of electron sharing in covalent bonds.

10.2 Half-reactions

Since all redox reactions involve both an oxidation and a reduction, we can describe such reactions by two **half-equations** or **half-reactions.** In the reaction above representing the formation of sodium chloride, the sodium atom has been oxidised and the chlorine molecule reduced. We can write half-equations to represent these two reactions:

$Na \rightarrow Na^+ + e^-$ (i) oxidation of sodium

$\frac{1}{2}Cl_2 + e^- \rightarrow Cl^-$ (ii) reduction of chlorine

If we add together the right-hand sides and the left-hand sides of equations (i) and (ii) the electrons cancel out and we obtain the overall equation for the reaction.

$Na + \frac{1}{2}Cl_2 \rightarrow Na^+ Cl^-$ (i) + (ii) overall equation

Consider a biological redox couple – the cellular conversion of malate to oxaloacetate. The reaction is catalysed by the enzyme malate dehydrogenase and involves the coenzyme NAD^+ (nicotinamide adenine dinucleotide). The reaction can be written as:

$$\text{malate} + NAD^+ \longrightarrow \text{oxaloacetate} + NADH + H^+$$

Considering the structures of malate and oxaloacetate (Fig. 73), malate has lost two protons and *two electrons* overall and so has been oxidised in the conversion to oxaloacetate. The reaction actually involves the removal of a hydride and a proton from malate; a hydride (H^-) is a proton with two electrons ($H^+ + 2e^-$).

Figure 73. Oxidation of malate to oxaloacetate

malate ⇌ oxaloacetate + $H^- + H^+$

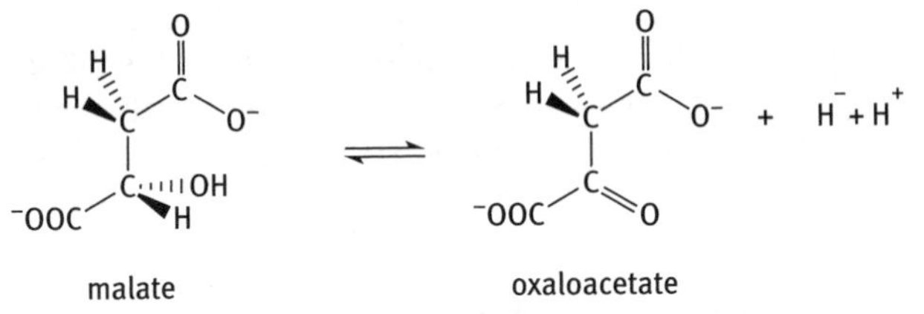

Question

Why has malate lost two electrons?

A proton H^+ has been removed from the hydroxyl group, and a hydride H^- has been removed from the -C-H. The 'H^-' has taken both electrons from the C-H bond with it

In the second half-reaction (Fig. 74), NAD^+ is reduced to NADH. NAD^+ *accepts two electrons* (from the hydride, H^-) and one proton and so is clearly reduced.

Figure 74. Reduction of NAD^+ to NADH

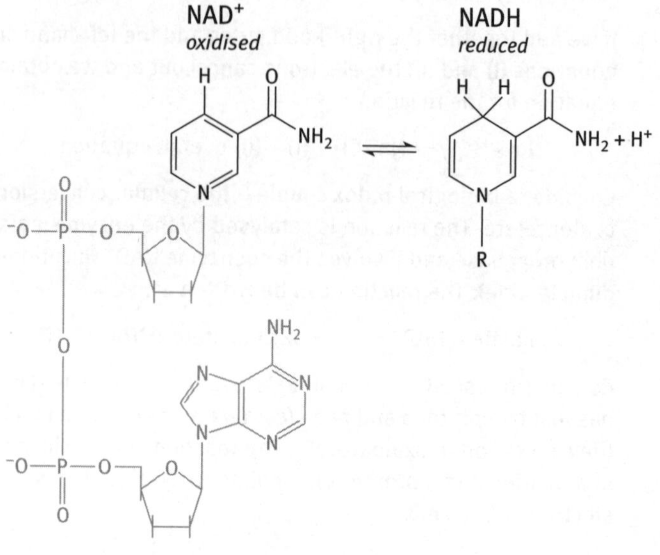

NAD^+ *oxidised* ⇌ NADH *reduced* $+ H^+$

The two half-reactions are clearly linked, and together they constitute a **redox couple**; we can add them together to get an overall equation. The malate–oxaloacetate reaction is reversible, as indeed are most redox reactions.

10.3 Redox potential

As we have previously seen, the thermodynamic potential of a chemical reaction is calculated from a knowledge of both its equilibrium constant, derived from the concentrations of reactants and products, and from the free energy change (ΔG) of the reaction. All redox reactions on the other hand involve the loss and gain of electrons. It is clearly impractical to measure electron concentrations directly. Instead, we measure the **redox potential** of each half-reaction. The redox potential is a measure of the tendency for a species to either gain or lose electrons.

The redox potential of a half-reaction must be measured relative to a reference; that reference, a standard half-reaction, is the hydrogen electrode, which is arbitrarily set at 0 volts. Redox potentials are measured in volts and chemists designate this by $\boldsymbol{E}^\circ$; in biological systems we measure redox potentials at physiological pH and use the symbol $\boldsymbol{E}^{\circ\prime}$ to denote this difference. Furthermore, at pH 7, the redox potential of the hydrogen electrode converts to −0.42 V; this is the value biologists use as their standard redox potential ($E^{\circ\prime}$). A measured redox potential which has a negative value for $E^{\circ\prime}$ indicates the tendency of the reaction to proceed in the direction of oxidation, i.e. to lose electrons. A positive value for $E^{\circ\prime}$ suggests that the reaction is likely to be one of reduction, i.e. a gain in electrons.

In the table below, redox potentials are given for a number of common biological half-reactions. Notice that all of these are shown as reductions. This does not imply that this is the preferred direction of each reaction, but is simply a convention that chemists use in reporting such data.

Redox potentials of some common biological half-reactions

Redox half-reaction	$E^{\circ\prime}$/volt	
$2H^+ + 2e^- \longrightarrow H_2$	–0.42	low
ferredoxin(Fe^{3+}) + e^- $\longrightarrow$ ferredoxin(Fe^{2+})	–0.42	redox
$NAD^+ + 2H^+ + 2e^- \longrightarrow NADH + H^+$	–0.32	potential
$S + 2H^+ + 2e^- \longrightarrow H_2S$	–0.274	↑
$SO_4^{2-} + 10H^+ + 8e^- \longrightarrow H_2S + 4H_2O$	–0.22	
acetaldehyde + $2H^+ + 2e^-$ $\longrightarrow$ ethanol	–0.20	
pyruvate + $2H^+ + 2e^-$ $\longrightarrow$ lactate	–0.185	
$FAD + 2H^+ + 2e^- \longrightarrow FADH + H^+$	–0.18	
oxaloacetate + $2H^+ + 2e^-$ $\longrightarrow$ malate	–0.17	
fumarate + $2H^+ + 2e^-$ $\longrightarrow$ succinate	0.03	
cytochrome *b*(Fe^{3+}) + e^- $\longrightarrow$ cytochrome *b*(Fe^{2+})	0.075	
ubiquinone + $2H^+ + 2e^-$ $\longrightarrow$ ubiquinone H_2	0.10	
cytochrome *c*(Fe^{3+}) + e^- $\longrightarrow$ cytochrome *c*(Fe^{2+})	0.254	
$NO_3^- + 2H^+ + 2e^- \longrightarrow NO_2^- + H_2O$	0.421	↓
$NO_2^- + 8H^+ + 6e^- \longrightarrow NH_4^+ + 2H_2O$	0.44	high
$Fe^{3+} + e^- \longrightarrow Fe^{2+}$	0.771	redox
$O_2 + 4H^+ + 4e^- \longrightarrow 2H_2O$	0.815	potential

A reaction which we have used before, namely the reduction of pyruvate to lactate, is an example of a redox reaction.

$$O{=}C{-}O^- \;|\; C{=}O \;|\; CH_3 \;+\; NADH + H^+ \rightleftharpoons O{=}C{-}O^- \;|\; H{-}C{-}OH \;|\; CH_3 \;+\; NAD^+$$

pyruvate lactate

The two half-reactions of this redox couple are also given in the table. That for pyruvate reduction is:

$$\text{pyruvate} + 2H^+ + 2e^- \longrightarrow \text{lactate} \qquad (1)$$

The redox potential for this half-reaction is –0.185 V (read from the table).

For NADH the half-reaction, which is an oxidation reaction, is written as:

$$NADH + H^+ \longrightarrow NAD^+ + 2H^+ + 2e^- \quad (2)$$

The half-reaction for NAD^+/NADH shown in the table is a reduction reaction (by convention). To obtain the redox potential for the oxidation of NADH, we simply reverse the sign of the $E^{\circ\prime}$ given in the table. Therefore, the redox potential for the oxidation of NADH is +0.32 V (rather than –0.32 V). When we add the redox potentials for both half-reactions together, to find the overall redox potential of the enzyme-catalysed reaction, we get:

$$-0.185\ V + 0.32\ V = +0.265\ V$$

When we add the right-hand sides and the left-hand sides of equations (1) and (2) the electrons cancel and we get:

$$\text{pyruvate} + NADH + H^+ \rightleftharpoons \text{lactate} + NAD^+ \quad E^{\circ\prime} = 0.265\ V$$

10.4 Free energy and redox potentials

As mentioned earlier, it is impractical to measure an electron concentration in order to derive an equilibrium constant for a redox reaction. However, by derivation we find that the free energy of a redox reaction can be calculated directly from its $E^{\circ\prime}$ by the **Nernst equation**. The Nernst equation is shown as:

$$\Delta G^{\circ\prime} = -nF\,\Delta E^{\circ\prime}$$

where n is the number of electrons transferred in the reaction and F is the Faraday constant (23.06 kcal V^{-1} mol^{-1} or 96.5 kJ V^{-1} mol^{-1}).

Using determinations of equilibrium constants (K_{eq}) and redox potentials ($E^{\circ\prime}$), biologists can therefore determine the $\Delta G^{\circ\prime}$ for a variety of reactions with a view to assessing their feasibility and directionality.

10.5 Obtaining energy for life

Biological molecules which are capable of accepting and donating electrons, being alternately reduced and oxidised, are referred to as **electron carriers**. NADH is an example of an electron carrier. Dependent on their redox potentials, we could arrange a series of different electron carriers to form an **electron transport chain**. Electrons could pass down such a chain, passing from electron carriers with a low redox potential (more negative and therefore more likely to become oxidised by losing an electron) to ones with an increasingly high redox potential (more positive). Electron carriers with a low (more negative) redox potential have a lower affinity for the electron and will pass it on to a carrier with a higher (more positive) redox potential and greater affinity for the electron.

In Section 8.7 we indicated that the reduction of oxygen (or the oxidation of hydrogen) in the reaction

$$2H_2 + O_2 \rightarrow 2H_2O$$

was highly exergonic and spontaneous, liberating some 447 kJ mol^{-1} of energy, mostly as heat and sound. The subcellular organelle, the mitochondrion, carries out an analogous process whereby a series of electron carriers provide a carefully controlled pathway in which free energy can be extracted in small packages. At one end of the **'electron transport chain'** NADH , which is generated in the cell by catabolic pathways, is oxidised and the two electrons 'lost' in this oxidation are passed down the chain, from one electron carrier to the next, finally ending in a reaction which results in the reduction of oxygen to water (which is why you need to breathe oxygen to survive!).

The oxidation of NADH, with a relatively low redox potential, and the reduction of oxygen, with a relatively high redox potential, creates a potential difference from 'top to bottom' of the chain of some +1.13 V

$$\tfrac{1}{2}O_2 + 2H^+ + 2e^- \rightarrow H_2O \qquad E^{\circ\prime} = +0.815\ \text{V}$$

$$NADH + H^+ \rightarrow NAD^+ + 2H^+ + 2e^- \qquad E^{\circ\prime} = +0.32\ \text{V}$$

The overall reaction is:

$$\tfrac{1}{2}O_2 + NADH + H^+ \rightarrow H_2O + NAD^+ \qquad \Delta E^{\circ\prime} = +1.13\ \text{V}$$

Inserting this value of $E^{\circ\prime}$ into the Nernst equation gives a free energy change of −218 kJ mol^{-1} .

$$\Delta G^{\circ\prime} = -nF\,\Delta E^{\circ\prime} = -2(96.5\ \text{kJ V}^{-1}\ \text{mol}^{-1})\ (1.13\ \text{V}) = -218\ \text{kJ mol}^{-1}$$

Note that n, the number of electrons involved in the reaction, is 2 in this case.

Not only is this series of reactions highly exergonic, releasing a considerable amount of free energy, but the passage of electrons 'down' this chain is driven by a significant potential difference of +1.13 V.

A redox couple such as these two half-equations, which together comprise the biological reaction between oxygen and hydrogen to produce water, form a link in an electron transport chain. Such a link can be represented by the following convention.

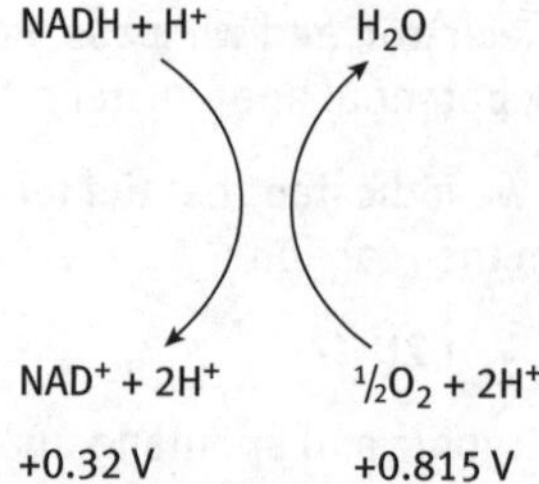

The curved arrows represent the movement of electrons down the chain as each species is oxidised.

10.6 What happens to this free energy?

The mitochondrial electron transport chain is depicted in Fig. 75; the various electron carriers are shown in a linear fashion. Each redox couple is associated with a redox potential ($\Delta E^{\circ\prime}$) and a calculated $\Delta G^{\circ\prime}$. As we pass along the chain from left to right, the redox potential becomes progressively more positive (with a greater tendency to become reduced), whereas we note that there are three redox couples in particular which are associated with much higher free energy changes; these are the oxidation of NADH and reduction of FAD, oxidation of cytochrome *b* and reduction of cytochrome *c*, and the oxidation of cytochrome *a* and reduction of oxygen. The large negative free energy change at these three sites is used to do work; ultimately this results in the synthesis of ATP. ATP is the cells' energy currency; the free energy stored in this molecule can be coupled to unfavourable endergonic reactions to drive anabolic processes.

TAKING IT FURTHER: Free energy and metabolic pathways (p. 111)

Figure 75. Redox and free energy steps in the mitochondrial electron transport chain

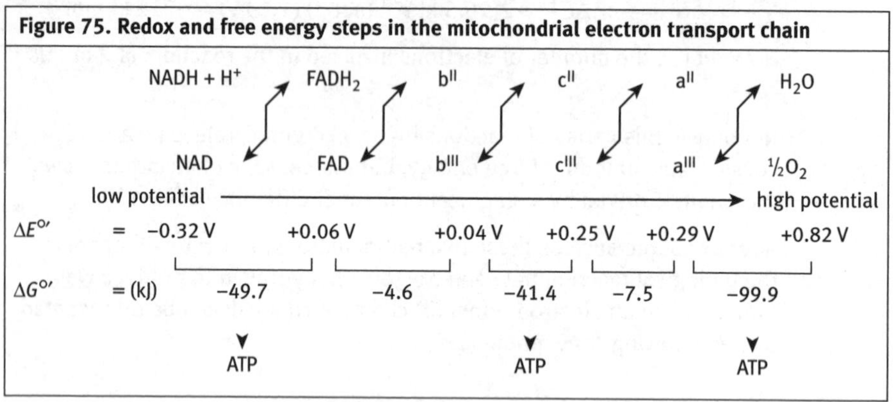

[Key to the mitochondrial electron transport chain: NAD = nicotinamide adenine dinucleotide; FAD = flavin adenine dinucleotide; b = cytochrome *b*; c = cytochrome *c*; a = cytochrome *a*; b^{II}, c^{II} and a^{II} represent the reduced cytochrome, and b^{III}, c^{III} and a^{III} the oxidised cytochrome]

Mitochondria are the principal site for ATP generation in higher organisms. This aerobic process, referred to as oxidative phosphorylation, is essential for energy provision in all higher organisms.

Reminder

The oxidative phosphorylation process is a series of oxidation steps which eventually bring about the synthesis of ATP through the phosphorylation of ADP

10.7 Summing up

1. Oxidation always involves the removal of electrons from a molecule. Whenever one molecule is oxidised another must be reduced. Thus we speak of redox reactions.
2. All redox reactions are made up from two half-reactions, one half-reaction involving an oxidation and the other involving a reduction. For each half-reaction we may measure a redox potential; the redox potential is a measure of the degree to which a reaction will either gain or lose electrons. It is measured relative to a reference; that reference, a standard half-reaction, is the hydrogen electrode, which is arbitrarily set to 0 volts. Redox potentials are therefore measured in volts and are designated by E°.
3. A negative value for $E^{\circ\prime}$ indicates the tendency of the reaction to proceed in the direction of oxidation, i.e. to lose electrons.
4. The free energy change of a redox reaction is related to the redox potential by the Nernst equation, $\Delta G^{\circ\prime} = -nF\Delta E^{\circ\prime}$
5. Electrons will transfer from a low redox potential (more negative $E^{\circ\prime}$) to a higher redox potential (more positive $E^{\circ\prime}$). This transfer of electrons is associated with a free energy change.
6. The subcellular organelle, the mitochondrion, uses an arrangement of electron carriers to mediate the transfer of electrons from a low to a higher redox potential, and utilises the free energy released to do work and eventually generate ATP in the process of oxidative phosphorylation.

10.8 Test yourself

The answers are given on page 162.

Question 10.1

Write half-reactions for the following redox reactions

(a) $Zn + Cu^{2+} \rightarrow Zn^{2+} + Cu$

(b) $Fe^{2+} + Cu^{2+} \rightarrow Fe^{3+} + Cu^{+}$

Question 10.2

The following reaction shows the reduction of acetaldehyde to form ethanol, catalysed by the enzyme alcohol dehydrogenase.

acetaldehyde + NADH + $H^{+} \rightleftharpoons$ ethanol + NAD^{+}

(a) Write the two half-reactions for this redox couple.

(b) Using the redox potentials tabulated on page 133, decide whether the above reaction will proceed in a forward (left to right) or backward direction.

Question 10.3

Using the redox potentials tabulated on page 133, decide in which direction the following redox couple will proceed.

ubiquinone $_{\text{(oxidised)}}$ + cytochrome c $_{\text{(reduced)}}$ $\rightleftharpoons$ ubiquinone $_{\text{(reduced)}}$ + cytochrome c $_{\text{(oxidised)}}$

Question 10.4

Using the equation $\Delta G^{\circ\prime} = -nF\Delta E^{\circ\prime}$, calculate the free energy change for the reaction:

succinate + FAD $\rightleftharpoons$ fumarate + $FADH_2$

(use tabulated values on page 133 for the redox potentials; number of electrons transferred in this reaction = 2; take F as 96485 J V^{-1} mol^{-1})

Question 10.5

Using the equation $\Delta G^{\circ\prime} = -nF\Delta E^{\circ\prime}$, calculate $\Delta G^{\circ\prime}$ for the following reaction:

malate + $NAD^{+} \rightleftharpoons$ oxaloacetate + NADH

Taking it further

10.9 Further oxidation

Just as electrons in atoms occupy atomic orbitals, such that the lowest energy orbitals are filled first, so in molecules electrons occupy molecular orbitals. Molecular orbitals are regions of space in which there is a certain probability of finding an electron. Electrons in molecules possess energy and fill molecular orbitals such that a minimum energy state is adopted. The higher in energy the molecular orbital, the more energy the electron possesses.

We have defined **oxidation** as the loss of electrons. Stated slightly differently, oxidation is the movement of electrons *away from* an atom (or, more precisely, away from an atom's nucleus). Thus, when oxygen oxidises something, it pulls the electrons away from that something and towards itself (in the process oxygen serves as an oxidising agent and is itself reduced). Remember, oxygen is an electronegative atom. Note that oxidation refers to this movement of electrons even when oxygen atoms are not involved in the process.

In Sections 2.5 and 2.6 we considered **polar covalent bonds** which arise when a more **electronegative** atom, such as oxygen, is bonded to a less electronegative atom such as hydrogen or carbon. The more electronegative atom draws electrons towards it (and is reduced) whilst the other atom in the bond, which 'loses' electrons, is oxidised.

Consider the succession of compounds in Fig. 76.

Figure 76. Oxidation of high energy compounds

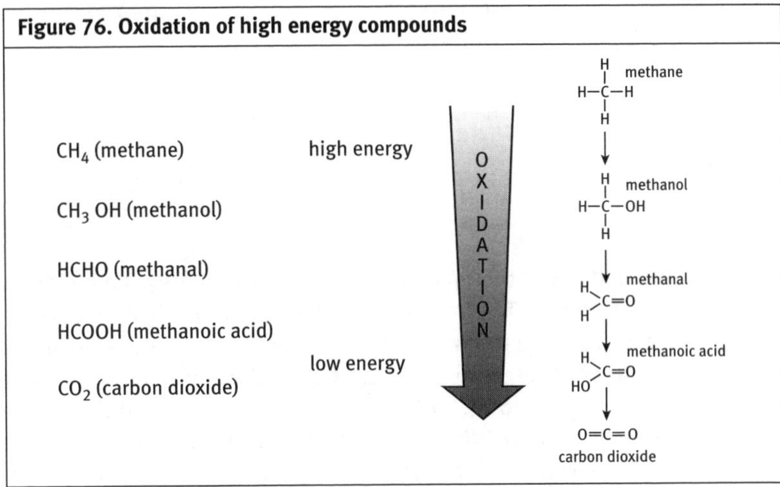

In carbon–hydrogen bonds (e.g. methane), the electronegativity of carbon and hydrogen are approximately equal. Therefore the electrons in this symmetrical covalent bond are as far from the nuclei of the two atoms as is possible. They are more or less shared evenly between carbon and hydrogen. As we pass down this series, hydrogen atoms are progressively replaced by electronegative oxygen, and the carbon atom is progressively oxidised by the formation of polar covalent bonds (in which electrons are drawn away from the carbon towards the oxygen). It must follow therefore that the electrons in the methane covalent bonds possess more energy than those in carbon dioxide (since through oxidation they are found progressively closer to the oxygen atomic nucleus). Energy is stored in the carbon–hydrogen bonds of methane and methane is a **high energy** compound, whereas carbon dioxide is a **low energy** compound.

In the process of photosynthesis, plants capture light energy and incorporate this primarily into carbohydrates. Carbohydrates produced by photosynthesis, such as glucose, mainly consist of carbon–carbon and carbon–hydrogen covalent bonds. We might consider these symmetrical covalent bonds as '**energy-rich bonds**', in so much as the electrons in them possess a high energy. Captured solar energy is stored in carbon compounds, such as cellulose in trees; over time these have become buried and now provide us with our hydrocarbon reserves of oil, gas and coal so-called 'fossil fuels'. **Hydrocarbons** are so called because they consist of only carbon and hydrogen. Some 'energy-rich' compounds are shown in Fig. 77.

Figure 77. High-energy glucose and hydrocarbons

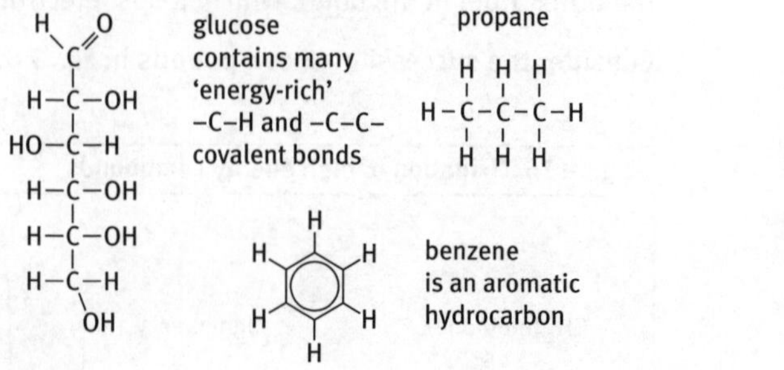

glucose contains many 'energy-rich' –C–H and –C–C– covalent bonds

propane

benzene is an aromatic hydrocarbon

The first law of thermodynamics states that energy can neither be created nor destroyed. The transfer of an electron from a less electronegative atom to a more electronegative atom must therefore result in some transfer of energy from that electron to the surrounding environment. When we burn fossil fuels in air, we oxidise them and the energy is released as heat. When the mitochondria in our cells 'burn' fuels they likewise oxidise them, but the energy is released step-wise, harnessed and stored in ATP for future use. It is worth pointing out that this mitochondrial transfer of energy to ATP is far from being 100% efficient, indeed much is lost as heat. As warm-blooded animals we owe most of our heat to this oxidation process.

11 Reactivity of biological molecules

Basic concepts:
The presence of functional groups on biomolecules provides reactive sites, at which such molecules may be transformed or linked to other molecules. Reactive sites may be nucleophilic or electrophilic, and the reactions which occur at such sites can be classified as addition, substitution or elimination reactions. Functional groups are essential in conferring a 'site of attack' for enzymes; reactions at a nucleophilic or electrophilic centre often generate intermediates or transition states which can be stabilised by an enzyme. It is by these means that enzymes provide alternative and feasible reaction routes which underline their role as catalysts.

Understanding reaction mechanisms allows biologists and chemists to optimise the yield of a chemical in a biotechnological industrial process, or understand catalysis by metals and by enzymes, and to design inhibitors of enzymes, such as drugs and pesticides.

The metabolic reactions which occur in organisms consist of many apparently very complex transformations. Biomolecules consist of a large number of atoms and bonds at any one of which a chemical reaction could occur. Fortunately, however, we know that the **reactive sites** on such molecules invariably involve their **functional groups**. The functional group behaves differently from the rest of the molecule, usually because it contains an electronegative atom and consequently a polar covalent bond. As we have previously discussed in Chapter 3, functional groups are vitally important in the provision of intermolecular interactions which stabilise molecular interactions and conformations (shapes) of molecules. But in providing reactive sites, functional groups also provide the means for molecular transformations, as well as the means for molecular building and construction of biological **macromolecules**.

Reactive sites on a molecule may contain a **nucleophilic centre** or an **electrophilic centre**. Nucleophilic centres are electron rich; they may have a negative charge, have lone pairs of electrons, or possess an increased electron density typical of double covalent bonds. A nucleophilic centre will attract positively charged groups, i.e. an electrophile (for example, a positively charged proton). Electrophilic centres are electron deficient and seek out negative charges.

Take for example the carbonyl group. Carbonyl groups (C=O) are found in many biological molecules containing:

- RCOOH (carboxylic acid), in acids
- RCHO (aldehyde), in sugars
- R_2CO (ketone), in sugars
- $RCONH_2$ (amide), in proteins

The C=O group is polar, with a small positive charge (δ+) on C and a small negative charge (δ–) on O. Nucleophiles are attracted to the slightly positive C, forcing electrons upon it. Electrophilic protons are attracted to the slightly negative O, which can donate electrons.

Reminder

An electronegative atom is frequently found as part of a functional group, which generates a charge dipole in that group, and subsequently results in both a nucleophilic and an electrophilic centre.

Nucleophilic and electrophilic reactive sites provide the basis of a number of different types of reaction mechanisms.

11.1 Addition reactions

The reaction sequence in Fig. 78 shows a ketone called propanone (acetone) forming a hydrate under acidic conditions. The nucleophilic OH^- attacks the electrophilic carbon of the ketone group, whilst an electrophilic proton attacks the nucleophilic oxygen of the ketone group. The large curved arrow in this diagram indicates the first point of attack by the nucleophile; the short curly arrow is used to indicate the movement of a pair of electrons, in this case two electrons from the pi covalent bond to form a new sigma covalent bond with the proton.

Figure 78. An addition reaction to a ketone

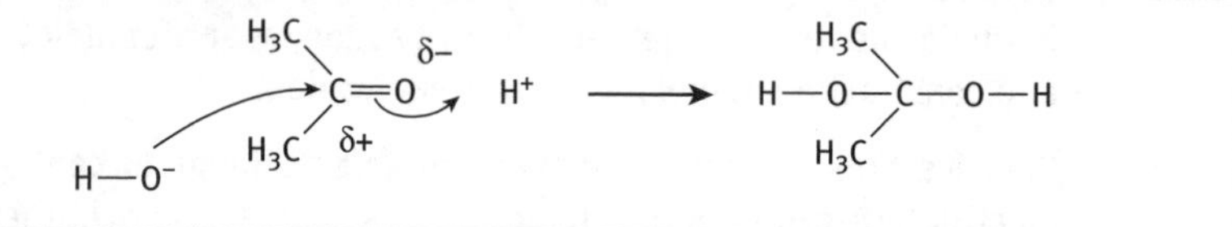

H_3C δ–
H_3C–C=O δ+ H^+ ⟶ H—O—C(H_3C)(H_3C)—O—H
H—O^-

This type of reaction is referred to as an **addition reaction** (OH^- and H^+ have both been added to the ketone).

When reactions involve either nucleophiles or electrophiles, then the electrons involved in forming new bonds or breaking existing bonds move in pairs.

11.2 Substitution reactions

Shown in Fig. 79 is the mechanism of the reaction of bromoethane in alkaline solution to produce ethanol. In bromoethane the carbon is an electrophilic centre, because of the polar covalent bond with bromine (bromine is a more electronegative atom). A nucleophilic hydroxide ion is attracted to the electrophilic carbon centre, causing the electron pair of the C-Br polar covalent bond to move onto the bromine, resulting in a substitution of Br by OH and formation of the alcohol and a Br^- ion.

Figure 79. Formation of an alcohol by a substitution reaction

$$CH_3{-}\overset{\delta+}{CH_2}{-}\overset{\delta-}{Br} \longrightarrow CH_3{-}CH_2{-}OH \quad + \quad Br^-$$

$$\bar{O}H$$

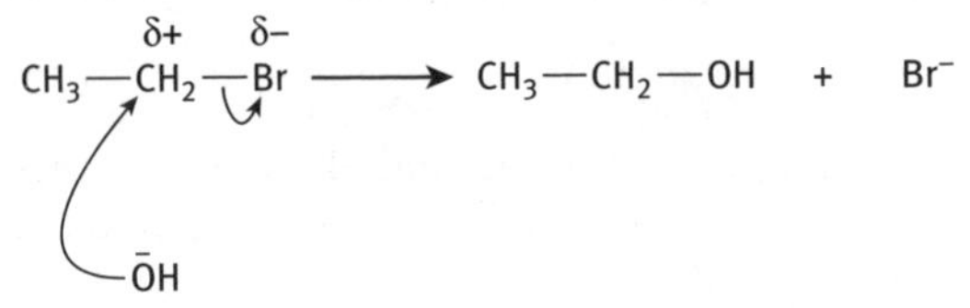

This is an example of a **nucleophilic substitution**.

11.3 Elimination reactions

Consider the dehydration (removal of water) from ethanol in acidic solution to produce ethene (Fig. 80).

The lone pair of electrons on the oxygen atom moves to form a coordinate covalent bond with an H^+ ion (both electrons in the bond are provided by the oxygen). This momentarily imparts a positive charge on the oxygen, causing an electron pair to move from the C-O bond to reinstate the lone pair of electrons on the oxygen, in the process resulting in the formation and loss of water from the ethanol molecule, but also forming an electrophilic centre on the carbon atom. This unstable intermediate molecule rearranges further, with an electron pair moving from a C-H bond of the methyl (CH_3) group to the C^+, so satisfying the tetravalency of carbon and resulting in the elimination of H^+ and formation of ethene.

Figure 80. The dehydration of ethanol by an elimination reaction

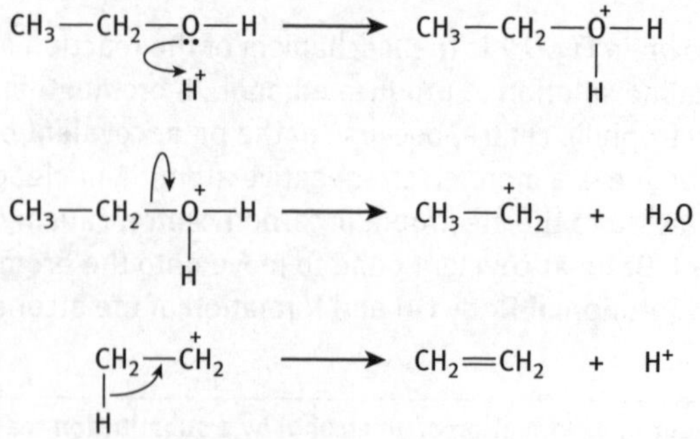

$$CH_3—CH_2—\ddot{O}—H \xrightarrow{H^+} CH_3—CH_2—\overset{+}{O}(H)—H$$

$$CH_3—CH_2—\overset{+}{O}(H)—H \longrightarrow CH_3—\overset{+}{C}H_2 + H_2O$$

$$CH_2(H)—\overset{+}{C}H_2 \longrightarrow CH_2{=}CH_2 + H^+$$

Reminder

In addition, substitution and elimination reactions, the electrons involved in forming new bonds or breaking existing bonds always move in pairs

11.4 Free radical reactions

Normally, bonds do not split in a way that leaves an atom or molecule with an odd, unpaired electron. However, **free radicals** do contain a single unpaired electron, and are produced by the **homolytic** fission of a covalent bond. Free radicals are indicated by a single dot to the side and middle of the atom concerned; for example, the chlorine free radical is shown as **Cl·**

In Fig. 81, the first reaction shown is a **heterolytic** fission, the 'full head' curly arrow denoting the movement of a **pair** of electrons; the products formed are **ions.** The second reaction is a homolytic reaction, the 'half head' curly arrows denoting the movements of single electrons; the products formed are free radicals.

Figure 81. Heterolytic and homolytic fission of covalent bonds

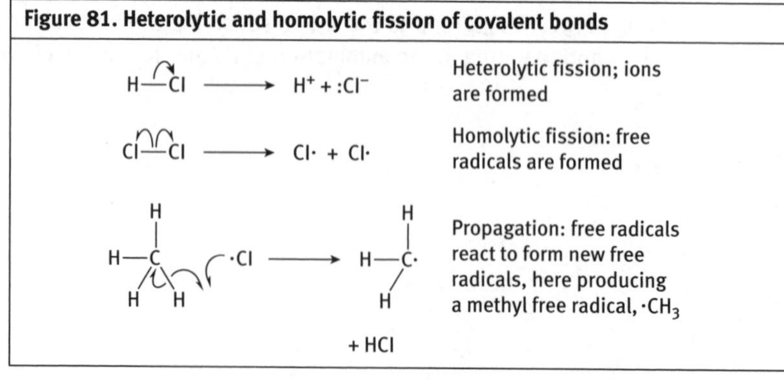

Free radicals are very unstable and react quickly with other compounds, trying to capture the needed electron to gain stability. Generally, free radicals attack the nearest stable molecule, 'stealing' an electron. When the 'attacked' molecule loses an electron, it becomes a free radical itself, beginning a chain reaction and **propagating** the reaction to other molecules. Once the process is started, it can cascade resulting in damage and disruption to the living cell.

Some free radicals arise normally during metabolism. Indeed, the body's immune system cells purposefully create them to neutralise viruses and bacteria. However, environmental factors such as pollution, radiation, cigarette smoke and herbicides can also spawn free radicals. **Antioxidants** include a range of different chemicals which can **terminate** free radical propagation; they do this by donating an electron, so ending the 'electron stealing' reaction chain. Vitamins C and E are examples of biological antioxidants.

Reminder

Free radicals are generated by the homolytic cleavage of a covalent bond and contain a single unpaired electron

11.5 Pi bonds and addition reactions

Perhaps not obviously so, the pi bond may also be considered to be a functional group, across which addition reactions can occur. Remember that a covalent double bond consists of a sigma and a pi bond. The pi bond is 'electron rich' and can effectively act as a nucleophilic site (Fig. 82). The pi bond often breaks and the electrons in it are used to join other atoms or groups in an addition reaction.

Figure 82. The 'electron-rich' pi bond is a nucleophilic site

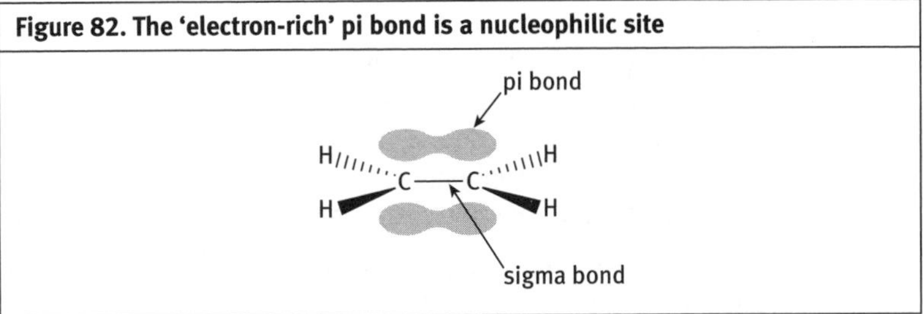

Consider the reaction scheme in Fig. 83 involving a carbon–carbon double bond.

In stage **1**, the pi bond is approached by the molecule X-Y; Y is highly electronegative and so X is an electrophile. As $X^{\delta+}$ approaches the pi bond, electrons in the X-Y bond are further pushed onto Y (shown by ··).

Reminder

Pi bonds provide an 'electron rich' site which effectively acts as a nucleophilic reactive site

Figure 83. The addition reaction across a pi bond

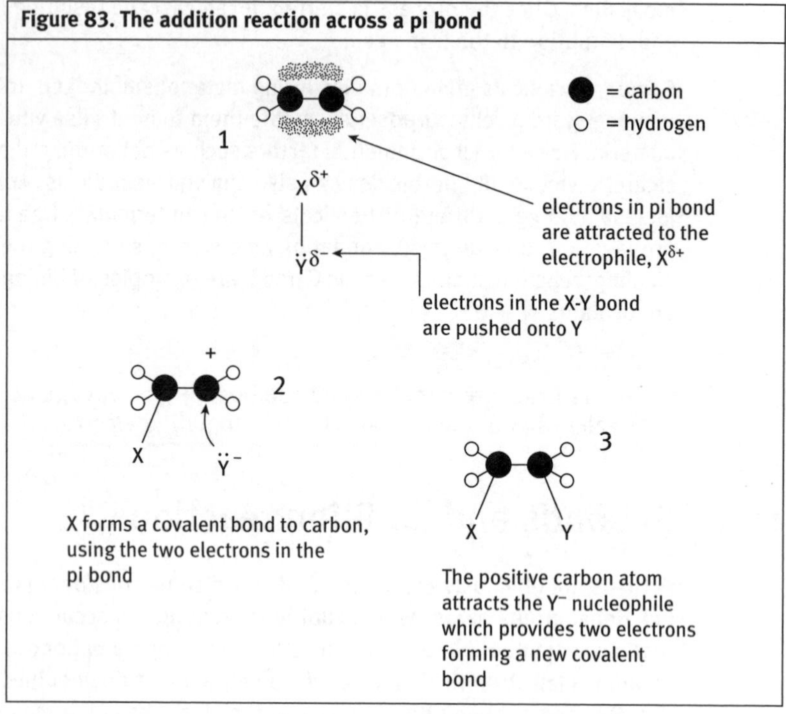

In stage **2,** X forms a new covalent bond to a carbon atom using the two electrons from the pi bond. The pi bond was originally made using an electron from each carbon atom, but both of these electrons have now been used to make a bond to the X atom. This leaves the right-hand carbon atom an electron short – hence positively charged. In stage **3**, Y has an electron pair, originally used to bond X to Y; this pair of electrons can form a new covalent bond with the 'electron deficient' carbon.

Thus, an addition reaction has been made across the carbon double bond. Pi covalent bonds are a useful 'source' of electrons for addition reactions to occur.

11.6 Functional groups link molecules together

The hallmark of organisms is their ability to link simple molecules together to form **macromolecules.** Carbohydrates, proteins and nucleic acids are **polymers,** large macromolecules formed through the linkage of numerous **monomers**. Macromolecules can be structural (e.g. collagen in bone and connective tissues, cellulose in plants), functional (e.g. enzymes and hormones), deposits of information (e.g. DNA), as well as deposits of energy (e.g. glycogen and starch).

The hydroxyl functional groups on sugar monomers can react through a condensation reaction (dehydration, loss of water) to form sugar linkages, eventually building to macromolecules such as starch. For example amylose, a constituent of starch, consists typically of 200 to 20 000 glucose units (Fig. 84).

Figure 84. First step in amylose formation through linking of glucose monomers

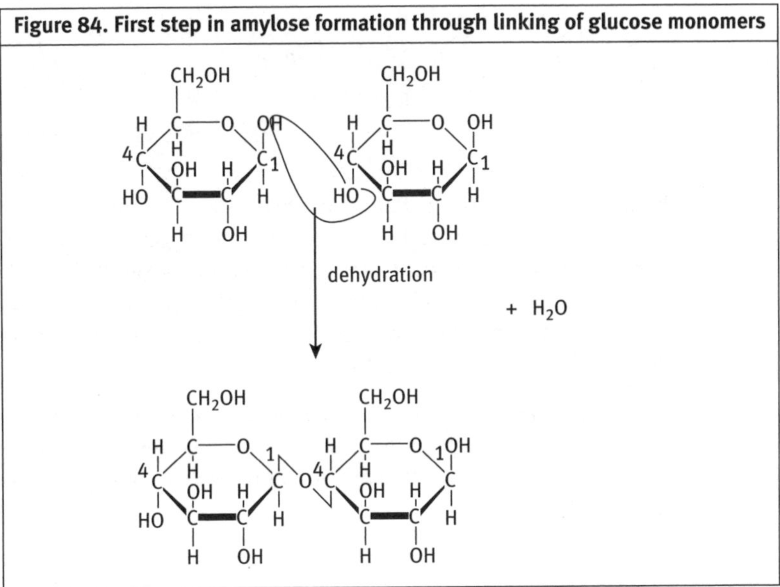

The bond linkages, between carbon atom 1 of one glucose molecule, and carbon atom 4 of another glucose molecule, are referred to as 1,4-glycosidic bonds.

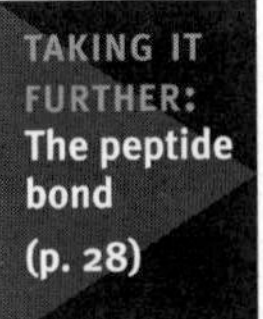

Similarly, as we have seen previously, amino acids are linked together by peptide bonds through a condensation reaction to form those polymers which are proteins. Again, the reaction involves a functional hydroxyl group, on the 'carboxyl end' of one amino acid, with the elimination of water (Fig. 85).

Figure 85. Amino acid monomers are linked together through peptide bonds

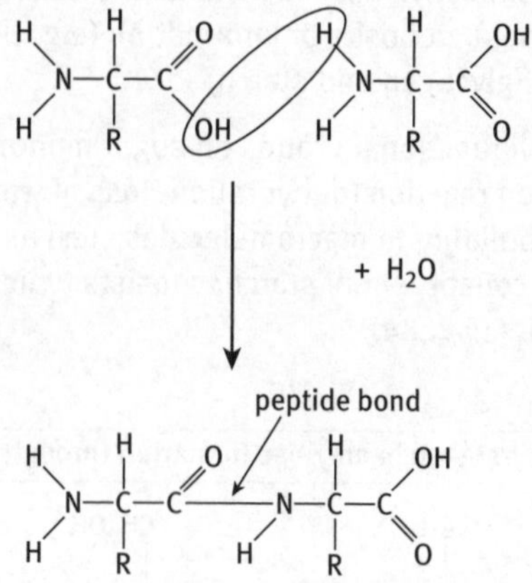

In nucleic acids, the 'backbone' of each single nucleic acid strand is a polymer of nucleotides, linked through dehydration reactions to form phosphodiester bonds between the phosphate group and the 3′ and 5′ carbons of adjacent nucleotides. Once again, functional hydroxyl groups at the 3′ and 5′ positions of the sugar molecule provide the means for such linkages to occur (Fig. 86).

Through these 'polymerisation' reactions, nucleic acids are formed (Fig. 87), the single polymer strands being held together in a double-stranded structure by hydrogen bonding between opposite bases (see Chapter 3).

Reminder

Carbohydrates, proteins and nucleic acids are polymers, linked together through the functional groups of their monomers

Figure 86. Hydroxyl groups on sugar molecules provide the means for constructing nucleotide polymers

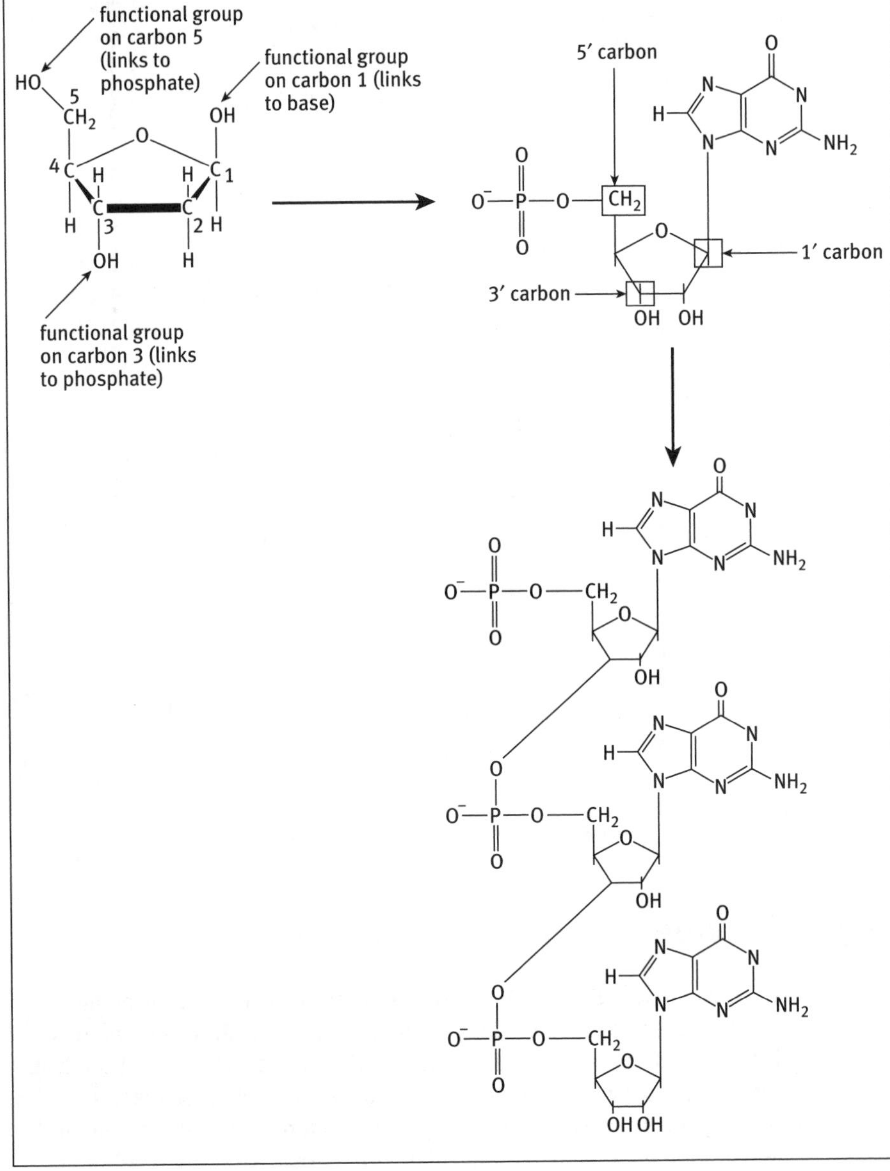

Figure 87. Double-stranded nucleic acid

5′ end | 3′ end

3′ end | 5′ end

Base A = adenine, C = cytosine, G = guanine, T = thymine

11.7 Enzyme-catalysed reactions

We considered in Section 8.2 how an enzyme was able to provide an alternative reaction route, and therefore a lower activation energy; it is this ability of enzymes which underlines their roles as biological catalysts. But exactly how is an enzyme able to provide an alternative reaction route? A clue is provided earlier in the elimination reaction scheme for the dehydration of ethanol to ethene (Fig. 80). In this scheme a number of intermediates are shown; such intermediates constitute **transition states** which are generally very unstable and short-lived. Enzymes are able to bind and stabilise transition states; indeed enzymes can make possible different, and often multiple, transition states providing alternative reaction routes from reactant to product with lower activation energies.

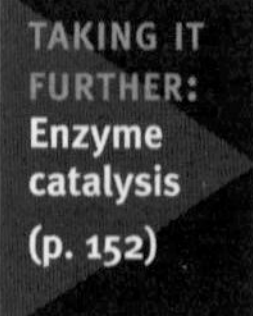

11.8 Summing up

1. Functional groups provide reactive sites on biomolecules; these may be nucleophilic or electrophilic centres.
2. Reactions at active sites can be classified as addition, substitution or elimination.
3. Free radicals are produced by the homolytic fission of a covalent bond, as opposed to heterolytic fission which is most common for covalent bond breakage.
4. A pi molecular orbital provides an additional nucleophilic reactive site; the two electrons of the pi bond are used to add additional atoms or groups across the double bond.
5. Functional groups are utilised to join simple molecules (monomers) together to form large macromolecules (polymers). Typical reactions in organisms involve condensation reactions (a dehydration reaction where there is a loss of water), linking monomers through their hydroxyl groups to form macromolecules such as polysaccharides, proteins and nucleic acids.

11.9 Test yourself

The answers are given on p. 162.

Question 11.1
Explain what is meant by a nucleophilic and an electrophilic centre on a molecule.

Question 11.2
Why is a pi bond considered to be a nucleophilic site?

Question 11.3
Explain what is happening in the following sequence.

$$H_2C{=}CH_2 + A^{\delta+}{-}B^{\delta-} \longrightarrow H_2C(A){-}\overset{+}{C}H_2 + :B^- \longrightarrow H_2C(A){-}CH_2(B)$$

Question 11.4
Explain the relationship between enzymes, activation energy and transition states.

Question 11.5
Explain the difference between an ion and a free radical.

Taking it further

11.10 Enzyme catalysis

Enzymes are biological catalysts. They make possible many hundreds of chemical reactions in the cell which might otherwise not be thermodynamically feasible. They are highly specific with regard to which **substrate** they will interact with, and they are controllable making possible the highly complex integration of the cell's metabolism.

Enzymes are proteins; large three-dimensional amino acid polymers which adopt specific conformations (shapes) and recognise only certain substrates. The high specificity of substrate recognition is a consequence of a three-dimensional **'active site'** on each enzyme. An active site is a region in the three-dimensional structure that will accommodate only a particular shape and size of substrate. An analogy may be drawn with a 'lock and key' – the enzyme is the 'lock' and only one type of 'key', a specific substrate, will fit it!

It is relatively straightforward to list those properties of enzymes which make them good catalysts:

- Enzymes **bind** and hold 'still' their substrate (this is far more effective than relying on random collisions between molecules). This is often how chemical catalysts work, by binding and 'presenting' molecules or atoms on their surface.
- Enzymes make reactions feasible by providing an alternative reaction route with a lower activation energy.

Easy to say, but exactly how do enzymes **bind** their substrates, and how do they make reactions feasible?

Binding of substrate

The secret lies in the sequence of amino acids, linked together through peptide bonds, to form the polymeric enzyme (protein) molecule. The general structure of an amino acid is shown below.

```
        H
        |
H2N—C—COOH
        |
        R
```

In a protein (polypeptide) chain, the amino (NH_2) and carboxyl (COOH) groups of each amino acid are involved in peptide bonding. The R group is specific for each amino acid.

Some R groups contain a carboxylate or an amino group which is charged (dissociated or protonated) at physiological pH; examples include:

- R group of glutamic acid, $HOOC(CH_2)_2$- dissociates to $^{-}OOC(CH_2)_2$-
- R group of lysine, $H_2N(CH_2)_4$- is protonated to $H_3{}^{+}N(CH_2)_4$-

Some R groups are polar, but not dissociated, because they contain a relatively highly electronegative atom, such as the sulfhydryl group in cystine, HS-CH_2-

R groups may be apolar, such as the methyl groups in leucine, $(CH_3)_2$-CH-CH_2-

In addition to the various R groups, each peptide bond contains a carbonyl (-C=O) and an amine (-NH) group. These are all **functional** groups; many are dissociated and charged at physiological pH, or contain an electronegative atom and polar covalent bond. Arrangements of such groups about the three-dimensional 'surfaces' of the **active site** of an enzyme provide specific possibilities for charge–charge interactions or hydrophobic interactions with a substrate, which itself possesses interactive functional groups. The specific arrangement of amino acids at the active site leads to specific interactions with specific substrates. In other words, enzymes bind their substrates through intermolecular interactions. In Chapter 3 we learnt that such intermolecular interactions are relatively weak and reversible (as opposed to the intramolecular interactions which are strong covalent bonds). This is the basis of **binding**. Binding is a central concept in biology. Enzymes bind their substrates, hormones bind their receptors, antibodies bind antigens. Biological events, at the molecular level, are always preceded by binding, which results in a recognition and from which follows an action.

Providing an alternative reaction route

The second property of enzymes, namely that of providing an alternative reaction route with a lower activation energy, is also a property of the various R groups on the different amino acids. We can demonstrate this comprehensively by considering the mechanism of action of the enzyme **chymotrypsin**. Chymotrypsin is an enzyme which is present in our small intestine and which catalyses the hydrolysis of peptide bonds in proteins; in other words this enzyme helps to digest proteins into smaller peptides and amino acids which can be absorbed more readily into the body.

In the laboratory, the chemical hydrolysis of a peptide bond is a very slow process, unless a strong acid catalyst is added.

$$R{-}C({=}O){-}N(R){-}R \;+\; H{-}O{-}H \;\rightleftharpoons\; R{-}C({=}O){-}O{-}H \;+\; R{-}N(R){-}H$$

amide water carboxylic acid amine

However, in the small intestine, where the pH is essentially neutral, the reaction catalysed by chymotrypsin occurs very quickly.

There are six stages in the mechanism of chymotrypsin catalysis which are explained in the following sequence.

Stage 1: A protein substrate approaches the active site of the enzyme. The active site of chymotrypsin allows that part of the enzyme to bind a portion of the protein which has a non-polar side chain, like those found in phenylalanine (consequently, those amino acid R groups lining this portion of the active site will be apolar). Once the protein is in place at the active site, an H^+ ion moves from the serine amino acid at position 195 (Ser-195) of the enzyme's amino acid sequence, to the histidine amino acid at position 57 (His-57). The oxygen atom in serine's hydroxyl group then forms a covalent bond to the carbon of one of the substrate's peptide bonds, in turn 'pushing' the two electrons from the pi bond on to the carbonyl oxygen. This is one transition state in the substrate transformation which is stabilised by the enzyme active site.

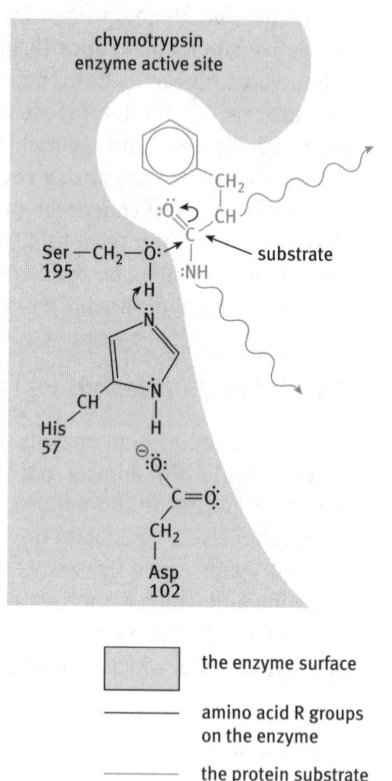

Stage 2: The positive charge formed on His-57 is stabilised by the negative charge on the aspartic acid at position 102 (Asp-102). When the double bond on the carbonyl group of the peptide bond is reformed, the bond between the carbon and the nitrogen in the peptide bond is broken. The nitrogen-containing group is stabilised by the formation of a bond to a hydrogen atom from His-57. This is a second transition state in the substrate transformation.

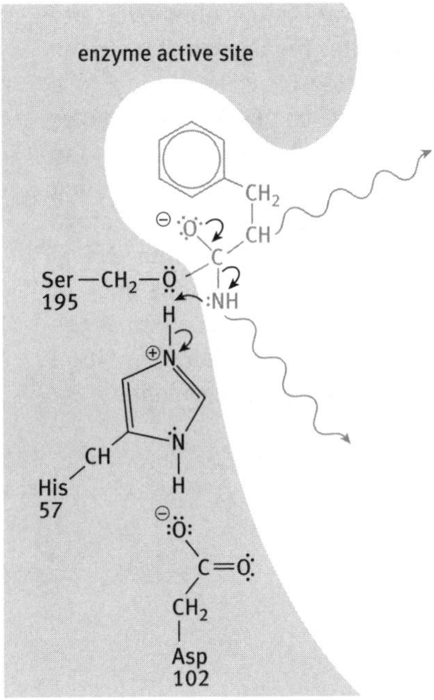

Stage 3: The portion of the polypeptide which contains the nitrogen atom from the broken peptide bond moves out of the active site. This represents a third transition state in the substrate transformation.

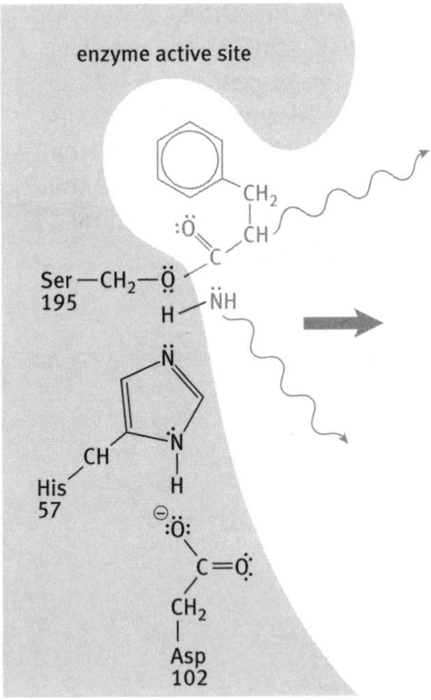

Stage 4: A water molecule moves into the active site. The oxygen atom in the water molecule loses an H^+ ion to a nitrogen atom on His-57. This allows water's oxygen atom to form a bond to the carbon atom on the remaining portion of the substrate. Like in stage 1, the pair of electrons from the pi bond moves to form a lone pair of electrons on the oxygen atom of the carbonyl group. Another transition state in the substrate is evident.

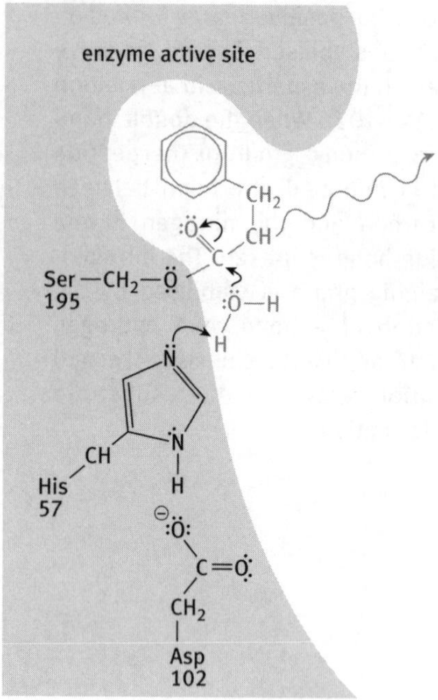

Stage 5: When the carbonyl double bond reforms, the bond between carbon and the oxygen of Ser-195 is broken. The –OH group on Ser-195 is restored with a transfer of an H^+ ion from His-57. With this step, the Ser-195 and His-57 are both returned to their original forms.

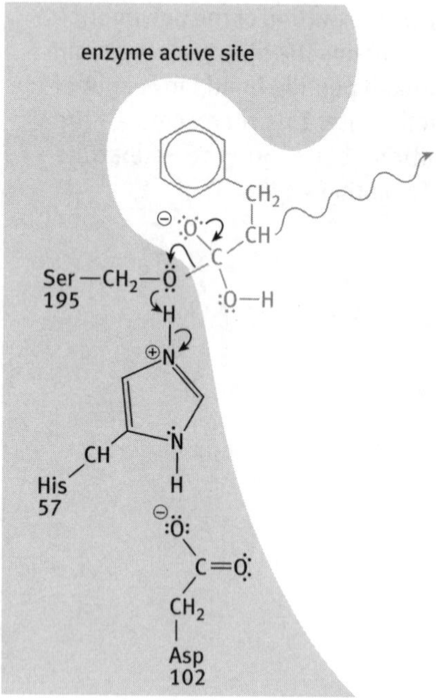

Stage 6: The remaining portion of the substrate moves out of the active site, leaving the active site in its original form, ready to repeat stages 1–5 with another protein molecule.

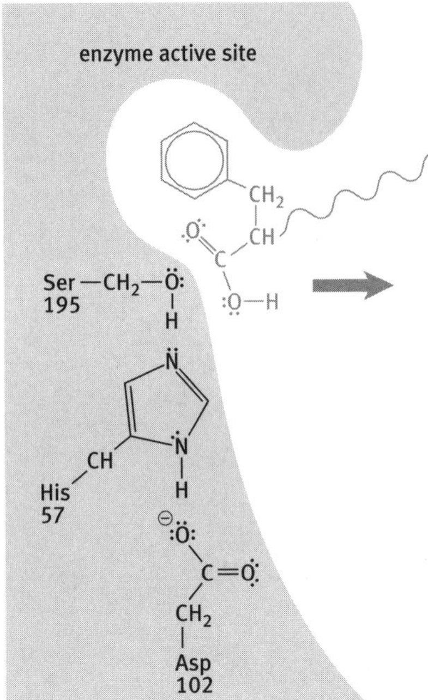

Even this reaction scheme has been simplified! Nevertheless, it can be seen that a number of transition states must occur during this transformation, each being stabilised through the intervention of different functional groups in the enzyme active site.

So, functional groups on biomolecules such as proteins are responsible for specific binding events, mediate catalytic processes and, of course, make possible the intermolecular interactions within such molecules which define the molecules' active and functional three-dimensional shape.

Answers to "test yourself" questions

Answer 1.1
Hydrogen-1 (^1_1H) has just one proton and a mass number of 1. Hydrogen-2 (^2_1H), or deuterium, has an additional neutron and a mass number of 2 (1 proton + 1 neutron). Hydrogen-3 (^3_1H), or tritium, has two additional neutrons and a mass number of 3 (1 proton + 2 neutrons).

Answer 1.2
(a) Just the one, i.e. 1s.
(b) In energy level 2 we can find two types of atomic orbital, namely 2s and 2p; a total of four atomic orbitals may be present, namely $2s + 2p_x + 2p_y + 2p_z$.

Answer 1.3
A region in space about the atomic nucleus where there is a high probability of finding an electron.

Answer 1.4
(a) 10; two in 1s, two in 2s, and two in each of three 2p atomic orbitals.
(b) This would be an unreactive element since there are no unpaired electrons in the outer energy level which can participate in covalent or ionic bonding; the 2s and 2p atomic orbitals in energy level $n = 2$ are full (this element is in fact neon, one of the inert gases).

Answer 1.5
(a) 1s, 2s, ($2p_x$, $2p_y$, $2p_z$); all 2p atomic orbitals have equivalent energy.
(b) Atomic orbitals closest to the nucleus have lowest energy.

Answer 2.1
The atomic orbitals of two atoms merge to form one or more molecular orbitals in which electrons are shared in the formation of a covalent bond.

Answer 2.2
A sigma molecular orbital may be formed through the 'head-to-head' merging of s or p atomic orbitals, whereas a pi molecular orbital is formed by the 'side-to-side' merging of p atomic orbitals.

Answer 2.3
An 'asymmetric' covalent bond is one in which the electrons in the bond are drawn towards the more electronegative atom, creating a dipole (a separation of charge – hence the term 'polar'). The partial charges on the atoms allow for intermolecular charge–charge interactions with water.

Answer 2.4
Nothing really! A dative covalent (or coordinate) bond is the same as any other similar covalent bond, except for the fact that the two electrons constituting the bond are both provided by just one of the atoms in the bond, rather than the usual case of one electron being contributed by each of the two atoms forming the bond.

Answer 2.5
The 'octet rule' is explained in the 'Taking it further' section on the periodic table. For most of the lighter elements a complete outer, or valence, shell requires eight electrons, which leads to a more stable and unreactive state (Group 18 of the periodic table). These observations are said to obey the octet rule.

Answer 3.1
(a) A hydrogen bond (b) Common functional groups which participate in hydrogen bonding include hydroxyl (-OH) , carbonyl (-C=O), and amino ($-NH_2$).

Answer 3.2
The hydrogen bond is (i) a relatively strong form of intermolecular interaction, (ii) of a relatively 'fixed' distance, (iii) strongly directional.

Answer 3.3
Intermolecular interactions (i) are much weaker than covalent bonds, (ii) 'make and break' relatively quickly in water, (iii) tend to be rather dependent on pH.

Answer 3.4
Van der Waals short range intermolecular interactions.

Answer 3.5
(c) and (e); in (c) the methyl group (CH_3) shows no polarity since C and H have similar electronegativities, in (e) the carbon ring structure is particularly hydrophobic. The ring structure in (f) is made polar by the attachment of a hydroxyl group.

Answer 4.1
Remember, no. of moles = mass in grams/molar mass. Therefore, we have $0.01/6000 = 1.66 \times 10^{-6}$, or 1.66 micromoles of insulin.

Answer 4.2
The M_r of ethanoic acid is $(12 \times 2) + (4 \times 1) + (16 \times 2) = 60$. In other words, 60 grams of ethanoic acid is 1 mole. A 1 M solution of ethanoic acid would contain 1 mole (60 grams) in 1 litre. A 0.1 M solution of ethanoic acid would contain 1/10 moles (6 grams). It therefore follows that to make 10 litres of a 0.1 M solution of ethanoic acid, you would need $10 \times 6 = 60$ grams.

Answer 4.3
10 grams of glucose is equivalent to $10/180 = 0.055$ moles. There are therefore 0.055 moles of glucose present in the 50 ml of solution. The molarity is defined as the number of moles in a litre. If there are 0.055 moles in 50 ml, then by extrapolation there would be $0.055 \times 1000/50$ (assuming no dilution) in a litre = 1.1 M (remember, the molarity does not change with volume [unless diluted], but the number of moles will).

Answer 4.4
If the stock solution of glycine is 0.02 M, then by definition this would contain 0.02 moles of glycine in 1 litre. Therefore, 1 ml of this solution must contain $0.02/1000 = 2 \times 10^{-5}$ moles. Of course, 1 ml of this solution is still 0.02 M! The total volume of the enzyme assay is 3 ml, 1 ml of which is from the added glycine solution. In this total volume there would still be 2×10^{-5} moles of glycine, but the concentration has been diluted. The molarity is now (1 in 3 dilution) $0.02/3 = 0.0066$ M (6.6×10^{-3} M).

Answer 4.5
1.2 grams of glycine is equivalent to $1.2/79 = 0.015$ moles. (a) In 1 ml of this 100 ml there must be $0.015/100 = 0.00015$ moles. (b) One litre of this solution would contain 1000×0.00015 moles = 0.15 mol l^{-1}, i.e. a 0.15 M solution. (c) 1 ml of this solution must contain $0.00015 \times 6.022 \times 10^{23}$ (since 1 mol is equivalent to 6.022×10^{23} molecules = Avogadro's number) $= 9.033 \times 10^{19}$ molecules.

Answer 5.1
In an sp^3 hybridised carbon atom, there are four hybridised atomic orbitals derived from the 2s and three 2p atomic orbitals. In an sp^2 hybridised carbon, there are three hybridised atomic orbitals leaving one 2p atomic orbital unchanged.

Answer 5.2
In methane the carbon is sp^3 hybridised; its four hybridised atomic orbitals each produce a covalent bond with hydrogen. The four C-H bonds arrange themselves as far apart from one another as possible; consequently the three-dimensional shape of methane is that of a tetrahedron.

Answer 5.3
In (a), (b) and (d) the carbon must be forming four sigma covalent bonds, i.e. the carbon is sp^3 hybridised.

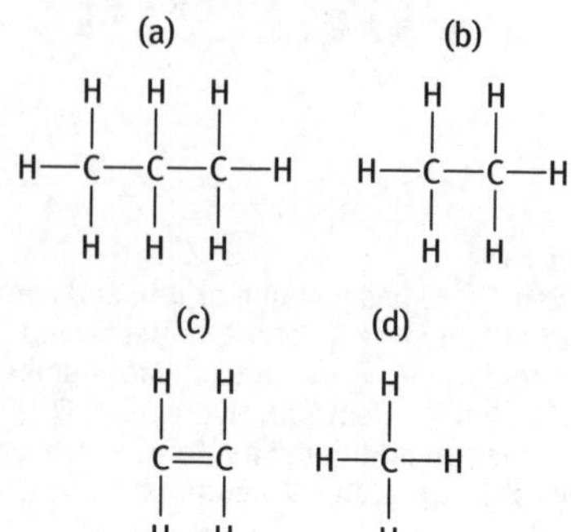

In (c) however, the only way that carbon could satisfy its valency and form four covalent bonds is if it were sp^2 hybridised and therefore able to form a pi bond through the unhybridised p orbital, so creating a carbon–carbon double bond.

Answer 5.4
B and C. In B, four of the six carbons are sp^2 hybridised (those either side of the two double bonds), while in C every carbon is sp^2 hybridised and the ring is aromatic. In A, each carbon is sp^3 hybridised, forming four covalent bonds; one to each of two adjacent carbons, and two to each of two hydrogens (not shown).

Answer 6.1
(i); the two molecules are structural isomers.

Answer 6.2
Both pairs of molecules are stereoisomers, but enantiomers are mirror images of each other; diastereomers are not mirror images.

Answer 6.3
A chiral centre (most often an asymmetric carbon atom) has four different groups attached to it.

Answer 6.4
The number of possible stereoisomers is given by 2^n, where n is the number of chiral centres. The answer is therefore $2^6 = 64$.

Answer 6.5
By observing their effect on the rotation of plane polarised light. D-glucose will rotate plane polarised light to the right (it is dextrorotatory), whereas L-glucose will rotate it to the left (it is levorotatory).

Answer 7.1
Water is a dipolar molecule. The electronegative oxygen atom draws electrons to itself, causing the hydrogen atoms to have a partial positive charge. In the diagram as shown the water molecules would repel each other.

Answer 7.2
(a) An acid is a substance which produces H^+ ions by dissociation
(b) A strong acid will essentially completely dissociate in solution (its dissociation constant K_a will be very large), whereas a weak acid will only partially dissociate and its K_a will be smaller
(c) The pH scale is a log scale, from 0 to 14
(d) Remember, each pH unit represents a 10-fold increase (or decrease) in $[H^+]$; so an increase in pH from 9 to 4 is five pH units, which equals a $10 \times 10 \times 10 \times 10 \times 10$, i.e. a 100 000 fold increase in $[H^+]$.

Answer 7.3
(a) Assuming the strong acid is completely dissociated, then $[H^+] = 0.05$ M, so using the equation:

$$\begin{aligned} pH &= -\log[H^+] \\ &= -\log[0.05] \\ &= -(-1.30) \\ &= 1.30 \end{aligned}$$

(b) For a solution of pH 6.2, the $[H^+]$ is given by:
$6.2 = -\log[H^+]$
the negative antilog of 6.2 is 6.3×10^{-7} *[on the calculator enter 'shift' 10^x-6.2]*
Therefore, $[H^+] = 6.3 \times 10^{-7}$ M

Answer 7.4
$pK_a = -\log_{10} K_a$
therefore,

$$\begin{aligned} pK_a &= -\log_{10}(1.8 \times 10^{-5}) \\ &= -(-4.74) \\ &= 4.74 \end{aligned}$$

Answer 7.5
Use the Henderson–Hasselbalch equation.

$$\begin{aligned} pH &= pK_a + \log \frac{[\text{base}]}{[\text{acid}]} \\ &= 4.75 + \log \frac{0.05}{0.10} \\ &= 4.75 + (-0.30) \quad \textit{[enter log (0.05 divided by 0.10)]} \\ &= 4.45 \end{aligned}$$

Answer 7.6
Using the Henderson–Hasselbalch equation:

$$\begin{aligned} pH &= pK_a + \log \frac{[\text{base}]}{[\text{acid}]} \\ &= 8.08 + \log \frac{0.186}{0.14} \\ &= 8.08 + (0.123) \\ &= 8.20 \end{aligned}$$

Answer 7.7
That molecule with a pK_a of 4.2 would give the most acid solution. Using the equation $pK_a = -\log_{10}[K_a]$, i.e. taking the negative antilog of each pK_a value, the $[H^+]$ are respectively 6.31×10^{-5}, 1.58×10^{-7} and 6.31×10^{-9} M.

Answer 8.1
(c) is correct; there are no heat gradients in cells to do work.

Answer 8.2
(b) is correct; this is the only catabolic process, all the other processes are synthetic (anabolic) and will require a net input of energy to make them feasible.

Answer 8.3
(c) is correct; lowering the activation energy.

Answer 8.4
Exergonic describes reactions/processes which result in the release of free energy, i.e. a negative ΔG. Endergonic processes require an input of free energy (ΔG is positive), usually through being coupled to an exergonic process.

Answer 8.5
Firstly, to calculate the entropy change, use the equation:
$\Delta G = \Delta H - T\Delta S$
$-3089.0 \text{ kJ mol}^{-1} = -2807.8 \text{ kJ mol}^{-1} - (310 \times \Delta S)$ *[T=273 + 37 =310 K]*
To simplify, if we add 2807.8 kJ mol^{-1} to both sides, we get:
$-281.2 \text{ kJ mol}^{-1} = -310 \text{ K} \times \Delta S$

$$\Delta S = \frac{-281.2 \text{ kJ mol}^{-1}}{-310 \text{ K}} \quad \textit{[dividing two minuses gives a plus!]}$$

$$= 0.907 \text{ kJ mol}^{-1} \text{ K}^{-1}$$

The thermodynamic data provided for this reaction would suggest it is spontaneous, i.e. ΔG is negative, as is ΔH (reaction is exothermic). Furthermore, the calculated ΔS is positive. We would assume the reaction to be feasible and both 'enthalpy-driven' and 'entropy-driven'.

Answer 9.1
The equilibrium constant, K_{eq}, is given by the expression:

$$K_{eq} = \frac{[\alpha\text{-ketoglutarate}][CO_2][NADH]}{[\text{isocitrate}][NAD^+]}$$

Answer 9.2
An equilibrium constant of 3.18×10^{520} (which is very high!) would indicate that the reaction goes almost to completion with almost no reactants left.

Answer 9.3
Using the equation:
$\Delta G = -RT \ln K_{eq}$, we have
-3.7×10^3 Jmol^{-1} = -8.314 JK^{-1}mol^{-1} $\times$ 310 K $\times$ $\ln K_{eq}$
(multiplying ΔG by 10^3 because R is in Joules and ΔG is in kilojoules; and using the absolute temperature T as 273 + 37 = 310 K)
therefore,

$$\ln K_{eq} = \frac{-3.7 \times 10^3}{-8.314 \times 310} = 1.44$$

and so $K_{eq} = 4.20$ M *(use 'shift' e^x on the calculator to give e1.44 = 4.20)*
ΔG is relatively small for this reaction, and the value of K_{eq} is likewise low, suggesting that there are significant amounts of reactant still present.

Answer 9.4
Using the equation:
$\Delta G^{o\prime} = -RT \ln K_{eq}$
and since R (the gas constant) is given in joules, substituting values gives:
13000 Jmol^{-1} = 8.314 JK^{-1}mol^{-1} $\times$ 310 K $\times$ $\ln K_{eq}$

$$\ln K_{eq} = \frac{-13\,000}{-8.314 \times 310}$$

$\ln K_{eq} = -5.04$
$K_{eq} = 0.00645 = 6.45 \times 10^{-3}$M

Answer 10.1
The two half-reactions for (a) are: $Zn \rightarrow Zn^{2+} + 2e^-$ and $Cu^{2+} + 2e^- \rightarrow Cu$, and for (b) are: $Fe^{2+} \rightarrow Fe^{3+} + e^-$ and $Cu^{2+} + e^- \rightarrow Cu^+$

Answer 10.2
(a) The two half-reactions are written as:
acetaldehyde + $2H^+ + 2e^- \rightarrow$ ethanol
$NADH + H^+ \rightarrow NAD^+ + 2e^- + 2H^+$
(b) Using the tabulated redox potential values given in the text, for acetaldehyde/ethanol, $\Delta E^{o\prime} = -0.2$ V and for NAD^+ / NADH. $\Delta E^{o\prime} = -0.32$ V. However, in the reaction as shown, NADH is being oxidised; therefore we reverse the sign of the redox potential (which by convention is given as a reduction reaction), and so the $\Delta E^{o\prime}$ is +0.32 V.
Remember that electrons will tend to flow from a 'more negative' to a 'more positive' redox potential. Here we have electrons flowing from the oxidation of NADH ($\Delta E^{o\prime} = 0.32$ V) to the reduction of acetaldehyde ($\Delta E^{o\prime} = -0.2$ V). The reaction therefore is favoured and will tend to proceed from left to right.

Answer 10.3
Using tabulated redox values, $\Delta E^{o\prime}$ for reduction of ubiquinone (ubiquinone $_{(oxidised)}$ $\rightarrow$ ubiquinone $_{(reduced)}$) is 0.10 V, and $\Delta E^{o\prime}$ for reduction of cytochrome *c* (cytochrome *c* $_{(oxidised)}$ $\rightarrow$ cytochrome *c* $_{(reduced)}$) is 0.254 V. The reaction as written shows a reduction of ubiqinone ($\Delta E^{o\prime} = 0.10$ V), electrons being supplied by the oxidation of cytochrome *c* (so $\Delta E^{o\prime} = -0.254$ V, reversing the 'sign'). Since electrons tend to flow from a 'more negative' to a 'more positive' redox potential, this reaction is likely to proceed from right to left; therefore the reaction as shown is unfavourable.

Answer 10.4
Using tabulated redox potential values; succinate is being oxidised (donating two electrons to FAD), therefore the $\Delta E^{o\prime}$ is -0.03 V. For FAD, which is being reduced, $\Delta E^{o\prime} = -0.18$ V. The change in redox potential ($\Delta E^{o\prime}$) is therefore: $-0.03 + (-0.18) = -0.21$ V. Inserting this value into the equation gives:
$\Delta G^{o\prime} = -nF\Delta E^{o\prime}$
$\Delta G^{o\prime} = -2 \times 96485 \times -0.21$
$\Delta G^{o\prime} = +40524$ J mol^{-1} *(remember, two minuses make a plus)*
The $\Delta G^{o\prime}$ value for this reaction as written is positive, suggesting the reaction is unfavourable and is more likely to proceed from right to left.

Answer 10.5
In this reaction as written, NAD^+ is being reduced to NADH ($\Delta E^{o\prime} = -0.32$ V), and malate is being oxidised to oxaloacetate ($\Delta E^{o\prime}$ is therefore +0.17 V). The $\Delta E^{o\prime}$ for this reaction is therefore $-0.32 + 0.17 = -0.15$ V. Inserting this into the equation gives:
$\Delta G^{o\prime} = -nF\Delta E^{o\prime}$
$\Delta G^{o\prime} = -2 \times 96485 \times -0.15$
$\Delta G^{o\prime} = 28945$ J mol^{-1} *(remember, two minuses make a plus)*
A positive $\Delta G^{o\prime}$ suggests this reaction as written is unfavourable and is more likely to proceed from right to left.

Answer 11.1
A nucleophilic site on a molecule is an area which is 'electron rich' and one which will attract a nucleophile, such as a proton. Nucleophilic sites may result from electronegative atoms forming polar covalent bonds, pi bonds, lone electron pairs on an atom, or dissociated groups. An electrophilic centre is one which carries a positive charge and therefore attracts an electrophile.

Answer 11.2
Pi bonds, which are formed by the side-to-side merger of p atomic orbitals, effectively produce an 'electron-rich cloud' above and below the plane of the sigma covalent bond. The two electrons in the pi bond readily 'add' to an electrophile in an addition-type reaction.

Answer 11.3
The hydrocarbon molecule (ethene) contains a carbon–carbon double bond. In the molecule A-B, B is a more electronegative atom, indicated by the negative charge on B, the positive charge on A, and the proximity of the two electrons (··) in the bond to B. A is therefore an electrophile and is attracted to the pi bond. As A-B approaches the pi bond, electrons in the A-B bond are pushed further onto atom B (shown by the small arrow in the diagram). Eventually, the A-B bond becomes so polarised that it breaks; A forms a new bond to carbon using the two electrons in the pi bond (a form of coordinate bonding). This leaves a positive charge on the other carbon atom (now an electrophilic site) which attracts the ion B^- with its lone pair of electrons. B 'adds' to C, providing the two electrons needed to form a new covalent bond.

Answer 11.4
Enzymes are biological catalysts which make reactions feasible by providing alternative reaction routes with lower activation energies. Intermolecular interactions at the active site of the enzyme with its substrate are able to stabilise different transition states which are the basis of the alternative reaction route (with a lower activation energy).

Answer 11.5
When covalent bonds are broken, they normally do so in a heterolytic fashion, i.e. the two electrons in the bond go to one or other of the two atoms involved; one atom will effectively gain an electron (becoming negatively charged) while the other atom effectively loses an electron, becoming positively charged. Again, two ions are formed, although they are likely to exist only transiently. In both cases the normal event is for the two electrons comprising the covalent or ionic bond to go to one or other of the two atoms when the bond is broken. However, a free radical is formed when a covalent bond is broken homolytically, i.e. the bond is split equally such that the two electrons in the bond are shared equally between the two atoms. This produces a free radical, containing an atom with a single unpaired electron. This is an unstable state and that atom will be very reactive, looking for an additional electron to pair with. It will often gain this electron by 'stealing' from another atom, thereby generating another radical, then another, in a chain reaction. This can cause serious damage to biological molecules.

Appendix 1. Some common chemical formulae

Formula	Name	Type of compound
HCl	Hydrochloric acid	Strong acid in aqueous solution
H_2SO_4	Sulfuric acid	Strong acid in aqueous solution
HNO_3	Nitric acid	Strong acid in aqueous solution
H_3PO_4	Phosphoric acid	Weak acid in aqueous solution
NH_3	Ammonia	Weak base in aqueous solution
CO_2	Carbon dioxide	Gas at STP (standard temperature and pressure) Weak acid in aqueous solution
CO	Carbon monoxide	Gas at STP
CH_4	Methane	Alkane
CH_3OH	Methanol	Alcohol
HCHO	Methanal	Aldehyde
HCOOH	Methanoic (formic) acid	Carboxylic acid (weak acid)
CH_3CH_3	Ethane	Alkane
CH_2CH_2	Ethene	Alkene
C_6H_6	Benzene	Aromatic hydrocarbon
CH_3CH_2OH	Ethanol	Alcohol
CH_3CHO	Ethanal	Aldehyde
CH_3COOH	Ethanoic (acetic) acid	Carboxylic acid (weak acid)

Appendix 2. Common anions and cations

Cations		Anions	
Formula	Name	Formula	Name
Na^+	Sodium	F^-	Fluoride
K^+	Potassium	Cl^-	Chloride
Ca^{2+}	Calcium	Br^-	Bromide
Mg^{2+}	Magnesium	SO_4^{2-}	Sulfate
Fe^{2+}	Iron(II) - ferrous	CO_3^{2-}	Carbonate
Fe^{3+}	Iron(III) - ferric	NO_3^-	Nitrate
NH_4^+	Ammonium	OH^-	Hydroxide
		HCO_3^-	Hydrogen carbonate
		PO_4^{3-}	Phosphate
		HPO_4^{2-}	Hydrogen phosphate
		$H_2PO_4^-$	Dihydrogen phosphate

Appendix 3. Common functional groups

Group	Name
$>C=C<$	Alkene (pi double bond)
$-O-H$	Hydroxyl
$>C=O$	Carbonyl (ketone)
$-C(=O)H$	Aldehyde
$-C(=O)OH$	Carboxyl
$-NH_2$	Amine
$-C(=O)NH_2$	Amide
$-C(=O)-N(R)-$	Peptide linkage
R	General representation of an alkyl group (C_nH_{2n-1})

Appendix 4. Notations, formulae and constants

Atoms

${}^{A}_{Z}X$ where A = the mass number and Z = the atomic number

Molecules

A line connecting atoms in a structural formula denotes a covalent (sigma bond), a double line denotes a double bond, i.e. a sigma bond plus a pi bond. A solid wedge denotes a bond coming out of the page, a hatched wedge denotes a bond going in to the page.

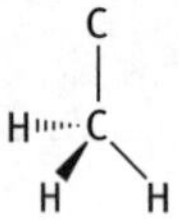

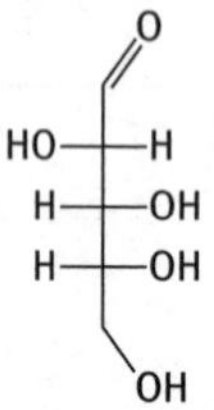

can be written as

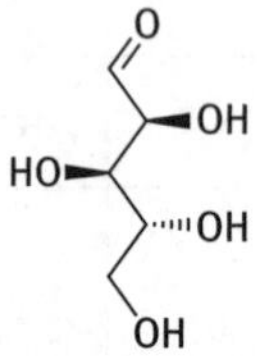

Carbon atoms are not normally shown in chain or ring structures, nor indeed the hydrogen atoms attached to them.

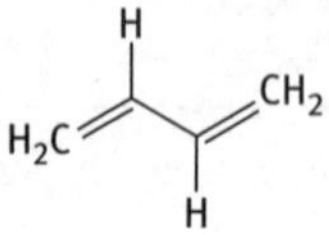

can be written as

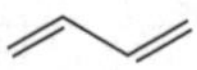

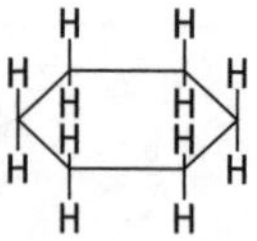

can be drawn as

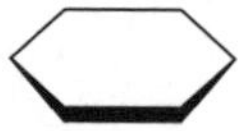

Atoms other than carbon, that form part of a ring structure, are always shown.

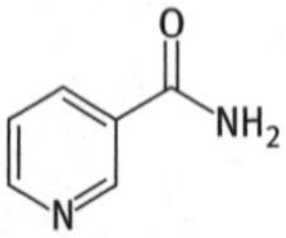

Amounts and concentrations

Avagadro's number (constant) = 6.022×10^{23}, the number of atoms, molecules or particles in 1 mole of a substance. A mole is an amount; the amount that contains a number of molecules equal to Avagadro's number.

One **mole** (mol) of a compound is also the amount of the substance equal to its molecular mass expressed in grams. For example, the molecular formula of glucose is $C_6H_{12}O_6$. Its molecular mass is given by the sum of the atomic masses of its atoms, which is approximately $(6\times12) + (12\times1) + 6\times16) = 180$; therefore we can say that 180 g of glucose is equivalent to 1 mole.

Molarity (M) is a concentration; 1 mole of a solute dissolved in 1 litre of solvent makes a 1 M solution. In other words, 180 g of glucose dissolved in 1 l of water is a 1 M glucose solution. If we take 1 ml of that solution we have not changed the concentration (the ratio of glucose to water is the same), it is still 1 M, but that 1 ml only contains 1 mmol, one thousandth of a mole.

Acids and bases

Acidity is measured according to the hydrogen ion concentration, denoted by $[H^+]$ or $[H_3O^+]$. Hydrogen ion concentrations ($[H^+]$) are generally very small and so the pH scale is used to denote hydrogen ion concentration on a scale of 0–14.

$$\text{pH} = -\log_{10}[H^+]$$

Water dissociates weakly (the auto-ionisation of water) to give a neutral solution.

K_w is the dissociation constant of water:

$$K_w = [H^+]\,[OH^-] = [1 \times 10^{-7}]\,[1 \times 10^{-7}] = 1 \times 10^{-14}\ \text{M}^2\ (\text{or mol}^2\ \text{dm}^{-6})$$

Acids dissociate in solution; a weak acid dissociates partly and a strong acid almost completely: K_a = acid dissociation constant for the reaction.

In $AH + H_2O \rightleftharpoons A^- + H_3O^+$ (the acid AH dissociates to form its conjugate base, A^-)

$$K_a = \frac{[A^-]\,[H_3O^+]}{[AH]}$$

Values of acid dissociation constants are usually very small and so, as in the case of the hydrogen ion concentration, we take the log and express the result as a pK_a.

$$pK_a = -\log_{10}K_a$$

To calculate the pH of a solution of a weak acid in equilibrium with its conjugate base we can use the Henderson–Hasselbach equation:

$$pH = pK_a + \log_{10}\frac{[base]}{[acid]}$$

where [base] = concentration of conjugate base and [acid] = concentration of conjugate acid.

This equation also allows us to calculate the pH of buffer solutions.

Thermodynamics

Standard Conditions: The symbol Δ is used to denote a change in a quantity such as the enthalpy or entropy of a system. This is because we cannot measure absolute values but only a change in such quantities. In order to ensure we are comparing like with like we need to define the conditions under which we are measuring these changes to ensure they are similar throughout a reaction or other change. The reference conditions which chemists use to compare these changes are called the conditions of *standard-state* and are represented by the superscript ° following the quantity. For example ΔH^o means the enthalpy change when reactants in their standard states are converted to products in their standard states. The standard state of a substance is the pure form of the substance at a pressure of 1 atm pressure. For solutions this refers to a 1 M concentration of the substance. Standard enthalpy changes are usually reported at a temperature of 25°C or 298 K. There is one further condition which is important for biological changes and this is the pH. Biologists use the symbol $\Delta G^{o\prime}$ to denote a free energy change under biological conditions which defines the pH as being 7.0.

ΔH = change in enthalpy (kJ mol^{-1}).

ΔS = change in entropy (kJ $mol^{-1}K^{-1}$).

ΔG = change in Gibbs free energy (kJ mol^{-1}).

Gibbs equation is $\Delta G = \Delta H - T\Delta S$, where T is the absolute temperature in Kelvin (K).

ΔG^o = the standard free energy change at 1 atmosphere and 298 K.

$\Delta G^{o\prime}$ = the standard free energy change which refers to the standard state in an ideal solution (commonly used for biological systems and measured at 25°C, pH = 7.0, 1 atmosphere pressure and 1 M concentration).

A positive ΔG indicates an endergonic process.

A negative ΔG indicates an exergonic process.

A positive ΔH indicates an endothermic process.

A negative ΔH indicates an exothermic process.

Kinetics

The rate of a reaction is given by the change in concentration of the reactant [R] with time:

$$\text{rate} = \frac{-\Delta[\text{R}]}{\Delta t}$$

Or the rate of increase in the concentration of the product [P]:

$$\text{rate} = \frac{\Delta[\text{P}]}{\Delta t}$$

For the reaction A + B $\rightarrow$ C + D:

$$\text{rate} = k[\text{A}]^x[\text{B}]^y$$

where k is the rate constant (for a stated temperature), and x and y are orders of reaction.

If x and y are zero, i.e. the reaction rate does not depend on the concentration of any of the reactants (common for enzyme-catalysed reactions), then the rate of the reaction is equal to the rate constant:

rate = k.

This arrow symbol $\rightleftharpoons$ denotes a reversible reaction.

At equilibrium there is no net change in the concentration of products or reactants; the forward and reverse rates are equal.

For the reaction:

$$\text{pyruvate} + \text{NADH} + \text{H}^+ \rightleftharpoons \text{lactate} + \text{NAD}^+$$

the equilibrium constant, K_{eq}, is given by the expression:

$$K_{eq} = \frac{[\text{lactate}][\text{NAD}^+]}{[\text{pyruvate}][\text{NADH}][\text{H}^+]}$$

Free energy and equilibrium

Free energy and equilibrium constants are related by the expression:

$$\Delta G^{\circ\prime} = -RT\ln K_{eq}$$

where R is the gas constant and T the absolute temperature.

Free energy and redox potential

Free energy and redox potentials are related by the Nernst equation:

$$\Delta G^{\circ\prime} = -nF\Delta E^{\circ\prime}$$

where F is a constant, the Faraday, n is equal to the number of electrons transferred in the reaction and $E^{\circ\prime}$ is the redox potential (V).

Reactivity

In describing reaction mechanisms, a short 'full head' curly arrow is used to indicate the movement of a pair of electrons. A 'half-head' curly arrow describes the movement of a single electron (with the formation of a radical).

for **pairs** of electrons
(more common)

for a **single** electrons
(*i.e.* radical reactions)

Two 'dots' against an atom denotes a lone pair of electrons. For example, oxygen (in the carbonyl group) has two lone pairs of electrons:

$$\rangle C{=}\ddot{\underset{\cdot\cdot}{O}}$$

One 'dot' against an atom denotes a single unpaired electron (a radical).

Cl· is the chlorine radical.

Appendix 5. Glossary

Acid (Brønsted-Lowry definition): a substance which can donate a proton or hydrogen ion.

Acid dissociation constant (K_a): the equilibrium constant for the dissociation of a proton from an acid.

Activation energy: the amount of energy that reactants must absorb before a chemical reaction will start (to promote the reactants from the ground state to the transition state).

Active site: the area of an enzyme where the substrate binds by means of weak chemical bonds. It is often located in a cleft or pocket in the protein's tertiary structure.

Addition reaction: the addition of a small molecule (e.g. H_2) to a double or triple carbon bond.

Alcohol: an organic molecule containing a –OH group attached to a carbon atom that is not part of a carbonyl group or an aromatic ring.

Aldehyde: an organic molecule containing the –CHO group.

Amine: an organic molecule derived from ammonia by replacing various numbers of H atoms with organic groups. Amines contain the group $-NH_2$, -NH or –N.

Amino acid: an organic molecule possessing both carboxyl and amino groups. Amino acids serve as the monomers of proteins.

Amino group: a functional group that consists of a nitrogen atom bonded to two hydrogen atoms. It can act as a base in solution, accepting a hydrogen ion and acquiring a charge of +1.

Amphipathic (also amphiphilic): a molecule that has both a hydrophilic region and a hydrophobic region.

Anabolic pathway: a metabolic pathway that synthesises a complex molecule from simpler compounds.

Angstrom (Å): a unit of length equal to 1×10^{-10} m.

Anion: a negatively charged ion.

Anomeric carbon: the most oxidised carbon atom of a cyclised monosaccharide. The anomeric carbon has the chemical reactivity of a carbonyl group.

Anomers: isomers of a sugar molecule that have different configurations only at the anomeric carbon atom.

Apolar: opposite of polar. See *Polar*.

Aromatic hydrocarbon: an aromatic hydrocarbon is one which incorporates one or more planar sets of six carbon atoms that are connected by delocalised electrons numbering the same as if they consisted of alternating single and double covalent bonds. The simplest aromatic hydrocarbon is benzene; this configuration of six carbon atoms is known as a benzene ring.

Asymmetric carbon: a carbon atom which is attached to four different atoms or groups.

Atom: the smallest unit of matter that retains the properties of an element.

Atomic mass unit (amu): the unit of atomic mass equal to one-twelfth the mass of the ^{12}C isotope of carbon.

Atomic nucleus: an atom's central core that contains protons and neutrons.

Atomic orbital: a region in space around an atom in which there is a high probability of finding an electron.

ATP: adenosine triphosphate. An adenosine-containing nucleoside triphosphate that releases free energy when its phosphate bonds are hydrolysed. The 'energy currency' of the cell. The free energy released is used to drive endergonic reactions in the cell.

Auto-ionisation of water: a reaction in which a proton is transferred from one molecule of water to another to form an H_3O^+ ion and an OH^- ion.

$$2H_2O \rightleftharpoons H_3O^+ + OH^-$$

Autotroph: autotrophic organisms use energy from the sun or from the oxidation of inorganic compounds to make organic molecules.

Avagadro's number: formally defined as the number of carbon-12 atoms in 0.012 kg of carbon-12. Avogadro's number is 6.022×10^{23}.

Base (Brønsted-Lowry definition): a substance which can extract a proton, e.g. OH^-, NH_3, RNH_2.

Beta-pleated sheet: one form of the secondary structure of proteins.

Bond energy: an average value for the energy required to break a covalent bond in such a way that each participating atom retains an unpaired electron.

Buffer: a substance that consists of conjugate acid and base forms in a solution that minimises changes in pH when extraneous acid or base is added to the solution.

Calorie: the amount of heat energy required to raise the temperature of 1 g of water by 1°C. The Calorie, with a capital C, usually used to indicate the energy content of food, is a kilocalorie.

Carbohydrates: a sugar (monosaccharide), or one of its dimers (disaccharide) or polymers (polysaccharide), in which the ratio C: H: O is usually 1: 2: 1.

Carbonyl group: the functional group >C=O found in aldehydes and ketones.

Carboxyl group: the functional group –COOH found in carboxylic acids.

Carboxylic acid: an organic molecule containing the –COOH group. They are weak acids.

Catabolic pathway: a metabolic pathway that releases energy by breaking down complex molecules to simpler compounds.

Catalyst: a chemical agent that changes the rate of a reaction without itself being consumed in the reaction.

Cation: an ion with a positive charge, produced by the loss of one or more electrons.

Chemical bond: an attraction between two atoms resulting from a sharing of outer energy level electrons, or the presence of opposite charges on ions.

Chemical energy: energy stored in the chemical bonds of molecules; a form of potential energy.

Chemical equilibrium: a stage in a chemical reaction in which the rate at which the reactants are converted to products is equal to the rate at which products are being converted back to reactants. A chemical equilibrium is a dynamic equilibrium.

Chemical reaction: a process that results in chemical changes to matter, and involves the making and/or breaking of chemical bonds.

Chemo-autotroph: an organism that needs only carbon dioxide as a carbon source but that obtains energy by oxidising inorganic compounds.

Chiral (or asymmetric) molecule: a molecule which cannot be superimposed upon its mirror image.

Cholesterol: a steroid that forms an essential component of animal cell membranes and which acts as a precursor for a number of other biologically important steroids.

Condensation reaction: a reaction in which two molecules become covalently bonded to each other through the loss of a small molecule, usually water (a dehydration reaction).

Coordinate bond: a chemical bond formed by the sharing of a pair of bonding electrons between two atoms, when the electron pair originates from one of the atoms. See also *Dative covalent bond.*

Covalent bond: a chemical bond formed by the sharing of a pair of electrons.

Conjugate acid/base: the product resulting from the gain of a proton by a base; or, the product resulting from the loss of a proton by an acid.

Conjugated system: a chain of atoms (usually carbon) connected by alternating single and double bonds.

Cytochrome: an iron-containing protein component of electron transport chains in mitochondria and chloroplasts.

Dalton: a unit of atomic mass equivalent to one-twelfth of the mass of the atom ^{12}C.

Dative covalent bond: a chemical bond formed by the sharing of a pair of bonding electrons between two atoms, when the electron pair originates from one of the atoms. See also *Coordinate bond.*

Dehydration reaction: a chemical reaction in which two molecules covalently bond to each other with the removal of a water molecule.

Delocalisation of electrons: the spread of electrons over several atoms within a molecule.

Deoxyribonucleic acid (DNA): a double-stranded helical nucleic acid molecule that is capable of replicating and determining the inherited structure of a cell's proteins.

Deoxyribose: the sugar component of DNA, having one less hydroxyl group than ribose, the sugar component of RNA.

Dipole: two equal but opposite charges, separated in space, resulting from the unequal distribution of charge within a molecule or chemical bond. See also *Polar.*

Dissociation:

i. breaking of a bond

ii. separation into ions by an ionic compound on dissolving in water.

Dissociation constant (of an acid), K_a: the equilibrium constant which indicates the degree to which an acid is dissociated into ions in water.

Electron: a negatively charged sub-atomic particle found outside the nucleus of an atom.

Electron configuration: the way in which the electrons of an atom are arranged in orbitals in an atom or molecule, e.g. C $1s^2 2s^2 2p^2$

Electron energy level: a permitted value for the energy of an electron in an atom or molecule.

Electron shell: 'shell' is used interchangeably with energy level. See *Electron energy level.*

Electron transport chain: a sequence of electron carrier molecules that possess redox components and collectively shuttle electrons along their length.

Electrophile: species that are attracted to negative centres. Electrophiles are typically cations such as the hydroxonium ion, H_3O^+.

Electronegativity: the attraction of an atom for the electrons in a covalent bond.

Electrostatic interactions: a general term for the electronic interaction between particles. Electrostatic interactions include charge–charge interactions, hydrogen bonds, and van der Waals forces.

Element: a substance which cannot be separated into simpler substances by chemical techniques. All atoms of the same element have the same atomic number and electron configuration.

Elimination reaction: a chemical reaction in which a molecule forms two different molecules.

Enantiomer: one of a pair of optical isomers (stereoisomers) that are non-superimposable mirror images. D- and L-glucose are enantiomers.

Endergenic reaction: a chemical reaction in which there is a net input of free energy.

Endothermic reaction: a chemical reaction in which heat energy is absorbed from the surroundings (i.e. $\Delta H > 0$).

Enthalpy: a thermodynamic term that is a measure of the heat content of a system.

Entropy : a thermodynamic term that is a measure of the extent of disorder of a system.

Equilibrium: the position in a chemical reaction where the forward rate of formation of products is equal to the reverse rate of formation of reactants; the reaction is said to be in equilibrium.

Equilibrium constant (K_c , K_{eq}): the equilibrium constant K_c relates to a chemical reaction at equilibrium and can be calculated from a knowledge of the concentration of reactants and products at equilibrium. For enzyme-catalysed reactions the equilibrium constant is usually depicted as K_{eq}. For the reaction aA + bB $\rightleftharpoons$ cC + dD, the equilibrium constant is given by the expression:

$$K_c = \frac{[C]^c[D]^d}{[A]^a[B]^b}$$

Exergonic reaction: a spontaneous chemical reaction in which there is a net release of free energy.

Exothermic reaction: a chemical reaction in which there is a net release of heat energy (i.e. $\Delta H < 0$).

Fatty acid: a long chain carboxylic acid. Fatty acids vary in length and the number and location of double bonds. Double bonds allow for *cis* and *trans* geometric isomers. Three fatty acids linked through a glycerol molecule form a triacylglycerol (fat); two fatty acids linked to glycerol that has a phosphate group form a phospholipid, the major component of biological membranes.

Feasibility: the extent to which (a chemical process) will be successful.

Feedback inhibition: a method of metabolic control in which the end product of a metabolic pathway (or reaction therein) acts as an inhibitor to an enzyme in that pathway.

First law of thermodynamics: the principle of conservation of energy. The internal energy of a system is a constant; energy can be neither created nor destroyed.

Free energy: the (Gibbs) free energy is energy that is available to do work. It is related to enthalpy and entropy by the Gibbs equation:

$$\Delta G = \Delta H - T\Delta S$$

Free energy that is released in a catabolic reaction is often coupled to an anabolic reaction, such as ATP synthesis.

Free radical: a molecule or atom with an unpaired electron that has resulted from the homolytic fission of a covalent bond. Radicals are highly reactive.

Functional group: a specific configuration of atoms, often containing a relatively highly electronegative atom, commonly attached to the carbon skeleton of organic molecules, and usually involved in chemical reactions, e.g. –OH, >C=O.

Geometric isomers: compounds that have the same chemical formula but differ in the spatial arrangement of their atoms. *Cis-trans* isomers are isomers which have different arrangements of atoms about a double bond.

Gibbs free energy: See *Free energy*.

Gluconeogenesis: the synthesis of 'new' glucose from non-sugar precursors such as lactate. Applied more specifically to the synthesis of glucose in the liver.

Glycolysis: the splitting of glucose into pyruvate. Glycolysis is an oxidative metabolic pathway that occurs in all living cells and is an essential provider of cellular energy.

Glycosidic link: a covalent bond formed between two monosaccharides by a dehydration reaction. The most commonly encountered glycosidic bonds are formed between the anomeric carbon of one sugar and a hydroxyl group of another sugar.

Half-life:

i. the time taken for the concentration of a substance to fall by a half (in chemical kinetics)

ii. the time taken for half the original number of radionuclides to disintegrate (in radioactivity).

Half-reaction: an equation which shows the electron loss or gain in an oxidation or reduction.

Heterolytic fission: fission is the process by which a molecule splits into two constituent parts. This occurs when one of the bonds between atoms in the molecule is broken. In heterolytic fission, the two electrons from the broken bond go to the same species. This occurs when one species is significantly more electronegative than the other. Heterolytic fission results in a negatively charged anion, which received both electrons, and a positively charged cation, which received neither.

Homolytic fission: in homolytic fission, the two electrons from the broken bond are shared between the resulting species. This means that each species contains an unpaired electron in an outer shell. They are therefore highly reactive, and are known as free radicals. Homolytic fission occurs when the two atoms being separated have a similar or identical electronegativity; that is, they have roughly the same ability to attract electrons to themselves.

Hybridisation: a model for chemical bonding in which hybrid orbitals are formed.

Hydration shell: a layer of water molecules surrounding a central ion or other species.

Hydrocarbon: an organic molecule consisting of only hydrogen and carbon.

Hydrogen bond: an electrostatic interaction between a hydrogen atom in one molecule and strongly electronegative atom (O, N or F) in a nearby molecule. The interaction may be between two different molecules or within the same molecule.

Hydrogen ion: a hydrogen atom which has lost its single outer electron.

Hydrolysis: the reaction of a substance with water resulting in the cleavage of the compound into two parts, each of which combines with a fragment (H^+ or OH^-) from the water molecule.

Hydrophilic: having an affinity for water. From 'hydro' referring to water and 'philic' meaning 'liking'.

Hydrophobic: having an aversion to water. From 'hydro' referring to water and 'phobic' meaning 'fearing'.

Hydrophobic interaction: a repulsive interaction between a molecule and water; the coalescing of hydrophobic groups caused by their exclusion by water.

Hydroxyl group: a functional group consisting of a hydrogen atom linked to an oxygen atom by a polar covalent bond. Molecules possessing this group are soluble in water and are called alcohols.

Intermolecular forces: the attractive and repulsive forces which exist between molecules.

Intramolecular forces: the interactions which exist between different fragments of the same molecule.

Ion: an electrically charged atom or group of atoms.

Ionic product of water (K_w): the product of the concentrations of hydroxonium ions and hydroxide ions in an aqueous solution, equal to $1 \times 10^{-14}\ M^2$.

Ionic bond: a chemical bond resulting from the attraction between oppositely charged ions.

Isomer: one of several organic compounds with the same molecular formula but different structures and therefore different properties. The major types of isomerism are structural, geometric and stereoisomers (enantiomers).

Isotope: one of several atomic forms of an element, each containing a different number of neutrons (but the same number of protons) and therefore differing in atomic mass. Isotopes may be stable or unstable (radioisotopes).

Joule: a unit of energy. 1 J = 0.239 cal; 1 cal = 4.184 J.

K_a: See *Acid dissociation constant.*

K_{eq}: See *Equilibrium constant.*

K_w: See *Ionic product of water.*

Kilocalorie: a unit of energy equal to 4.184 kJ; 1000 calories, or 1 Calorie (see *Calorie*).

Kinetics: the study of the rates and mechanisms of chemical reactions.

Kinetic energy: the energy of a particle due to its motion.

Ketone: an organic compound with a carbonyl group of which the carbon atom is covalently bonded to two other carbon atoms.

Le Chatelier's principle: states that when a stress is applied to a system at equilibrium, the equilibrium will shift so as to minimise the effects of the stress.

Lone pair: a pair of electrons in the valence shell of an atom, not involved in bonding.

Macromolecule: a large molecule (polymer) formed by the joining together of smaller molecules (monomers) usually by a condensation reaction. Polysaccharides, proteins and nucleic acid are macromolecules.

Metabolic pathway: a connected series of chemical reactions which take place in a living cell.

Metabolism: the totality of an organism's chemical reactions, consisting of catabolic and anabolic pathways.

Molarity: a common measure of solution concentration referring to the number of moles per volume (l or dm^3) of solution.

Molar mass: the mass in grams of one mole of a substance.

Mole: the amount of substance which contains the same number or particles as there are atoms in 12.0 g of the isotope ^{12}C.

Molecule: two or more atoms held together by covalent bonds.

Molecular orbital: a region in space within a molecule that represents the probability of finding an electron.

Monomer: the sub-unit that serves as the building block of a polymer.

Monosaccharide: the simplest carbohydrate (simple sugar), active alone or serving as a monomer in disaccharides or polysaccharides. The molecular formulae of monosaccharides are generally some multiple of CH_2O.

M_r: See *Relative molecular mass.*

Mutarotation: the change in specific rotation that occurs when an α (alpha) or β (beta) hemiacetal form of carbohydrate is converted into an equilibrium mixture of the two forms.

NAD^+ (NADH): nicotinamide adenine dinucleotide is a coenzyme that participates in redox reactions. NAD^+ is the oxidised form that gains two electrons in being reduced to NADH.

Nernst equation: an equation that relates the free energy change in a redox reaction to the change in standard reduction potential ($\Delta E^{\circ\prime}$) of a reaction.

$$\Delta G^{\circ\prime} = -nF\Delta E^{\circ\prime}$$

Neutron: an electrically neutral particle found in the nucleus of an atom. Different numbers of neutrons in the same element give rise to isotopes.

Nucleic acid: a polymer consisting of nucleotide monomers (base–sugar–phosphate). In DNA the sugar is deoxyribose, in RNA the sugar is ribose.

Nucleotide: molecule consisting of a purine or pyrimidine base, a pentose sugar (ribose or deoxyribose) and a phosphate group.

Nucleophile: species that are attracted to positive centres. Typical nucleophiles are negative ions or neutral atoms with a lone pair of electrons.

Nucleus (of an atom): the small positively charged centre of the atom in which most of the mass of the atom is concentrated.

Octet rule: the tendency of atoms to form ionic or covalent bonds in order that their outer (valence) energy level attains a full (8) complement of electrons. Atoms or ions that attain this state are generally stable and unreactive, e.g. the inert gases such as helium and neon.

Optical isomerism: a form of isomerism (specifically stereoisomerism) whereby the two isomers are the same in every way except that they are non-superimposable mirror images of each other. Optical isomers are known as chiral molecules.

Orbital hybridisation: the rearrangement of electrons in the outer energy level of an atom in order to maximise the number of unpaired electrons available to participate in covalent bonding.

Order of reaction: the order of a reaction refers to the number of components involved in the rate-determining step. A first order reaction depends on only one component in the reaction mixture.

Oxidation: combination with oxygen or loss of electrons by an atom or group of atoms in a reaction.

Oxidation–reduction (redox) reaction: a simultaneous oxidation and reduction reaction.

Oxidative phosphorylation: the synthesis of ATP using energy derived from the redox reactions of an electron transport chain.

Oxidizing agent: a species which causes oxidation by accepting electrons from another species.

Peptide bond: the covalent secondary amide linkage that joins the carbonyl group of one amino acid to the amino nitrogen of another in peptides and proteins.

Periodic table: an arrangement of the known elements according to their atomic number.

pH: defined as $-\log_{10}[H^+]$ this is a measure of the acidity of a species, solution or system. $pH < 7$ indicates an acidic solution, $pH = 7$ indicates a neutral solution, and $pH > 7$ indicates an alkaline solution.

pK_a: a logarithmic value that indicates the strength of an acid. pK_a is defined as the negative logarithm of the acid dissociation constant K_a.

Photon: a quantum of light energy.

Phospholipid: a molecule consisting of a polar hydrophilic glycerol + phosphate head-group, and a non-polar hydrophobic 'tail' consisting of two fatty acids. The molecule is amphipathic. Phospholipids are major constituent of biological membranes.

Pi bonds: the covalent bond formed between two atoms that have an unpaired electron in a p orbital, and formed by the 'side-to-side' merging of the p orbitals, above and below the bond axis. Always found in addition to a sigma covalent bond between the two atoms, hence the description of a 'double bond'. Double bonds are considered as a functional group since they constitute a region of increased electron density that may be 'attacked' by an electrophilic reagent.

Polar: a term used to describe species which have an uneven distribution of charge – often indicated by $\delta+/\delta-$.

Polar molecule (group): a molecule that possesses an electrical dipole; for example, those functional groups containing a relatively highly electronegative atom are said to be polar. Polar molecules (groups) are generally soluble in water. Apolar molecules do not possess such groups and are not soluble in water.

Polar covalent bond: a covalent bond in which the two atoms differ in their electronegativities. The more electronegative atom draws electrons towards itself becoming slightly negatively charged, with the other atom becoming slightly positively charged.

Polymer: a large molecule formed from many monomers.

Potential energy: the energy an object has because of its position, rather than its motion. The bond energy between two atoms is potential energy until that bond is broken.

Protein: a large polymer constructed from more than 50 α-amino acid units (monomers) linked through peptide bonds.

Racemic mixture: an equimolar mixture of enantiomers of a molecule, e.g. a mixture of D- and L-glucose.

Radical: a molecule or atom with an unpaired electron that has resulted from the homolytic fission of a covalent bond. Radicals are highly reactive.

Radioactivity: the emission of α or β particles or γ radiation by an atom (or a combination thereof).

Radioisotope: an unstable form of an element in which the nucleus decays spontaneously emitting particles (α or β particles) and/or energy (γ radiation).

Rate constant: the constant of proportionality in a rate equation.

Rate equation: an expression of the observed relationship between the rate of a reaction and the concentration of each reactant.

Rate-limiting step: a number of intermediate states may be involved in the conversion of a reactant to a product. The intermediate that is formed at the slowest rate constitutes the rate-limiting step for the overall reaction.

Rate of reaction: the rate of any chemical reaction is defined as the change in concentration of a component of the reaction over a period of time.

Reactant: a substance that participates in a chemical reaction.

Reaction mechanism: the reaction steps and the nature of the intermediates involved in the conversion of reactant to product.

Reaction order: See *Order of reaction*.

Reactive site: a site on a molecule, in or on a functional group, which is either electron deficient or electron rich, that will attract attack by a nucleophilic or electrophilic agent respectively.

Redox reaction: two coupled half-reactions, one an oxidation and the other a reduction.

Redox potential: a measure of an atom's tendency to gain or lose electrons. Atoms with a high redox potential (a positive value) gain electrons from atoms with a lower redox potential (a more negative value).

Reducing agent: a substance which causes reduction by donating electrons.

Reduction: the gain of electrons by a substance.

Reduction potential (E): a measure of the tendency of a substance to reduce other substances. The more negative the reduction potential, the greater the tendency to donate electrons.

Relative molecular mass (M_r): the mass of a molecule relative to one-twelfth the mass of ^{12}C.

Salt:

i. an ionic compound that can be formed by replacing one or more of the hydrogen ions of an acid with another positive ion.

ii. the product, formed along with water, from the reaction of an acid with a base.

Saturated: referring to a biological molecule that does not contain a carbon–carbon double bond. Also: unable to take up further material; can refer to compounds which can take up no further hydrogen or solutions which can take no further solute.

Second law of thermodynamics: the principle whereby energy transfers or transformations increase the disorder of the universe. A spontaneous change is always accompanied by an increase in entropy of the system. Ordered forms of energy (in covalent bonds) are at least partly converted to heat energy during metabolic reactions, leading to an increase in disorder (an increase in entropy) of the surroundings.

Sigma bond: a covalent bond formed from the overlap of atomic orbitals along the bond axis ('head-to-head'), as opposed to pi bonds where the orbital overlap is below and above the bond axis.

Solute: a substance that is dissolved in a solution.

Solution: a homogeneous liquid mixture of two or more substances.

Solvent: the dissolving agent of a solution. Water is the most versatile solvent.

Spontaneous change: a change which occurs naturally and does not need to be driven.

Standard state (standard free energy change, $\Delta G^{\circ\prime}$) (standard reduction potential, $\Delta E^{\circ\prime}$): A set of reference conditions for a chemical reaction. In biochemistry the standard state is defined as that at a temperature of 298 K (25°C), a pressure of one atmosphere, a solute concentration of 1 M, and a pH of 7.0.

Stereoisomer: a molecule that is a mirror image of another molecule with the same molecular formula; these isomers have atoms with the same connectivity but are arranged differently in space.

Stoichiometry: the ratio of number of moles of reactants consumed to products formed in the chemical equation for a reaction.

Structural formula: a molecular notation in which the individual atoms of a molecule are shown joined together by lines representing covalent bonds.

Structural isomers: compounds that have the same molecular formula but differ in the covalent arrangement of their atoms.

Substitution reaction: a reaction, normally involving organic compounds, whereby an atom, or group of atoms, is replaced by another atom, or group of atoms.

Substrate: the reactant on which an enzyme works.

Sulfhydryl group: a functional group consisting of a sulfur atom bonded to a hydrogen atom.

Tetrahedral: the geometry adopted around a saturated carbon atom; four atoms (or groups) covalently bonded to one carbon atom will tend to point towards the corners of a tetrahedron.

Transition state: an unstable high energy arrangement of atoms in which chemical bonds are being formed or broken. Transition states have structures between those of the substrate (reactant) and product of the reaction.

Unpaired electron: a single electron in an outer (valence) energy level that may participate in covalent bond formation.

Unsaturated fatty acid: a fatty acid with at least one carbon–carbon double bond. In general the double bonds of an unsaturated fatty acid are of the *cis* configuration.

Valence: the bonding capacity of an atom; generally equal to the number of unpaired electrons in an atom's outer electron energy level.

Valence electron: an electron in a valence shell which can undergo bonding with other valence electrons.

Valence shell: the outermost energy shell (energy level) of an atom containing the valence electrons involved in the chemical reactions of that element.

van der Waals forces: weak charge–charge attractions or repulsions between molecules (intermolecular) or parts of molecules in close proximity. Such forces are the result of temporary transient dipoles on molecules leading to localised charge fluctuations.

Index

Maths & Stats

For the life and medical sciences

Michael Harris

Lecturer in General Practice, Royal United Hospital, Bath, UK

Gordon Taylor

Senior Research Fellow in Medical Statistics, University of Bath, Bath, UK

and

Jacquelyn Taylor

Mathematics Teacher, Writhlington School, Radstock, Bath, UK

Scion

First published 2005

A CIP catalogue record for this book is available from the British Library.

ISBN 1 904842 11 9

Scion Publishing Limited
Bloxham Mill, Barford Road, Bloxham, Oxfordshire OX15 4FF
www.scionpublishing.com

Important Note from the Publisher

The information contained within this book was obtained by Scion Publishing Limited from sources believed by us to be reliable. However, while every effort has been made to ensure its accuracy, no responsibility for loss or injury whatsoever occasioned to any person acting or refraining from action as a result of information contained herein can be accepted by the authors or publishers.

Typeset by Phoenix Photosetting, Chatham, Kent, UK
Printed by Biddles Ltd, King's Lynn, UK, www.biddles.co.uk

Contents

Applications of mathematics section

Statistics section

Foreword

A generation ago, training in the life and medical sciences almost completely ignored mathematics and statistics. But today's students and practitioners face developments such as analysis of complex genome data, predictive modelling of the spread of diseases, nonlinear interactions in ecology, a wide range of quantitative physiological data, etc. etc. – all of these require a firm grounding in mathematics and statistics, and yet the maths background of many life and medical science students is getting weaker.

In my research work at the interface between mathematics and biology, I collaborate with ecologists, biologists and clinicians who are keen to learn basic mathematics and statistics in order to make full use of these new developments. But when they ask me about introductory books, I have had to shrug my shoulders.

Similarly, when I moved to Heriot-Watt I developed two new teaching modules on Mathematics for Biologists, and was frustrated by the lack of textbooks suitable for those students with a weaker background in maths.

Therefore when I was sent details of this book, I was thrilled. Both content and style are tailored to people who have a background in the life and medical sciences, but who need to equip themselves for today's world of quantitative biology.

It is my hope and belief that greater quantitative training will enable the next generation of biologists and clinicians to make full use of the exciting recent and ongoing research in quantitative biology and medicine. This book by Harris, Taylor and Taylor is a major contribution to this process.

Jonathan A Sherratt FRSE
Professor of Mathematics
Heriot-Watt University

Preface

This book is designed for life and medical science students and professionals who need a basic knowledge of mathematics and statistics.

Whether you love or hate maths and stats, you need to have some working knowledge of the subjects if you want to work in the life or medical sciences.

This book assumes that you have nothing more than a very basic maths or stats knowledge. However basic your knowledge, you will find that everything is clearly presented and explained.

A few readers will find some of the sections very simplistic; others will find that some need a lot of concentration. Start with concepts that suit your level of understanding.

All the sections have worked examples, and you can check your understanding of what you have learnt by going through the "test yourself" questions, then comparing your answers with those of the authors.

Michael Harris
Gordon Taylor
Jacquelyn Taylor
Bath, April 2005

About the authors

Dr Michael Harris MB BS FRCGP MMEd is a GP and senior lecturer in postgraduate medicine in Bath. Until recently he was an examiner for the Royal College of General Practitioners. He has a special interest in the design of educational materials.

Dr Gordon Taylor PhD MSc BSc (Hons) is a senior research fellow in medical statistics at the University of Bath. His main role is in the teaching, support and supervision of health-care professionals involved in non-commercial research.

Mrs Jacquelyn Taylor MSc BSc (Hons) PGCE has worked in both secondary and higher education in mathematics and science.

Acknowledgements

We would like to thank all our reviewers, whether expert or enthusiastic amateur.

We are very grateful to Professor Jonathan Sherratt of Heriot-Watt University for his comments and for having given us permission to use some of his material.

Thank you also to our publisher, Dr Jonathan Ray, for his patience and helpful advice.

Finally, special thanks go to Sue Harris for her forbearance and support.

1 How to use this book

If you want a maths and stats course

- Work through from start to finish for a complete course in the mathematics and statistics relevant to the life and medical sciences.
- The first page starts with the assumption that you want to go right back to basics.
- If you already know some maths or stats, start with concepts that suit your level of understanding.
- Each chapter will build on what you have learnt in previous chapters.
- All the chapters have worked examples that illustrate what you have read. These examples will help you reinforce your learning.
- We have cut down the jargon as much as possible. All new words are put in bold and explained.

If you're in a hurry

- Choose the chapters that are relevant to you. Each chapter is designed so that it can be read in isolation.

If you want a reference book

- You can use this as a reference book. The index is detailed enough for you to find what you want in a hurry.

Test your understanding

- Use the "test yourself" questions at the end of the chapters to check your understanding of what you have just read, then compare your answers with those of the authors.
- You will be able to answer most questions by using mental arithmetic. Some questions will be easier to answer if you use a calculator.

Study advice

- Try not to cover too much at once.
- Go through difficult sections when you are fresh.
- You may need to read some sections a couple of times before the meaning sinks in. You will find that working through the examples helps you to understand the principles.

2 Handling numbers

This chapter goes right back to basics, with reminders of the principles of handling numbers.

2.1 Factors

The **factors** of a number are all the whole numbers that divide into it without a remainder.

The numbers 1, 2, 3, 4, 6 and 12 all divide into 12 exactly. These numbers are called the factors of 12.

$$1 \times 12 = 12$$

$$2 \times 6 = 12$$

$$3 \times 4 = 12$$

Example

The factors of 15 are 1, 3, 5 and 15.

2.2 Common factors

Common factors are numbers that will divide into two or more numbers.

Example

1 and 3 are the common factors of 12 and 15.

The **highest common factor** is 3.

2.3 Use of brackets

We use **brackets** to change the order of mathematical calculation and ensure correct interpretation of mathematical **expressions**.

Example

The expression

$$3 \times 8 - 5$$

is calculated as follows:

$3 \times 8 = 24$; subtract 5; answer: 19.

Putting brackets round the 8 – 5 expression changes the order of calculation: the contents of brackets need to be worked out *before* doing the rest of a calculation.

$$3(8-5)=3\times 3=9$$

Note that when we write "3 times (8 – 5)" we don't need to write the "×" sign.

2.4 BODMAS

Calculations in a mathematical expression should be in **BODMAS** order:

- Brackets
- "Of": this relates to powers of numbers
- Division
- Multiplication
- Addition
- Subtraction

Example

To calculate

$$9+6-8(7+5)\div 4$$

First we work out the sum in brackets, giving:

$$9+6-8(12)\div 4$$

After the division we get:

$$9+6-8\times 3$$

After multiplying:

$$9+6-24$$

After the addition:

$$15-24$$

Finally, the subtraction gives:

$$-9$$

2.5 Absolute values

Absolute value bars around a number turn it into a positive value. Absolute value does nothing to a positive number or zero.

Example

$$|-3|=3$$

$$|3|=3$$

2.6 Prime numbers

A whole number with only two factors, 1 and itself, is called a **prime number.**

Example

7 has only two factors, 1 and 7, so it is a prime number.

2.7 Square numbers

Square numbers are formed by multiplying a whole number by itself, for example:

9 is 3×3, 25 is 5×5

3×3 can be written as 3^2, and can be spoken as "3 squared" or "3 to the power of 2".

When two negative numbers are multiplied the answer is positive, so the square of a negative number is positive.

Examples

$(-3)^2 = -3 \times -3 = 9$

$5 \times 5 = 5^2 = 25 =$ "5 squared", or "5 to the power of 2".

A **quadrat** is an area used as a sample unit.

So, a 5 by 5 metre square quadrat of a field is 25 m^2.

2.8 Square roots

Because $9 = 3 \times 3$, 3 is called the **square root** of 9. The square root of a number is written using the $\sqrt{\ }$ symbol, so $\sqrt{9} = 3$.

However, we know that $(-3)^2$ also equals 9, so $\sqrt{9}$ could also be -3.

The square root of any number can be positive or negative, so $\sqrt{16} = \pm 4$.

The $\pm$ symbol means "plus or minus".

2.9 Cube numbers

Cube numbers are formed by multiplying a number by itself and then by itself again.

Example

$5 \times 5 \times 5 = 5^3 = 125 =$ "5 cubed", or "5 to the power of 3".

Thus a block of plant tissue 5 by 5 by 5 millimetres is 125 mm^3.

2.10 Cube roots

Because $27 = 3 \times 3 \times 3$, 3 is the **cube root** of 27.

This is written as $\sqrt[3]{27} = 3$.

Test yourself

The answers are given on page 169.

Question 2.1
What are the factors of 18, 21 and 24?
What is their highest common factor?

Question 2.2
Calculate
$7(4 + 3)(5 - 2)$

Question 2.3
Work out
$16(9 \div 3 + 1) - 10 \div 5$

Question 2.4
Which of these are prime numbers: 21, 22, 23?

Question 2.5
What is the surface area of a 7 by 7 m square quadrat of a field?

Question 2.6
What are the lengths of the sides of a square quadrat that is 64 m^2?

Question 2.7
What is the volume of a 40 by 40 by 40 mm cube of soil sample?

Question 2.8
A cube of tissue from a biopsy is 64 mm^3. What are its dimensions?

3 Working with fractions

As soon as we work with anything other than whole numbers, we need to use fractions.

3.1 Fractions

The **fraction** 3/5 means 3 parts out of 5. The top number in a fraction is known as the **numerator**, the bottom number is called the **denominator**.

To calculate a fraction of an amount, multiply by the numerator and divide by the denominator.

Example

$$\frac{3}{5} \text{ of } 20 = (3 \times 20) \div 5 = 60 \div 5 = 12$$

3.2 Simplifying fractions

Fractions can be **simplified** if the numerator and denominator have a common factor.

With fractions, whatever we do to the numerator, we also have to do to the denominator.

Example

To simplify 12/15, the 12 and the 15 have a common factor of 3, so we can divide the numerator and denominator by 3.

$$\frac{12}{15} = \frac{12 \div 3}{15 \div 3} = \frac{4}{5}$$

Where there is no common factor, the fraction is already in its simplest form.

3.3 Reciprocals

The **reciprocal** of a number or a mathematical expression is 1 divided by that number or expression.

To get the reciprocal of a fraction, flip it upside down.

Examples

The reciprocal of $\frac{5}{6}$ is $\frac{6}{5}$, or $1\frac{1}{5}$

The reciprocal of 7 is $\frac{1}{7}$

3.4 Multiplying fractions

To **multiply fractions**, multiply straight across the top, and straight across the bottom.

Example

$$\frac{3}{4} \times \frac{5}{6} = \frac{3 \times 5}{4 \times 6} = \frac{15}{24} = \frac{5}{8}$$

3.5 Dividing fractions

To divide a fraction by another, flip the second fraction (i.e. take its reciprocal) and multiply the two fractions.

Example

$$\frac{3}{4} \div \frac{5}{6} = \frac{3}{4} \times \frac{6}{5} = \frac{3 \times 6}{4 \times 5} = \frac{18}{20} = \frac{9}{10}$$

3.6 Adding fractions

Where the denominator is the same in both fractions, i.e. where there is a **common denominator**, add across the top.

Example

$$\frac{3}{7} + \frac{2}{7} = \frac{3+2}{7} = \frac{5}{7}$$

Where there is *no* common denominator, the simplest way is to convert the fractions so that they have the **lowest common denominator**, and then add across the top.

The lowest common denominator is the smallest number that has all the denominators as factors.

Example

$$\frac{2}{3} + \frac{4}{5}$$

The lowest common denominator is 15, as it is the smallest number that has 3 and 5 as factors.

To convert 2/3 to have 15 as a denominator, we need to multiply the numerator and denominator by 5.

To convert 4/5 to have 15 as a denominator, we need to multiply the numerator and denominator by 3.

$$\frac{2\times 5}{3\times 5}+\frac{4\times 3}{5\times 3}=\frac{10}{15}+\frac{12}{15}=\frac{10+12}{15}=\frac{22}{15}=1\frac{7}{15}$$

3.7 Subtracting fractions

The process is analogous to that for adding fractions.

Example

$$\frac{3}{7}-\frac{2}{7}=\frac{3-2}{7}=\frac{1}{7}$$

Again, where there is no common denominator, convert each fraction to the lowest common denominator and subtract across the top.

3.8 Changing fractions to decimals

To change a fraction into its **decimal** equivalent, divide the numerator by the denominator.

Example

$$\frac{1}{2}=1\div 2=0.5$$

Test yourself

The answers are given on page 169.

Question 3.1

Calculate $\frac{5}{6}$ of 72.

Question 3.2

Simplify $\frac{20}{24}$

Question 3.3

Give the reciprocal of $\frac{24}{28}$

Question 3.4

Multiply $\frac{2}{5}$ by $\frac{9}{10}$

Question 3.5

Divide $\frac{2}{5}$ by $\frac{9}{10}$

Question 3.6

Add $\frac{6}{7}$ to $\frac{9}{14}$

Question 3.7

Subtract $\frac{7}{12}$ from $1\frac{3}{8}$

Question 3.8

What is the decimal equivalent of $1\frac{5}{8}$?

4 Percentages

A percentage is another way of describing a fraction and it can be easier to visualise.

4.1 Percentages

A **percentage** is a fraction out of 100.

$$15\% \text{ is the same as } \frac{15}{100}$$

So, calculating 15% of a number is the same as calculating 15/100 of the number.

It is also the same as multiplying the number by 0.15, as $0.15 = 15 \div 100$.

Example

$$15\% \text{ of } 480 = \left(\frac{15}{100}\right) \times 480 = 72$$

Also, $15\% \text{ of } 480 = 0.15 \times 480 = 72$

4.2 Converting decimals into percentages

To change a decimal into a percentage, multiply it by 100.

Example

$$0.05 = (100 \times 0.05)\% = 5\%$$

4.3 Calculating percentages using decimals

To calculate a percentage increase or decrease, convert the percentage to decimals.

Example

A stalk of common wheat, *Triticum aestivum*, measures 625 mm in height. In 1 week it grows by 12%.

An increase *of* 12% is the same as an increase *to* 112%; multiplying a number by 112% is the same as multiplying it by 1.12.

$112\% \text{ of } 625 = 1.12 \times 625 = 700$

So, after 1 week the stalk has grown to 700 mm.

When a number increases (or decreases), we can calculate the increase (or decrease) as a percentage of the original number.

Work out the change as a fraction of the original number, then convert it to decimals. Multiply this by 100 to get the percentage.

Example

The weight of a baby has increased from 1.3 kg to 1.56 kg.

It has therefore increased by 0.26 kg, or $\frac{0.26}{1.30}$ of the original weight.

$$\frac{0.26}{1.30} \times 100 = 20\%$$

So, the baby has gained 20% in weight.

4.4 Tabulating data using percentages

We use percentages when **tabulating** data to give a scale on which to assess or compare the data.

Example

We can use a table to compare data for body and tail length in 10 mice.

Table comparing body and tail length in 10 mice

Body length (mm)	Tail length (mm)	$\frac{\text{Tail length}}{\text{Body length}}$ (%)
92	31	34
97	32	33
96	35	36
99	36	36
100	40	40
111	43	39
109	44	40
115	49	43
120	49	41
122	52	43

Note how tabulating the percentages makes the trend clear: in this group of mice, longer mice have proportionately longer tails.

We can also tabulate **frequencies** (the number of times events occur) and compare them with percentages.

Example

We wish to compare the ages of 80 patients referred for heart transplantation.

Table comparing ages of 80 patients referred for heart transplantation

Years	Frequency	Percentage
0–9	2	2.5
10–19	5	6.25
20–29	6	7.5
30–39	14	17.5
40–49	21	26.25
50–59	20	25
≥60	12	15
Totals	80	100

The first column gives age in 10-year ranges.

The ≥ symbol means "more than or equal to", in this case "more than or equal to 60 years old".

The second column gives the frequency, i.e. the number of patients in each 10-year range.

The last column gives the percentage of patients in each age range. For example, in the 30–39 year range there were 14 patients and we know the ages of 80 patients, so:

$$\frac{14}{80} \times 100 = 17.5\%$$

Take care when interpreting percentages, though.

To say that 50% of a sample meets certain criteria when there are only four subjects in the sample is clearly not providing the same level of information as 50% of a sample based on 400 subjects.

So, percentages should be used as an additional help when interpreting data, rather than replacing the actual data.

Test yourself

The answers are given on page 169.

Question 4.1
Drying a 375 g sample of soil has reduced the mass by 40%. What mass of water was there in the sample?

Question 4.2
A patient's peak flow rate (a measure of respiratory airflow) is 400 litres per minute during an attack of asthma.
Twenty minutes after treatment, his peak flow has increased to 560 litres per minute. What percentage increase is this?

Question 4.3
The mean mass of leaf litter per square metre in an area of woodland is 900 g. The mass reduces by 18% in a month. What is the resulting mean mass?

Question 4.4
In 100 ml of cell culture at 37°C, the concentration of *E. coli* is 24 million cells per ml immediately after inoculation.
Three hours later the cell concentration has increased to 912 million cells per ml.
What is the percentage increase?

5 Powers

Powers are used throughout the life and medical sciences, whether for describing very large and very small numbers, to describe exponential and other relationships, or for their use in calculus and statistical analysis.

5.1 Indices

The **power** of a number is the same as the **index** of a number (plural: **indices**).

3^2 is known as "3 to the power of 2", or "3 squared",

3^3 is "3 to the power of 3", or "3 cubed",

$3 \times 3 \times 3 \times 3$ is 3^4, or "3 to the power 4",

and so on.

The power of a number is also its **logarithm**. This is explained in detail in Chapter 16.

5.2 Powers of 10

We use **powers of 10** to describe very large and very small numbers.

Table of powers of 10

Power of 10	Spoken as	Calculated by	Ordinary form
10^4	"10 to the power of 4"	$10 \times 10 \times 10 \times 10$	10 000
10^3	"10 to the power of 3" or "10 cubed"	$10 \times 10 \times 10$	1000
10^2	"10 to the power of 2" or "10 squared"	10×10	100
10^1	"10 to the power of 1"	10	10
10^0	"10 to the power of zero"	$\frac{10}{10}$	1
10^{-1}	"10 to the power of minus 1"	$\frac{1}{10}$	$\frac{1}{10}$ (or 0.1)
10^{-2}	"10 to the power of minus 2"	$\frac{1}{10 \times 10}$	$\frac{1}{100}$ (or 0.01)

We also use powers of 10 to describe other large numbers:

$$2\,380\,000 = 2.38 \times 1\,000\,000 = 2.38 \times 10^6$$

The 2 380 000 format is known as **ordinary form.**

The 2.38×10^6 format, where there is only one digit before the decimal point, is known as **standard form.**

Note that, for each move of the decimal point to the left, the power increases by 1.

In this example, to get standard form the decimal point has been moved six places to the left, so it results in 10 to the power of 6.

Small numbers can be written in standard form as well:

$$0.0056 = 5.6 \times 0.001 = 5.6 \times 10^{-3}$$

Here, for each move of the decimal point to the right the power reduces by 1.

So, to get standard form the decimal point has been moved three places to the right, resulting in 10 to the power of –3.

5.3 Multiplying or dividing powers

When we multiply numbers with powers, we have to *add* the powers.

Example

$$3^3 \times 3^2 = 3^{3+2} = 3^5 = 3 \times 3 \times 3 \times 3 \times 3 = 243$$

$$10^4 \times 10^2 = 10^{4+2} = 10^6 = 10 \times 10 \times 10 \times 10 \times 10 \times 10 = 1\,000\,000$$

Similarly, to divide a number with a power by another, we *subtract* the powers.

Example

$$3^6 \div 3^2 = 3^{6-2} = 3^4 = 3 \times 3 \times 3 \times 3 = 81$$

$$10^7 \div 10^3 = 10^{7-3} = 10^4 = 10 \times 10 \times 10 \times 10 = 10\,000$$

5.4 Multiplying or dividing numbers in standard form

When multiplying or dividing numbers in standard form, group the numbers together, and group the powers of 10 together.

Example

A sample of water has 2200 (2.2×10^3) bacteria per litre. To calculate how many bacteria there are in 36 000 litres (3.6×10^4 l) of water, we need to multiply 3.6×10^4 by 2.2×10^3.

Grouping the numbers (3.6×2.2) and also the powers ($10^4 \times 10^3$) together, we get:

$$(3.6 \times 2.2)\,(10^4 \times 10^3) = 7.92 \times 10^{4+3} = 7.92 \times 10^7$$

Note that $(3.6 \times 2.2)\,(10^4 \times 10^3)$ means $(3.6 \times 2.2) \times (10^4 \times 10^3)$.

So there are 7.92×10^7 bacteria in the 36 000 litres of water.

5.5 Addition and subtraction of numbers in standard form

To add or subtract numbers in standard form, if the powers are the same:

- the two numbers can be added or subtracted;
- the power remains the same.

Example

Two samples of tissue weigh 2.3×10^{-3} and 5.6×10^{-3} kg.

Their total mass is:

$$(2.3 + 5.6) \times 10^{-3} = 7.9 \times 10^{-3}\,\text{kg}$$

If the powers are different, convert one away from standard form so that the powers are the same and then add or subtract the numbers, leaving the power the same. This can then be converted back to standard form.

Example

Two samples of seawater are 4.41×10^7 and $7.9 \times 10^5\,\text{mm}^3$.

It doesn't matter which sample is converted away from standard form, the end result will be the same.

If we convert the first sample,

$$4.41 \times 10^7 = 441 \times 10^5$$

Their total volume is therefore:

$$(441 + 7.9) \times 10^5 = 448.9 \times 10^5\,\text{mm}^3$$

Converting back to standard form gives:

$$4.489 \times 10^7\,\text{mm}^3$$

Test yourself

The answers are given on page 170.

Question 5.1
An onion leaf epidermal cell, *Allium cepa*, is 0.00045 m long. Give this in standard form.

Question 5.2
Human genomic DNA is made up of approximately 3×10^9 base pairs. Give this in ordinary form.

Question 5.3
It is estimated that in a rural area there is a mean of 150 people per square kilometre. Using standard form, calculate how many people will there be in a square plot of 40 by 40 km.

6 Approximation and errors

We sometimes need to **approximate** numbers and give the **degree of accuracy.**

6.1 Approximation

One way of approximating is to give the **nearest whole number**.

32.543716 is nearer 33 than 32, so it is "33 to the nearest whole number".

Another way of giving the degree of accuracy when approximating is to state the number of **decimal places.**

32.543716 is 32.54 to two decimal places, and 32.5437 to four decimal places.

A third way of giving the degree of accuracy when approximating is to give the number of **significant figures**, i.e. the number of digits quoted.

32.543716 is 32.54 correct to four significant figures.

Example

28 365 is 28 000 correct to two significant figures.

6.2 Significant figures and handling zeros

When there are zeros *within* a number, the zeros are considered as significant figures.

Example

10.54 counts as four significant figures.

If the *last* figures in a *whole* number are zeros, the zeros are not counted as significant figures.

Example

6 754 000 counts as four significant figures.

If the zeros are *before* a decimal number, the zeros are not counted as significant figures.

Example

0.0004832 counts as four significant figures.

However, zeros *after* a decimal number are counted as significant figures.

Example

0.8760 counts as four significant figures.

6.3 Rounding numbers

When approximating numbers, the last significant digit stays as it is, i.e. the number is "rounded down", if the next digit is below 5.

Example

6340 is 6300 to two significant figures.

The last significant digit is "rounded up" if the next digit is above 5.

Example

6360 is 6400 to two significant figures.

When the next digit is exactly 5, the common convention is to round the last significant digit up.

Example

6350 is 6400 to two significant figures.

6.4 Choosing the number of significant figures

When two or more measurements are taken, the result is only as reliable as the least reliable value.

So, to work out how many significant figures to use, use the same number of significant figures as the least precise value that the result was derived from.

If you need to adjust the number of significant figures, always do so at the end of your calculations.

Example

An animal is noted to have run 47.81 metres in 8.5 seconds. We want to know its velocity.

The least precise value was the time, which was measured to two significant figures. So, the velocity can also only be given to two significant figures.

$$\frac{47.81}{8.5} = 5.6247 = 5.6\,\mathrm{m\,s^{-1}}$$ to two significant figures.

Note that "metres per second" can be written as $\mathrm{m\,s^{-1}}$ or as m/s.

6.5 Errors

Where we use approximations, there will be some **errors** in the calculations.

If the height of a plant is given as 4 m, the lack of a decimal place implies that the measurement is correct to the nearest metre. There is an error of $\pm$ 0.5 m, so the actual height could be anywhere between 3.5 and just below 4.5 m (remember that a height of exactly 4.5 m would have been rounded up to 5 m).

If the plant's height is given as 4.29 m, the error is $\pm$ 0.005 m, and the actual height could be anywhere between 4.285 and just below 4.295 m.

6.6 Precision and accuracy

Measuring instruments may be precise, in that they can give results to many significant figures, but inaccurate, in that they may be incorrectly calibrated.

Example

A pH meter may measure pH to three decimal places of precision. However, if it has been set up incorrectly it will always give an inaccurate value.

Test yourself

The answers are given on page 170.

Question 6.1
The height of a child is 1.050 m. How many significant figures is this?

Question 6.2
58.44 g of NaCl is dissolved in 0.137 m^3 of water.
Use a calculator to work out the concentration, and state your answer to the appropriate number of significant figures.

Question 6.3
The mass of a hen's egg is given as 56 g to the nearest gram. Within what range could the actual mass be?

7 Introduction to graphs

One way to present data is in the form of a table. However, often we can understand and interpret data more easily by plotting them on a graph.

Graphs also help to define the relationship between two variables.

7.1 The *x*- and *y*-axes

To compare two variables we use a two-dimensional graph. It uses an ***x*-axis** and a ***y*-axis**.

The horizontal axis is known as the *x*-axis, and the vertical axis is called the *y*-axis.

The axes of a graph

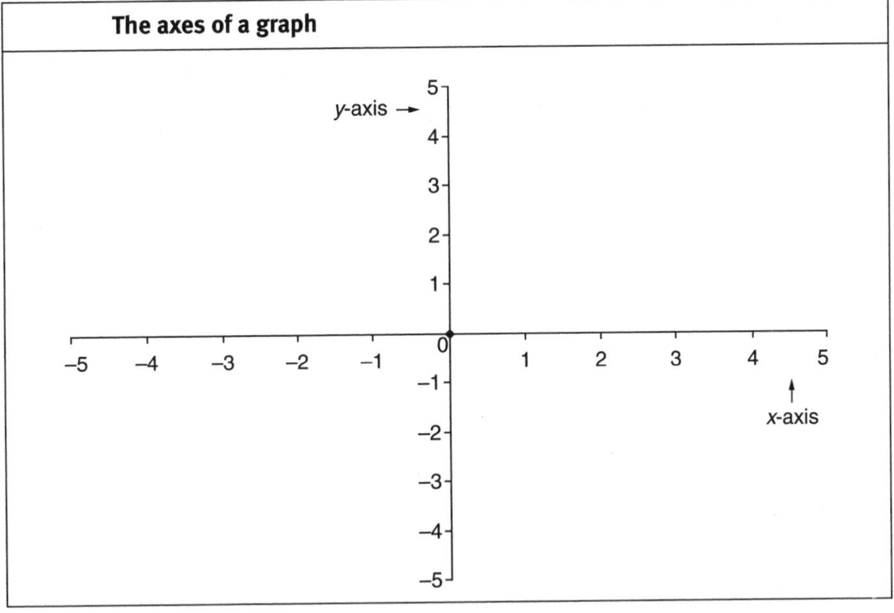

The variable that is "controlled" by the researcher is usually plotted on the *x*-axis. We call this the **independent variable**, as it is not dependent on the other variable, and can therefore be determined by the researcher.

We plot the **dependent variable** on the *y*-axis. This is the variable which can be determined or predicted if the *x* value is known.

> **Example**
>
> When measuring a reaction over time, time is plotted on the x-axis as the researcher will decide at what times to measure the reaction.

7.2 Plotting values on a graph

Say that we want to **plot** the values in the table below on a graph.

Table of *x* and *y* values

Value for x	0	2	4	6
Value for y	1	3	5	7

First we need to draw the *x*- and *y*-axes to a suitable scale. In this example *x*- and *y*-axes can be drawn from 0 to 8.

We then plot the values.

For example, with the first pair of values, where $x = 0$ and $y = 1$, we find 0 on the *x*-axis and then move vertically up until we reach the value of 1 on the *y*-axis, and mark that point.

While computer software, books and journals will use dots, when drawing a graph by hand it is best to use a cross, as this pinpoints the exact place with more accuracy than a dot.

Plot of *x* and *y* values

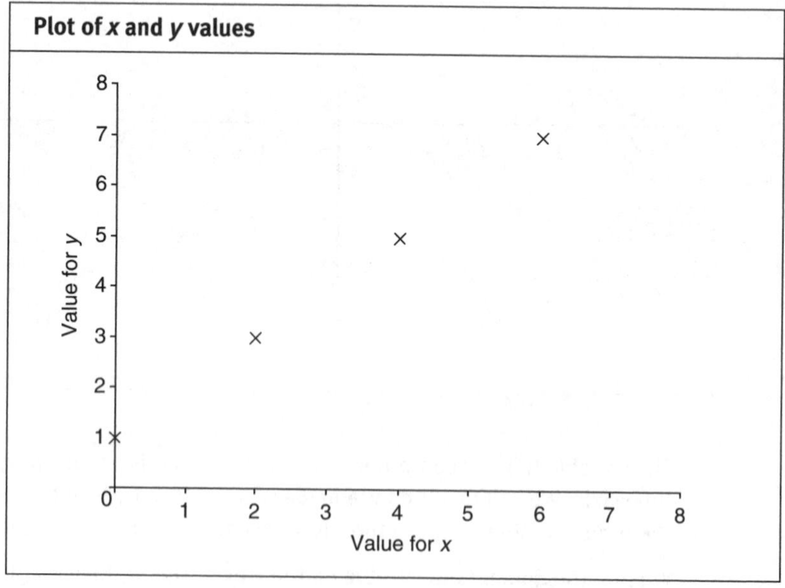

7.3 Co-ordinates

We can define a point on a graph by its **co-ordinates.**

A co-ordinate is written like this:

(*x*-value, *y*-value)

Therefore the co-ordinates of the points in the plot above are:

(0,1) (2,3) (4,5) and (6,7)

The co-ordinate (0,0) is known as the **origin** of a graph.

7.4 Direct proportion

Variables are said to be in **direct proportion** if:

- when one variable is zero, the other is also zero;
- when one variable changes, the other changes in the same ratio.

Example

The number of plankton in a water sample is in direct proportion to the size of the water sample. Doubling the volume of water doubles the number of plankton. If there is no water, there are no plankton.

We use the symbol $\propto$ to indicate direct proportion. So, $x \propto y$ indicates that the value of variable x is directly proportional to the value of variable y.

This can be shown in the form of a graph.

Example

This linear (straight line) graph shows how the number of plankton in a sample of seawater relates to the size of the sample.

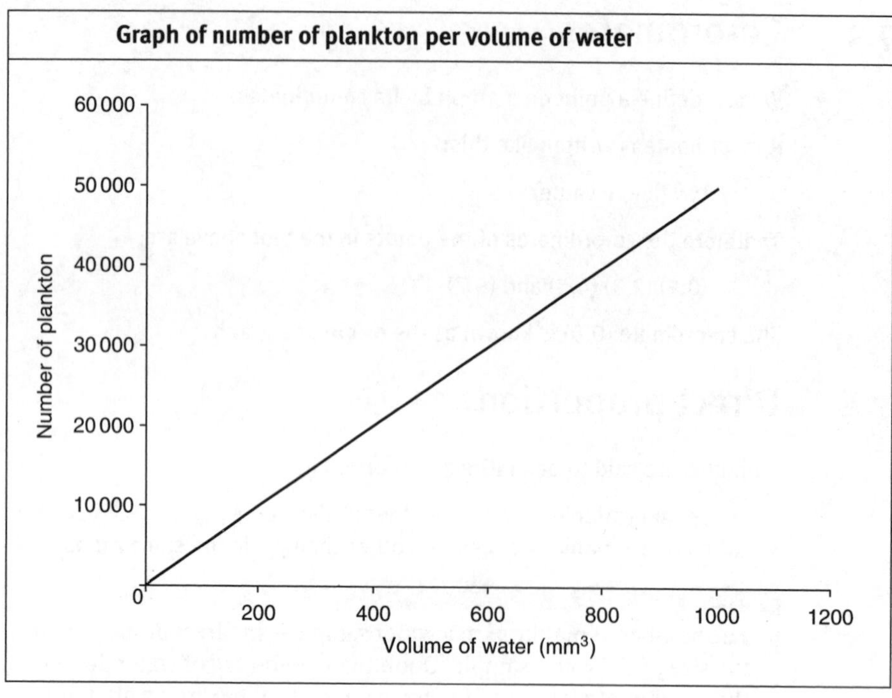

A linear (straight line) graph that goes through the origin can be written as an equation:

$$y = mx$$

where m is the gradient (or slope) of the graph.

When two variables are in direct proportion, if we know both variables we can calculate the gradient of the graph.

If we know one variable and the gradient, we can calculate the value of the other variable.

Test yourself

The answer is given on page 170.

Question 7.1

In a maze-learning experiment, the number of errors made by a rat in a maze is tabulated against the number of trips.
Plot these values on a graph.

Table of number of errors made by a rat in a maze

Trip number	1	2	3	4	5	6
Error score	31	18	15	6	7	3

8 The gradient of a graph

The gradient of a graph describes the steepness of the graph.

8.1 The gradient of a straight line

To calculate the gradient of a graph, we take a section of the graph that we have drawn, then divide the number of units that the graph has moved up the y-axis by the number of units it has moved along the x-axis.

The **gradient** (sometimes known as the **slope**) is therefore $\frac{\text{Change in } y}{\text{Change in } x}$.

Example

In this graph, for every two units up the y-axis, the line goes one unit along the x-axis.

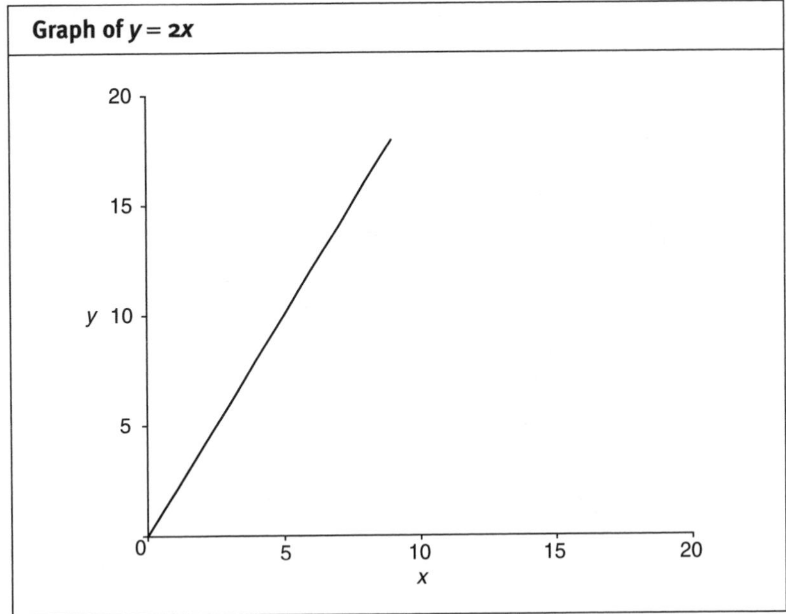

The gradient Change in y/Change in x is therefore 2/1, or 2, so the equation for the graph is $y = 2x$.

The steeper the graph, the larger value for the gradient.

This is the graph for $y = x/3$ plotted using the same scale axes as the previous graph.

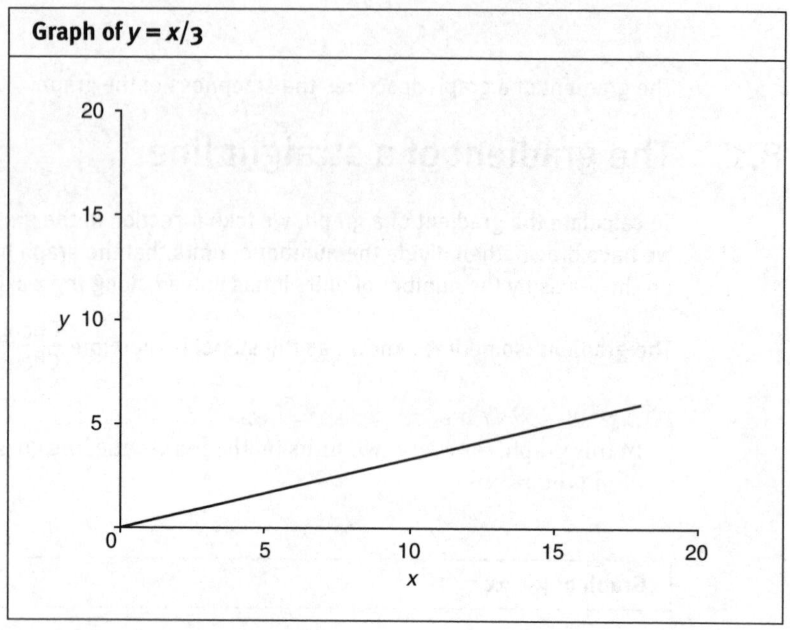

The gradient can be given as a decimal. The gradient of this graph is 0.3333, correct to four decimal places.

8.2 The formula for the gradient

The formula that describes the gradient

$$\frac{\text{Change in } y}{\text{Change in } x}$$

is

$$\frac{y_2 - y_1}{x_2 - x_1}$$

Example

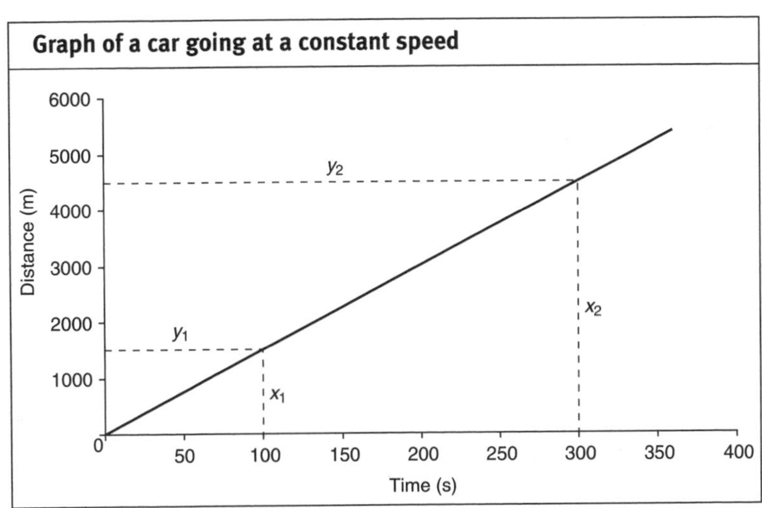

This graph shows the distance that a car travels in any given time.

Between 100 and 300 seconds, it has travelled from 1500 to 4500 m.

$$\text{Gradient} = \frac{y_2 - y_1}{x_2 - x_1} = \frac{4500 - 1500}{300 - 100} = \frac{3000}{200}\,\text{m s}^{-1}$$

So, the gradient of the line is 15, making the speed of the car $15\,\text{m s}^{-1}$.

The equation for the graph is therefore $y = 15x$

8.3 The symbol for the gradient

For a straight-line graph, the gradient is symbolised by the constant *m*.

$$\text{Gradient} = \frac{\text{Change in } y}{\text{Change in } x} = m$$

m is the **rate of change** of *y* with respect to *x*.

8.4 Negative gradients

In this graph, for each unit along the *x*-axis, the graph goes two units *down* the *y*-axis:

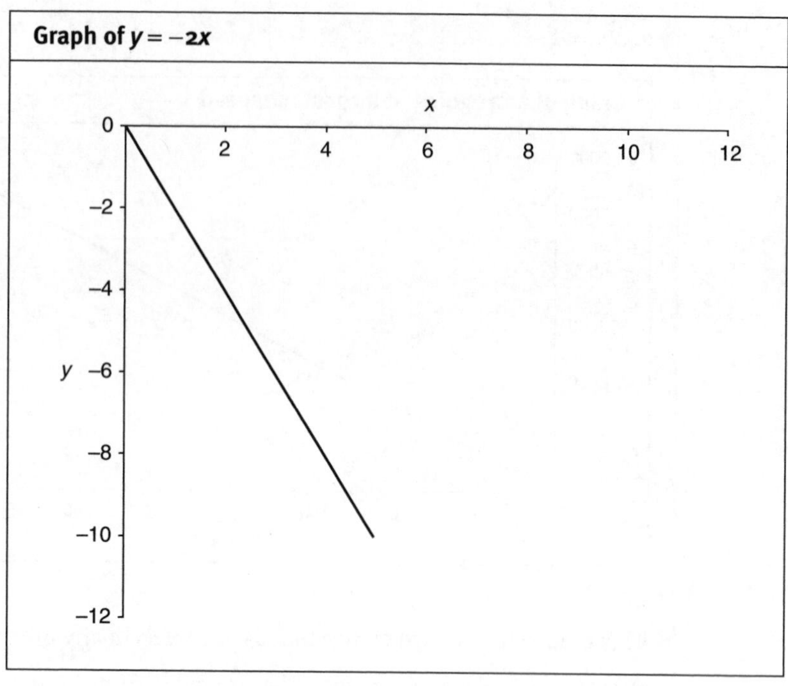

The gradient is therefore –2/1, or –2, so the equation for the graph is $y = -2x$.

So, a line that runs up to the right has a *positive* gradient . . .

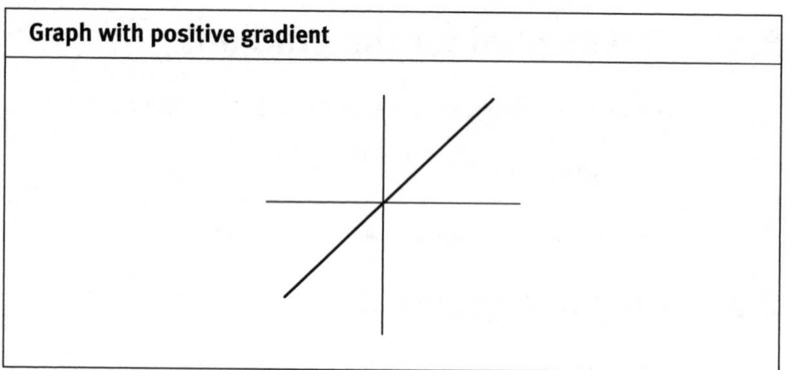

. . . while one that runs down to the right has a *negative* gradient:

Graph with negative gradient

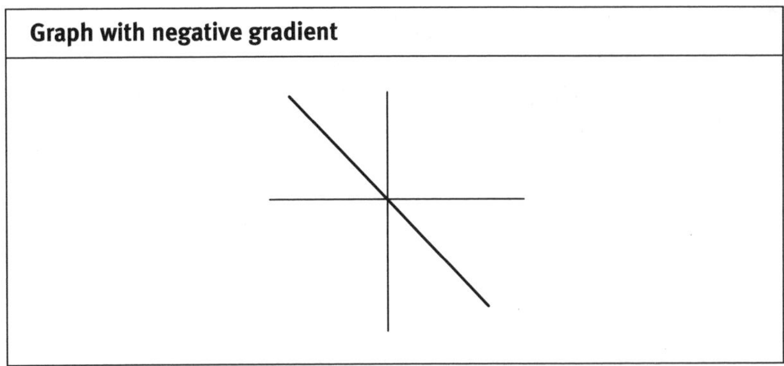

8.5 Graphs that don't go through the origin

The next graph has a gradient of 4 but the line doesn't go through the origin, i.e. it does not go through (0,0).

Graph of $y = 2x - 4$

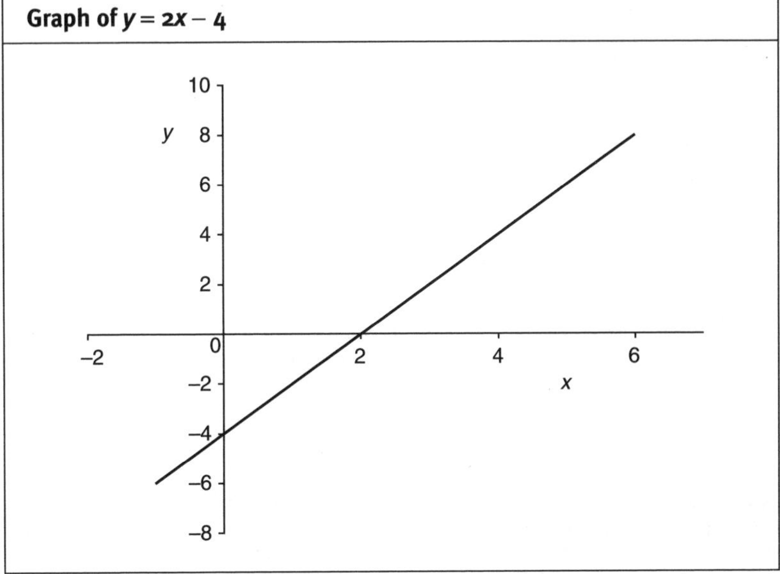

The equation for a graph that does not go through the origin is $y = mx + c$, where m is the gradient (as before), and c is the point at which the line crosses the y-axis (i.e. when $x = 0$ then $y = c$). This point is called the "y-intercept".

In this case, the gradient is 2, and the line crosses the y-axis at –4, so the equation for this line is $y = 2x - 4$.

8.6 Linear equations

The equation $y = mx + c$ is called **linear** as it describes a straight-line graph: there is a straight-line relationship between the two variables x and y.

Test yourself

The answers are given on page 170.

Question 8.1

The following graph shows the distance that a cyclist covers over 50 minutes. Calculate the speed that the cyclist is going in kilometres per hour.

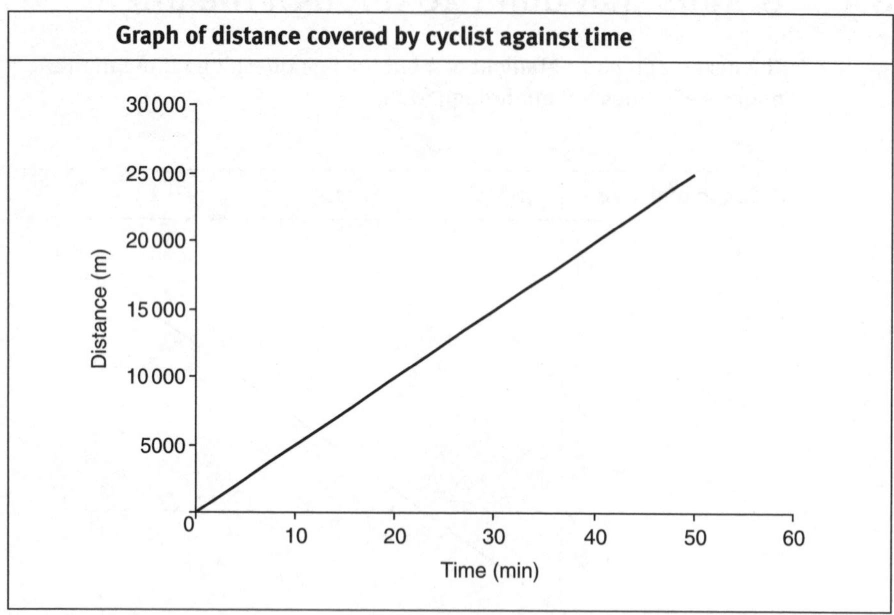

Question 8.2

A baby girl is 500 mm long when born. She grows 10 mm per week. Give the equation for this relationship and draw a graph that shows how she grows over the first 6 weeks.

Question 8.3

A yew tree, *Taxus baccata*, was planted as a sapling. One year later it is found to be 200 mm high. Two years after that it is found to be 400 mm high. State this growth as an equation and, assuming that the rate of growth is constant, calculate how high it was when planted.

9 Algebra

Algebra is the area of mathematics that uses symbols to represent numbers. It lets us investigate relationships between quantities.

We need to be competent in algebra to be able to handle ("manipulate") the many equations and formulae used in the life and medical sciences.

9.1 Using symbols

In arithmetic, 2, 5, 7, 9, 10, etc. each have a fixed value.

In algebra, *a*, *b*, *c*, *x*, *y*, *z*, etc. stand for *any* values.

For instance, we may use the letter "*x*" as a symbol to represent an unknown, variable number. We could equally well use *a*, *b*, *c* or any other letter as a symbol.

9.2 Simplifying expressions

Collecting algebraic terms together helps to **simplify** expressions.

Examples

$4a + 3a$ can be simplified to $7a$.

$4a + 6b + 3a + b$ can be simplified to $7a + 7b$. This can be further simplified to $7(a + b)$.

Identical powers can be collected together.

Example

$2a^2$ is the same as $a^2 + a^2$.

$3a^2$ is the same as $a^2 + a^2 + a^2$.

So, $3a^2$ plus $2a^2$ is the same as $a^2 + a^2 + a^2 + a^2 + a^2$, which can be simplified to $5a^2$.

$4a^4$ plus $5a^4$ can be simplified to $9a^4$.

Different powers cannot be collected together.

Example

$2a^2$ is the same as $a^2 + a^2$.

$3a^4$ is the same as $a^4 + a^4 + a^4$.

So, $3a^4$ plus $2a^2$ becomes $a^4 + a^4 + a^4 + a^2 + a^2$. It cannot be simplified and remains as $3a^4 + 2a^2$.

Collect different powers separately. Conventionally we state the larger powers first.

$3a^2 + 6a^4 + 2a^2 + a$ can be simplified to

$6a^4 + (3 + 2)a^2 + a$, which further simplifies to

$6a^4 + 5a^2 + a$

9.3 The factors of an expression

In section 2.1 we learnt that the factors of a number are all the whole numbers that divide into it without a remainder.

Algebraic expressions also have factors.

Example

The factors of $30ab^2$ include 2, 3, 5, a and b.

$$30ab^2 = 2 \times 3 \times 5 \times a \times b \times b$$

This can be stated in different ways, for example:

$$30ab^2 = 6(5ab^2)$$

$$30ab^2 = 3b(10ab)$$

9.4 Cancelling in fractions

Cancelling is another way to simplify fractions.

Common factors are those factors that are common to two or more numbers. In order to cancel in algebraic fractions, we find common factors in the numerator and the denominator of the fraction and then we can cancel.

Example

$\frac{a^4b^2}{a^3c}$ can be simplified by cancelling a^3 from the top and the bottom.

$$\frac{a^4b^2}{a^3c} = \frac{ab^2}{c}$$

If you are unsure how to do this, write the powers out in full.

$$\frac{a^4b^2}{a^3c} = \frac{\cancel{a \times a \times a \times} a \times b \times b}{\cancel{a \times a \times a \times} c} = \frac{ab^2}{c}$$

9.5 Cancelling expressions

We can cancel expressions in the same way that we can cancel single variables.

Example

$\frac{d^2(ab+c)^5}{e(ab+c)^3}$ can be simplified by cancelling $(ab+c)^3$ from top and bottom.

$$\frac{d^2(ab+c)^5}{e(ab+c)^3} = \frac{d^2(ab+c)^2\cancel{(ab+c)^3}}{e\cancel{(ab+c)^3}} = \frac{d^2(ab+c)^2}{e}$$

9.6 When not to cancel in fractions

If we cannot find a common factor for the whole of the top and the whole of the bottom of the fraction, we cannot cancel in the fraction.

Example

$\frac{a^4b^2+d}{a^3c}$ cannot be simplified by cancellation, because there is no common factor for a^4b^2+d and a^3c.

9.7 Multiplying out expressions

Expressions with brackets can be **multiplied out**.

Example

$3(4a+2b)$ multiplies out to $12a+6b$.

Make sure that you multiply everything inside the bracket with everything outside.

9.8 The highest common factor

Factorising an algebraic expression is the opposite of multiplying out. To do this, we pull out a common factor, and put the remainder in brackets.

To factorise an algebraic expression, pull out the **highest common factor** – the largest factor that is common to each term.

Example

We wish to factorise $12a$ and $6b$.

Separating the components, for 12a and 6b the highest common factor is 6.

Pulling out the 6 and putting the remainder in brackets gives:

$$6(2a + b).$$

So, $12a + 6b$ factorised gives $6(2a + b)$.

9.9 The difference of squares

Where there is a **difference of squares**, i.e. when one square is subtracted from another, an expression can be factorised.

Example

$$a^2 - b^2 = (a - b)(a + b)$$

(Note that $(a - b)(a + b) = a^2 - b^2 + ab - ab = a^2 - b^2$)

However, we can't factorise a *sum* of squares.

Example

$a^2 + b^2$ cannot be factorised.

Test yourself

The answers are given on page 171.

Question 9.1
Simplify $15a^5 + 12a^5 + 2a^3 + 4a^2 + a^2 + 7a$

Question 9.2
Simplify $\frac{a^2b}{a^3} \times \frac{a^4b^2}{b^3}$

Question 9.3
Simplify by cancelling:
$\frac{a^3b^3\,(c + 2d)^4}{a^2b^4\,(c + 2d)}$

Question 9.4
Which of these fractions can be simplified by cancelling?

1) $\frac{a^2 - b^4}{b}$

2) $\frac{c^4d^2 + b^2c^2d^2}{c^2 + b^2}$

3) $\frac{e^3d^2 - cf}{cf}$

Question 9.5
Multiply out $5a(2a - b^2)$

Question 9.6
Factorise $6a^3b^2 + 9a^2b^4$

Question 9.7
Factorise $a^2 - 4b^2$

10 Polynomials

In science, some relationships are linear, like the equation $y = mx + c$ for a straight-line graph.

Polynomials describe relationships that contain *powers* of numbers and therefore are not linear.

10.1 The definition of a polynomial

Polynomials are expressions where all the terms have a variable raised to a positive "integer" (i.e. whole number) power.

Example

$5x^4 + 6a^2 - 4x + 3$ is a polynomial.

The **degree** of a polynomial is its highest power.

Example

$5x^4 + 6a^2 - 4x + 3$ is a 4th degree polynomial.

The zeroth degree polynomial is x^0 and is equal to 1.

10.2 Other names for polynomials

A **binomial** expression is a polynomial with two terms.

A **trinomial** expression is a polynomial with three terms.

A **quadratic** expression is a second degree polynomial: the highest power is 2.

A **cubic** expression is a third degree polynomial: the highest power is 3.

Example

$6a^2 - 4x + 3$ has three terms, so is a trinomial. Its highest power is 2, so it is a quadratic expression.

10.3 Adding and subtracting polynomials

In order to add or subtract polynomials, keep each term separate and only add or subtract terms with the same power.

Example

When adding $4x^2+3x+6$ to $8x^3+2x+4$, we collect terms which are raised to the same power:

$$\begin{array}{r} 4x^2+3x+\ 6 \\ 8x^3 \qquad\ \ +2x+\ 4 \\ \hline 8x^3+4x^2+5x+10 \end{array}$$

We can use the same process to subtract x^3+6x^2-3 from $4x^2+2x+1$:

$$\begin{array}{l} \qquad\quad 4x^2\ +2x\ +1 \\ -(x^3)-(6x^2) \qquad -(-3) \\ \hline -x^3-2x^2\ +2x\ +4 \end{array}$$

10.4 Multiplying polynomials

When multiplying polynomials, every term in the first expression must be multiplied by every term in the second expression.

Examples

Multiplying the first degree polynomials $x+2$ and $x+3$ can be represented by $(x+2)(x+3)$.

We need to multiply both x and 2 in the first expression by both x and 3 in the second expression.

$$(x+2)(x+3)=x(x+3)+2(x+3)$$

This gives:

x^2, $3x$, $2x$ and 6.

Adding these together gives:

x^2+5x+6.

We wish to multiply the terms x^5+3x^4+2 and $6x^2+3x-5$.

Multiplying x^5 with each component of the second term gives $6x^7$, $3x^6$ and $-5x^5$.

Multiplying $3x^4$ with each component of the second term gives $18x^6$, $9x^5$ and $-15x^4$.

Multiplying 2 with each component of the second term gives $12x^2$, $6x$ and -10.

Adding all these together:

$$\begin{array}{l} 6x^7+\ 3x^6-5x^5 \\ \qquad +18x^6+9x^5-15x^4 \\ \qquad\qquad\qquad\qquad\qquad +12x^2+6x-10 \\ \hline 6x^7+21x^6+4x^5-15x^4+12x^2+6x-10 \end{array}$$

10.5 Factorising polynomials

The previous chapter explained how to factorise an algebraic expression, by pulling out a common factor and putting the remainder in brackets.

Factorising a polynomial like $4x^3 - 6x^2 + 2x - 3$ means reversing the multiplication and taking it back to $(2x^2 + 1)(2x - 3)$.

Look for the common factors. This can be a bit tricky but it gets easier with practice.

If we represent the constant with the symbol *a*, many polynomials follow one of the following patterns on the left hand side of the following table. Follow each polynomial across to the right hand side to see how it can be factorised.

Table showing factorisation of common polynomials

	Polynomial multiplied	↔	Intermediate step	↔	Polynomial factorised
1	$x^2 + 2xa + a^2$	↔	$(x + a)(x + a)$	↔	$(x + a)^2$
2	$x^2 - 2xa + a^2$	↔	$(x - a)(x - a)$	↔	$(x - a)^2$
3	$x^2 - a^2$	↔		↔	$(x + a)(x - a)$
4	$x^3 + 3x^2a + 3xa^2 + a^3$	↔	$(x + a)(x + a)(x + a)$	↔	$(x + a)^3$
5	$x^3 - 3x^2a + 3xa^2 - a^3$	↔	$(x - a)(x - a)(x - a)$	↔	$(x - a)^3$

Example

The polynomial

$$x^2 + 8x + 16$$

can be factorised to

$$x^2 + 2(4x) + 4^2$$

When $a = 4$, this is directly equivalent to the first polynomial in the table above:

$$x^2 + 2xa + a^2$$

which we know from the table factorises to

$$(x + a)^2$$

So,

$x^2 + 8x + 16$ factorises to $(x + 4)^2$.

You can check this by multiplying out $(x + 4)(x + 4)$.

Test yourself

The answers are given on page 171.

Question 10.1
What is the degree of the following polynomial?
$6a^5 + 5a^3 + 2a^2 - 12$

Question 10.2
Subtract $6x^4 + 9x^3 - x^2 + 5$ from
$2x^5 + 7x^4 + 5x^3 + 4$

Question 10.3
Multiply out $(4x^4 - x^2 + 5)(2x^5 + 3x^2 + 6)$

Question 10.4
Factorise the polynomial $x^2 - 6x + 9$

11 Algebraic equations

Many relationships in science can be generalised into algebraic equations. Anything from the growth of a child to the photosynthesis rate of a plant can be written as an algebraic equation.

11.1 Balancing the sides of an equation

In an equation, the equals sign can be represented as a balance.

Whatever is done to one side of the equation must also be done to the *whole* of the other side. So, if we add, subtract, multiply or divide something to one side, we need to do the same to the whole of the other side.

Example

$$3x + 2 = 11$$

Subtracting 2 from each side keeps the equation balanced:

$$3x + 2 - 2 = 11 - 2$$

This can be simplified to:

$$3x = 9$$

Then we can divide both sides by 3:

$$\frac{3x}{3} = \frac{9}{3}$$

So $x = 3$

11.2 Manipulating equations with different powers

We can also manipulate equations with different powers.

Example

We wish to solve the equation $ay^2 - b = x$ for y. This means that we wish to manipulate the equation to show what y equals.

We call this "making y the subject of the equation".

Adding b to each side gives

$$ay^2 = x + b$$

Dividing each side by a gives

$$y^2 = \frac{x + b}{a}$$

Taking the square root of each side gives

$$y = \sqrt{\frac{x + b}{a}}$$

Test yourself

The answers are given on page 172.

Question 11.1
Solve $3x^2 = 12$

Question 11.2
Make y the subject of the equation $x = 4y^3 + 1$.

12 Quadratic equations

Some scientific relationships can be represented by quadratic equations. For example, the equation underlying the Hardy-Weinberg equilibrium in population genetics is a quadratic equation.

12.1 Solving quadratic equations

Quadratic equations have expressions containing a square power, e.g. x^2.

They have two solutions, i.e. x can have two possible values.

Quadratic equations have the form:

$$ax^2 + bx + c = 0$$

where a, b, and c are constants.

Example

$x^2 + 2x - 15 = 0$ is an example of a quadratic equation.

See how it relates to the quadratic formula $ax^2 + bx + c = 0$.

In this example, a is 1, b is 2, and c is -15.

x can equal $+3$ and it can also equal -5.

12.2 Different ways to solve quadratic equations

We will describe four ways to solve quadratic equations:

- graphically
- by factorisation
- by using the quadratic formula
- by completing the square

12.3 Graphical solution

If we plot the quadratic equation on a graph, the solutions are the points at which the graph crosses the x-axis.

Example

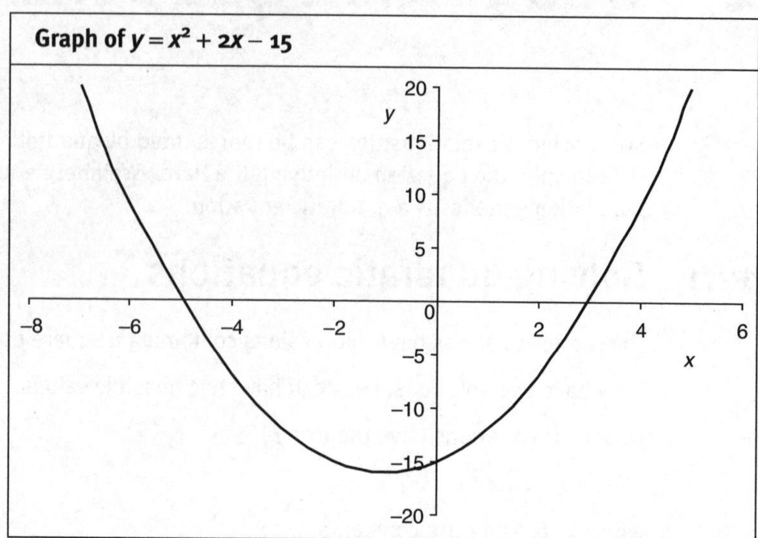

In this graph of $y = x^2 + 2x - 15$, the graph crosses the x-axis at + 3 and – 5, i.e. when $y = 0$, $x = +3$ and – 5.

So, + 3 and – 5 are the solutions to the quadratic equation $x^2 + 2x - 15 = 0$.

12.4 Solution by factorisation

If the quadratic equation is factorised (factorising can be thought of as reversing a multiplication; Chapter 9 explains the basics), the solution is given if each factor is set to zero.

Example

$$x^2 + 2x - 15 = 0$$

can be factorised to:

$$(x + 5)(x - 3) = 0$$

If either one of the expressions in brackets is zero, the equation will still equal zero.

For example

$$0(x - 3) = 0 \text{ or } (x + 5)0 = 0$$

To set the first expression, $(x + 5)$, to zero, x must equal –5.

To set the second expression, $(x - 3)$, to zero, x must equal +3.

Thus $x = -5$ or $x = 3$

12.5 Solution by using the quadratic formula

Given the quadratic equation

$$ax^2 + bx + c = 0,$$

the solution can be calculated from the **quadratic formula:**

$$x = \frac{-b \pm \sqrt{b^2 - 4ac}}{2a}$$

Put the numbers into the equation and calculate the two possible values of x.

Example

We noted above that for the quadratic equation $x^2 + 2x - 15 = 0$, a is 1, b is 2, and c is -15.

$$x = \frac{-b \pm \sqrt{b^2 - 4ac}}{2a} = \frac{-2 \pm \sqrt{2^2 - (4 \times 1 \times -15)}}{2 \times 1} =$$

$$\frac{-2 \pm \sqrt{4 - (-60)}}{2} = \frac{-2 \pm \sqrt{64}}{2} = \frac{-2 \pm 8}{2}$$

The two results are therefore:

$$\frac{-2+8}{2} = 3 \text{ and } \frac{-2-8}{2} = -5$$

12.6 Solution by completing the square

This involves creating a square trinomial that we can solve by taking its square root.

Example

To solve $3x^2 + 24x - 27 = 0$

Put the x^2 and x terms on one side and the constant on the other:

$$3x^2 + 24x = 27$$

Divide both sides by the "coefficient" (the multiplier) of x^2 (in this case 3):

$$\frac{3x^2}{3} + \frac{24x}{3} = \frac{27}{3}, \text{ therefore}$$

$$x^2 + 8x = 9$$

We take half the coefficient of x, (in this case half of 8, giving 4), square it (giving 16), and add it to both sides.

This gives a square trinomial:

$$x^2 + 8x + 16 = 9 + 16 = 25$$

We can now factorise the left-hand side of the equation:

$$(x + 4)^2 = 25$$

Now we need to take the square root of both sides:

$$\sqrt{(x + 4)^2} = \sqrt{25}$$

As a square root can be positive or negative, there needs to be a ± on the right side of the equation.

$$x + 4 = \pm 5$$

We can then solve the equation.

$$x = +5 - 4 = 1 \text{ and } x = -5 - 4 = -9$$

Test yourself

The answers are given on page 172.

Question 12.1
Solve by factorisation:
$x^2 + 6x + 8 = 0$

Question 12.2
Solve, using the quadratic formula:
$x^2 + 6x + 8 = 0$

Question 12.3
Solve by completing the square:
$x^2 + 6x + 8 = 0$

13 Simultaneous equations

Simultaneous equations can be used to calculate when two relationships coincide.

In the life and medical sciences, the relationship between two variables may directly reflect on the relationship of two other variables. For example, a change in numbers of an animal species over time may affect the numbers of another species higher up the food chain. We can establish the link by studying the population growth or decay of both simultaneously.

Mathematically, we do this by solving simultaneous equations.

13.1 Two equations with a single solution

Two equations that can be solved by one value of *x* and one value of *y* are called **simultaneous equations.**

Example

$$x - y = 5 \text{ and } x + 2y = -4$$

can both be solved if

$$x = 2 \text{ and } y = -3$$

13.2 Three ways to solve simultaneous equations

We will show three ways to solve simultaneous equations: graphically, by substitution, and by elimination.

13.3 The graphical solution

If we plot the two equations on a graph, the co-ordinates of where they intersect give the solution.

Example

To solve the simultaneous equations $x + 3y = 4$ and $6x - 5y = 1$ by graph, we need to plot both lines on a graph. The solution is the point where they intersect.

First, we need to rearrange the two equations to the form $y = mx + c$, so that they can be plotted on the graph.

The equation $x + 3y = 4$ manipulates to $y = -\frac{x}{3} + 1.333$.

We found in Chapter 8 that this means that the gradient of the line is $-1/3$, and that the line crosses the y-axis at $+1.333$.

Manipulating the equation $6x - 5y = 1$ gives

$y = 1.2x - 0.2$

so the gradient of this line is 1.2 and it crosses the y-axis at -0.2

Graph showing the solution of simultaneous equations

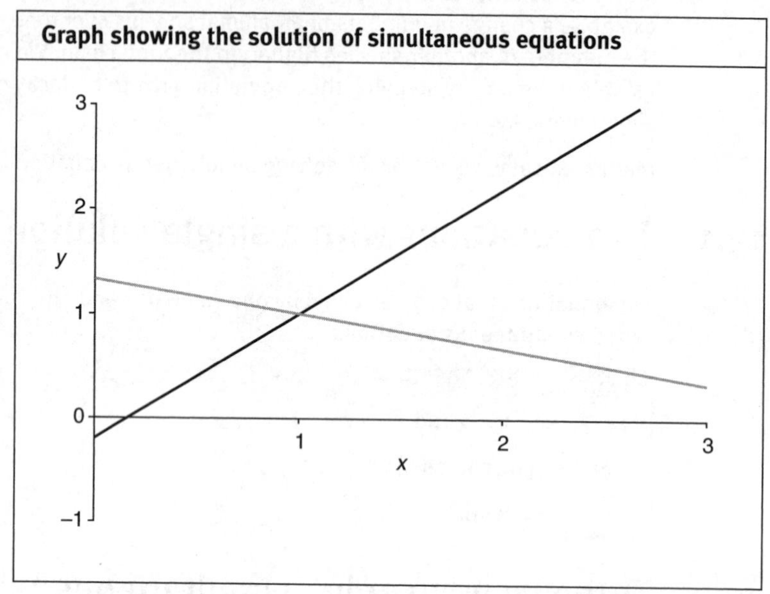

The lines intersect each other at (1,1), i.e. where $x = 1$ and $y = 1$, so the solution is $x = 1$, $y = 1$.

13.4 Solution by substitution

For solution by substitution, we manipulate one of the equations so that y is written as an expression of x. This value of x is then substituted into the second equation, and the solution can be calculated.

Example

To solve the simultaneous equations $x + 3y = 4$ and $6x - 5y = 1$ by substitution, we manipulate the first equation by subtracting $3y$ from each side:

$$x = 4 - 3y$$

We now substitute this into the second equation, giving:

$$6(4-3y)-5y=1$$

This multiplies out to:

$$24-18y-5y=1$$

Subtract 24 from each side:

$$-18y-5y=1-24$$

This simplifies to:

$$-23y=-23$$

Dividing both sides by 23 gives:

$$-y=-1$$

So $y=1$

Now we know the value of y, we can substitute it into either of the original equations to calculate the value of x:

$$x+3y=4$$

$$x+(3\times 1)=x+3=4$$

$$x=4-3=1$$

So the solution is $x=1$, $y=1$.

13.5 Solution by elimination

Another way of solving simultaneous equations is by eliminating one of the unknowns (either the x or the y).

Example

To solve the simultaneous equations

$$x+3y=4 \quad (1)$$

and

$$6x-5y=1 \quad (2)$$

by elimination, we can manipulate equation (1) by multiplying both sides by 6:

$$6(x+3y)=6\times 4$$

This multiplies out to:

$$6x + 18y = 24$$

So, the equation (1) can be manipulated to:

$$6x = 24 - 18y$$

By adding $5y$ to both sides of equation (2), we get:

$$6x = 1 + 5y$$

Both sides have now been manipulated to equal $6x$, so both equations are now equal:

$$24 - 18y = 6x = 1 + 5y$$

i.e.:

$$24 - 18y = 1 + 5y$$

This simplifies to:

$$-18y - 5y = 1 - 24$$

and again to:

$$-23y = -23$$

Once more, dividing both sides by 23 gives:

$$-y = -1$$

So $y = 1$

As in the solution by substitution example above, now we know the value of y, we can substitute it into either of the original equations and find that $x = 1$.

Test yourself

The answers are given on page 172.

Question 13.1
Solve $2x + y = 8$ and $3x + 2y = 14$ by substitution

Question 13.2
Use elimination to solve $2x + y = 8$ and $3x + 2y = 14$

Question 13.3
There are 1000 people living in a village. Each year the number is expected to increase by 50. The population in another village is 1600, declining by 50 per year.
In how many years will they have an identical population size?
What will the population be at that time?
Solve this in two ways: graphically, and by calculation.

14 Sequences and series of numbers

In nature we can often identify patterns. These patterns can frequently be identified as one of two distinct groups: **arithmetic series** and **geometric series**.

14.1 Sequences

Where we have a **sequence** of numbers, we can use a letter to represent a position in the sequence. Typically, for a simple variable we use the letter n.

So, for the sequence 2, 4, 6, 8, 10 . . ., the nth term is $2n$.

To work out the 30th term, we substitute 30 for n, so that $2n = 2 \times 30 = 60$.

14.2 Arithmetic series

In the example above, the difference between each number is constant.

This means that it is an **arithmetic series** (also known as an arithmetic progression).

An arithmetic series is one in which each number or term is obtained from the previous number or term, by adding (or subtracting) a constant quantity.

This constant is known as the **common difference**.

Example

For the 7, 12, 17, 22, 27... sequence, the common difference is +5:

Term	1st		2nd		3rd		4th		5th
Number	7		12		17		22		27
Common difference		+5		+5		+5		+5	

An arithmetic series can be symbolised by:

$a, (a+d), (a+2d), (a+3d)$ . . . . . . to n terms,

where a is the first term (7 in the example above), d is the common difference (5 in the example above), and n symbolises the "term" number (the number that we want to count to).

The nth term for any arithmetic series is given by:

$$a_n = a + (n - 1)d$$

Example

For the sequence 13, 16, 19, 22, 25 . . ., we wish to calculate the 78th term.

The first term is 13, so $a = 13$; the common difference is 3, so $d = 3$; we want the 78th term, so $n = 78$.

Substituting these into the equation

$$a_n = a + (n - 1)d$$

we get

$$13 + (78 - 1)3 = 244$$

So the 78th term is 244.

14.3 The sum of an arithmetic series

We use the Σ ("summation") symbol to indicate that we are adding a list of terms together. We then put a comma between each term, rather than a plus sign.

Adding the arithmetic series

$$a + (a + d) + (a + 2d) + (a + 3d) \ldots \text{ to } n \text{ terms}$$

can therefore be symbolised by:

$$\sum a, (a + d), (a + 2d), (a + 3d) \ldots (a + (n - 1)d)$$

The formula that calculates this total is:

$$S_n = \frac{n}{2}\,[2a + (n - 1)d]$$

where S_n is the sum of the first n terms.

While this may look complex, note that it is the formula for the nth term for any arithmetic series

$$a + (n - 1)d$$

plus the first number in the series, a, the total being multiplied by $\frac{1}{2}n$.

Try not to confuse the S_n notation (meaning the sum of the first n terms) with the log base notation, $\log_n$.

Example

Given the sequence 13, 16, 19, 22, 25 . . ., we wish to calculate the sum of the first 78 terms.

The first term is 13, so $a = 13$; the common difference is 3, so $d = 3$; $n = 78$.

Substituting these into the formula

$$S_n = \frac{n}{2}\,[2a + (n-1)d]$$

gives

$$S_{78} = \frac{78}{2}\,[(2 \times 13) + (78-1)3] = 39(26 + 231) = 10\,023$$

So the sum of the first 78 terms is 10 023.

14.4 Geometric series

A **geometric series**, also known as a geometric progression, is formed by multiplying each term by a constant. This constant is known as the **common ratio** and can be any value except 0, 1, or –1.

Example

For the 3, 6, 12, 24, 48 . . . sequence, each number is multiplied by 2:

Term	1st		2nd		3rd		4th		5th
Number	3		6		12		24		48
Common ratio		×2		×2		×2		×2	

A geometric series can be symbolised by:

a, ar, ar^2, ar^3 . . . to n terms,

where:

a is the first term (3 in the example above),

r is the common ratio (2 in the example above), and

n is the number of terms.

The nth term for any geometric series is given by:

$$a\,r^{n-1}$$

Example

For the geometric series 5, 20, 80, 320, 1280 . . ., we wish to calculate the 9th term.

Here, the first term is 5, so $a = 5$. The common ratio is 4, so $r = 4$. The nth term is the 9th, so $n = 9$.

Substituting these into the formula

$$a\,r^{n-1}$$

we get

$$5 \times 4^{9-1} = 5 \times 4^8 = 5 \times 65\,536 = 327\,680$$

14.5 The sum of a geometric series

The sum of any geometric series can be written as

$$S_n = \frac{a(r^n - 1)}{r - 1}$$

You may also see it written as

$$S_n = \frac{a(1 - r^n)}{1 - r}$$

which gives the same answer, but is easier to use if $r < 1$.

Example

In calculating the sum of the first nine terms of the series 5, 20, 80, 320, 1280 . . ., again $a = 5$, $r = 4$, and $n = 9$.

Substituting these into the formula above gives:

$$S_9 = \frac{5(4^9 - 1)}{4 - 1} = \frac{5 \times 262\,143}{3} = 436\,905$$

So the sum of the terms is 436 905.

Test yourself

The answers are given on page 173.

Question 14.1
There are initially 12 robins in an area of parkland. The yearly count of robins increases as follows: 16, 20, 24, 28. Assuming that the rate of increase remains constant, how many robins would be expected in the 10th year?

Question 14.2
If the life expectancy of the robins in the previous question is 1 year, how many robins in total would have inhabited the parkland in the 10 years?

Question 14.3
One year after a forest fire, in one quadrat (measured area) there are 250 plants which are greater than 100 mm high. In the succeeding years the numbers go up to 750, 2250 and 6750.
Assuming the numbers increase in the same pattern each year, how many plants would we expect to find 7 years after the forest fire?

Question 14.4
If the plants studied in the previous question were all annuals, what is the total number of plants that would have grown in the plot during the 7 years?

15 Working with powers

This chapter describes a number of rules that can help us when working with powers.

15.1 The power of zero

In Section 5.2, we found that $10^0 = 1$.

This applies for any value (except zero itself: $0^0 = 0$). Any value to the power of zero is 1.

Example

$$x^0 = 1$$

15.2 Useful rules for working with powers

Other useful rules for working with powers are as follows:

$$x^{-2} = \frac{1}{x^2}$$

$$x^{1/2} = \sqrt{x}$$

$$x^{1/3} = \sqrt[3]{x}$$

$$x^{2/3} = \sqrt[3]{x^2} = \left(\sqrt[3]{x}\right)^2$$

$$x^2 \times x^3 = x^{2+3} = x^5$$

$$\frac{x^5}{x^3} = x^{5-3} = x^2$$

$$\left(x^2\right)^3 = x^6$$

$$(xab)^2 = x^2a^2b^2$$

$$\left(\frac{x}{a}\right)^2 = \frac{x^2}{a^2}$$

15.3 Adding and subtracting powers

As seen in Section 5.4, numbers in standard form can be added or subtracted if the powers are the same. The same applies to algebraic expressions.

Example

$$5x^3 - 2x^3 = 3x^3$$

However,

$$x^2 + x^5$$

cannot be added, as the powers are different.

15.4 Working with roots

All roots can be converted into powers.

Examples

$$\sqrt[3]{x} = x^{1/3} = x^{0.333}$$

$$\sqrt[4]{x^2} = x^{2/4} = x^{1/2} = x^{0.5}$$

In basic mathematics, we can't have a square root (or any even-numbered root) of a negative value.

$\sqrt[4]{-x}$ (crossed out)

You may find these root rules helpful:

$$\sqrt[3]{x} \times \sqrt[3]{a} = \sqrt[3]{xa}$$

$$\frac{\sqrt[3]{x}}{\sqrt[3]{a}} = \sqrt[3]{\frac{x}{a}}$$

$$\sqrt[2]{\sqrt[3]{x}} = \sqrt[2\times3]{x} = \sqrt[6]{x}$$

Test yourself

The answers are given on page 174.

Question 15.1

Calculate $\dfrac{(a+2)^7}{(a+2)^5}$

Question 15.2

Calculate $\sqrt[3]{(2a-1)^6}$

16 Logarithms

Many biological and biochemical systems are logarithmic. For example, population growth can show logarithmic properties; the pH measure of acidity is a logarithmic measure.

16.1 Introducing logarithms

We have already met powers, like 2^3.

The first number, 2 in this case, is called the "base".

The power, 3 in this case, is called the "exponent".

The exponent is the **logarithm** ("log") of the base.

Examples

$2^3 = 8$ is saying the same as $\log_2 8 = 3$ (spoken as "log base 2 of eight equals 3").

With 10^5, the exponent is 5, therefore the log of 10^5 is 5. This the same as stating that $\log_{10}$ of 100 000 is 5, i.e. $\log_{10} 100\,000 = 5$.

Conventionally, if the base is 10 we don't write the 10.

Example

$\log_{10} 100 = 2$ is written as $\log 100 = 2$ (spoken as "log 100 equals 2").

Scientific calculators and computer software will calculate logarithms for you: use the "log" key or function.

To convert a logarithm back to its original number, use the "inverse" (or "shift") key and then the log key.

16.2 The natural logarithm

Log base e (e is a constant ≈ 2.718) is called the "natural logarithm". It is conventionally written ln rather than $\log_e$.

e has the property that $\ln e^x = x$ (i.e. $\log_e e^x = x$).

So, $\ln e = 1$ (remember that ln e is the same as $\ln e^1$, and $\ln e^1 = 1$).

This is important because, if we have an **exponential** relationship, a graph of ln y against x will give a straight line.

Chapter 17 explains the use of e, natural logarithms and exponential relationships in more detail.

16.3 Logarithm rules

The following rules may prove useful:

$\log_a 1 = 0$ (this is the same as saying $a^0 = 1$)

$\log_a a = 1$ (this is the same as $a^1 = a$)

$\log_a (bc) = \log_a b + \log_a c$, i.e. to multiply two numbers, we *add* their logarithms. Remember that logarithms are indices, and we add indices when multiplying.

$\log_a (b/c) = \log_a b - \log_a c$, so to divide one number into another, in the same way that we subtract indices, we *subtract* their logarithms.

$\log_a b^c = c \log_a b$

$\log_a a^c = c$ (because $\log_a a^c = c \log_a a$, and $\log_a a = 1$)

Examples

$\ln 1 = 0$ (i.e. $\log_e 1 = 0$)

$\log(1546 \times 4326) = \log 1546 + \log 4326 \approx 3.189 + 3.636 = 6.825$

$\log_2 56^4 = 4 \log_2 56$

Test yourself

The answers are given on page 174.

Question 16.1

We will cover pH in detail in Chapter 24. It is a logarithmic scale of hydrogen ion concentration and is defined by:

$pH = -\log[H^+]$

where $[H^+]$ symbolises the concentration of hydrogen ions.

1) If $[H^+] = 1.2 \times 10^{-5}$, what is the pH?

2) If the pH is 6.3, what is $[H^+]$?

Use the “log” and “inverse” (or “shift”) keys of a calculator to help you.

Question 16.2

What is $\ln e^4$, where e is the constant ≈ 2.718?

Hint: you can do this without using your calculator.

17 Exponential growth and decay

Many life science systems have an **exponential** relationship with time.

For example:

- the number of bacteria in a culture may double every hour, an example of **exponential growth**;
- the radioactivity of technetium-99*m* halves every 6 hours, an example of **exponential decay**.

17.1 Formulae for exponential growth

A simple formula that describes exponential growth is

$$y = a^x$$

where a is a constant that depends on the system being studied, and x is the exponent (also called the power, or the index).

Taking the logarithm of both sides of the general equation gives

$$\log y = x \log a$$

So, if we plot $\log y$ against x, an exponential relationship will plot as a straight line with a gradient of $\log a$.

In this example, we have used log to the base 10. However, we will still get a straight line if the log to any base is used.

A relationship that is exponential will give a straight-line graph if plotted on semi-log graph paper (graph paper that has a logarithmic scale on one axis). There is an example of this in Section 21.8.

A more general formula for exponential growth can be written in the form

$$y = a\mathrm{e}^{bx}$$

Where a and b are constants that depend on the system.

Taking the natural logarithm of both sides of this equation gives:

$$\ln y = \ln(a\mathrm{e}^{bx})$$

Using logarithm rules for multiplication, this is the same as

$$\ln y = \ln a + bx \ln \mathrm{e}$$

However, we know that ln e = 1, so

$$\ln y = bx + \ln a$$

From this we can see that by plotting ln *y* against *x*, we will get a straight-line graph with a gradient of *b* and a *y*-axis intercept of ln *a*.

17.2 The growth–decay formula

You may see the formula for exponential growth and decay, $y = ae^{bx}$, stated as:

$$N = N_0 e^{kt}$$

where *N* is the exponentially changing quantity, *t* is time, N_0 is its value at time $t = 0$, *k* is the growth constant or decay constant, and e is the constant ≈ 2.718.

Example

100 bacteria ($N_0 = 100$) are plated out onto agar ($t = 0$ hours). Five hours later ($t = 5$ hours), there are 300 bacteria ($N_5 = 300$). If we assume exponential growth, we can calculate the growth constant.

$$300 = 100e^{5k}$$

This can be manipulated to:

$$\frac{300}{100} = 3 = e^{5k}$$

Now we take the natural logarithm of both sides of the equation. Remember that the natural logarithm of e^x is x, so $\ln e^{5k} = 5k$

$$\ln 3 = 5k$$

$$\frac{\ln 3}{5} = k$$

$$k = 0.22 \text{ per hour}$$

17.3 The doubling time

We can also use the growth–decay formula to calculate the growth constant if we know how often a population doubles.

Example

Moore's law suggests that the power of the fastest computer chips doubles every 18 months.

We could use any initial value for this, say $N_0 = 1$. After 18 months ($t = 18$ months), we have double the initial value, $N_{18} = 2$.

Substituting these values into $N = N_0 e^{kt}$, we get

$$2 = 1e^{18k}$$

This can be manipulated to:

$$\frac{2}{1} = 2 = e^{18k}$$

Taking the natural logarithm of each side gives:

$$\ln 2 = 18k$$

$$\frac{\ln 2}{18} = k$$

$$k = 0.039 \text{ per month}$$

17.4 Exponential decay

We also use the growth–decay formula for exponential decay.

Example

The radioisotope technetium-99*m* is a short-lived isotope used in nuclear medicine in the diagnosis of various disorders. It has a half-life of 6 hours.

Again, we can use any initial value for this, say $N_0 = 1$. After 6 hours ($t = 6$ hours), we have *half* the initial value, $N_6 = 0.5$.

Substituting these values into $N = N_0 e^{kt}$, we get

$$0.5 = 1e^{6k}$$

which is then manipulated to:

$$\frac{0.5}{1} = 0.5 = e^{6k}$$

$$\ln 0.5 = 6k$$

$$\frac{\ln 0.5}{6} = k$$

$$k = -0.116 \text{ per hour.}$$

Note that when there is exponential decay, the constant is negative.

17.5 Using the growth constant

Given the growth constant and an initial population size, we can use the growth–decay formula to calculate the population at any given time.

Example

Using the bacterial incubation growth constant of $k = 0.22$ per hour given above, we can calculate how many bacteria would be present after 24 hours ($t = 24$ hours), given an initial ($t = 0$ hours) inoculation of 5000 bacteria ($N_0 = 5000$).

$$N_{24} = N_0 e^{kt} = 5000e^{0.22 \times 24} = 5000e^{5.28} = 5000 \times 196 = 980\,000 \text{ bacteria}$$

Test yourself

The answers are given on page 174.

Question 17.1
A highly infectious virus had affected five patients when first diagnosed. Three weeks later there are 25 patients. Assuming exponential growth of the outbreak, what is the growth constant? Use your calculator to calculate the natural logarithm.

Question 17.2
If the outbreak continues to spread at the same rate, use your calculated growth constant to predict how many patients will be affected in another 4 weeks. Again, use your calculator.

Question 17.3
The radioisotope iodine-131 has a half-life of 8 days. Calculate the decay constant.

18 Circles and spheres

The mathematical formulae relating to circles and spheres can be helpful in many scientific calculations.

18.1 Pi

All the formulae for circles and spheres relate to the mathematical constant π. You may see π written as **pi**, and it is pronounced "pie".

π is an infinitely long number: $\pi = 3.1416$ to five significant figures.

18.2 Formulae for circles and spheres

For circles and spheres of radius r:

Circumference of a circle	$C = 2\pi r$
Area of a circle	$A = \pi r^2$
Volume of a sphere	$V = \frac{4}{3}\pi r^3$
Surface area of a sphere	$A = 4\pi r^2$

18.3 Formulae for cylinders and cones

For cylinders and cones of radius r and height H:

Volume of a cylinder	$V = \pi r^2 h$
Volume of a cone	$V = \frac{1}{3}\pi r^2 h$

Example

A glass cylinder of radius 120 mm contains solution up to a height of 85 mm. We wish to know the volume of solution.

First converting to standard form:

$$120\,\text{mm} = 1.2 \times 10^2\,\text{mm}$$

$$85\,\text{mm} = 8.5 \times 10\,\text{mm}$$

Then substituting them into the formula for volume of a cylinder:

$V = \pi r^2 h = 3.1416\,(1.2 \times 10^2)^2 \times 8.5 \times 10 = 3.8 \times 10^6\,\text{mm}^3$ to two significant figures.

Test yourself

The answer is given on page 174.

Question 18.1
An egg yolk to be used for cell culture is 24 mm diameter. Assuming it is spherical, calculate its volume?

19 Differential calculus

In this book we cover two types of calculus:

- differential calculus
- integral calculus

Also known as differentiation, **differential calculus** is concerned with rates of change. It calculates the gradient of a graph at a particular point.

19.1 Constant speed . . .

The speed of a car is the rate at which the distance the car travels changes with time, for instance miles per hour, or metres per second. This can be calculated by taking the gradient from a distance/time graph.

This is a graph for a car going at a constant speed:

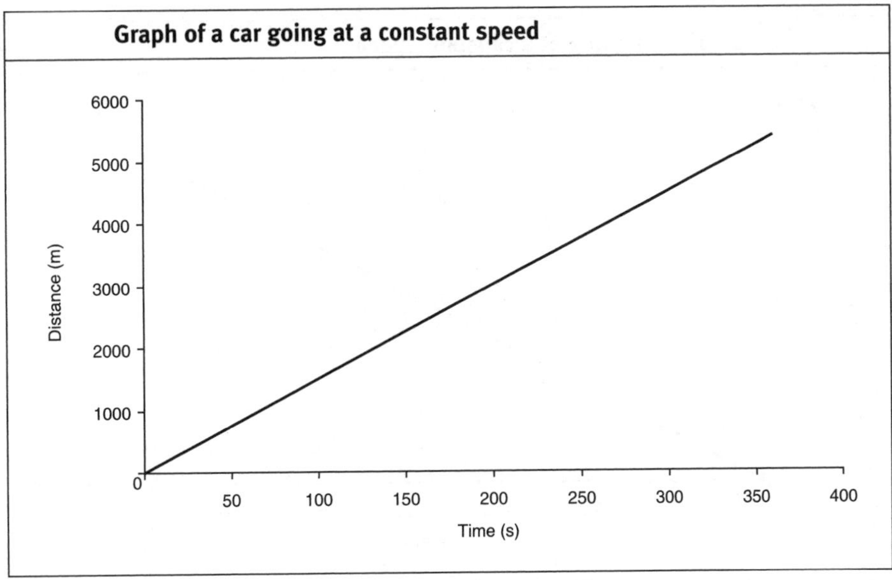

$$\text{The speed of the car in metres per second} = \frac{\text{Distance travelled (m)}}{\text{Time taken (s)}}$$

In Chapter 8, we learnt that the equation for a straight-line graph is $y = mx + c$, where m is the gradient, and c the point where the line crosses the y-axis.

In this example, the car travels 5400 metres in 360 seconds at a constant speed. The gradient m is therefore $\frac{5400}{360} = 15\,\text{m}\,\text{s}^{-1}$.

The intercept on the y-axis is 0, i.e. at the start of the journey the car hasn't moved, so the constant c = zero.

The equation for the line is therefore:

$$y = 15x + 0, \text{ or } y = 15x$$

where y is distance in metres, and x is time in seconds.

We also learnt that the gradient of a straight-line graph can be described as:

$$\frac{\text{Change in } y}{\text{Change in } x} = \frac{y_2 - y_1}{x_2 - x_1}$$

This is the rate of change of y with respect to x.

Graph of a car going at a constant speed

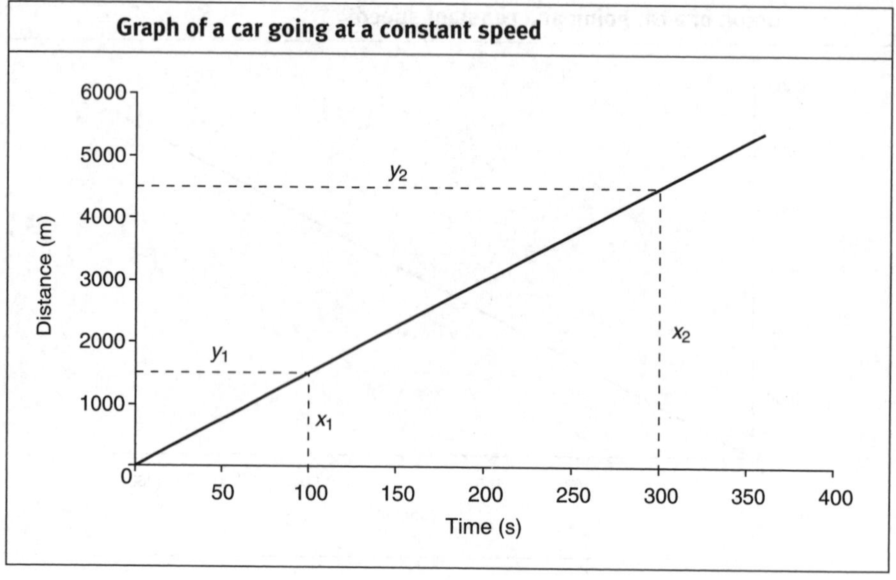

In this example, $\frac{y_2 - y_1}{x_2 - x_1} = \frac{4500 - 1500}{300 - 100} = \frac{3000}{200} = 15\,\text{m}\,\text{s}^{-1}$.

In calculus the gradient, $(y_2 - y_1)/(x_2 - x_1)$, is called the **derivative.**

So, the derivative $(y_2 - y_1)/(x_2 - x_1) = 15\,\mathrm{m\,s^{-1}}$. The graph is a straight line, so the car's speed, and the derivative $(y_2 - y_1)/(x_2 - x_1)$, is constant throughout the journey.

19.2 ... or acceleration

In the next graph, the car is accelerating, so the speed is increasing throughout the time-span shown.

To calculate the gradient of a curve isn't as easy as with a straight line – the car is constantly accelerating, so the car's speed is constantly changing with time.

Graph of a car accelerating

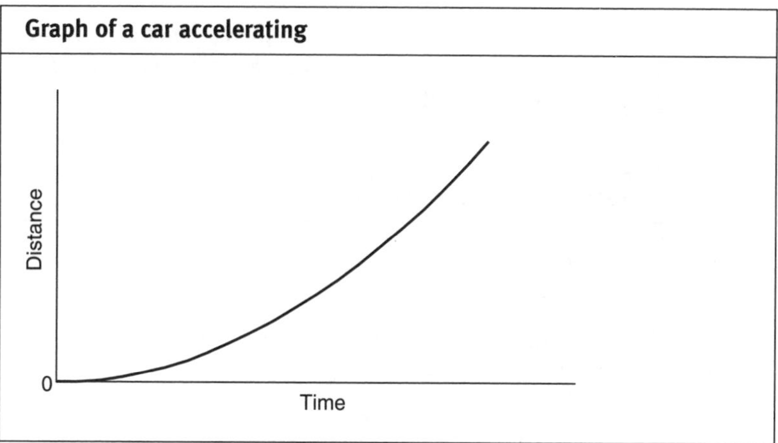

19.3 Calculating the gradient of a curve

The formula for the next curve is $y = \frac{1}{2}x^2$.

The gradient (or speed) is constantly changing, so it is different at each point of the graph.

Graph showing an increasing gradient

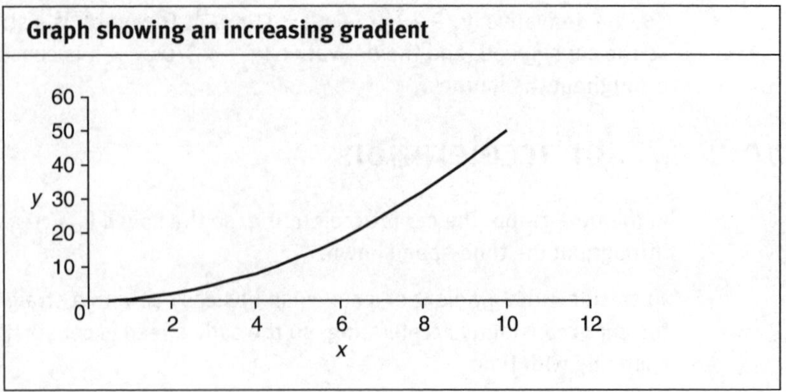

At any specific point we can calculate the gradient by drawing the **tangent** (a straight line touching the curve at that point) and calculating the gradient of this straight line.

Graph with a tangent drawn

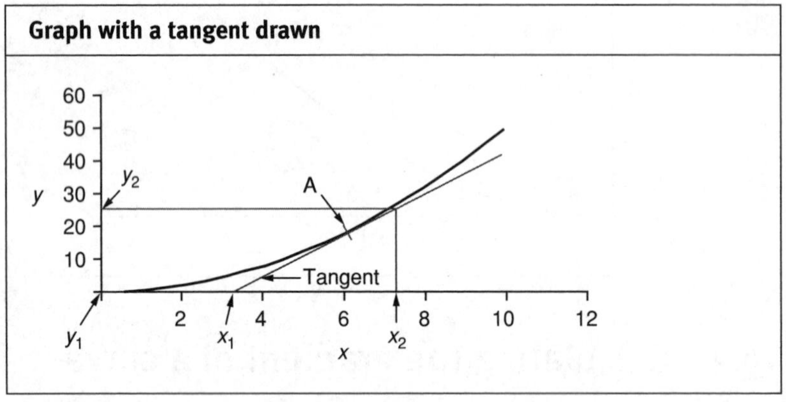

At point A, the gradient of the tangent, and therefore of the curve, is:

$$\frac{y_2 - y_1}{x_2 - x_1} = \frac{25 - 0}{7.5 - 3.5} = \frac{25}{4} = 6.25$$

Differential calculus, otherwise known as **differentiation**, is the mathematical approach to find this gradient (**differential**).

19.4 Functions

A **function** is a relationship between two or more things, where the value of one thing depends on the value of the other.

Examples

- The distance that a fish has swum depends on the time it has been swimming.
- The number of bacteria in a sample depends on the size of the sample.
- In the graph $y = \frac{1}{2}x^2$ the value of y depends on half the square of x.

In each of these relationships, there is a **dependent variable** (distance, number of bacteria, and y in these examples) and an **independent variable** (time, sample size, x).

Usually, we want to know what the value of the dependent variable is for a given independent variable. So, for $y = \frac{1}{2}x^2$ we might want to know the value of y for a given value of x.

19.5 Function notation

In an equation, one way to write the function $y = \frac{1}{2}x^2$ is to write $f(x) = \frac{1}{2}x^2$.

$f(x)$ is spoken as "*f* of *x*".

Both these functions state exactly the same thing.

$f(x)$ doesn't mean f times x. $f(x)$ is simply another way of writing y, i.e. it states that y is a function of x, for example the distance a fish has swum is a function of the time spent swimming.

19.6 Problems with curves

We stated that to be able to calculate a gradient at a particular point, we need to know the change in y for a given change in x, i.e. $\frac{(y_2 - y_1)}{(x_2 - x_1)}$.

To do this, we need to take a section of the graph that has a straight line – and the problem with a curve is that it has no straight line.

For a curve, we can get an *approximation* of the gradient at a particular point by taking a line that intersects the curve at two points (a **secant line**).

Example

Graph showing a curve with a secant

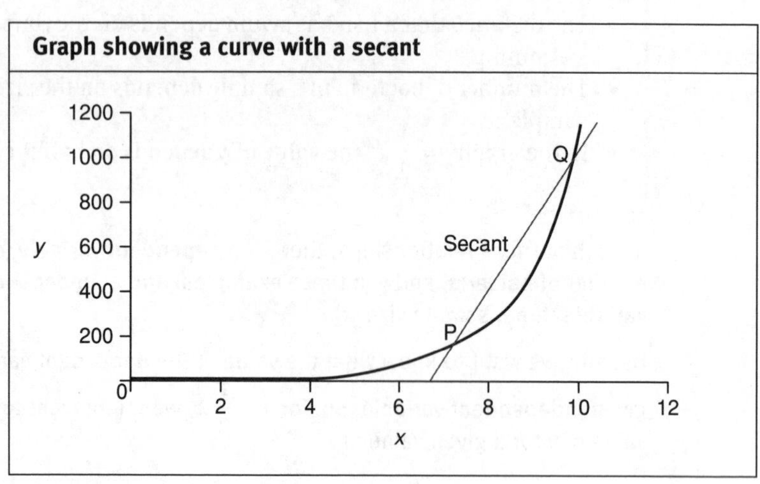

Here, to get an approximation of the gradient at point P, we have drawn a secant line that intersects the curve at point P and point Q.

$$\text{Secant line gradient} = \frac{y_2 - y_1}{x_2 - x_1} = \frac{1000 - 200}{9.75 - 7.5} = \frac{800}{2.25} = 355.55$$

However, if we add a tangent at point P, as in the following graph, we can see that the secant line gradient is steeper than the tangent line.

Graph showing a curve with a secant and a tangent

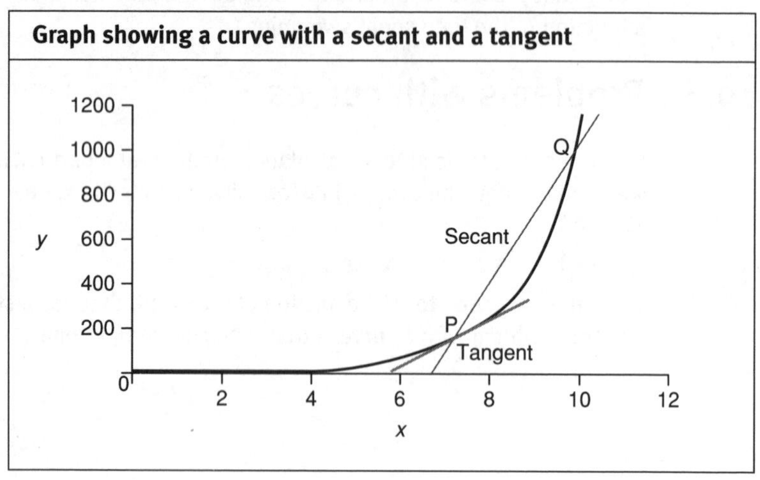

We can use a shorter secant line, from point P to point R:

Graph showing a curve with a shorter secant

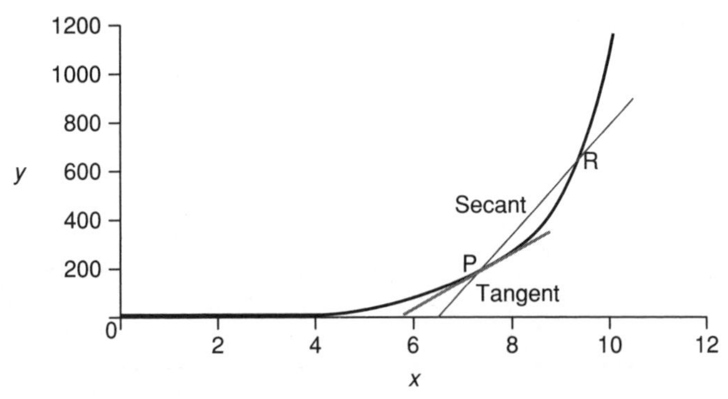

Now the secant line gradient $= \frac{y_2 - y_1}{x_2 - x_1} = \frac{650 - 200}{9 - 7.5} = \frac{450}{1.5} = 300.$

This is closer to the tangent's gradient, but still only an approximation of it.

You will see that the shorter the secant, the closer the secant's gradient gets to the gradient of the curve at P.

Now comes one of the key concepts underlying calculus – when the gradient of the secant reaches its **limit**, i.e. when the secant is infinitesimally small, it is, effectively, the gradient of the curve at that point. The notation for this is

$$\lim_{\delta x \to 0} \frac{y_2 - y_1}{x_2 - x_1}$$

In other words, as the difference in x (i.e. $x_2 - x_1$, symbolised here by δx) approaches its limit of zero (symbolised by $\lim_{\delta x \to 0}$), we reach the gradient of the tangent at P.

19.7 The gradient of the chord

With the following curve, the position of point A is given by its co-ordinates x, y.

$y_2 - y_1$ is the difference between y_2 and y_1 and it can also be written as δy (which stands for "difference in y"). It is pronounced "delta y".

So, the position of point B is given by its co-ordinates $(x + \delta x, y + \delta y)$.

Graph showing the gradient of a chord

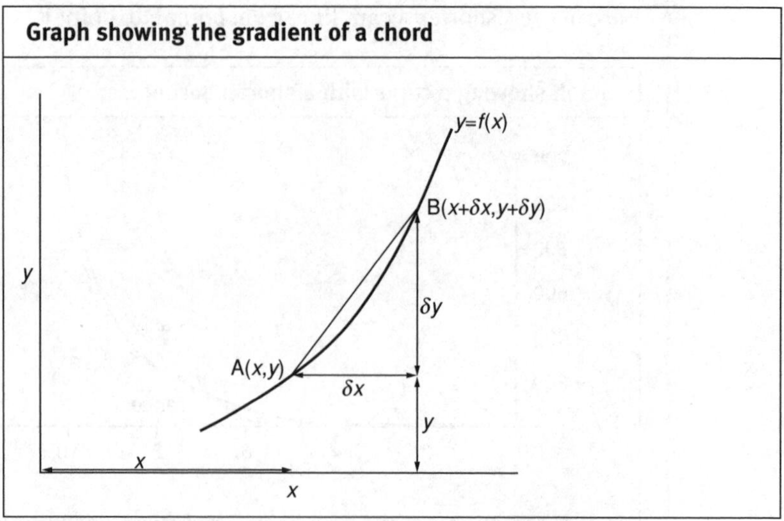

The gradient of the chord approaches the tangent of point A as we move point B closer and closer to point A.

At the same time, δy and δx approach their limit of zero.

So, at point A the gradient of the chord AB is:

$$\frac{(y+\delta y)-y}{(x+\delta x)-x}$$

which changes to the mathematical expression dy/dx as δx approaches its limit of zero.

dy/dx is known as the **differential coefficient.** It is the **derivative** of y with respect to x.

Differentiation is the process of calculating this derivative.

Note that dy/dx isn't a fraction and the d and y cannot be separated: the d implies a limit.

Putting these together,

$$\lim_{\delta x \to 0} \frac{y_2 - y_1}{x_2 - x_1} = \frac{dy}{dx}$$

dy/dx is known as the differential coefficient, the derivative of y, or y'.

As y is a function of x (i.e. $y = f(x)$) the derivative may also be written as $f'(x)$.

Because y is a function of x, the co-ordinates A and B can be written completely in terms of x, as follows:

the co-ordinates of A are (x, y).

So, as y is a function of x (i.e. $y = f(x)$), the co-ordinates of A can also be written $(x, f(x))$.

The co-ordinates of B can therefore be written:

$$(x + \delta x, f(x + \delta x))$$

As the chord between A and B gets smaller, δx gets smaller and moves towards its limit of zero, and A gets closer to B.

19.8 Calculating the differential of x^2

One common function is $y = x^2$.

We want to know the differential (differential coefficient) dy/dx for this function.

If $\quad y = x^2$

then $\quad y + \delta y = (x + \delta x)^2$

Multiplying out $(x + \delta x)^2$ gives:

$$y + \delta y = x^2 + 2x\,\delta x + \delta x^2$$

From this subtracting $\quad y = x^2$

gives $\quad \delta y = 2x\,\delta x + \delta x^2$

Dividing both sides by δx gives the differential coefficient:

$$\frac{dy}{dx} = 2x + \delta x$$

Now here's the calculus equivalent of sleight of hand – we're interested in an infinitesimally small value for δx, i.e. when it approaches zero.

When δx approaches zero,

$$\frac{dy}{dx} = 2x + 0 = 2x$$

So, for the equation $y = x^2$ we've calculated that the differential is $2x$.

19.9 Differentiating with a constant

In the equation for a graph, the constant has no effect on the gradient.

Graphs of $y = x^2$ and $y = x^2 + 5$

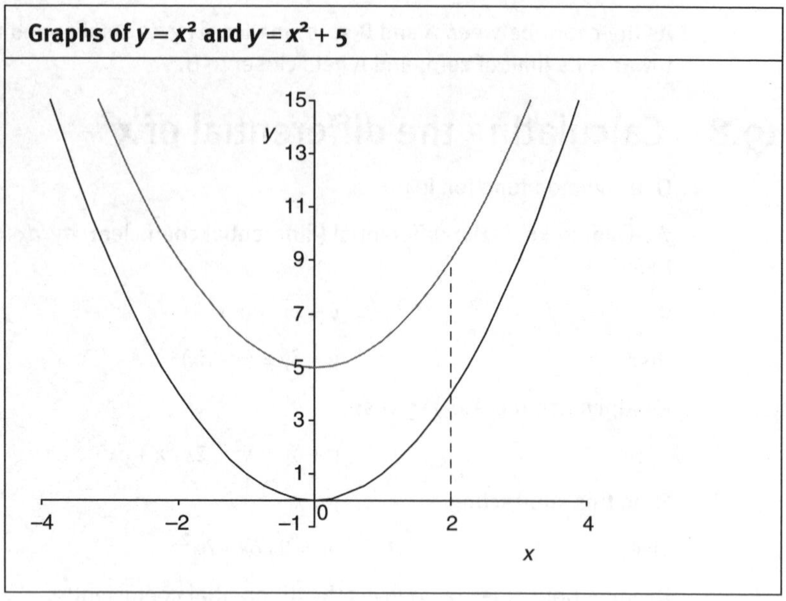

In these graphs of $y = x^2$ and $y = x^2 + 5$, for any given value of x the gradient is the same. For instance, where the line $x = 2$ crosses the graphs, the gradient is 4 for both graphs.

The differential of the constant is zero, so the value of the constant (how far up the y-axis the graph starts) doesn't affect, or appear in, the differential.

So, the differentials of $y = x^2$ and $y = x^2 + 5$ are both $dy/dx = 2x$.

19.10 Using the differential of x^2

The formula for the area of a circle is $A = \pi r^2$, where A is the surface area, π is a constant 3.1416 and r is the radius of the circle. We can differentiate this formula to calculate the rate of increase of the area at any given radius.

Example

In the previous section we have shown that when $y = x^2$, the differential is $2x$, so:

$$\frac{dA}{dr} = \pi(2r) \approx 3.1416(2r) = 6.2832r$$

At the point that the radius of a circle enlarges to 10 mm, the area of the circle is 314.16 mm^2.

At that point, the rate of increase of area is:

$$\frac{dA}{dr} = 6.2832 \times 10 = 62.832 \text{ mm}^2 \text{ per mm.}$$

19.11 The differential of x^3

For the equation $y = x^3$, $dy/dx = 3x^2$.

Example

The formula for the volume of a sphere is

$$V = \frac{4}{3}\pi r^3$$

where V is the volume, π is the constant 3.1416 and r is the radius of the sphere.

For the equation

$$y = x^3$$

the differential is

$$\frac{dy}{dx} = 3x^2$$

so for a sphere

$$\frac{dV}{dr} = \frac{4}{3} \times 3\pi r^2 = 4\pi r^2 \approx 12.57\ r^2$$

When the radius of the sphere enlarges to 10 mm, the volume is increasing by:

$$\frac{dV}{dr} = 12.57(10^2) = 12.57 \text{ mm}^3 \text{ per mm.}$$

19.12 The differential of x^n

We've stated that the differential of

x^2 is $2x$

and that the differential of

x^3 is $3x^2$.

The differential of

x^4 is $4x^3$

The differential of

$2x^4$ is $4 \times 2x^3 = 8x^3$

You may have spotted the pattern.

The equation to differentiate x to any power, i.e. $y = mx^n$ is as follows:

$$\frac{dy}{dx} = nmx^{n-1}$$

The constant c in an equation is effectively cx^0 (i.e. $c \times 1$), and the differential of

cx^0 is $0cx^{0-1} = 0$

so the differential of a constant is always zero.

Example

The differential of $y = 3x^{15} + 7$ is

$$\frac{dy}{dx} = 15(3x^{15-1}) + 0 = 45x^{14}$$

19.13 The differential of $y = 1/x$

$y = \frac{1}{x}$ is the same as writing $y = x^{-1}$.

We can use the $\frac{dy}{dx} = nmx^{n-1}$ formula for this.

$$\frac{dy}{dx} = -1x^{-1-1} = -x^{-2}$$

19.14 The differential of e^x

The differential of e^x is e^x, i.e. for e^x, $dy/dx = e^x$.

It is the only function (apart from zero) which is equal to its own derivative.

The number that satisfies this is approximately 2.718, so $e \approx 2.718$

You may recognise the $e \approx 2.718$ constant from the chapters on logarithms and exponential growth, Chapters 16 and 17.

Test yourself

The answers are given on page 174.

Question 19.1
The heat loss of an organism depends on the surface area of that organism. At a given ambient temperature, the heat loss in a species is found to be $y = 50x^2$, where y is the heat loss in watts and x is the length of the animal in metres.
What is the rate of increase of heat loss when the animal is growing through 1.2 m in length?

Question 19.2
A number of wasps in a colony increases with the cube of its radius, described by the formula $y = x^3/60$

where y = number of wasps and x = radius in mm.
What is the rate of increase in number of wasps when the colony has reached 30 mm in radius?

Question 19.3
What is the differential of $y = 3x^{20} - 8$ with respect to x?

Question 19.4
What is the differential of $y = 3/x^5$ with respect to x?

20 Integral calculus

Integral calculus is concerned with **integration.** Integration is the reverse process to differentiation.

For our purposes, there are two main uses of integration:

- to calculate the original equation for a curve when we only know its differential;
- to calculate the area underneath a portion of a curve.

20.1 Integration: the converse of differentiation

We learnt in Chapter 19 that the differential of $y = x^2$ is $\frac{dy}{dx} = 2x$.

So, the integral of $2x$ would appear to be x^2.

Simple. Except, we also learnt that the differential of $y = x^2 + 9$ is $dy/dx = 2x$, so the integral of $2x$ could also be $x^2 + 9$.

Thus, in the same way that we *lose* the constant (because the differential of the constant is zero) when differentiating, we have to *replace* the constant when integrating.

This unknown constant is symbolised by the capital letter C.

Therefore the integral of $2x$ is $x^2 + C$.

Example

The integral of $46x$ is the same as the integral of $23(2x)$.

The integral of $23(2x) = 23x^2 + C$.

Because we cannot give C as a number we call this the "indefinite integral". The integral becomes "definite" if a co-ordinate is given, and C can then be calculated.

20.2 The integration symbol

The symbol to indicate integration is $\int$. Think of it as an elongated S for "sum up".

The $\int$ symbol is always followed by *d(something)*, in these examples *dx*, meaning with respect to *x*.

So, using the example in the previous section, the integral of $2x$ with respect to x is written as $\int 2x\,dx$, so

$$\int 2x\,dx = x^2 + C.$$

The symbol $\int_2^5 f(x)dx$ means the integral of the function of x with respect to x between the values of $x = 2$ to $x = 5$.

Using function notation (i.e. $y = f(x)$) we can state $\int f'(x)\,dx = f(x)$ and we can see that integration is the opposite of differentiation. The differential of $f(x)$ with respect to x is $f'(x)$ (see Section 19.7) and the integral of $f'(x)$ with respect to x is the function $f(x)$.

20.3 The integral of x^n

The integral of x^n with respect to x is

$$\frac{x^{n+1}}{n+1} + C.$$

Using the integration symbol, this is written as

$$\int x^n\,dx = \frac{x^{n+1}}{n+1} + C$$

Example

We wish to integrate $y = 4x^3$ with respect to x.

$$\int 4x^3\,dx = 4\,\frac{x^{3+1}}{3+1} + C = 4\,\frac{x^4}{4} + C = x^4 + C$$

The integral is therefore $x^4 + C$.

20.4 Calculating areas by using integration

With a straight-line graph, it is easy to calculate the area under part of the line.

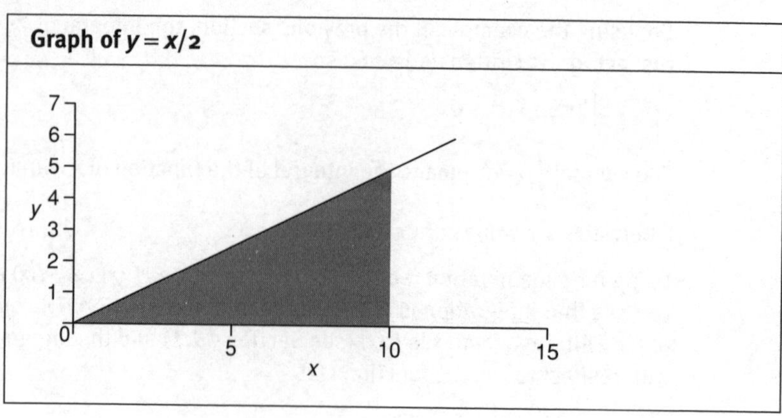

The equation for this graph is $y = \frac{x}{2}$.

The area under the line between the values $x = 0$ and $x = 10$ is calculated by:

$$\frac{\delta y \times \delta x}{2} = \frac{10 \times 5}{2} = 25$$

However, to calculate the area under a curve we need to use integration.

The area under a curve is the *integral* of the equation for the curve.

The area under the curve $y = x^n$ is the integral of x^n. We use the capital letter A to symbolise area.

A = integral with respect to x of

$$x^n = \int x^n \, dx = \frac{x^{n+1}}{n+1} + C$$

This integral is defined by cancelling the C when two values of x are given as boundaries, in this case between $x = 0$ and $x = 10$.

$$\text{For } y = \frac{x}{2}$$

$$\int_0^{10} \frac{x}{2} \, dx = \left[\frac{x^{1+1}}{2(1+1)} + C \right]_0^{10} = \left[\frac{x^2}{4} + C \right]_0^{10}$$

Notice that square brackets are the standard notation when using the limits for definite integration.

As the numbers at the top and bottom of the square brackets define the limits, substitute $x = 10$ into the equation, and then again $x = 0$ into the equation. Then find the area between by subtracting one from the other.

$$\left(\frac{10^2}{4}+C\right)-\left(\frac{0^2}{4}+C\right)=25-0+C-C=25$$

Note how the constant C is cancelled out.

Example

Graph of $y = x^2$

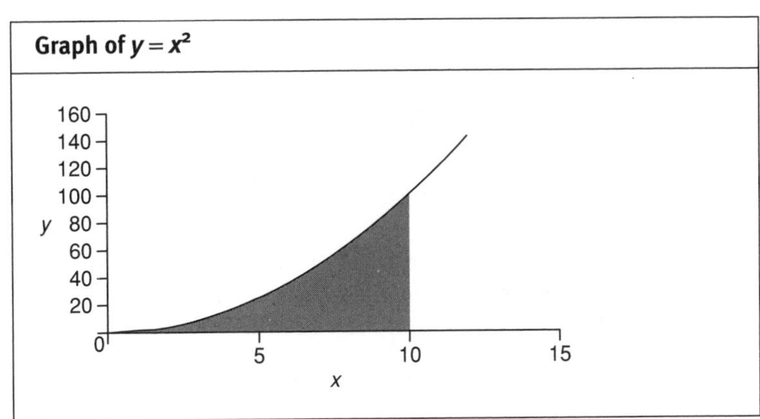

The equation for the area under the curve is the integral of x^2:

$$A=\int x^2\,dx=\frac{x^{2+1}}{2+1}+C=\frac{x^3}{3}+C$$

If we want to calculate the area between $x = 0$ and $x = 10$ we can write this as

$$A=\int_0^{10} x^2\,dx=\left[\frac{x^3}{3}+C\right]_0^{10}=\left(\frac{10^3}{3}+C\right)-\left(\frac{0^3}{3}+C\right)=\frac{10^3}{3}-0$$

$= 333.33$ to two decimal places.

If we want to calculate the area between $x = 1$ and $x = 10$ we can write this as

$$A=\int_1^{10} x^2\,dx=\left[\frac{x^3}{3}+C\right]_1^{10}=\left(\frac{10^3}{3}+C\right)-\left(\frac{1^3}{3}+C\right)=\frac{10^3}{3}-\frac{1}{3}=333$$

20.5 How does it work?

This section explains the mathematical proof that the area under a curve is equal to the integral of the formula for the curve.

Graph showing $y \times \delta x$

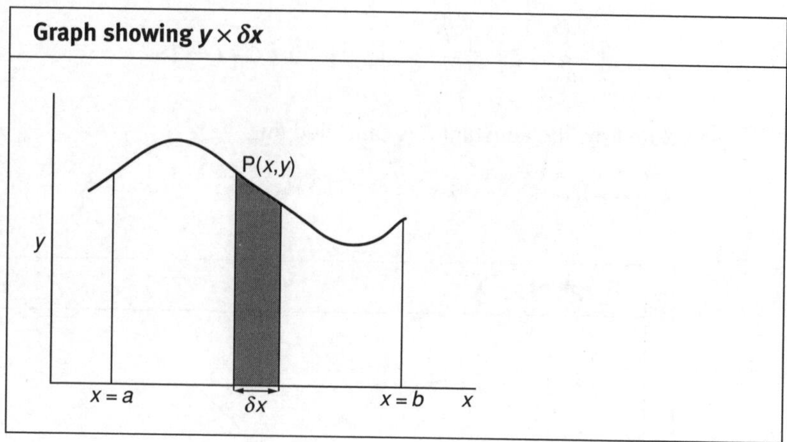

The approximate area of the thin strip under the curve in this graph is the height of the strip, y, multiplied by the difference between the strip's two points on the x-axis, which can be written as δx.

So, the area of the strip approximates to

$$y \times \delta x$$

i.e. $y\,\delta x \approx \delta A$

Note the use of the $\approx$ sign, meaning "approximately equal to".

If we fill the area from $x = a$ to $x = b$ with an infinite number of infinitely small strips, and add their areas, the mathematical notation is

$$A \approx \sum_{x=a}^{x=b} \delta A$$

Remember A symbolises the area, and that the sigma sign Σ means a summation of all that follows it. The $x = b$ and $x = a$ symbols above and below the Σ sign mean that all the strips from $x = b$ to $x = a$ are included in the expression.

We've already stated that $\delta A \approx y\,\delta x$.

Substituting $y\,\delta x$ for δA in $A \approx \sum_{x=a}^{x=b} \delta A$ gives:

$$A \approx \sum_{x=a}^{x=b} y\,\delta x$$

As the width of the strips, i.e. δx, gets smaller, the estimation of the area becomes more accurate.

We've met the lim symbol already in Section 19.6.

When the strips are infinitely small, i.e. $\delta x \to 0$, we have an accurate equation for the area under the curve, i.e.

$$A = \lim_{\delta x \to 0} \sum_{x=a}^{x=b} y\,\delta x$$

As $y = f(x)$ then A is also a function of x alone.

Therefore we can manipulate the equation $\delta A \approx y\delta x$ to

$$\frac{\delta A}{\delta x} \approx y$$

This also becomes more accurate as δx gets smaller:

$$\lim_{\delta x \to 0} \frac{\delta A}{\delta x} = y$$

But $\lim_{\delta x \to 0} \frac{\delta A}{\delta x}$ is $\frac{dA}{dx}$ so $\frac{dA}{dx} = y$

If $\frac{dA}{dx} = y$, then

$$A = \int y\,dx$$

so the area is the integral of y with respect to x.

So, with limits $x = a$ and $x = b$ we can say:

$$\text{Total area } (A) = \int_a^b y\,dx$$

Thus integration can be seen as calculating the area under a curve between the limits of $x = a$ and $x = b$.

Test yourself

The answers are given on page 174.

Question 20.1
Integrate $y = 7x^4$ with respect to x.

Question 20.2
Find the area under the curve from $x = 3$ to $x = 5$ for the curve $y = 2x^3$ shown in the following graph.

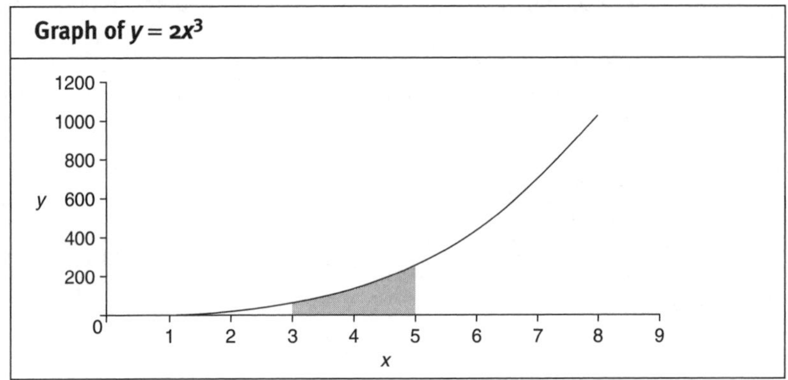

21 Using and recognising graphs

Many equations have characteristic curves – drawing a graph and recognising the curve can help us decide what the relationship between two variables is.

21.1 Labelling graphs

Some general comments on plotting graphs:

- always label the axes with the names of the variables;
- also, where relevant, give the units used in brackets after the label, for example: Time (min); and
- give the graph a title.

21.2 Scatter plots

Chapter 7 explained the rudiments of plotting a graph from a table of data.

One way to decide whether there is a relationship between two sets of data is by drawing a **scatter plot**, a graph showing all the data as individual points.

We can use a scatter plot when the units on the *x*- and *y*-axes are **continuous**, i.e. when we can use values in between the co-ordinates that we are plotting.

Examples of continuous variables are time and distance.

If there seems to be a pattern, we can join the points on the plot.

Example

This table gives the number of bacteria growing in a colony as a function of time.

Table showing growth of bacterial colony								
Time (min)	0	10	25	45	60	75	100	120
No. of bacteria	470	650	1030	1900	3040	4830	10 500	19 500

The data are effectively continuous:

- time is certainly continuous;
- although we can't have half a bacterium, the numbers of bacteria are so large that that they can be treated as being continuous.

Thus we can use a scatter plot and, as there is clearly a pattern, draw a line through the points.

Graph showing growth of bacterial colony

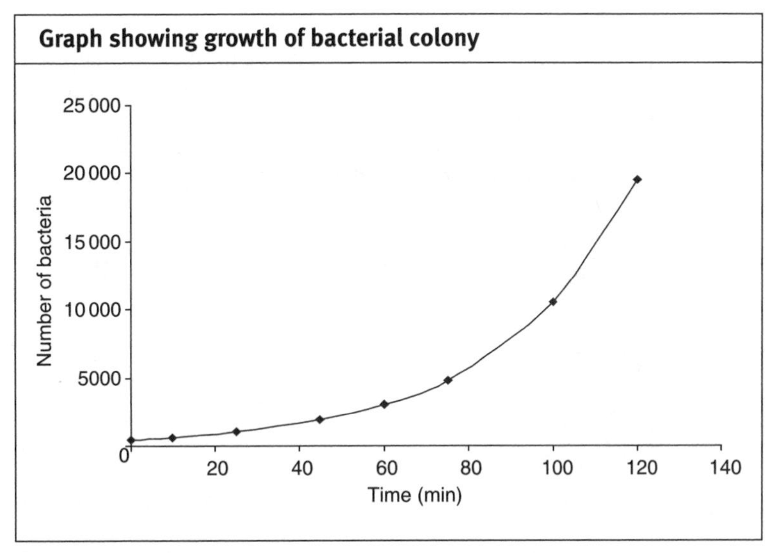

21.3 Graphs and types of variable

Because continuous variables are those which can take any value within a given range, we can get information from the line that we have drawn. In the example above, we can estimate the number of bacteria in the colony after 90 minutes even though no data were taken at that point.

For **categorical variables**, those where only certain values can exist, or **nominal variables**, those that have no ordering to their categories, we cannot join the points on a scatter plot.

If the data are **discrete**, meaning that the values in between the co-ordinates are meaningless, or **categorical variables**, where they represent different categories of the same feature, we cannot join the points on a scatter plot.

Example

The number of patients admitted to a hospital because of fractures is collected over 7 consecutive days.

Table of patients admitted to hospital over 1 week							
Day	1	2	3	4	5	6	7
Number of admissions due to fractures	5	4	6	7	8	5	4

As it is not possible to have a half a patient, the data are discrete, so we can use a scatter plot but we cannot join the points.

Graph of patients admitted to hospital over 1 week

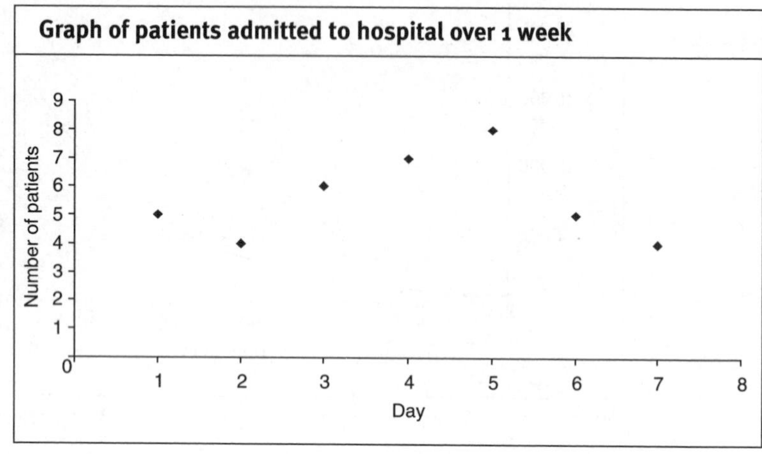

Because we cannot join the points, a clearer way of displaying these data is with a bar chart.

Bar chart of number of patients admitted to hospital over 1 week

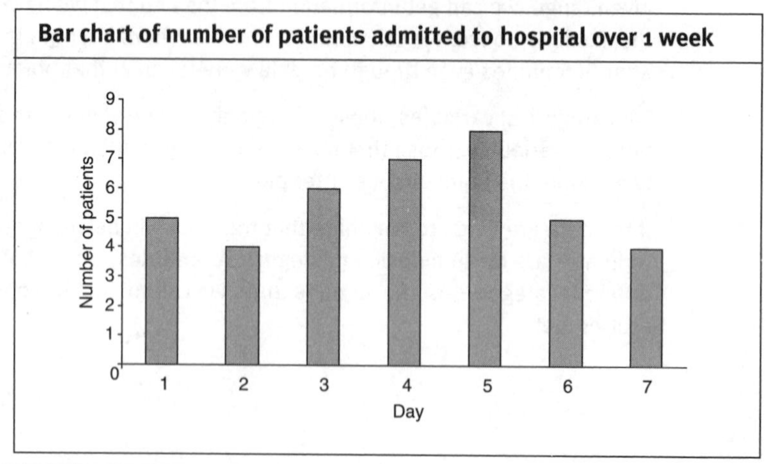

21.4 The line of best fit

When we collect and plot biological data, they don't usually follow an *exact* mathematical relationship (for instance, a straight line). This could be because of normal variations in the population or errors in collecting the data.

However, if the data seem close to a mathematical relationship, then we can demonstrate that relationship by drawing a **line of best fit** on the graph, rather than joining all the points.

The line of best fit is the line that best shows the trend of the plotted points.

If this is a straight line, we can use it to calculate the gradient.

Example

The tail and body length of a group of 10 mice are related as follows:

Table showing relationship between body and tail length in 10 mice

Body length (mm)	92	97	96	99	100	111	109	115	120	122
Tail length (mm)	31	32	35	36	40	43	44	49	49	52

Drawing a scatter plot of these data shows the relationship between the body and tail length.

Graph showing relationship between body and tail length in 10 mice

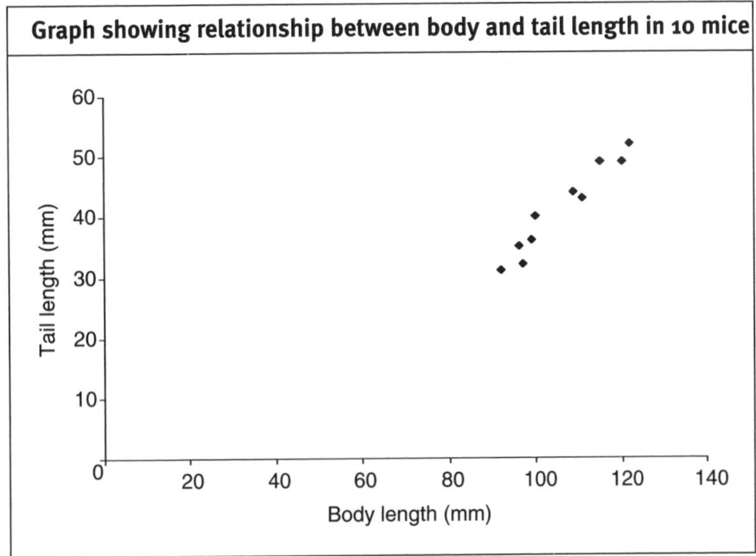

This suggests that there is a fairly linear (straight-line) relationship between the length of the body of a mouse and its tail, and therefore it follows the $y = mx + c$ pattern.

Example

Drawing the line of best fit to the scatter plot above gives the following.

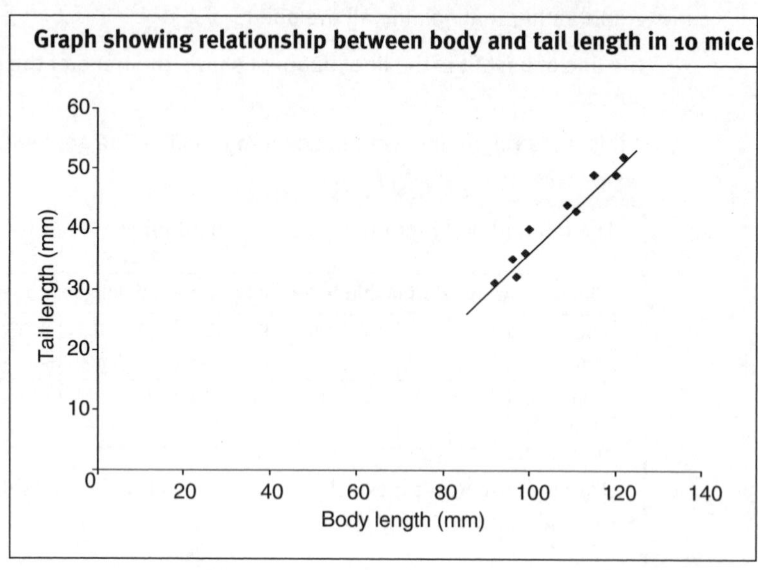

Here, a change in x (body length) of 30 mm results in a change in y (tail length) of approximately 21 mm.

Remember that

$$\text{Gradient} = \frac{\text{Change in } y}{\text{Change in } x}$$

or

$$\text{Gradient} = \frac{y_2 - y_1}{x_2 - x_1}$$

So here, the gradient is approximately

$$\frac{21}{30} = 0.7$$

so the equation for the graph is, $y \approx 0.7x + c$

If we continue the line of best fit until it crosses the y-axis we can also estimate the constant c.

Graph showing relationship between body and tail length in 10 mice

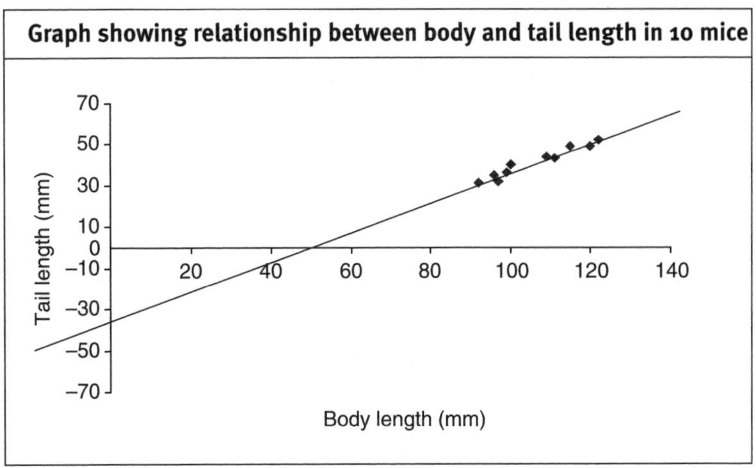

The line crosses the y-axis at $y \approx -35$, so we can now write out the equation that links body and tail length in this group of mice:

$$y \approx 0.7x - 35$$

Obviously, we cannot have mice with negative tail lengths, so we can only predict tail length within the range of the available data.

Chapter 42 explains an exact way of calculating the line of best fit.

21.5 The quadratic relationship

In Chapter 12, we learnt that **quadratic equations** have two solutions, and that they have the formula:

$$ax^2 + bx + c = 0$$

They have a symmetrical ∪ shape (or, for negative values like $-ax^2$, a ∩ shape).

Example

This graph does have a symmetrical ∪ shape so there may well be a quadratic relationship between x and y...

Graph of $y = 2x^2 + 4x - 20$

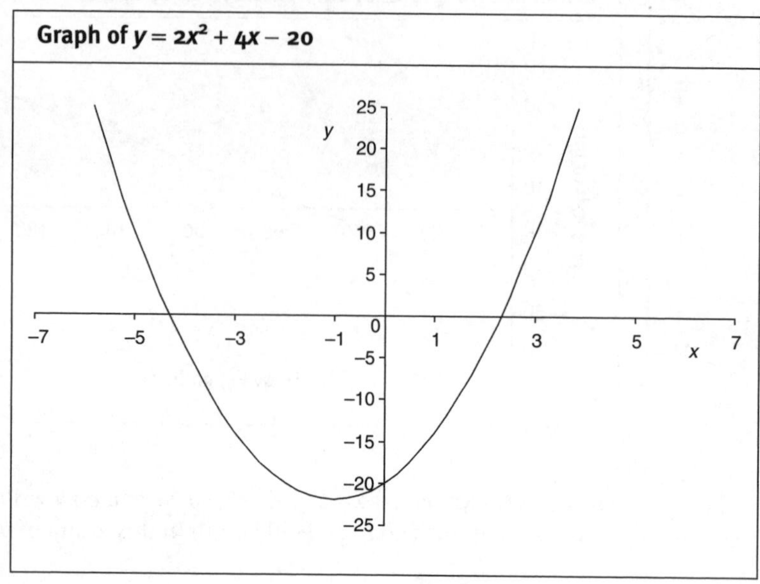

... and indeed there is. It is a graph of the quadratic equation:

$$y = 2x^2 + 4x - 20$$

Note that, as expected, the line crosses the y-axis at –20, the constant of the equation.

21.6 A cubic relationship

A polynomial with a **cube** as the highest power results in a characteristic tilted Ƨ or S shape.

Example

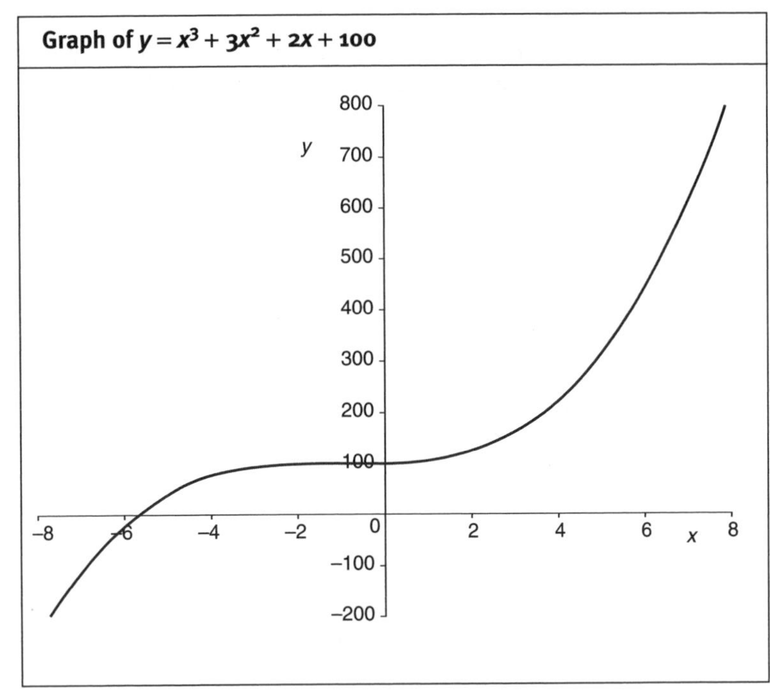

This is the graph of the cubic equation $y = x^3 + 3x^2 + 2x + 100$.

21.7 Graphs suggesting asymptotes

An **asymptote** is where a function gets infinitely close to a line without crossing it.

Plotting a graph that appears to have asymptotes gives information about the relationship between the two variables.

Example

We will meet the **Michaelis–Menten** equation for enzyme-catalysed reactions in Section 26.3.

At this stage, just note the pattern of its graph of substrate concentration, [S], against rate of reaction, V:

Graph of Michaelis–Menten equation

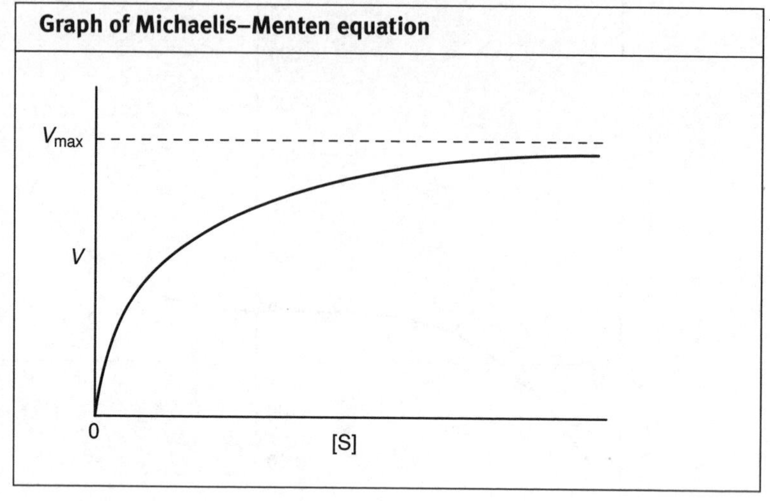

We can see that the reaction rate will never reach its maximum of V_{max} so the maximal rate is an asymptote. A graph of this pattern follows the equation $y = \frac{1}{ax} + b$

21.8 Exponential relationships

The chapter on **exponential** relationships showed that exponential growth can be described by the equation $y = a^x$.

The shape of an exponential growth graph is characteristic: the line continues to go up and to the right forever.

Example

Twenty seeds of rye grass are sown and grown under optimal conditions, and the numbers of grass seedlings is counted every month.

Table of number of grass seedlings over time

Time (months)	0	1	2	3	4	5	6
No. of grass seedlings	20	35	61	105	188	320	557

Plotted on a graph, we get the following curve:

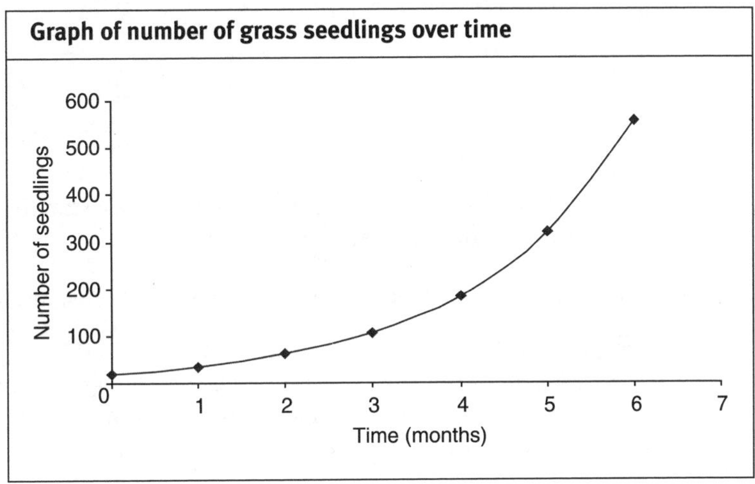

The graph looks as if it could be exponential. To assess this, the graph can be plotted either:

- on "semi-log" graph paper, where one scale is logarithmic, and the other scale is not; or
- as a graph of time against the natural logarithm of the number of seedlings, symbolised by ln(number of seedlings).

If, as we suspect, the relationship is exponential, the plot should be **transformed** (converted) into a straight line.

In the following graph, note the scale of the y-axis: each unit is 10 times that of the unit below, so it is a logarithmic scale. The x-axis is a normal, non-logarithmic scale. The graph paper is therefore semi-logarithmic.

Graph of number of grass seedlings over time

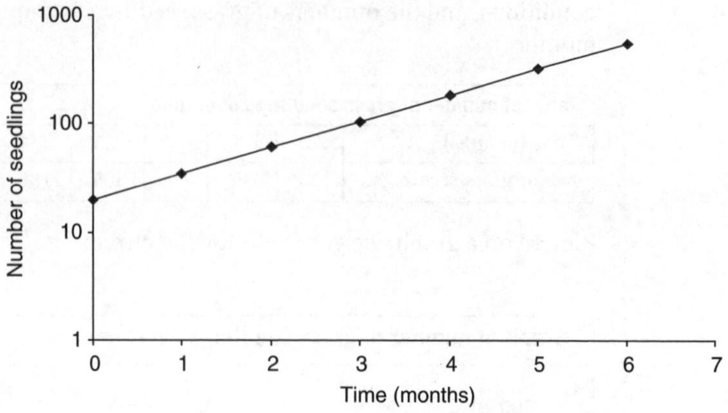

Sure enough, the semi-logarithmic plot shows a straight line, confirming the exponential relationship.

We can use the same techniques to look for exponential decay.

Example

Chicken meal is heated to 60°C to remove *Staphylococcus aureus* bacteria. The following are measurements of the surviving bacteria as a function of time:

Table showing bacterial numbers in chicken meal over time

Time (min)	0	3	6	9	12
Bacteria (cells mm^{-3})	3×10^6	8.4×10^5	1.9×10^5	5.4×10^4	1.4×10^4

This produces the following curve:

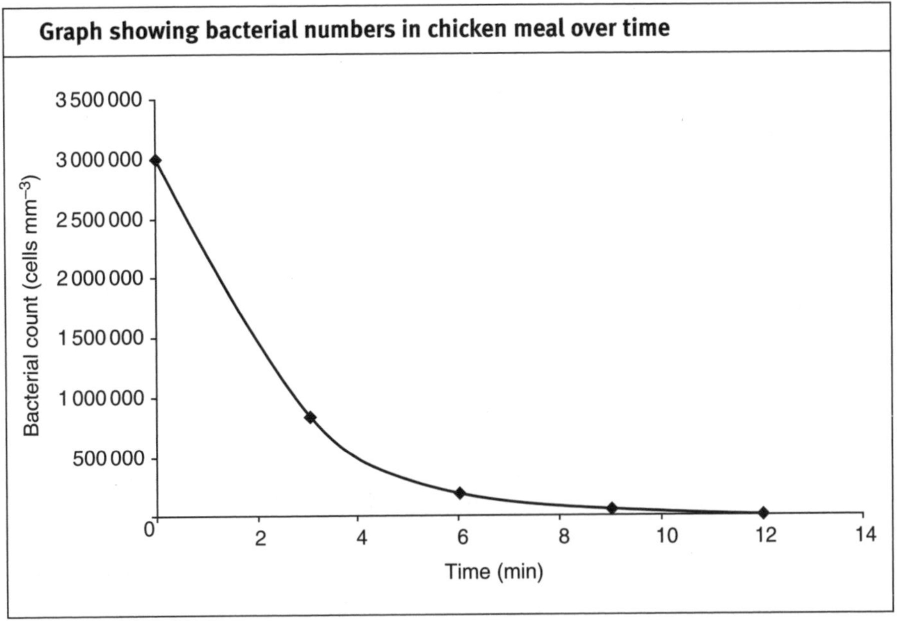

We want to know whether this is a negative exponential.

This time, instead of using semi-log graph paper, we will convert the bacterial counts to their natural logarithms, i.e. ln(bacterial count).

Table of natural log of bacterial count in chicken meal over time

Time (minutes)	0	3	6	9	12
Bacteria (cells mm^{-3})	3×10^6	8.4×10^5	1.9×10^5	5.4×10^4	1.4×10^4
ln(bacterial count)	14.91	13.64	12.15	10.90	9.55

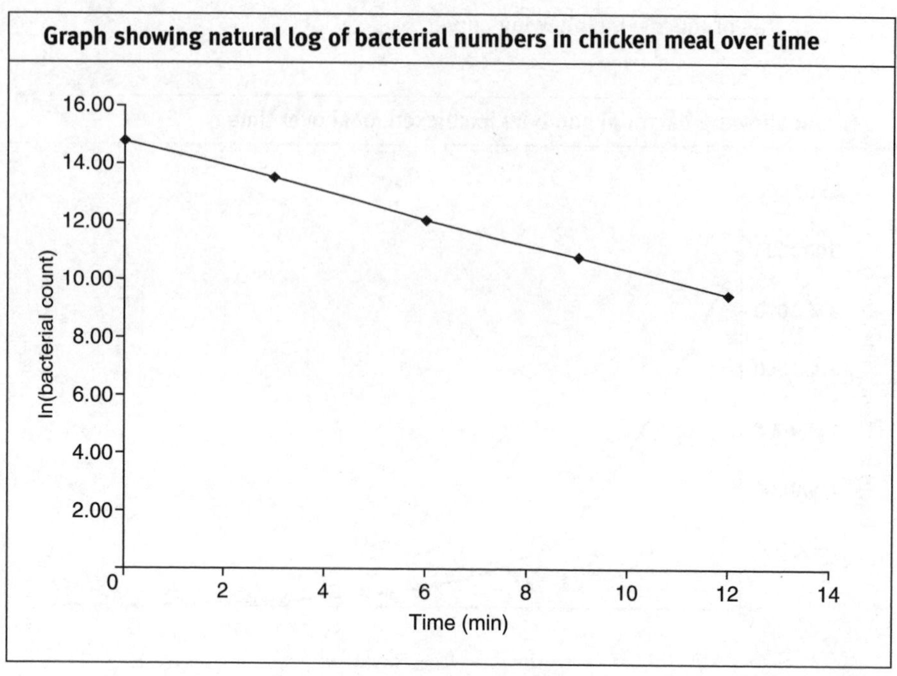

The plot is now straight and has a negative gradient, confirming that the relationship between the bacterial count and time is one of exponential decay.

22 SI units

Any data that we use should, ideally, be converted to **SI units** (Système International d'Unités) before working with them.

The base SI units most frequently used by life and medical scientists are shown with their symbols in the following table.

Table of the base SI units		
Quantity	SI unit	Abbreviation
Mass	kilogram	kg
Length	metre	m
Time	second	s
Temperature	kelvin	K
Amount of substance	mole	mol

All other SI units are derived from these.

Example

The unit for force (the Newton, N) is defined by $kg\,m\,s^{-2}$.

22.1 SI units for large and small numbers

We can use a **metric prefix** to describe very large or small numbers, as an alternative to expressing them in powers of 10. Each prefix indicates a multiplication factor of 1000.

However, for calculations, numbers need to be converted to SI units, using **standard form** (explained in Section 5.2).

Table of metric prefixes and abbreviations		
Metric prefix	Abbreviation	Power of 10
exa-	E	10^{18}
peta-	P	10^{15}
tera-	T	10^{12}
giga-	G	10^{9}
mega-	M	10^{6}
kilo-	k	10^{3}

Metric prefix	Abbreviation	Power of 10
milli-	m	10^{-3}
micro-	μ	10^{-6}
nano-	n	10^{-9}
pico-	p	10^{-12}
femto-	f	10^{-15}
atto-	a	10^{-18}

Examples

$$12\,\text{km} = 1.2 \times 10^4\,\text{m}$$

$$0.78\,\text{nm} = 0.78 \times 10^{-9}\,\text{m} = 7.8 \times 10^{-10}\,\text{m}$$

$$13.7\,\mu\text{g} = 13.7 \times 10^{-6}\,\text{kg} = 1.37 \times 10^{-5}\,\text{kg}$$

22.2 Use of non-SI units

Some units are accepted for use with the SI, including the following non-SI units.

Table of commonly used non-SI units		
Name	Symbol	Value in SI units
degree Celsius	°C	0°C = 273.15K
litre	l	$1\,\text{l} = 1 \times 10^{-3}\,\text{m}^3$
gram	g	$1\,\text{g} = 1 \times 10^{-3}\,\text{kg}$
angstrom	Å	$1\,\text{Å} = 10^{-10}\,\text{m}$
minute (time)	min	1 min = 60 s
hour	h	1 h = 3600 s
day	d	1 d = 86 400 s

The formula to convert from °C to K is °C = K − 273.15. However, note that a change in temperature of 1°C is the same as a change in temperature of 1 K.

Test yourself

The answers are given on page 174.

Question 22.1
An experiment to calculate the mass of DNA in a human cell gives a value of 5.5 pg. Give this value in SI units in standard form.

Question 22.2
An epidemiologist studies the population in an area 5 km long by 6 km wide. Give the area in SI units, in standard form.

Question 22.3
Some frozen cells are stored at −40.5°C. Give this temperature in kelvin.

23 Moles

The mole is a key unit of biochemical quantity.

23.1 Molecular mass

The **molecular mass**, symbol MM, of a substance is the mass of one molecule expressed in **daltons**, symbol Da. A dalton is one-twelfth the weight of an atom of ^{12}C, also known as carbon-12, the most common naturally occurring isotope of the element carbon.

Atomic mass is the mass of one atom expressed in daltons. It therefore also follows that a molecular mass is the sum of the relevant atomic masses.

Example

Atomic mass of sodium (Na)	22.99 Da
Atomic mass of chlorine (Cl)	+ 35.45 Da
Molecular mass of sodium chloride (NaCl)	58.44 Da

23.2 Relative molecular mass

The **relative molecular mass** (symbol M_r), is the ratio of the molecular mass to the mass of one-twelfth the weight of an atom of ^{12}C. Thus, M_r is a pure number with no units.

Example

The molecular mass of a particular protein is 10 000 Da, abbreviated as 10 kDa. Its relative molecular mass M_r, is therefore 10 000.

23.3 Moles

The strict definition of a **mole** is the amount of a substance that contains the same number of atoms, molecules, ions, or other elementary units as the number of atoms in 0.012 kg of ^{12}C. Its symbol is "mol".

0.012 kg of carbon-12 contains 6.022×10^{23} particles, so a mole of any substance also contains 6.022×10^{23} particles.

The value 6.022×10^{23} is known as "Avogadro's number".

Most frequently in the life and medical sciences, we refer to a mole in the context of molecules, so a mole of, say, glucose contains 6.022×10^{23}

molecules. One mole of a compound has a mass equivalent to its relative molecular mass in grams.

Example

The relative molecular mass of glucose is 180.18 so a mole of glucose weighs 180.18 g.

23.4 Calculating the number of moles of a sample

The formula to calculate the number of moles of a sample is:

$$\text{number of moles} = \frac{\text{mass of sample (g)}}{\text{relative molecular mass}}$$

Example

360.36 g of glucose is

$$\frac{360.36\,\text{g}}{180.18} = 2\ \text{mol}$$

23.5 Molarity

The **concentration** of a solution is defined as the amount of the substance in a set volume:

$$\text{concentration} = \frac{\text{amount}}{\text{volume}}$$

The **molarity,** M, of a solution is its concentration expressed as the number of moles per litre.

$$\text{M} = \frac{\text{no. of moles}}{\text{volume (l)}} = \frac{\text{mass (g)}}{\text{relative molecular mass} \times \text{solution volume (l)}}\ \text{mol}\,\text{l}^{-1}$$

When a solution contains 1 mole of a compound in 1 litre of solution, it is said to be a "1 molar" (1 M) solution.

Example

A 1 M solution of NaCl contains $58.44\,\text{g}\,\text{l}^{-1}$.

23.6 Preparing solutions of known molarity

We can calculate the mass of a substance required to prepare a solution of given molarity as follows:

Mass of substance (g) = relative molecular mass × molarity (M) × required volume of solution (l)

Example

To prepare 500 ml (i.e. 0.5 litres) of a 3 M solution of NaCl:

mass of NaCl needed = 58.44 × 3 M × 0.5 l = 87.66 g

While the mole is an SI unit, for historical reasons moles and molarity are derived from grams and litres, which are not SI units.

23.7 Diluting molar solutions

Calculations for diluting stock solutions can be made using this equation:

$$\frac{\text{required molarity of new solution (M)}}{\text{molarity of stock solution (M)}} = \frac{\text{required volume of stock solution (l)}}{\text{required total volume of new solution (l)}}$$

Example

From a 2 M copper sulphate ($CuSO_4$) solution, we wish to prepare 200 ml of 0.4 M $CuSO_4$ solution.

$$\frac{0.4\,\text{M}}{2\,\text{M}} = \frac{\text{required volume (ml)}}{200\,\text{ml}}$$

$$\text{Required volume} = \frac{0.4 \times 200}{2} = \frac{80}{2} = 40\,\text{ml}$$

40 ml of 2 M $CuSO_4$ solution, made up to 200 ml with water, will produce a 0.4 M solution.

Another approach to calculating stock dilutions is to use the following equation:

$$C_1V_1 = C_2V_2$$

where C_1 is the concentration of the stock solution, V_1 is the required volume of stock solution, C_2 is the required concentration of the new solution, and V_2 is the required volume of the new solution.

Example

To calculate the amount of 10% glucose solution needed to make 50 ml of 2% glucose solution:

$$10\% \times V_1 = 2\% \times 50\,\text{ml}$$

$$V_1 = \frac{2 \times 50}{10} = 10\,\text{ml}$$

We therefore need 10 ml of 10% glucose, made up to 50 ml with water.

23.8 Per cent solutions

When the term **per cent** (symbol "%") is applied to a solution, this may be **% w/w** (per cent weight/weight), **% w/v** (per cent weight/volume) or **% v/v** (per cent volume/volume).

If this is not defined, it usually represents % w/v, which is the number of grams per 100 ml.

Example

To calculate the concentration of a 2 M NaCl solution as a % w/v solution, we first need to know the relative molecular mass of NaCl, which is 58.44.

The amount of NaCl in 1 litre of a 2 M NaCl solution is therefore:

$$2 \times 58.44 = 116.88\,\text{g}$$

The amount in 100 ml is:

$$116.88 \times \frac{100}{1000} \approx 11.7\,\text{g}$$

A 2 M NaCl solution is thus equivalent to a 11.7% w/v NaCl solution.

We can also convert a per cent value to molarity.

Example

To express a 5% w/v NaCl solution by its molarity, first we calculate the mass per litre.

$$5\%\ \text{w/v NaCl} = \frac{5\,\text{g}}{100\,\text{ml}} = 50\,\text{g}\,\text{l}^{-1}$$

The relative molecular mass of NaCl is 58.44.

$$50\,\text{g}\,\text{l}^{-1} = \frac{50\,\text{g}}{58.44} \approx 0.86\,\text{M}$$

A 5% NaCl solution is therefore equivalent to a 0.86 M solution.

23.9 Normality

A **normal** solution (1 N) has one mole of hydrogen ions (H^+) for an acid, or for an alkali hydroxide ions (OH^-) in solution per litre.

Normality and molarity are related as follows:

$$\text{N} = n\,\text{M}$$

where N is the normality, n is the number of H^+ ions per molecule, and M is the molarity.

Example

Sulphuric acid has two replaceable H^+ ions per H_2SO_4 molecule, so to calculate the normality of a 3 M sulphuric acid solution:

$$\text{Normality} = n\,\text{M} = 2 \times 3 = 6\,\text{N}$$

A 3 M sulphuric acid solution therefore has a normality of 6 N.

Test yourself

The answers are given on page 175.

Question 23.1
The relative molecular mass of glucose is 180.18. How many moles are there in 450.45 g of glucose?

Question 23.2
Uric acid is $C_5H_4N_4O_3$
The component atomic masses are as follows:

Carbon (C)	12.01 Da
Hydrogen (H)	1.01 Da
Nitrogen (N)	14.01 Da
Oxygen (O)	16.00 Da

What is the molecular mass of uric acid?

Question 23.3
The molecular mass of glucose is 180.18 Da. What mass of glucose is there in 200 ml of a 0.5 M solution?

Question 23.4
How much 0.25 M solution can be made from 500 ml of stock 1 M NaCl solution?

Question 23.5
The molecular mass of glucose is 180.18 Da. What is the molarity of a 5% w/v solution?

24 pH

A measure of hydrogen ion concentration, pH is a scale of acidity.

24.1 pH and the concentration of hydrogen ions

In water, at any one moment a number of water molecules are splitting ("dissociating") to produce hydroxide (OH^-) and hydrogen (H^+) ions. Almost immediately they recombine ("reassociate") back to H_2O.

This is represented by the following equation:

$$H_2O \rightleftharpoons H^+ + OH^-$$

In fact, the hydrogen ions actually join other H_2O molecules to form oxonium ions (H_3O^+):

$$H_2O + H_2O \rightleftharpoons H_3O^+ + OH^-$$

However, it is conventional to think of the H^+ ion as being a dissociation product of water, rather than the H_3O^+ ion.

pH is a measure of the concentration of H^+ ions in a solution. It is a logarithmic scale of hydrogen ion concentration and is defined by:

$$pH = -\log[H^+]$$

Note that the square brackets indicate concentration measured in molarity, so $[H^+]$ symbolises the molarity of hydrogen ions.

Example

The concentration of H^+ ions in pure water is 10^{-7} M

$$pH = -\log[H^+] = -\log 10^{-7} = -(-7) = 7$$

Thus the pH of pure water is 7.

24.2 The ion product

In water, the product of the concentrations of hydrogen ions and hydroxide ions always remains the same:

$$[H^+][OH^-] = 10^{-14}$$

This constant is known as the **ion product of water,** K_w and is helpful when determining the pH of alkaline solutions.

Example

Sodium hydroxide, NaOH, is a very strong base that is almost completely ionised to OH^- ions.

In 1 M NaOH the concentration of OH^- is therefore also 1 M, so $[OH^-] = 1$.

Substituting this into $[H^+][OH^-] = 10^{-14}$ gives:

$$[H^+][OH^-] = [H^+] \times 1 = 10^{-14}$$

So

$$[H^+] = 10^{-14}$$

$$pH = -\log[H^+] = -\log 10^{-14} = 14$$

Thus the pH of 1 M NaOH is 14.

24.3 Acids and bases

There are various definitions of acids, but for our purposes the most useful one is that acids are substances that dissociate to produce hydrogen ions. Using this definition we can write the equation:

$$HA \rightleftharpoons H^+ + A^-$$

where HA is the acid and A^- is its conjugate base.

The equilibrium constant for this reaction, known as the "acid dissociation constant", is K_a:

$$K_a = \frac{[H^+][A^-]}{[HA]}$$

Strong acids have a high K_a as they reach equilibrium when they are fully dissociated (fully "ionised"). Weak acids have a lower proportion of ionised molecules, so have a lower K_a.

The pK_a of an acid is defined as the negative logarithm of K_a:

$$pK_a = -\log_{10} K_a.$$

Example

The K_a of acetic acid is 1.78×10^{-5} mol l^{-1}

Its pK_a is:

$$-\log K_a = -\log(1.78 \times 10^{-5}) = 4.75$$

The equation for a base, B^-, which accepts H^+ ions, is:

$$B^- + H^+ \rightleftharpoons BH$$

where B^- is the base, and BH its conjugate acid.

Here, the equilibrium constant K_a is represented by:

$$K_a = \frac{[H^+][B]}{[BH^+]}$$

Again, $pK_a = -\log K_a$.

Test yourself

The answers are given on page 175.

Question 24.1
Hydrochloric acid, HCl, nearly fully dissociates to H^+ and Cl^- ions. What is the pH of 0.1 M HCl?

Question 24.2
What is the pH of a 0.1 M solution of NaOH?

Question 24.3
The pK_a of ammonia is 9.25. What is its K_a? You will need a calculator for this.

25 Buffers

A pH **buffer** is a solution that keeps the pH at a given value, resisting pH change even when acids or bases are added.

25.1 Making a buffer

There are two ways to make a buffer.

One way is by partially neutralising a weak acid (or base) with a strong base (or acid). This is called a **titration**.

The other way to make a buffer is to mix a weak acid (or base) with its conjugate acid (base).

25.2 Calculating the pH of a buffered solution

The **Henderson–Hasselbach equation** relates pH to the composition of buffer solutions. The equation is:

$$\text{pH} = \text{p}K_a + \log\frac{[\text{A}^-]}{[\text{HA}]}$$

where $[\text{A}^-]$ is the concentration of the conjugate base, and [HA] is the concentration of the acid.

We can use the Henderson–Hasselbach equation to calculate the pH of a buffer solution.

Example

We wish to make a buffer by mixing Tris hydrochloride (Tris–HCl, the conjugate acid) with Tris base (the conjugate base).

The pK_a of Tris–HCl is 8.3.

If we mix 250 ml of 1 M Tris–HCl with 750 ml of 1 M Tris base, the resulting acid and base molarities are 0.25 M and 0.75 M.

$$\text{pH} = \text{p}K_a + \log\frac{[\text{A}^-]}{[\text{HA}]} = 8.3 + \log\frac{0.75}{0.25} = 8.3 + 0.477 = 8.777.$$

Thus the pH of the resulting buffer is 8.8, to two significant figures.

25.3 Calculating the titration needed for a given pH

We can also use the Henderson–Hasselbach equation to calculate the titration needed to achieve a given pH.

Example

Acetic acid is a weak acid with a pK_a of 4.75. Its conjugate base is the acetate ion. When the two are mixed, we get an "acetate buffer".

We wish to prepare an acetate buffer of pH 5.0

We use the Henderson–Hasselbach equation to calculate the necessary ratio of the acid and its conjugate base.

$$\text{Desired pH} = 5.0 = 4.75 + \log\frac{[A^-]}{[HA]}$$

So,

$$\log\frac{[A^-]}{[HA]} = 5.0 - 4.75 = 0.25$$

$$\frac{[A^-]}{[HA]} = 1.778$$

We therefore need the ratio of acetate and acetic acid to be 1.8 to 1, to two significant figures.

Note that the Henderson–Hasselbach equation does not tell us the exact amounts of acetic acid and acetate ion to use, only their ratio.

In this example, the ratio of 1.8 base to its acid could be made by mixing 1.8 litres of 1 M sodium acetate with 1 litre of 1 M acetic acid.

Test yourself

You will need to use your calculator to answer these questions.
The answers are given on page 175.

Question 25.1
The pK_a of Tris–HCl is 8.3. If we mix 300 ml of 1 M Tris–HCl with 200 ml of 1 M Tris base, what is the resulting pH?

Question 25.2
Diethylmalonic acid has a pK_a of 7.2. If we have 1 M diethylmalonic acid and 1 M conjugate base, how much of each do we need to make 1 litre of pH 7.4 buffer?

26 Kinetics

Kinetics is the science of measuring changes. In the life and medical sciences, it usually refers to **enzyme kinetics**, the study of the reactions carried out by enzymes.

Enzymes are **biological catalysts** responsible for supporting many of the chemical reactions that maintain homeostasis. Most significant life processes are dependent on enzyme activity.

26.1 Chemical reaction rates

The rate of a chemical reaction is described by the number of molecules of **reactant** that are converted to a **product** in a given time. The reaction rate depends on:

- the concentration of the chemicals involved in the process
- the **rate constant** for that reaction

A reaction in which chemical A (the **reactant**) is converted to B (the **product**) can be written as:

$$A \longrightarrow B$$

The rate of this forward reaction is equal to the product of the molar concentration of A (symbolised by [A]) and the **forward rate constant,** k_{+1}

At any one moment, some of chemical B is also being converted back to chemical A:

$$A \longleftarrow B$$

The rate of this reverse reaction is equal to the product of [B] and the **reverse rate constant,** k_{-1}

When the rate of the forward reaction is equal to the rate of the reverse reaction, the reaction is said to be in **equilibrium.** When the two chemicals, A and B, are in equilibrium, the ratio of the two concentrations gives the **equilibrium constant** of the reaction, symbolised by K_{eq}. It also equals the ratio of the two rate constants. So,

$$K_{eq} = \frac{[B]}{[A]} = \frac{k_{+1}}{k_{-1}}$$

With a chemical reaction, the higher the concentration of reactant, the faster the reaction, i.e.

$$\text{Rate of reaction} \propto [A]$$

The ∝ symbol means "directly proportional to" (see Section 7.4).

There is no upper limit to the rate of reaction.

26.2 Enzyme kinetics

In reactions catalysed by enzymes, reactants are called **substrates**. The substrate and product concentrations are typically thousands of times greater than the enzyme concentration. So, each enzyme molecule will catalyse the conversion of many substrate molecules.

A substrate binds with a specific site, the **active site**, of an enzyme, forming a transitional state called the **enzyme–substrate (ES)** complex. When the ES complex **dissociates** (breaks down), the reaction products are released, and the enzyme is free to link with another substrate molecule. This process can be represented by:

$$E + S \rightarrow ES \rightarrow E + P$$

Some of the ES will dissociate back to E + S, so a more accurate representation is:

$$E + S \rightleftharpoons ES \rightarrow E + P$$

The constants for these three reactions are k_1, k_{-1} and k_2.

$$E + S \underset{k_{-1}}{\overset{k_1}{\rightleftharpoons}} ES \overset{k_2}{\rightarrow} E + P$$

The **Michaelis constant** K_M is defined by:

$$K_M = \frac{(k_{-1} + k_2)}{k_1}$$

If an enzyme has a large K_M it means that it binds to its substrate very weakly – it has a low **affinity** for the substrate.

Conversely, an enzyme with a small K_M has a high affinity for its substrate.

26.3 The Michaelis–Menten equation

We have already learnt that, with a *non*-enzymatic reaction, the higher the concentration of reactant, the faster the rate of reaction. There is no upper limit.

However, when an enzyme's catalytic site is working as fast as it can, an increase in the concentration of substrate, [S], will not increase the rate of reaction, *V*. At this point, the reaction rate is at its maximum, V_{max}.

The **Michaelis–Menten** equation relates the rate of reaction, *V*, to the substrate concentration, [S], the Michaelis constant K_M, and the maximum rate of reaction, V_{max}, as follows:

$$V = \frac{V_{max}\,[S]}{K_M + [S]}$$

If the rate of reaction is half of V_{max} the equation becomes:

$$\frac{V_{max}}{2} = \frac{V_{max}\,[S]}{K_M + [S]}$$

This equation can be manipulated to:

$$K_M = [S]\left\{\left[\frac{2V_{max}}{V_{max}}\right] - 1\right\} = [S]\{2 - 1\} = [S]$$

So, the K_M of an enzyme is the substrate concentration at which the reaction goes at *half* the maximum rate, V_{max}.

Plotting substrate concentration, [S], against rate of reaction, *V*, using the Michaelis–Menten equation gives the following graph:

Graph of Michaelis–Menten equation

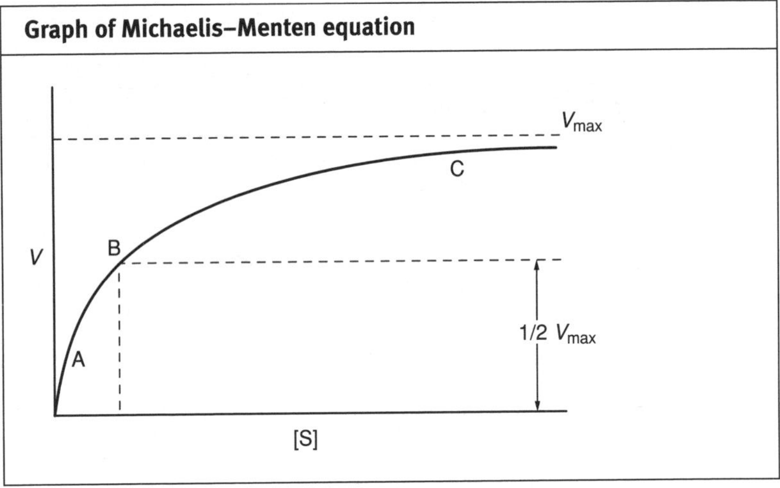

At point A, when levels of substrate are low, the limiting factor is availability of substrate. The rate of reaction is almost directly proportional to substrate concentration.

As more substrate is added, the rate of reaction increases rapidly.

At point B, 1/2 V_{max}, when [S] = K_M, 50% of the enzyme sites have substrate bound to them, i.e. are in an ES complex.

Point C, when [S] is high, is close to the point where all of the enzyme molecules have substrate bound to them. The reaction is moving towards its fastest possible rate, and adding more substrate has less and less effect on the rate of the reaction. This is indicated by the graph moving towards its asymptote of V_{max}

However, since V_{max} is approached only slowly as the substrate concentration is increased, it is difficult to calculate an accurate figure for V_{max} from a standard plot such as this. This problem is overcome by using a **Lineweaver–Burk** plot.

26.4 The Lineweaver–Burk plot

The Michaelis–Menten equation, explained above, can be rearranged by taking the reciprocal of both sides of the equation to give:

$$\frac{1}{V} = \frac{1}{V_{max}} + \left(\frac{K_M}{V_{max}} \times \frac{1}{[S]}\right)$$

A plot of $1/V$ against $1/[S]$ is called a **Lineweaver–Burk** plot. It gives a straight line with a gradient of K_M/V_{max}

V_{max} can easily be calculated from the intercept with the y–axis which is at $1/V_{max}$, and K_M can be calculated from the intercept on the x-axis which is at $-1/K_M$

Graph of Lineweaver–Burk plot

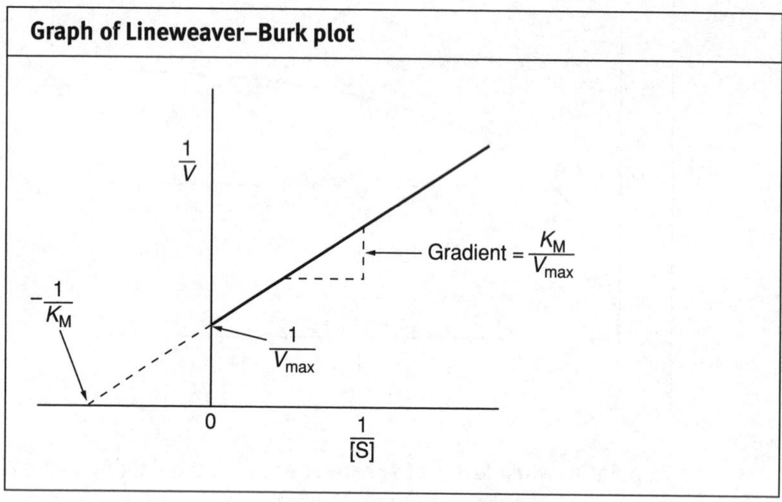

27 The language of statistics

This chapter explains some of the words that life and medical scientists need to know when analysing data.

27.1 Populations and samples

In statistics, the word **population** means all the individual items which could be studied.

A selection from a population is called a **sample**.

Example

A researcher wants to compare broad bean, *Vicia faba*, yield in two fields, one which has been artificially fertilised, the other organically fertilised.

The two populations being studied are all the broad beans in the two fields.

However, she wants to sample a total of $50\,m^2$ from each field.

Individuals in the sample may be called **sample units**, **subjects** or **cases**. The data collected from samples are known as **observations**.

The studied differences between individuals in a population are called **variables** or, sometimes, **fields**.

The observations are called **data**.

Example

The variables that the researcher wishes to study are the numbers of beans per pod and the mean mass of each sample unit, the individual beans.

The data are the resulting lists of measurements. (Note that the word "data" is a plural, so we say "data are . . ." rather than "data is . . .")

27.2 Bias

The aim is always that a sample should be representative of a population. If the sample doesn't truly represent the **parent population**, there is said to be **bias**.

An example of this is **observer bias**, where the researcher consciously or subconsciously biases the sample.

To avoid bias, the sample needs to be a **random selection** from the population.

Making a random selection usually involves using **random numbers**. These can be found in random number tables or can be generated by computer programs.

A **quadrat** is an area used as a sample unit.

Example

Because there is not time to count and weigh all the beans in both fields, the researcher needs to take samples.

She has a theory that organic farming reduces the numbers of beans per pod and the mass of the beans. She is aware that she could subconsciously choose smaller pods from the organic field, resulting in observer bias.

She also knows that the numbers and masses of the beans may vary in different areas of the same field, so she wants to avoid this sample bias as well.

She therefore tries to avoid these biases by using a random number generator to select five $10\,m^2$ quadrats from each field.

27.3 Variables

Continuous variables are those which can take any value within a given range.

There are two types of continuous variable: ratio and interval.

A **ratio variable** can take any value within a range, i.e. the value needn't be a whole number, but the zero must be a true zero.

Example

The length of a bean pod is a ratio variable.

An **interval variable** can take any value within a range, but a value of zero does not indicate an absolute zero.

Example

Temperature in °C is an interval variable.

A temperature of 0°C is not a true zero, as the true zero for temperature is −273.15°C (0 K).

A temperature of 10°C is therefore not twice the temperature of 5°C.

Categorical variables are those where only certain values can exist.

Categorical variables can be nominal or ordinal.

A **nominal variable** has no ordering to the categories.

> **Example**
>
> Broad bean varieties are categorical variables: we can't have half a variety, and there is no obvious numerical way to put different varieties in order.

An **ordinal variable** has an ordering to the categories.

> **Example**
>
> Rows of broad bean may be **ranked** (put in rank order). Ranks can be numbered (1st, 2nd, 3rd rank, etc.). However, the numbers used to label the ranks of ordinal data can't be manipulated mathematically: rank 3 isn't three times rank 1.

A **binary** variable is one that has only two categories, e.g. male and female.

27.4 Descriptive and inferential statistics

Descriptive statistics are those which *describe* the data in a sample. They include means, medians, standard deviations and quartiles. They are designed to give the reader an understanding of the data. They are described in detail in Chapters 28 and 29.

Inferential statistics are statistical methods which make "inferences" about the population, based on the sample of data that has been collected. They can estimate whether the results suggest that there is a real difference in the populations, or how well aspects of a sample are likely to represent the population as a whole. Chapter 32 gives an introduction to inferential statistics.

> **Example**
>
> A researcher may give the mean bean yield per square metre in a series of samples. This is a descriptive statistic.
>
> If she then wishes to test whether there is a difference in the mean bean yield per square metre in two fields, she would use inferential statistics.

28 Describing data: measuring averages

When faced with a lot of data, it helps to know what the average is.

There are three ways of giving an average: mean, median and mode. Which of these we need to use depends on the pattern of the data.

28.1 Mean

The **mean** is also known as an arithmetic mean.

It is the commonest measure of a "mid-point", so it's important to have an understanding of how it is calculated.

It is used when the spread of the data is fairly similar on each side of the mid-point, for example when the data are **normally distributed**.

The normal distribution (also known as the "Gaussian distribution") is referred to a lot in statistics. It's the symmetrical, bell-shaped distribution of data shown in the following graph.

The normal distribution

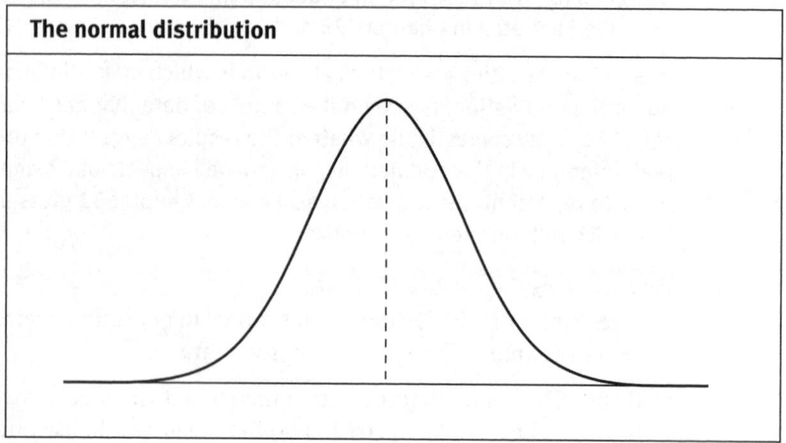

The dotted line shows the mean of the data.

The mean is the sum of all the values, divided by the number of values.

This can be symbolised by:

$$\bar{x} = \frac{\sum x}{n}$$

where $\bar{x}$ is the symbol for the sample mean (pronounced "x bar"), $\sum x$ is the sum of all the values, and n is the number of values in the sample.

Example

Five Scots pines, *Pinus sylvestris*, in a plantation are 3.5, 3.7, 3.8, 3.9 and 4.1 m high.

$$\bar{x} = \frac{\sum x}{n} = \frac{3.5 + 3.7 + 3.8 + 3.9 + 4.1}{5} = \frac{19}{5} = 3.8\,\text{m}$$

So the mean height is 3.8 m.

28.2 Population or sample mean?

If we have data on a whole population, then the symbol for the mean is μ, pronounced "mu".

However, if we only have data on a sample, then we use the $\bar{x}$ symbol to symbolise the mean.

Example

A researcher wanted to study the mass of all the mallard ducks, *Anas platyrhychos*, on a lake. He managed to capture and weigh all the ducks. The mean mass was 1.12 kg, so

$$\mu = 1.12\,\text{kg}$$

At a second lake he was unable to capture all the ducks, so could only measure a sample of them. In this sample, the mean mass was 1.39 kg, thus

$$\bar{x} = 1.39\,\text{kg}$$

28.3 Median

The **median** is the point which has half the values in a sample above, and half below.

If the data are not evenly distributed around their mean, the data are said to be **skewed**. In that case, the mean will not give a good picture of the typical value, so we use the median.

A skewed distribution

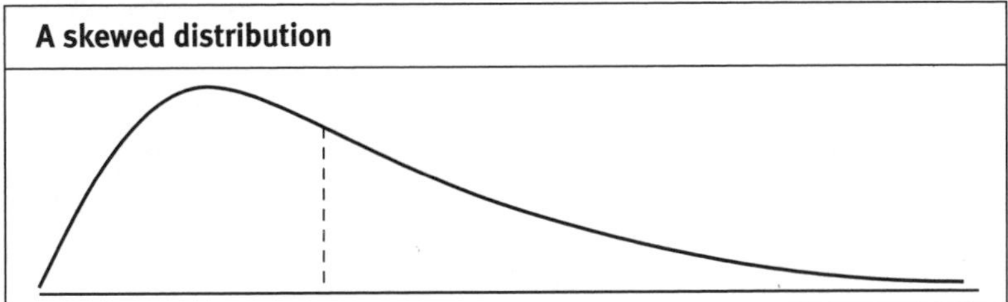

The dotted line shows the median.

Compare the shape of the graph with the normal distribution shown in Section 28.1.

Example

Let's take the same five Scots pines, *Pinus sylvestris*, as above, measuring 3.5, 3.7, 3.8, 3.9 and 4.1 m.

If there is a 6th pine that is 2.0 m high, then the mean height would be 3.5 m even though only one pine is less than 3.5 m tall. For this skewed sample, the *median* is a more suitable mid-point to use.

Where there are two "middle" values, the convention is that the median is half-way between these.

Example

Using the first example of five Scots pines measuring 3.5, 3.7, 3.8, 3.9 and 4.1 m, the median height is 3.8 m, the same as the mean – half the pines are taller, half are shorter.

However, in the second example with six pines measuring 2.0, 3.5, 3.7, 3.8, 3.9 and 4.1 m, there are two "middle" heights, 3.7 m and 3.8 m. The median is half-way between these, i.e. 3.75 m. This gives a better idea of the mid-point of these skewed data than the mean of 3.5 m.

The median may be given with its **quartiles**. The 1st quartile point has ¼ of the data below it; the 3rd quartile has ¾ of the sample below it. The inter-quartile range contains the middle ½ of the sample, i.e. the data between the 1st and 3rd quartiles. This can be shown in a **"box and whisker"** plot.

Example

A researcher measured the wing span of a population of 50 small brown bats, *Myotis lucifugus*. The median wing span was 245 mm, 1st quartile 238 mm and 3rd quartile 257 mm. The smallest wingspan was 221 mm, the largest 271 mm. This distribution is represented by the box and whisker plot in the following figure.

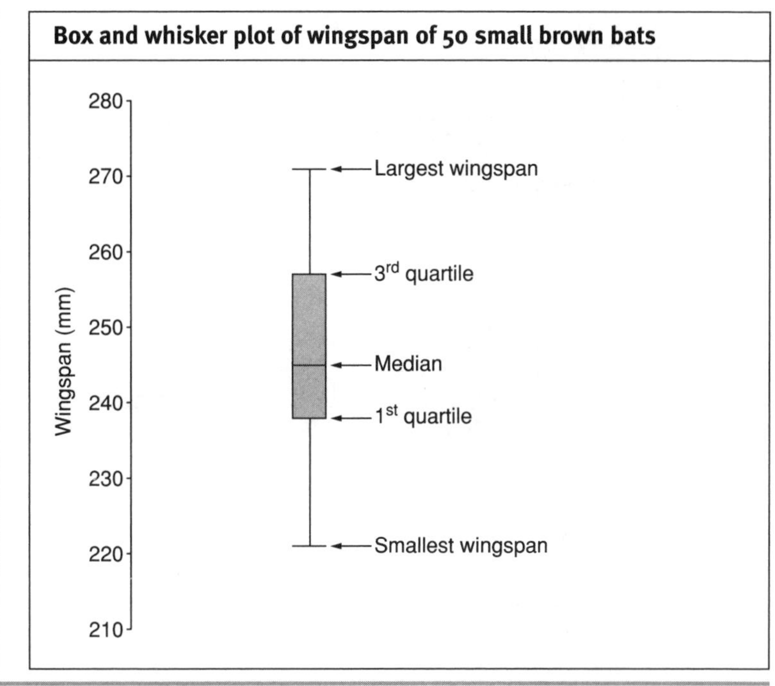

The ends of the whiskers represent the maximum and minimum values, excluding any extreme results.

28.4 Mode

The **mode** is the name for the most frequently occurring event.

This is usually only used when we have nominal variables, i.e. those which represent different categories of the same feature and where the categories are not ordered.

Example

A researcher noted the eye colour of 100 students. The results are shown here:

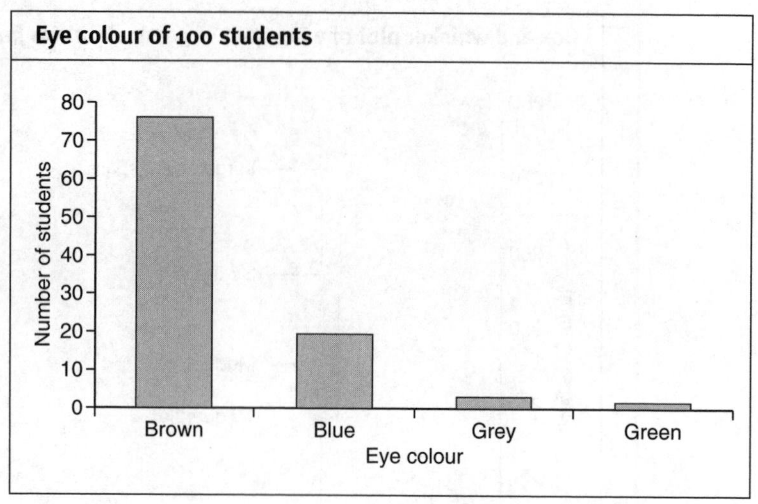

We can't use the mean or median here – there isn't an average of two eye colours. The commonest eye colour, and therefore the mode, is brown.

The mode can also be used when there is no one average value, for instance in a **bi-modal** distribution.

Example

In this graph there are two "peaks" to the data, i.e. it has a bi-modal distribution.

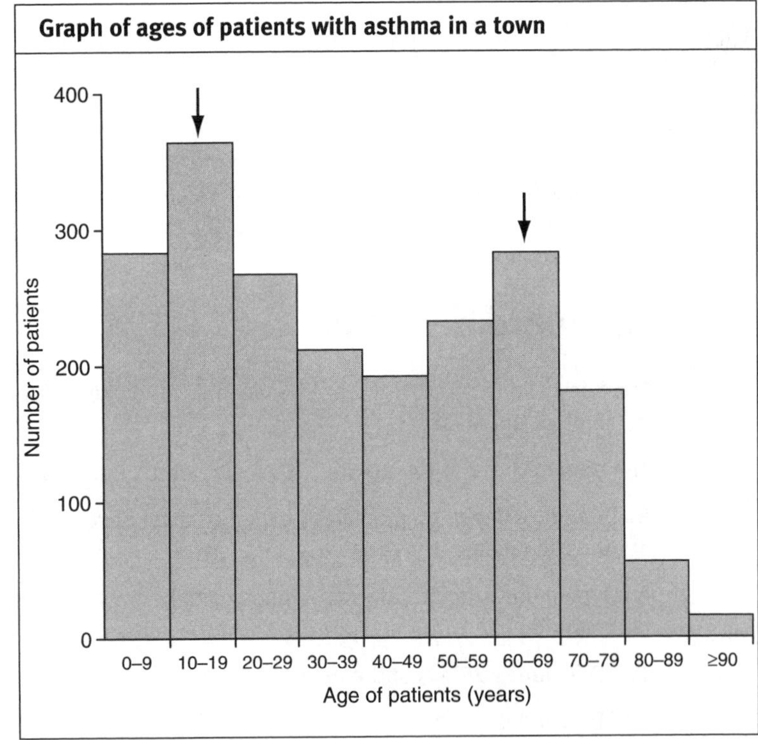

The arrows point to the modes at ages 10–19 and 60–69.

Bi-modal data usually suggest that two populations are present that are mixed together, so a mean or median is not a suitable measure for the distribution.

Test yourself

The answers are given on page 175.

Question 28.1
The haemoglobin levels in the blood of five female volunteers are 11.7, 11.9, 12.2, 12.7 and 13.0 $g\,dl^{-1}$.
What is the mean haemoglobin level?

Question 28.2
If the haemoglobin levels in the blood of eight female volunteers are 11.1, 11.7, 11.9, 12.3, 12.7, 13.3, 15.2 and 17.4 $g\,dl^{-1}$, what is the median?

29 Standard deviation

Standard deviation (SD) is used for data which are normally distributed (see Section 28.1). It provides an indicator of how much the data are spread around their mean.

29.1 Standard deviation

The "deviation" is the difference between any individual reading and the mean of all the readings.

The **standard deviation** is a kind of average of the individual deviations.

So, standard deviation indicates how much a set of values is spread around the mean of those values.

A range of one standard deviation above and below the mean (abbreviated to ±1 SD) includes 68.2% of the values.

±2 SD includes 95.4% of the data.

±3 SD includes 99.7%.

Example

Let's say that a group of patients has a normal distribution for weight. The mean weight of the patients is 80 kg. For this group, the SD is calculated to be 5 kg.

- 1 SD below the average is 80 – 5 = 75 kg.
- 1 SD above the average is 80 + 5 = 85 kg.

±1 SD will include 68.2% of the subjects, so 68.2% of patients will weigh between 75 and 85 kg.

95.4% will weigh between 70 and 90 kg (±2 SD).

99.7% of patients will weigh between 65 and 95 kg (±3 SD).

See how this relates to this graph of the data.

Graph showing normal distribution of weights of patients, mean 80 kg, SD 5 kg

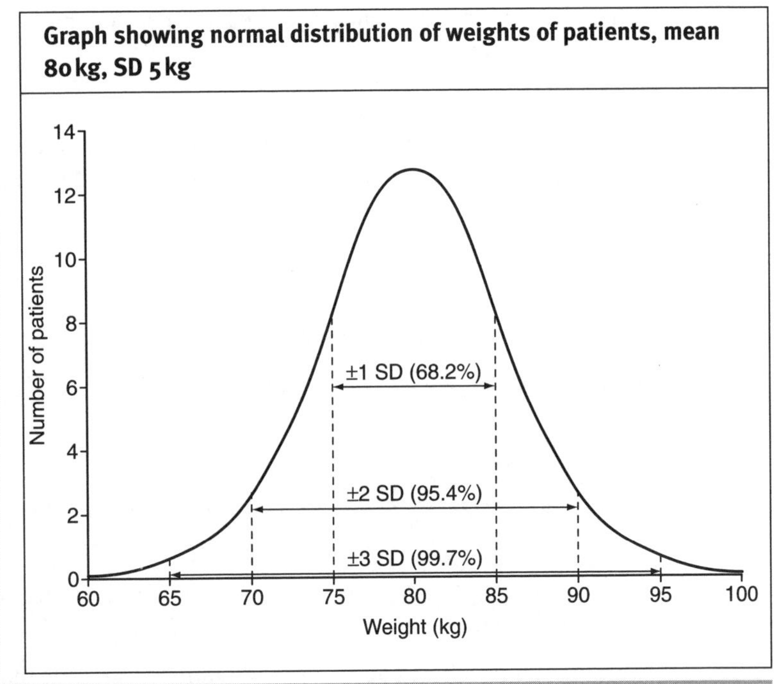

29.2 Calculating the standard deviation of a whole population

Calculating the SD of a *whole population* involves five steps.

1) First we calculate the mean, μ, of the population, by adding all the values and dividing the result by the number of values, as in Section 28.1.

2) Then we need to work out the difference (deviation) between each individual value and the mean, by subtracting the mean from each value, symbolised by:

$$x - \mu$$

3) Next we calculate the square of each deviation, multiplying each value by itself. Remember that any negative deviations become positive when squared.

The process so far is:

$$(x - \mu)^2$$

4) Then we calculate the sum of those squared deviations, known as the **sum of squares.**

$$\Sigma (x - \mu)^2$$

5) Then we need the mean of the sum of squares, so we divide the sum by the number of observations. This gives a value called the **population variance,** σ^2**:**

$$\sigma^2 = \frac{\Sigma (x - \mu)^2}{N}$$

6) The SD is the square root of the variance.

The whole process can be symbolised by:

$$\sigma = \sqrt{\frac{\Sigma (x - \mu)^2}{N}}$$

σ, pronounced "sigma", is the symbol for standard deviation.

Example

All 10 salmon, *Salmo salar*, in a lake weigh 1.6, 1.7, 1.8, 1.8, 2.3, 2.4, 2.6, 2.8, 3.1 and 3.3 kg.

NB: normally, whole populations will number far more than 10. We are using a small number so that it is easier to follow the calculations.

1) The sum of the values is 23.4 kg, giving a mean of 2.34 kg.

$$\mu = 2.34\,\text{kg}$$

2) The difference (deviation) between each individual value and the mean is as follows:

1.6 − 2.34	=	−0.74
1.7 − 2.34	=	−0.64
1.8 − 2.34	=	−0.54
1.8 − 2.34	=	−0.54
2.3 − 2.34	=	−0.04
2.4 − 2.34	=	0.06
2.6 − 2.34	=	0.26
2.8 − 2.34	=	0.46
3.1 − 2.34	=	0.76
3.3 − 2.34	=	0.96

3) Calculating the square of each deviation gives:

-0.74^2	=	0.5476
-0.64^2	=	0.4096
-0.54^2	=	0.2916
-0.54^2	=	0.2916
-0.04^2	=	0.0016
0.06^2	=	0.0036
0.26^2	=	0.0676
0.46^2	=	0.2116
0.76^2	=	0.5776
0.96^2	=	0.9216

4) The sum of these squares is 3.324.

5) The mean of the squares is the sum of squares divided by the number of salmon in the population, giving the variance, σ^2:

$$\sigma^2 = \frac{\Sigma\,(x-\mu)^2}{N} = \frac{3.324}{10} = 0.3324$$

6) The square root of the variance gives the SD:

$\sigma = \sqrt{0.3324} = 0.577$ kg to three significant figures.

29.3 Calculating standard deviation ranges

Having calculated the SD, we can work out the ranges that include 68.2% (± 1 SD), 94.5% (± 2 SD) and 95.4% (± 3 SD) of a population.

Example

The mean fasting triglyceride level of 183 patients is found to be 2.2 mmol l^{-1}.

The SD is calculated to be 0.3 mmol l^{-1}.

1 SD below the average is $2.2 - 0.3 = 1.9$ mmol l^{-1}.

1 SD above the average is $2.2 + 0.3 = 2.5$ mmol l^{-1}.

±1 SD will include 68.2% of the data, so we expect 68.2% of the population to have a triglyceride level between 1.9 and 2.5 mmol l^{-1}.

95.4% will be between 1.6 and 2.8 mmol l^{-1} (±2 SD).

99.7% of triglyceride levels will be between 1.3 and 3.1 mmol l^{-1} (±3 SD).

See how this relates to the following graph of the data.

Graph showing normal distribution of fasting triglyceride levels with mean 2.2 mmol l^{-1} and SD 0.3 mmol l^{-1}

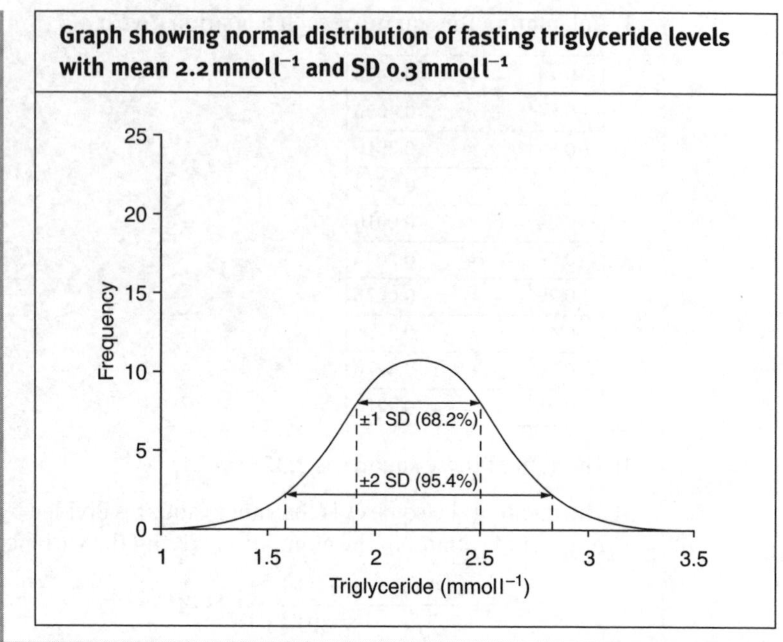

29.4 Comparing different standard deviations

If we have two populations with the same mean but different standard deviations, then the population with the larger SD has a wider spread than the population with the smaller SD.

Example

If another population of patients has the same mean fasting triglyceride level of 2.2 mmol l^{-1} but an SD of only 0.2 mmol l^{-1}, ±1 SD will include 68.2% of the subjects, so 68.2% of the patients will have a fasting triglyceride level of between 2.0 and 2.4 mmol l^{-1}.

Graph showing a normal distribution of fasting triglyceride levels with mean 2.2 mmoll^{-1} and SD 0.2 mmoll^{-1}

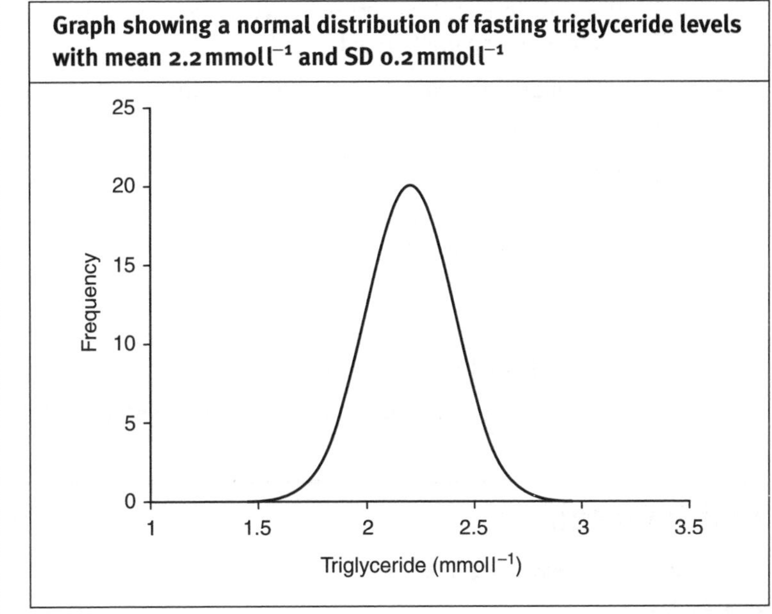

Compare this graph with the graph of the previous example which is drawn to the same scale. A smaller SD gives a taller, narrower normal distribution.

29.5 Calculating the standard deviation of a sample

When calculating the SD of a *sample*, the variance (denoted as V for a sample but σ^2 for a population) is calculated by dividing the sum of the squared deviations by *one fewer* than the number in the sample, i.e. $n - 1$. The value $n - 1$ is used rather than n as this gives a better estimate of the SD.

$n - 1$ is called the number of "degrees of freedom", and is explained in more detail in Chapter 31.

Example

Let's say that the 10 salmon in the example in Section 29.2 were a *sample* of salmon in a lake. Compare each step with the equivalent calculation above for a population of 10.

1) The sum of the values is still 23.4 kg, giving the same mean of 2.34 kg.

   However, this is now symbolised by $\bar{x} = 2.34$. The use of $\bar{x}$ (rather than μ) denotes that this a sample mean, not the mean of a whole population.

2) The difference (deviation) between each individual value and the mean is also unchanged.

3) Similarly, the calculation for the square of each deviation is unchanged.

4) The sum of squares is also unchanged at 3.324.

5) However, we now divide the sum of squares by (number of salmon – 1) to get the sample variance, V:

$$V = \frac{\Sigma(x - \bar{x})^2}{n - 1} = \frac{3.324}{9} = 0.3693$$

6) The square root of the variance gives the SD:

$$\text{SD} = \sqrt{0.3693} = 0.607 \text{ kg to three significant figures.}$$

Test yourself

The answer is given on pages 175–176.

Question 29.1
The haemoglobin levels in the blood of five volunteers are 11.7, 11.9, 12.2, 12.7 and 13.0 g l^{-1}.
What is the standard deviation?

30 Checking for a normal distribution

Standard deviation should only be used when the data have a normal distribution. However, means and standard deviations are often wrongly used for data which are not normally distributed.

If given only the mean and the SD, a simple check for a normal distribution is to see if 2 SD away from the mean is still within the possible range for the variable.

Example

If we have some data on the mass of a group of men that suggests a mean mass of 75 kg and a standard deviation of 40 kg, then:

$$\text{Mean} - (2 \times \text{SD}) = 75 - (2 \times 40) = -5\,\text{kg}$$

This is clearly an impossible value for the mass of a person, so the data cannot be normally distributed. The mean and standard deviations would therefore not be appropriate measures to use for this sample, as there should be 4.6% of the sample below the 2 SD level.

30.1 *z*-scores

The number of standard deviation units that a sample unit is away from the population mean is the ***z*-score**.

If an observation has a value above the population mean, it has a positive *z*-score, so

+1 SD gives a *z*-score of 1.

A value below the mean has a negative *z*-score, thus

−1 SD gives a *z*-score of −1.

The formula for the *z*-score is:

$$z = \frac{(x - \mu)}{\sigma}$$

where z is the *z*-score, x is the value of the observation, μ is the population mean and σ is the standard deviation.

Example

For a population of men, the mean mass is found to be 75 kg, SD 5 kg.

The mass of one man is 65 kg.

$$z = \frac{(x-\mu)}{\sigma} = \frac{(65-75)}{5} = -2$$

30.2 A tip

As well as knowing how to calculate the standard deviation, it's worth memorising how much data are included in each standard deviation, so a reminder:

±1 SD includes 68.2% of the data, ±2 SD includes 95.4%, ±3 SD includes 99.7%.

Keeping the "normal distribution" curve in Section 28.1 in mind may help.

Test yourself

The answer is given on page 176.

Question 30.1

The mean blood haemoglobin level in a female population is 12.5 g dl^{-1}, SD 1.2 g dl^{-1}.
One woman has a haemoglobin of 15.5 g dl^{-1}.
What is the *z*-score for her haemoglobin?

31 Degrees of freedom

Statisticians use **degrees of freedom**, df, in many of their calculations. The concept is difficult to grasp and, surprisingly, there is no neat definition to explain it.

However, the general concept can be shown as follows.

Let's say that we have three observations (A, B and C) and we want to know their values.

If we know nothing about them other than their existence, then each observation has the freedom of being unknown. There are three degrees of freedom: 3 df.

If we are given the value for the mean, 2.7 say, as soon as we know the values of two of the variables, we can calculate the value of the third. Thus there would be two degrees of freedom: 2 df.

Say we are now told that the standard deviation is 1.2 and the mean 2.7, we can now calculate the values of all the variables as soon as we know the value of any one of them. There is now one degree of freedom: 1 df.

Examples

If we have two yoghurt pots and we know that one is strawberry and one is chocolate, we don't have to taste both of the yoghurts before we can label both pots. We only need to taste one of them.

So, $\text{df} = N - 1 = 2 - 1 = 1$

where N equals the number of yoghurt pots.

If we have 10 yoghurt pots and have a list of their 10 different flavours, we only need to try nine before we can work out the flavour of each pot.

Here, $\text{df} = N - 1 = 10 - 1 = 9$

Similarly, if we have a sample of 28 babies and know their mean length, we need to know the length of 27 of them to be able to calculate the length of all of them.

$$\text{df} = N - 1 = 28 - 1 = 27$$

If we also know the standard deviation of their length, then we have again lost a degree of freedom.

$$\text{df} = N - 2 = 28 - 2 = 26$$

32 How to use statistics to make comparisons

Section 27.4 introduced us to the concept of inferential statistics, those that make "inferences" about a population, based on the sample of data that has been collected.

Inferential statistics estimate whether the results suggest that there is a real difference in the populations, or how well aspects of the sample are likely to represent the population.

This chapter explains the process we need to use when performing inferential statistics.

32.1 The null hypothesis

Inferential statistics involve testing a theory, known as a **hypothesis**.

Paradoxically, for statistical analysis the hypothesis is usually that there is *no* (null) difference between the populations being studied, the **null hypothesis**. The result of the test either supports or rejects that hypothesis.

The null hypothesis is generally the opposite of what we are actually interested in finding out. If we are interested in whether there is a difference between two groups, then the null hypothesis is that there is no difference, and we would try to disprove this.

Example

Chapter 27 gave an example of a researcher wanting to compare broad bean, *Vicia faba*, yield in two fields. The null hypothesis is that there is *no* difference between the yields in the two fields based on 5 quadrats from each field.

32.2 Choosing and using the right statistical test

Having collected research data for a comparative study, we need to use a statistical test to compare the data. The **decision-making flowchart** in Appendix 1 will help you decide which test to use.

Applying and calculating the test will give us a **test statistic**, which is a number that quantifies the difference between the samples.

In general, the larger the test statistic result, the larger the difference between the two samples.

> **Example**
>
> Using the flowchart in Appendix 1, the researcher decides that the best way to compare the broad bean yield per square metre in the two fields is with the unpaired *t* test. This is described in detail in Chapter 38.
>
> She calculates that the *t* test statistic is 2.51, df 18.

32.3 Significance levels

We now need to work out how likely it is that any difference between the data is due to chance.

We use tables like those in Appendices 2 and 3 to calculate a **significance level** from the test statistic.

If the difference is likely to be due to chance, the difference is said to be **non-significant** and the null hypothesis cannot be rejected.

However, if the difference is not likely to be due to chance, that difference is **significant** and the null hypothesis is rejected.

> **Example**
>
> In the table of critical values for *t* distribution in Appendix 2, the *t* test statistic of 2.51 for the bean yield comparison when df is 18 is more than the critical value of 2.10 for 5%, so the difference is unlikely to be due to chance.
>
> If, however, the *t* test statistic for the bean yield comparison has a low significance level, the null hypothesis cannot be rejected – the difference between the two fields could easily be due to chance.

32.4 Are there any confounding factors?

Finally, and outside the scope of this book, if significant differences *are* found, we need to consider whether anything else, any **confounding factors**, could have caused these differences.

> **Example**
>
> During her fieldwork, the researcher noticed that the two fields of broad beans differed in various respects: one was on a south-facing slope and was well drained; the other was on level ground and waterlogged.
>
> She realised that these confounding factors could have accounted for the difference in mass per bean, rather than the farming method.

33 The standard error of the mean

If we take measurements from a random sample of a large population, we can calculate the mean of that sample.

Because of chance, the sample mean is likely to vary from the mean of the whole population (the "population mean", also known as the "true value").

33.1 The standard error of the mean

The **standard error of the mean**, SEM, gives an indication of how close a sample mean might be to the population mean.

The standard error of a sample mean is given by dividing the standard deviation by the square root of the sample size, i.e.

$$\text{SEM} = \frac{\text{SD}}{\sqrt{n}}$$

The population mean ±1.96 SEM will include 95% of the sample means.

The population mean ±2.58 SEM will include 99% of the sample means.

The population mean ±3.29 SEM will include 99.9% of the sample means.

Example

Let's say that we are interested in the mean velocity of fibroblasts at a certain temperature. The "true value" for that mean will be the mean speed of all fibroblasts cultured under those conditions.

We study this by taking a sample of 10 fibroblasts and measuring their mean velocity. Because of random variation, the mean of this sample is likely to vary from the "true value". We don't know for certain how close it is to the true value, but the larger the sample, the closer it is likely to be.

Another sample would be likely to give slightly different results. If we took 10 separate samples, each would give a different result. However, we can see that the means would be distributed around a central area – this area would probably contain the true population mean.

This distribution of means around the population mean will be a normal distribution and can be described by the standard error of the mean.

If the sample sizes are larger (30 fibroblasts, say, instead of 10), then the standard error will be smaller, and the distribution will be narrower.

33.2 How SEM works

Even with only one sample, we know that the sample mean is part of that distribution around the population mean, we just don't know where it is. The standard error of the mean is the standard error of that distribution. It takes into account the sample size and the variation within that sample.

Example

In Section 29.5 we calculated the mean mass of 10 salmon to be 2.34 kg. The standard deviation was 0.607 kg.

$$\text{SEM} = \frac{\text{SD}}{\sqrt{n}} = \frac{0.607}{\sqrt{10}} = 0.1929\,\text{kg}$$

33.3 The effect of a larger sample size

Increasing sample size reduces the standard error.

Example

If we quadruple the sample size of salmon and the standard deviation is still 0.61 kg, then

$$\text{SEM} = \frac{\text{SD}}{\sqrt{n}} = \frac{0.607}{\sqrt{40}} = 0.0960\,\text{kg}$$

So, quadrupling the sample size halves the standard error.

33.4 Standard deviation or standard error?

Standard deviation tells us how much the data in *samples* vary around their own mean.

We use standard error when we want to know how much the sample *means* vary around their population mean.

Test yourself

The answer is given on page 176.

Question 33.1
The mean blood haemoglobin level in 16 pregnant women is 11.6 g dl^{-1}, SD 0.4 g dl^{-1}. What is the SEM?

34 Confidence intervals

If we have the mean value of a sample and want the range that is likely to contain the true population value, we can use the standard error to calculate the **confidence interval**, CI.

The confidence interval is the range (interval) in which we can be fairly sure (confident) that the population mean lies, i.e. the mean value that we would get if we had data for the whole population that we have sampled.

34.1 The 95% confidence interval

The population mean ±1.96 standard errors will include 95% of the sample means.

It follows that, for any given sample, there is a 95% chance (we can be 95% confident) that the sample mean is within ±1.96 standard errors of the population mean.

This is often taken to mean that there is a 95% chance that the population mean is within ±1.96 standard errors of the sample mean. Although technically not correct, the concept is commonly used.

34.2 The effect of sample size and SD on the confidence interval

The size of a confidence interval is related to the sample size and the size of the standard deviation:

- larger studies are likely to have narrower confidence intervals;
- the smaller the SD, the narrower the confidence interval.

34.3 Calculating the 95% confidence interval

The range from 1.96 standard errors below the sample mean to 1.96 standard errors above it is called the 95% confidence interval.

To calculate the 95% confidence interval, we first need to multiply the standard error by 1.96.

Subtracting the result from the sample mean gives the lower 95% **confidence limit**; adding it to the sample mean gives the upper 95% confidence limit.

$$95\%\ \mathrm{CI} = \bar{x} \pm (\mathrm{SEM} \times 1.96)$$

Example

In Section 33.2 we found that the mean mass of a sample of 10 salmon was 2.34 kg, with a standard error of the mean of 0.1929 kg.

To calculate the 95% confidence interval:

$$\text{SEM} \times 1.96 = 0.1929 \times 1.96 = 0.3781$$

$$\bar{x} - 0.3781 = 2.34 - 0.3781 = 1.9619$$

$$\bar{x} + 0.3781 = 2.34 + 0.3781 = 2.7181$$

This is usually interpreted as being 95% confident that the true population mean for this sample of fish is between 1.96 and 2.72 kg, to three significant figures.

This is written as:

mean mass 2.34 kg, 95% CI 1.96–2.72 kg.

34.4 Calculating other confidence intervals

The sample mean ±2.58 SEM gives the 99% confidence interval.

The sample mean ±3.29 SEM gives the 99.9% confidence interval.

Example

We wish to calculate the 99% confidence interval for the salmon in the example above.

$$\text{SEM} \times 2.58 = 0.1929 \times 2.58 = 0.4977$$

$$\bar{x} - 0.4977 = 2.34 - 0.4977 = 1.8423$$

$$\bar{x} + 0.4977 = 2.34 + 0.4977 = 2.8377$$

This is usually interpreted as being 99% confident that the true population mean for these fish is between 1.84 and 2.84 kg.

Similarly, to calculate the 99.9% confidence interval:

$$\text{SEM} \times 3.29 = 0.1929 \times 3.29 = 0.6346$$

$$\bar{x} - 0.6346 = 2.34 - 0.6346 = 1.7054$$

$$\bar{x} + 0.6346 = 2.34 + 0.6346 = 2.9746$$

This is usually interpreted as being 99.9% confident that the true population mean for the salmon is between 1.71 and 2.97 kg.

Test yourself

The answers are given on page 176.

Question 34.1

The mean blood haemoglobin level in a sample of 20 women is 12.8 g dl^{-1}. The SEM is 0.8 g dl^{-1}. Give the 95%, 99% and 99.9% confidence intervals.

35 Probability

An understanding of **probability** is key to many statistical analyses.

The concept of chance is something that we can intuitively understand. The many ways of describing probability can be confusing, however.

35.1 Five possible ways of stating a probability

A numerical value for a probability can be given in different ways.

Example

If we toss a coin, there is an equal chance that it will land heads or tails.

So, there is a 1 in 2 probability that it will land heads up.

Probability can be written as a fraction. We divide the number of nominated outcomes by the total number of possible outcomes.

Example

With a coin, the probability of throwing heads is 1 divided by 2: we are likely to get heads 1/2 the number of times that we toss the coin.

Probability is expressed on a scale from 0 to 1:

- a rare event has a probability close to 0;
- a very common event has a probability close to 1.

Example

$$\text{Probability of throwing a head} = \frac{1}{2} = 0.5$$

Probability can also be written using the abbreviation "P" or "p"

Example

For a coin, the chance of throwing a head is $P = 0.5$

Additionally we can state chance as a percentage.

Example

When tossing a coin, there is a 50% chance of throwing a head.

So, 1 in 2, 1/2, 0.5, $P = 0.5$ and 50% all give the same information about the chance of a coin landing heads up.

Example

The 95% confidence interval for the size of a strain of *E. coli* bacteria is 1.9 to 2.1 μm.

This is usually interpreted as meaning that we are 95% confident that the true population mean is between 1.9 and 2.1 – and there is a 5% chance that it isn't.

We can also describe the 95% probability as

- a 19 in 20 chance
- a 19/20 chance
- a 0.95 probability
- P = 0.95

Test yourself

The answers are given on page 176.

Question 35.1
Use five different ways to describe the chance of throwing two sixes with one roll of a pair of dice.

36 Significance and P values

Significance is an important concept related to probability.

When comparing measurements of two or more groups, we may make a hypothesis on the difference between them. The P (probability) value is used when we wish to see how likely it is that the hypothesis is true.

As described in Section 32.1, the hypothesis is usually that there is *no* difference between the groups, known as the "null hypothesis".

The significance level describes the likelihood that there actually *is* a difference.

36.1 What significance means

If we take two groups of patients who have been given different treatments, and the resulting mean cure rates are different, we may want to know whether there is a significant difference between the two means. Has the difference happened by chance, or might there truly be a difference between the two groups?

The hypothesis is that there is *no* difference between treatments, known as the **null hypothesis**.

Example

Two hundred adult patients with bronchopneumonia have been randomised to receive one of two possible antibiotics. Five days later, they are reassessed.

The researchers wish to know how likely it is that any difference between the effects of the two treatments could have happened by chance, or whether there is a significant difference.

The null hypothesis is that there is no difference between the effects of the two treatments.

36.2 The P value

The P value gives the probability of any observed difference in measurement between the two groups having happened by chance.

P = 0.5 means that the probability of the difference having happened by chance is 0.5 in 1.

P = 0.05 means that the probability of the difference having happened by chance is 0.05 in 1, i.e. 1 in 20. It is the figure frequently quoted as being "statistically significant", i.e. unlikely to have happened by chance and therefore important. However, this is an arbitrary figure. If we look at 20 studies, even if there is no difference in any of the groups studied, one of the studies is likely to have a P value of 0.05 and so appear significant!

The lower the P value, the less likely it is that the difference happened by chance, and so the higher the significance of the finding.

P = 0.01 is often considered to be "highly significant". It means that the difference will only have happened by chance 1 in 100 times. This is unlikely, but still possible.

P = 0.001 means the difference will have happened by chance 1 in 1000 times; even less likely, but still just possible. It's usually considered to be "very highly significant".

Example

An epidemiologist finds that in one town, out of 50 babies, 35 are female.

She wants to know the probability that this difference from the usual 50:50 male:female ratio in the rest of the county happened by chance.

The null hypothesis is that in this area the chance of having a female baby *hasn't* been altered.

The P value gives the probability that the null hypothesis is true.

The P value in this example is 0.007. Other sections show how this is calculated, but at the moment just concentrate on what it means.

P = 0.007 means the result would only have happened by chance in 0.007 in 1 (or 1 in 140) times if the choice of the area didn't actually affect the sex of the babies. This is highly unlikely, i.e. "highly significant", so we can reject our hypothesis and conclude that there is a highly significant difference between the sex ratio in this town and the rest of the county.

36.3 Does statistical significance always equal relevance?

Try not to confuse statistical significance with relevance. If a sample is too small, the results are unlikely to be statistically significant even if there really is a difference between them. Conversely a large sample may find a statistically significant difference that is too small to have any relevance.

Test yourself

The answers are given on page 176.

Question 36.1
We wish to study the effect of two different temperatures on the germination rate of common wheat seeds, *Triticum aestivum*. Set up a null hypothesis to test this.

Question 36.2
The germination rate of a sample of wheat seeds at 10°C is found to be 92%. The germination of another sample of wheat seeds, kept at 14°C, is 96%; $P = 0.25$.
How significant is this difference?

37 Tests of significance

There is a large array of significance tests, and it is not always easy to know which should be used when.

The **decision-making flowchart** in Appendix 1 is designed to help you decide which test to use.

There are two main groups of statistical tests: **parametric** and **non-parametric**. The choice of which to use depends on the distribution of the data.

37.1 Parametric tests

Generally, parametric tests compare means and variances. They should be used only when the data follow a **normal distribution**, the bell-shaped curve shown in Section 28.1.

For large samples (above 50, say) the sample *means* will usually be normally distributed even if the samples themselves aren't, so it may be possible to use parametric tests.

Some skewed data can be **transformed** to normally distributed data, which can then be analysed using the more accurate parametric testing. For instance, a skewed distribution might become normally distributed if the logarithms of the values are used.

You may see reference to the **Kolmogorov Smirnov** test. This tests the hypothesis that data are from a normal distribution, and therefore assesses whether parametric statistics can be used.

Statisticians prefer to use parametric tests when possible because:

- for parametric data, parametric tests are more powerful than non-parametric tests, and
- there are far more parametric tests readily available.

However, sample populations that don't meet (or cannot be transformed to meet) parametric criteria need to be analysed with non-parametric tests.

Most non-parametric tests work by **ranking** the data and then comparing the ranks.

37.2 The commonly used parametric tests

***t* tests** are used to compare sample means. They are described in detail in Chapter 38.

> **Example**
>
> We have two sets of 3-month-old pigeons that have been fed on different grain. We can use a *t* test to compare their mean masses.

When looking for an association between two categorical variables we analyse the data using the **chi-squared** test (see Chapter 40).

> **Example**
>
> If we studied the possible effect of a pollutant on the severity of asthma, we could label asthma severity as mild, moderate or severe. The effect of the presence or absence of the pollutant on severity would be analysed with the chi-squared test.

Most tests of significance compare the variance between samples. We can test the hypothesis that samples come from the same population by looking at the variation *within* each sample, and comparing that with the variance (see Section 29.2) *between* the sample means. This is called **analysis of variance** (ANOVA), and is particularly useful when comparing multiple variables. It is discussed in Chapter 39.

> **Example**
>
> A study of the effect of five different fertilisers on a variety of crops needs analysis of variance.

Correlation studies the strength of a linear (straight-line) relationship between two variables. This is explained in Chapter 41.

> **Example**
>
> If we wish to know the strength of a link between the incidence of obesity and diabetes in different age groups, we use Pearson's correlation test.

Regression analysis, explained in Chapter 42, quantifies how one set of data relates to another when one of the variables is dependent on the other, independent, variable.

> **Example**
>
> The relationship of hours of daylight (an independent variable) to plant growth (the dependent variable) can be quantified with regression analysis.

37.3 Non-parametric tests

When data cannot satisfy (or be transformed to satisfy) the requirements for analysis using a parametric test, we need to use a non-parametric test.

In general, non-parametric tests compare medians (see Section 28.3).

Rather than comparing the values of the raw data, the tests put the data in **ranks** and compare the ranks.

The non-parametric equivalents to parametric tests are given in the following table.

Table of parametric tests and their non-parametric equivalents	
Parametric test	**Non-parametric equivalent**
Mean	Median or mode
Standard deviation	Quartiles and interquartile range
One-sample *t* test	Wilcoxon test, sign test
Paired *t* test	Wilcoxon test, sign test
Unpaired *t* test	Mann–Whitney U test
One-way ANOVA	Kruskal–Wallis test or ANOVA on ranked data
Repeated measures ANOVA	Friedman test or ANOVA on ranked data
Pearson's correlation test	Spearman's rank correlation coefficient

A detailed description of the non-parametric tests is outside the scope of this book.

37.4 Which to use when

We cannot make comparisons between samples until we have decided whether to use parametric or non-parametric tests. However, the decision on which to use is surprisingly controversial – different statisticians can give different advice.

A simplified decision-making chart is given here, but you may get other equally valid advice.

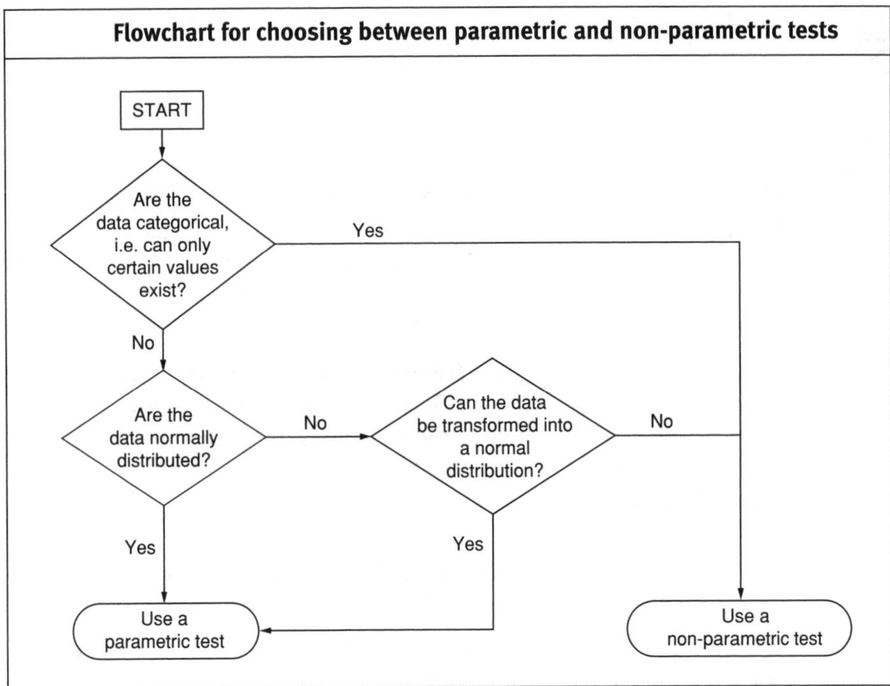
Flowchart for choosing between parametric and non-parametric tests
START
Are the data categorical, i.e. can only certain values exist?
Yes
No
Are the data normally distributed?
No
Can the data be transformed into a normal distribution?
No
Yes
Yes
Use a parametric test
Use a non-parametric test

38 *t* tests

Like other parametric tests, the ***t* test** (correctly termed the **Student's *t* test**) is used to compare samples of normally distributed data (see Section 28.1) with similar standard deviations. *t* tests are typically used to compare one or two samples. They test the probability that the samples come from a population with the same mean value.

For small samples, the *z*-score (see Section 30.1) does not provide a good estimate of the distribution of differences between groups. However, the *t* score has been developed to provide a better estimate that overcomes this.

38.1 *t* test tables

The table of *t* values in Appendix 2 gives us the significance level of the difference between two means for a given sample size and *t* score.

We reject the null hypothesis if the calculated value of *t* is larger than the value on the table for a chosen significance level.

Example

We wish to compare the masses of two samples of 10 eggs from different chicken species. The null hypothesis is that there is no difference between the masses.

A *t* test on the two samples gives a *t* score of 2.62, written as $t = 2.62$.

The two samples of 10 eggs means 18 degrees of freedom, df (see Chapter 31). Using the table in Appendix 2, we can see that for 18 df, at the 5% significance level, $t = 2.10$.

Our *t* score of 2.62 is larger than this value, so the significance level is less than 5%, i.e. $P < 0.05$.

We can be sure that the chance of getting results as extreme as these when there is actually no difference between the species is less than 5%, so can reject the null hypothesis.

38.2 One-tailed and two-tailed tests

In Section 34.1 we said that under a normal curve 95% of observations are within 1.96 standard deviations of the mean.

The remaining 5% are equally divided between the **tails** of the normal distribution, as shown below.

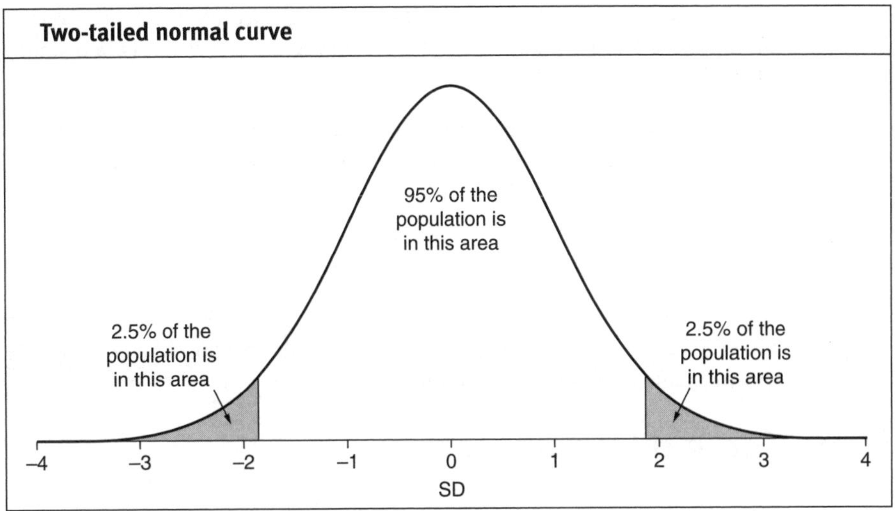

When trying to reject a null hypothesis (Section 32.1), we are generally interested in two possibilities: either we can reject it because the mean value of one sample is higher than that of the other sample, or because it is lower.

By allowing the null hypothesis to be rejected from either direction we are performing a **two-tailed test** – we are rejecting it when the result is in either tail of the test distribution.

However, if we know that a measurement in a population is larger (or smaller) than another, then we know that the residual 5% will be in the upper (or lower) tail of the normal distribution, as shown below:

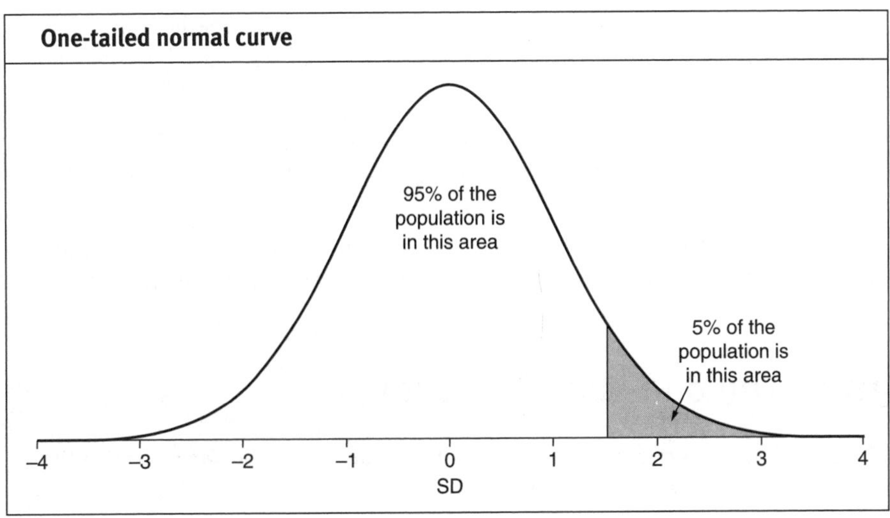

In this case, the critical values for *t* tests will be lower, and we use a **one-tailed test**, rejecting the null hypothesis only when the result is in a single tail of the test distribution.

Statistical software can calculate for both one- and two-tailed tests, and tables like the table of critical values for *t* distribution in Appendix 2 give significance levels for both.

In practice, usually there is the possibility of both improvement and deterioration, and therefore rarely the need to use one-tailed tests.

Also, a P value that is not quite significant on a two-tailed test may become significant if a one-tailed test is used. Researchers have been known to use this to their advantage!

38.3 Three different *t* tests

There are three different *t* tests.

When we have a single sample and wish to compare its mean with a fixed value, for example the population mean, we use the **one-sample *t* test**.

Where two observations are made on the same sample subjects, we need the **paired *t* test**. (Confusingly, it is also known as the related *t* test and paired-samples *t* test.)

Where we have two samples and have measured the same variable on each of them, the **unpaired *t* test** (also known as the two-sample *t* test and independent-samples *t* test) is used.

Examples

We are interested in the mean mass of day-old chicken eggs at a farm.

We have a sample of 10 eggs and wish to know whether their mean mass is significantly different to a national standard. For this, we need the one-sample *t* test.

If we weigh the *same* sample of eggs a day later and wish to know whether their mass has changed significantly, we use the paired *t* test.

If, however, we weigh a *different* sample of day-old eggs and want to know whether their mass is significantly different, we need the unpaired *t* test.

38.4 The one-sample *t* test

The one-sample *t* test compares the mean of a single sample with a fixed value, for example the population mean.

The *t* value is the number of standard errors that the sample mean is away from the population mean.

The formula is:

$$t = \frac{\bar{x} - E}{\text{SEM}}$$

where $\bar{x}$ is the sample mean, *E* is the fixed value, and SEM is the standard error of the mean for the sample (how to calculate these values was explained in Chapters 28 and 33).

Example

Using the example above, let's say that the mean mass of our sample of 10 eggs is 60 g, and the SEM of the mean is 1.6 g.

A national standard mass for day-old eggs is 55 g.

$$t = \frac{\bar{x} - E}{\text{SEM}} = \frac{60 - 55}{1.6} = 3.125$$

In the table of *t* values in Appendix 2, for 9 df a value of $t = 3.125$ is more than the 2.26 needed for the 5% significance level.

Thus the difference between the observed mean and the national standard is unlikely to have happened by chance.

38.5 The paired *t* test

The paired *t* test compares the means of a variable in the same sample under different conditions, or at two different times.

It is calculated by dividing the mean of the differences between each pair by the standard error of the difference.

Its formula is:

$$t = \frac{\bar{d}}{\text{SE}_d}$$

where $\bar{d}$ is the mean difference between each pair and $\overline{\text{SE}_d}$ is the standard error of the differences.

The paired samples *t* test is equivalent to a one-sample *t* test on the difference between the two samples and giving *E* a value of zero.

Example

With our sample of 10 eggs, on day 1 their mean mass was 60 g; the next day it was 58 g. We wish to know whether there is a significant difference between these two masses.

The standard error of their differences is calculated to be 1.05 g. The mean of the differences is 60 – 58 = 2 g.

$$t = \frac{\bar{d}}{SE_d} = \frac{2}{1.05} = 1.905$$

Using the table of *t* values in Appendix 2, for 9 df the critical value of *t* for 5% is 2.26. Our calculated value of 1.905 is therefore less than needed for the 5% significance level, so the difference may well have happened by chance.

38.6 The unpaired *t* test

The unpaired *t* test compares the means of the same variable in two different samples.

It is calculated by dividing the difference between the means of the two samples by the standard error of the differences.

The formula is:

$$t = \frac{\bar{x}_a - \bar{x}_b}{\overline{SE_d}}$$

where $\bar{x}_a$ and are $\bar{x}_b$ the means of the two samples, and $\overline{SE}_d$ and is the standard error of the differences.

Example

Our first sample of day-old eggs has a mean mass of 60 g. A different sample of day-old eggs has a mean mass of 51 g. We want to know if there is a significant difference in their masses.

The standard error of their differences is calculated to be 2.24 g.

$$t = \frac{\bar{x}_a - \bar{x}_b}{\overline{SE_d}} = \frac{60 - 51}{2.24} = 4.018$$

From the table of *t* values in Appendix 2, we can see that for 18 df a value of $t = 4.018$ is more than the 3.92 needed for the 0.1% significance level, so the difference is very unlikely to have happened by chance.

Test yourself

The answers are given on page 176.

Question 38.1

The germination rate for foxglove seeds, *Digitalis* spp., produced by a large nursery is 70%.

Twelve samples of seeds have been stored for 1 year. The mean germination rate of the samples is 62%, SEM 8%.

What is the *t* value? Use the table on *t* values in Appendix 2 to work out how likely it is that the difference is due to chance.

Question 38.2

The mean core temperature of a group of 20 volunteers is 36.80°C when the ambient temperature is 30.0°C.

Their mean core temperature is 36.90°C after exposure to an ambient temperature of 40.0°C.

The standard error of their differences is calculated to be 0.04°C.

What is the *t* value? Use the table in Appendix 2 to work out how likely it is that the difference is due to chance.

Question 38.3

The mean height of a sample of 30 common sunflower plants, *Helianthus annuus*, is 1.5 m.

Another sample has been grown in soil of lower humidity, and has a mean height of 1.2 m.

The standard error of their difference is 0.1 m.

What is the *t* value? Use the table on *t* values in Appendix 2 to work out how likely it is that the difference is due to chance.

39 Analysis of variance

Analysis of variance, also known by its acronym ANOVA, is used to compare multiple samples.

It is a powerful statistical technique that can be used to compare, for example, multiple samples, from multiple groups, under the influence of multiple factors.

39.1 ANOVA or *t* test?

Where we want to compare two means we use the *t* test.

Comparing three or more means could be done by multiple use of the *t* test. Comparing the sample means $\overline{A}$, $\overline{B}$ and $\overline{C}$, for example, could be done by using the *t* test three times: comparing $\overline{A}$ with $\overline{B}$, $\overline{A}$ with $\overline{C}$, and $\overline{B}$ with $\overline{C}$.

The larger the number of samples, the more *t* tests we would need to do.

ANOVA has the advantage that a single test covers all the comparisons, and it should therefore be used when comparing more than two samples.

ANOVA also confers the advantage of being able to consider the effect of multiple factors on a variable of interest.

Example

If we wish to test the null hypothesis that there is no difference in the yield of seven varieties of tomato, we would need to do 21 *t* tests.

A single ANOVA analysis would take the place of the 21 *t* tests.

We could also use ANOVA to consider the effect of both variety and field position on the yield of the crop.

39.2 Problems with multiple testing

Further to the above, when a *t* test gives a P value of 0.05, there is still a 5% possibility that we should not have rejected the null hypothesis and therefore a 5% chance that we have come to the wrong conclusion.

If we do lots of independent *t* tests, then this chance of making a mistake will be present each time we do a test. Therefore, the more tests we do, the greater the chances of drawing the wrong conclusion.

Because ANOVA is a single test, the problems of multiple testing don't apply.

Example

Continuing the example above, if we do 21 t tests, even if there is no difference between the varieties, each of the t tests would have a 5% chance of showing a "significant" difference. Taking the 21 tests as a whole, there is a 66% possibility that, by chance, one of them would indicate a "significant" difference.

39.3 How ANOVA works

Calculating ANOVA is, however, a complex process, so we shall simply give an outline of how it works.

ANOVA compares the variability *between* samples with the variability *within* samples.

Example

This plot shows the data from two samples.

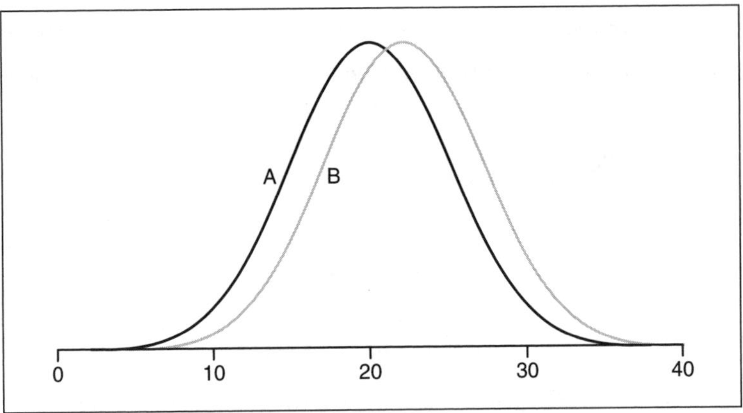

Samples A and B have different means, but a wide variability (scatter) within them. They could have come from the same population.

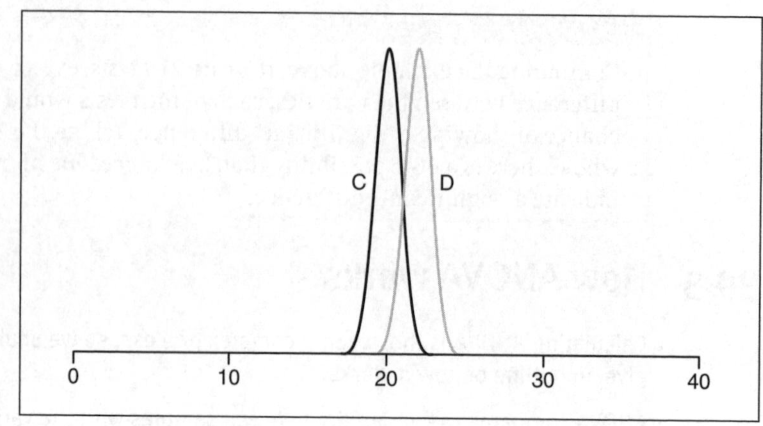

However, while samples C and D have the same means as in the previous plot, they each have a low variability, so probably come from different populations.

One-way ANOVA is used when we wish to compare means from more than two samples. It may therefore be considered as an extension of the *t* test.

Repeated measures ANOVA is used when there are repeated measurements on the same sampling unit.

39.4 The *F* value

The ANOVA test statistic, *F*, is derived from the ratio of the mean between-sample variations and the mean within-sample variations.

$$F = \frac{\text{between-sample variance}}{\text{within-sample variance}}$$

Calculating the *F* value is complex and best left to statistical software. The software will also calculate the P value for you.

Example

A researcher is interested in assessing whether three different samples of patients on different lipid-lowering medication have significantly different fasting cholesterol levels. Analysing the results using statistical software gives the following results.

Table of ANOVA results

	Sum of squares	df	Mean square	*F*	Sig.
Between groups	2604.205	2	1302.102	1.761	0.192
Within groups	19 228.002	26	739.539		
Total	21 832.207	28			

> The higher the F value, the more likely it is that the difference between the samples is statistically different. In this example we can see that we have an F value of 1.761 on two degrees of freedom. From the table we can see that the significance level (Sig.) is 0.192, and therefore there is a 19.2% possibility that the observed differences between the samples have arisen by chance.

39.5 Finding out which samples are different

ANOVA tells us whether there is a significant difference between samples, but it doesn't tell us *which* of those samples have that difference.

For this we need further tests, known as **post hoc tests**. Again, post hoc tests such as the **Bonferroni correction** and **Dunnett's, Scheffe** and **Tukey** tests are available on statistical software.

40 The chi-squared test

The **frequency** of an event is the number of times that it occurs.

Chi-squared is a measure of the difference between actual and expected frequencies.

Usually written as χ^2, chi is pronounced as in "sky" without the s.

40.1 Expected frequency

The **expected frequency** is the frequency if there is *no* difference between sets of results (the null hypothesis).

We can use a **contingency table** to compare expected and actual frequencies.

Example

In one area of forest, 15 out of 31 chimpanzees (Sample A) are found to be male. In another area, 36 out of 60 are male (Sample B). We wish to know whether this is a statistically significant difference.

Contingency table for sex of chimpanzee offspring

	Sample A	Sample B	Totals
Males	15	36	51
Females	16	24	40
Totals	31	60	91

40.2 Calculating chi-squared

This is given by:

$$\chi^2 = \Sigma \frac{(O-E)^2}{E}$$

where $(O - E)$ is the difference between observed and expected frequencies, and E is the expected frequency; the Σ symbol means "sum of".

Example

In the chimpanzee example, the expected frequency, E, of males in sample A is given by:

$$E = \frac{\text{Total number of males} \times \text{total number in sample A}}{\text{Total number of chimpanzees}}$$

So, $E = 51 \times 31/91 = 17.374$ for males in Sample A; $E = 40 \times 31/91 = 13.626$ for females in Sample A; $E = 51 \times 60/91 = 33.626$ for males in Sample B; $E = 40 \times 60/91 = 26.374$ for females in Sample B.

To calculate $\sum \frac{(O-E)^2}{E}$ we need to:

- subtract the expected from the observed frequencies,
- take the square of each of these differences,
- divide each squared difference by the expected values,
- take the sum of the results.

We have shown the calculations in the table below:

Table calculating χ^2 for chimpanzee example

	Observed frequency, O	Expected frequency, E	$O - E$	$(O - E)^2$	$\frac{(O-E)^2}{E}$
Males in Sample A	15	17.374	−2.374	5.636	0.3244
Females in Sample A	16	13.626	2.374	5.636	0.4136
Males in Sample B	36	33.626	2.374	5.636	0.1676
Females in Sample B	24	26.374	−2.374	5.636	0.2137
				$\sum \frac{(O-E)^2}{E} =$	1.1193

So $\chi^2 = 1.12$, to three significant places.

40.3 Calculating the significance level

Once we know the χ^2 value, we can work out the significance level.

The significance level for χ^2 depends on the number of degrees of freedom, df, Explained in Chapter 31.

The degrees of freedom for this test is the number of rows in the table less 1, multiplied by the number of columns in the table less 1, i.e.

$$\text{df} = (\text{rows} - 1)(\text{columns} - 1)$$

Critical values for χ^2 (usually written as X^2) are given in Appendix 3.

Example

For the chimpanzee example above, there are two rows (males and females) and two columns (samples A and B), so:

$$df = (2 - 1)(2 - 1) = 1$$

The table in Appendix 3 shows that the critical value for the 5% significance level is 3.84 for 1 df.

But, in our example, χ^2 is only 1.12, so the result is not significant and the null hypothesis stands.

40.4 Other tests for contingency tables

Instead of the χ^2 test, **Fisher's exact test** can be used to analyse contingency tables. Fisher's test is the best choice for 2 by 2 tables, with expected frequencies of fewer than 5.

The χ^2 test is simpler to calculate but gives only an approximate P value and is inappropriate for small samples. We can apply **Yates' continuity correction**, or other adjustments, to the χ^2 test to improve the accuracy of the P value.

The **Mantel–Haenszel** test is an extension of the χ^2 test that is used to compare several two-way tables.

Test yourself

The answers are given on page 176.

Question 40.1

A bacteriologist inoculates 480 Petri dishes with *E. coli* bacteria; 240 of the Petri dishes contained a standard culture medium, 240 had a new type of culture medium. After 3 days she checks whether or not there are any bacterial colonies. The results are shown below.

Table of effects of type of culture medium on *E. coli* growth

	Type of culture medium		
	Standard	New	Total
Bacterial colony present at 3 days	144	160	304
No bacterial colony at 3 days	96	80	176
Total	240	240	480

Calculate the χ^2 value. You will find a calculator helpful.

Question 40.2

For your calculated χ^2 value for the *E. coli* example, use the table in Appendix 3 to work out whether it is significant.

41 Correlation

Where there is a linear relationship between two variables there is said to be a **correlation** between them.

41.1 Positive or negative?

A **positive** correlation coefficient means that as one variable is increasing, the value for the other variable is also increasing – the line on the graph slopes up from left to right.

Example

Day length and plant growth have a positive correlation: plants grow faster as days get longer.

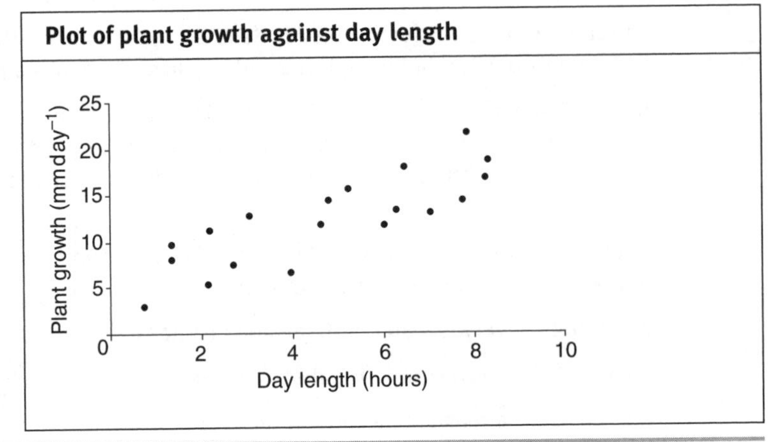

A **negative** correlation coefficient means that as the value of one variable goes up, the value for the other variable goes down – the graph slopes down from left to right.

Example

Higher pollution levels are associated with reduced crop yields, giving a negative correlation between the two variables.

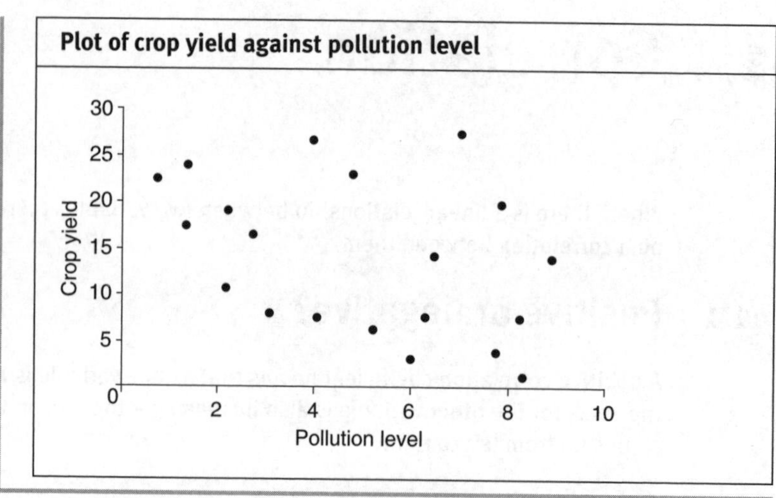

41.2 The correlation coefficient

The strength of a correlation is measured by the **correlation coefficient.**

The correlation coefficient is usually denoted by the Greek letter "ρ" (rho), for example $\rho = 0.8$, but is replaced with "*r*" in some books and journals.

If there is a perfect relationship between the two variables then $\rho = 1$ (a positive correlation) or $\rho = -1$ (a negative correlation). If there is no correlation at all (the points on the graph are randomly scattered) then $\rho = 0$.

Interpreting the size of the correlation coefficient is subjective but the following may be a useful rule of thumb:

$\rho = 0–0.2$	very low and probably meaningless
$\rho = 0.2–0.4$	a low correlation that might warrant further investigation
$\rho = 0.4–0.6$	a reasonable correlation
$\rho = 0.6–0.8$	a high correlation
$\rho = 0.8–1.0$	a very high correlation. Possibly too high! Check for errors or other reasons for such a high correlation

This guide also applies to negative correlations.

Example

With the day-length and plant-growth data above, $\rho = 0.8$, indicating a high correlation.

However, the correlation between crop growth and pollution levels is lower: $\rho = -0.39$.

41.3 Calculating the correlation coefficient

We need to calculate the mean for both sets of data:

$\bar{x}$ and $\bar{y}$

For each value of x and y we then calculate:

$x - \bar{x}$ and $y - \bar{y}$

These are used for the correlation coefficient formula:

$$\rho = \frac{\Sigma (x-\bar{x})(y-\bar{y})}{\sqrt{\Sigma (x-\bar{x})^2 \, \Sigma (y-\bar{y})^2}}$$

Example

A medical biochemist is interested in how closely blood glucose levels in humans relate to HbA1c (a measure of how much glucose is attached to haemoglobin molecules). The values, and their means, in a sample of eight people with diabetes mellitus are given in the table below.

Table of blood glucose and HbA1c in eight patients with diabetes mellitus

Patient	Blood glucose ($mmol l^{-1}$)	HbA1c (%)
A	5.1	5.8
B	4.6	6.9
C	6.3	8.3
D	8.3	6.1
E	9.7	7.8
F	12.0	8.4
G	12.7	10.8
H	14.1	9.1
Mean	9.1 ($\bar{y}$)	7.9 ($\bar{x}$)

Comparing the pairs of measurements gives the following graph:

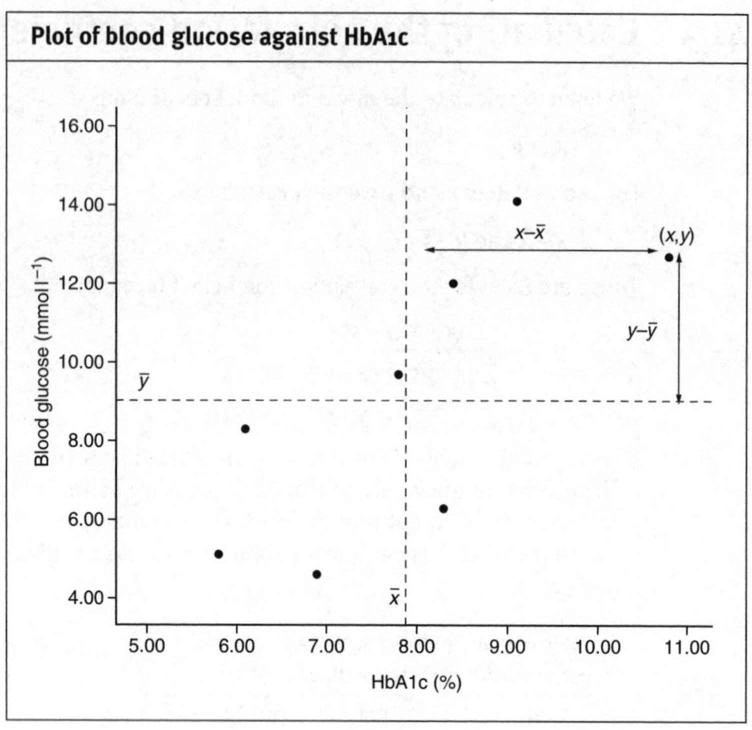

Also shown on the graph are dotted vertical and horizontal lines to indicate the mean values, $\bar{x}$ and $\bar{y}$, and the arrows show $x - \bar{x}$ and $y - \bar{y}$ for one of the patients.

Taking the vaues of x and y from the previous table, we can create a new data table to help solve the equation:

Table showing calculated values for $x - \bar{x}$, $y - \bar{y}$, $(x - \bar{x})^2$, and $(y - \bar{y})^2$

Patient	$x - \bar{x}$	$y - \bar{y}$	$(x - \bar{x})^2$	$(y - \bar{y})^2$
A	−2.1	−4.0	4.41	16.00
B	−1.0	−4.5	1.00	20.25
C	0.4	−2.8	0.16	7.84
D	−1.8	−0.8	3.24	0.64
E	−0.1	0.6	0.01	0.36
F	0.5	2.9	0.25	8.41
G	2.9	3.6	8.41	12.96
H	1.2	5.0	1.44	25.00

Substituting these values into the correlation coefficient formula gives the following result:

$$\rho = \frac{\Sigma\,(x-\bar{x})(y-\bar{y})}{\sqrt{\Sigma(x-\bar{x})^2\Sigma(y-\bar{y})^2}} = \frac{31.05}{\sqrt{91.46 \times 18.92}} = 0.7464$$

Thus $\rho = 0.75$, suggesting a high correlation between blood glucose levels and HbA1c.

41.4 Limitations of correlation

Correlation tells us about the strength of the association between the variables, but doesn't tell us about cause and effect in that relationship.

Take care when interpreting the significance of correlations. If a correlation is significant, we also need to consider the size of the correlation. If a study is sufficiently large, even a small correlation will have a high level of significance.

Also, bear in mind that a correlation only tells us about linear (straight-line) relationships between variables. Two variables may be strongly related but not have a straight line relationship, giving a low correlation coefficient.

Test yourself

The answer is given on page 177.

Question 41.1
A researcher measures the mean germination rates of delphinium seeds, *Delphinium cardinale*, planted at different times after harvesting.
Calculate the correlation coefficient, ρ.

Table of germination rates of *D. cardinale* seeds at different times after harvesting

	Germination rate (%)	Time after harvesting seeds (months)
	60	0
	53	4
	52	8
	38	12
	32	16
Mean	47	8

42 Regression

Regression analysis is used to quantify how one set of data relates to another. It is used when one of the variables is dependent on the other, independent, variable.

42.1 Linear regression

Linear regression is used where there is a linear (straight-line) relationship between the variables.

Example

In humans, HbA1c is a measure of how much glucose is attached to haemoglobin molecules. It is dependent on the blood glucose level.

The graph in Section 41.3 shows that the relationship is linear, so we can use linear regression to analyse it.

42.2 The line of best fit

In Section 21.4, we learnt how the **line of best fit** drawn on a scatter plot is the straight line that best shows the trend of the plotted points.

From that we can calculate the gradient. If we continue the line of best fit until it crosses the y-axis, we can also estimate the constant c for the formula of a straight-line graph, $y = mx + c$, explained in Chapter 8.

Regression analysis is the process of calculating this formula mathematically.

42.3 The regression line

A **regression line** is the line of best fit through the data points on a graph.

The **regression coefficient** gives the *gradient* of the graph, in that it gives the change in value of one outcome, per unit change in the other.

The **regression constant** gives the *position* of the line on the graph – it is the point where the line crosses the vertical axis.

So, the formula for the regression line is:

$$y = mx + c$$

where x is the independent variable, y is the dependent variable, m is the regression coefficient, and c is the regression constant.

42.4 Calculating the line of best fit

To fit a regression line to a scatter plot, we need the minimum vertical distance (deviation, d) from each point to the line.

Regression line fitted from scatter plot

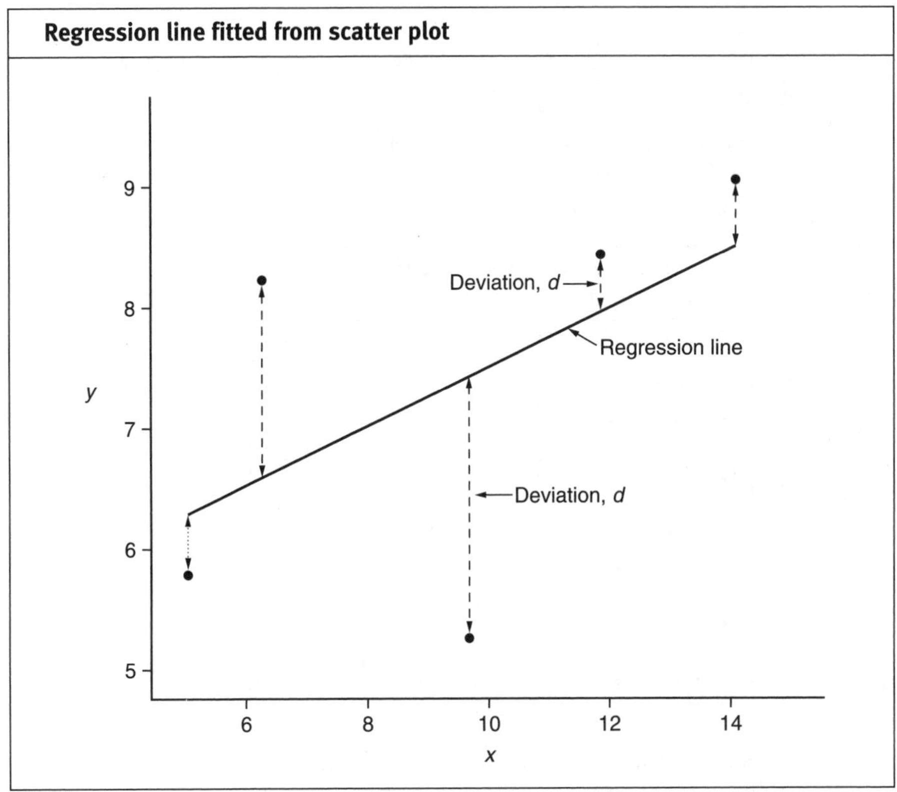

In this graph, the sum of the deviations above the line equals the sum of the deviations below the line.

For a straight-line graph, termed **linear regression**, this is usually calculated by the **method of least squares**, a calculation designed to give the smallest possible sum of the *squares* of the deviations.

Calculating the regression coefficient uses the formula:

$$m = \frac{\Sigma(x - \bar{x})(y - \bar{y})}{\Sigma(x - \bar{x})^2}$$

The regression line always goes through the mean values of x and y, $\bar{x}$ and $\bar{y}$. We can therefore substitute $\bar{x}$, $\bar{y}$ and the regression coefficient, m, into the regression line formula

$$\bar{y} = m\bar{x} + c$$

and calculate the regression constant, c.

Once we know the regression constant and coefficient, we can calculate the value of y for any given value of x.

Example

We want to quantify how altitude affects the diversity of tree and shrub species.

We find eight quadrats of woodland, of similar ages, that have had similar forestry management. The numbers of tree and shrub species at different altitudes are given below.

Table of number of tree and shrub species at different altitudes

Quadrat	Altitude (m)	Number of tree and shrub species
A	40	58
B	90	55
C	150	33
D	160	46
E	250	31
F	360	29
G	420	16
H	610	4
Mean	260	34

Substituting these values into the formula for the regression coefficient gives the following result:

$$m = \frac{\Sigma(x-\bar{x})(y-\bar{y})}{\Sigma(x-\bar{x})^2} = -\frac{23\,790}{257\,600} = -0.0924$$

Substituting into the regression line formula $\bar{y} = m\bar{x} + c$ gives:

$$34 = (-0.0924 \times 260) + c$$

so, $c = 58.0$

The graph for this regression is shown here:

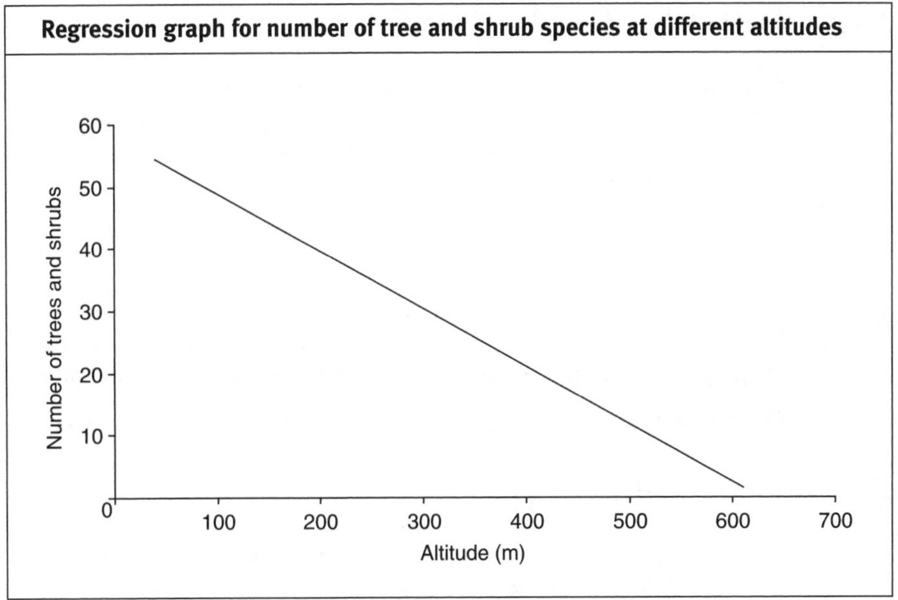

To predict the number of trees and shrubs for a given altitude we can now plug it into its regression formula $y = -0.0924x + 58$.

So, if we want to know the likely number of species at 300 m,

$$\text{Number of species} = y = (-0.0924 \times 300) + 58 \approx 30$$

42.5 Other values used with regression

We may wish to calculate the **standard error** of an estimate for a regression coefficient and constant. This indicates the accuracy that can be given to the calculations.

If there is considerable scatter, we may also wish to calculate the **significance level** of the regression – the probability that the calculated gradient is significantly different to zero.

Another useful value is the **R^2 value**. This shows how much a change in the dependent variable depends on a change in the independent variable.

Example

In the example above:

- the standard error of the estimate is 6.10;
- it is significant to $P < 0.001$ and it is therefore highly likely that the gradient is different to zero;
- the R^2 value is 0.91, so 91% of the variation in number of species is accounted for by variation in altitude.

42.6 Other types of regression

So far we have discussed linear regression, where the line that best fits the points is straight. Many biological relationships give curved graphs. These can often be "transformed" to a straight-line relationship, for instance by taking the logarithm of a set of data.

Other forms of regression include logistic regression and Poisson regression.

Logistic regression is used where each case in the sample can only belong to one of two groups (e.g. having disease or not) with the outcome as the probability that a case belongs to one group rather than the other.

Poisson regression is mainly used to study time between rare events.

42.7 Watch out for . . .

Regression should not be used to make predictions outside the range of the original data. In the example above, we can only make predictions from altitudes which are between 40 and 610 m.

42.8 Regression or correlation?

Regression and correlation are easily confused.

Correlation measures the *strength* of an association between variables.

Regression *quantifies* an association. It should only be used if one of the variables is thought to precede or cause the other.

Test yourself

The answer is given on page 177.

Question 42.1
Large warm-blooded animals have lower resting heart rates than small ones.
Turn the data into a linear regression by taking the log of each value, then calculate the regression coefficient and constant.

Use the results to calculate the expected resting heart rate of a warm-blooded animal of mass 15 kg.

Table comparing body mass and resting heart rate for different species		
	Mass (kg)	Resting heart rate (beats min^{-1})
Mouse	0.02	700
Rat	0.2	400
Cat	5	150
Dog	10	120
Man	70	70
Horse	450	40

43 Bayesian statistics

Bayesian analysis is a totally different statistical approach to the classical, "frequentist" statistics explained in this book.

It is being used increasingly often, for example in structural biology.

43.1 Prior and posterior distributions

In Bayesian statistics, rather than considering the sample of data on its own, a **prior distribution** is set up using information that is already available. For instance, a researcher may give a numerical value and weighting to previous opinion and experience as well as previous research findings.

One consideration is that different researchers may put different weighting on the same previous findings.

The new sample data are then used to adjust this prior information to form a **posterior distribution**. Thus these resulting figures have taken *both* the disparate old data *and* the new data into account.

Answers to "test yourself" questions

Answer 2.1
The factors of 18 are 1, 2, 3, 6, 9 and 18.
The factors of 21 are 1, 3, 7 and 21.
The factors of 24 are 1, 2, 3, 4, 6, 8, 12 and 24.
The highest common factor is 3.

Answer 2.2
$7(4+3)(5-2) = 7 \times 7 \times 3 = 147$

Answer 2.3
Calculating the sum in the brackets, we need to do the division before the addition, giving:
$16(4) - 10 \div 5$
Then the division and multiplication give:
$64 - 2$
So the sum works out to:
62

Answer 2.4
21 has 4 factors, 1, 3, 7 and 21, so it is *not* a prime number.
22 has 4 factors, 1, 2, 11 and 22, so it is *not* a prime number.
23 has only 2 factors, 1 and 23, so it *is* a prime number.

Answer 2.5
7 squared $= 7^2 = 7 \times 7 = 49\,\text{m}^2$

Answer 2.6
$8 \times 8 = 64$, so the square root of $64 = \sqrt{64} = 8$
The quadrat is 8 by 8 m wide.

Answer 2.7
40 by 40 by 40 mm cubed $= 40^3 = 40 \times 40 \times 40 = 64\,000\,\text{mm}^3$

Answer 2.8
$4 \times 4 \times 4 = 64$, so the cube root of $64 = \sqrt[3]{64} = 4$
The sample is 4 by 4 by 4 mm.

Answer 3.1
$\frac{5}{6} \times 72 = 60$

Answer 3.2
20 and 24 have a common factor of 4, so
$\frac{20}{24} = \frac{20 \div 4}{20 \div 4} = \frac{5}{6}$

Answer 3.3
The reciprocal of $\frac{24}{28}$ is $\frac{28}{24}$, or $1\frac{4}{24}$. This can be simplified to $1\frac{1}{6}$.

Answer 3.4
$\frac{2}{5} \times \frac{9}{10} = \frac{2 \times 9}{5 \times 10} = \frac{18}{50}$

This can be simplified to $\frac{9}{25}$.

Answer 3.5
$\frac{2}{5} \div \frac{9}{10} = \frac{2}{5} \times \frac{10}{9} = \frac{2 \times 10}{5 \times 9} = \frac{20}{45} = \frac{4}{9}$

Answer 3.6
Here, the lowest common denominator is 14.
$\frac{6 \times 2}{7 \times 2} + \frac{9}{14} = \frac{12}{14} + \frac{9}{14} = \frac{21}{14} = 1\frac{7}{14} = 1\frac{1}{2}$

Answer 3.7
Convert $1\frac{3}{8}$ completely to a fraction: $\frac{11}{8}$
The lowest common denominator of 12 and 8 is 24.
$\frac{11 \times 3}{8 \times 3} - \frac{7 \times 2}{12 \times 2} = \frac{33}{24} - \frac{14}{24} = \frac{33-14}{24} = \frac{19}{24}$

Answer 3.8
$\frac{5}{8} = 0.625$, so $1\frac{5}{8} = 1.625$

Answer 4.1
40% of $375 = \left(\frac{40}{100}\right) \times 375 = 150$
So, 150 g of the original sample was water.

Answer 4.2
The peak flow rate has increased by 160 litres per minute.
$\frac{160}{400} \times 100 = 40\%$
So, it has increased *by* 40%. However, note that it has increased *to* 140% of its original (the original rate of 100%, plus the growth of 40%).

Answer 4.3
A reduction of 18% is the same as 82% of the original (100% – 18% = 82%).
82% of 900 g is $0.82 \times 900 = 738\,\text{g}$

Answer 4.4
The cell concentration has increased by 888 million cells per ml, which is

$\frac{888 \text{ million}}{24 \text{ million}}$ of the original concentration.

Percentage increase $= \frac{888}{24} \times 100 = 3700\%$.

Answer 5.1
Moving the decimal point four places to the right gives 10 to the power of –4, so the measurement is 4.5×10^{-4} m.

Answer 5.2
Moving the decimal point nine places to the right gives 3 000 000 000 base pairs.

Answer 5.3
$150 = 1.5 \times 10^2$ people per km^2
40 by 40 km $= 1600\,\text{km}^2 = 1.6 \times 10^3\,\text{km}^2$

$(1.5 \times 10^2)(1.6 \times 10^3) = (1.5 \times 1.6)(10^{2+3}) =$
2.4×10^5 people
So we would expect a population of 2.4×10^5 people in the 40 by 40 km area.

Answer 6.1
1.050 m is stated to four significant figures.

Answer 6.2
$58.44 \div 0.137 = 426.569\,\text{g}\,\text{m}^{-3}$.
The least precise value is the volume of water, which is given to three significant figures. So, the concentration may only be given to three significant figures:
$427\,\text{g}\,\text{m}^{-3}$

Answer 6.3
The error is ± 0.5 g, so the mass could be anywhere between 55.5 and just below 56.5 g.

Answer 7.1

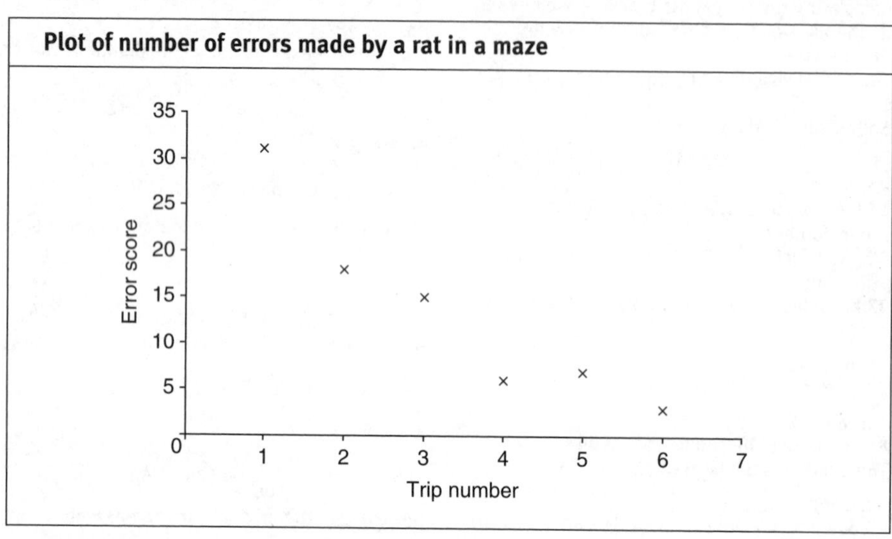

Answer 8.1
You could have used any portion of the graph. For example, when the *x* value changes from 10 to 50, the *y* value changes from 5000 to 25 000.

Gradient $= \frac{y_2 - y_1}{x_2 - x_1} = \frac{25\,000 - 5000}{50 - 10} =$

$500\,\text{m}\,\text{min}^{-1}$
$500\,\text{m}\,\text{min}^{-1} = 500 \times 60\,\text{m}\,\text{hour}^{-1} =$
$30\,000\,\text{m}\,\text{hour}^{-1} = 30\,\text{km}\,\text{hour}^{-1}$.

Answer 8.2
The formula for a straight line graph is:
$y = mx + c$
If we substitute in:
for y L = length of baby in millimetres
for m rate of growth = 10 mm per week
for x A = age in weeks
for c length at birth = 500 mm
Then the equation for the girl's growth is:
$L = 10A + 500$

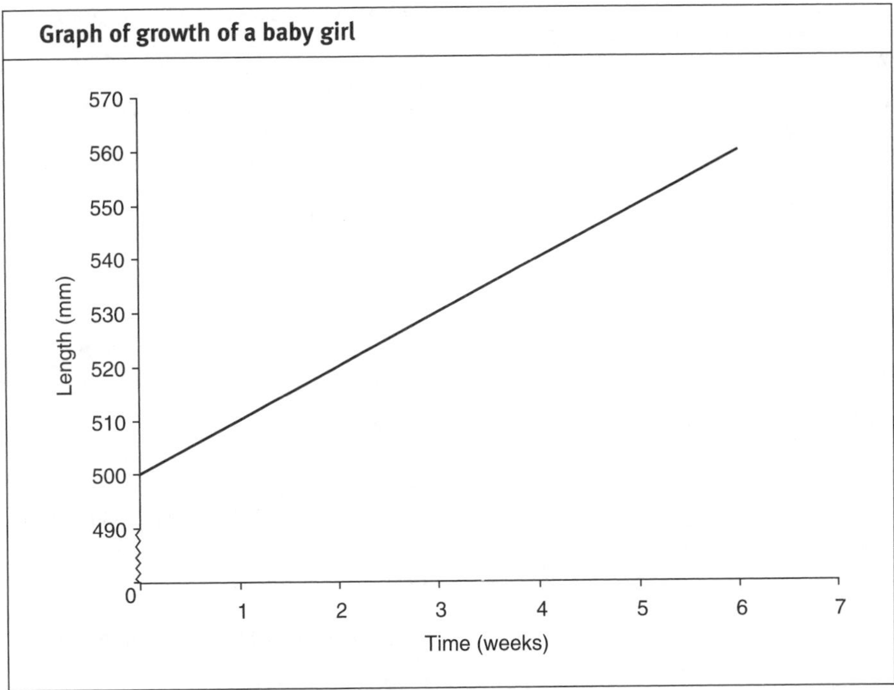

Answer 8.3

First calculate the gradient. From the end of the 1st to the end of the 3rd year it has grown from 200 to 400 mm, so

$$\text{Gradient} = \frac{y_2 - y_1}{x_2 - x_1} = \frac{400 - 200}{3 - 1} = 100$$

Thus the equation for the line is $y = 100x + c$.
Substituting in the height at year 1, we get
$200 = (100 \times 1) + c$
So, $c = 200 - (100 \times 1) = 100$
The sapling was 100 mm high when planted.

Answer 9.1

$27a^5 + 2a^3 + 5a^2 + 7a$
This can be further simplified to
$a(27a^4 + 2a^2 + 5a + 7)$

Answer 9.2

$$\frac{a^2b}{a^3} \times \frac{a^4b^2}{b^3} = \frac{a^{2+4}b^{1+2}}{a^3b^3} = \frac{a^6b^3}{a^3b^3} = a^{6-3}b^{3-3} =$$
$a^3b^0 = a^3$

Answer 9.3

Both top and bottom of the fraction can be divided by (i.e. have a common factor of) $a^2b^3\,(c + 2d)$

$$\frac{a^3b^3\,(c + 2d)^4}{a^2b^4\,(c + 2d)} = \frac{a(c + 2d)^3}{b}$$

Answer 9.4

1) No
2) Yes
$$\frac{c^4d^2 + b^2c^2d^2}{c^2 + b^2} = \frac{c^2d^2(c^2 + b^2)}{c^2 + b^2} = c^2d^2$$
3) No

Answer 9.5

$5a(2a - b^2) = (5a \times 2a) - (5a \times b^2) = 10a^2 - 5ab^2$

Answer 9.6

The highest common factor is $3a^2b^2$.
Pull this out and the remainder is $2a + 3b^2$.
The answer is therefore $3a^2b^2(2a + 3b^2)$.

Answer 9.7

$a^2 - 4b^2 = (a + 2b)(a - 2b)$

Answer 10.1

The highest power is 5, so it is a 5th degree polynomial.

Answer 10.2

$$\begin{array}{rrrrr} 2x^5 + & 7x^4 + & 5x^3 & & +4 \\ & -(6x^4) - & (9x^3) - & (-x^2) - & (5) \\ \hline = 2x^5 + & x^4 - & 4x^3 & +x^2 & -1 \end{array}$$

Answer 10.3

$$\begin{array}{rrrrrrr} +8x^9 & & +12x^6 & & +24x^4 & & \\ & -2x^7 & & & -3x^4 & -6x^2 & \\ & & & +10x^5 & & +15x^2 & +30 \\ \hline 8x^9 & -2x^7 & +12x^6 & +10x^5 & +21x^4 & +9x^2 & +30 \end{array}$$

Answer 10.4
$x^2 - 6x + 9$ can be factorised to $x^2 - 2(3x) + 3^2$.
When $a = 3$, this is directly equivalent to the 2nd polynomial in the table in Section 10.5:
$x^2 - 2xa + a^2$
which the table shows us factorises to
$(x - a)^2$
Thus,
$x^2 - 6x + 9$ factorises to $(x - 3)^2$.
You can check this by multiplying out
$(x - 3)(x - 3)$.

Answer 11.1

$$x^2 = \frac{12}{3} = 4$$

$$x = \sqrt{4} = 2$$

Answer 11.2
Subtracting 1 from each side gives:
$4y^3 = x - 1$
Divide both sides by 4:

$$y^3 = \frac{x-1}{4}$$

Taking the cube root of each side gives the solution:

$$y = \sqrt[3]{\frac{x-1}{4}}$$

Answer 12.1
When factorised, this becomes
$(x + 4)(x + 2) = 0$
To set the first factor to zero, x must be -4.
To set the second factor to zero, x must be -2.

Answer 12.2

$$x = \frac{-b \pm \sqrt{b^2 - 4ac}}{2a} = \frac{-6 \pm \sqrt{6^2 - (4 \times 1 \times 8)}}{2 \times 1}$$

$$= \frac{-6 \pm \sqrt{36 - (32)}}{2} = \frac{-6 \pm \sqrt{4}}{2} = \frac{-6 \pm 2}{2}$$

So the two solutions are:

$$\frac{-6+2}{2} = -2 \text{ and } \frac{-6-2}{2} = -4$$

Answer 12.3
Putting the x^2 and x terms on one side and the constant on the other we get:
$x^2 + 6x = -8$
The coefficient (multiplier) of x^2 is 1, and dividing both sides of an equation by 1 leaves it exactly the same.
We now take half the coefficient of x, i.e. 3 (half of 6), square it (giving 9), and add it to both sides.
$x^2 + 6x + 9 = -8 + 9 = 1$
Factorising the left side:
$(x + 3)^2 = 1$
Taking the square root of both sides gives:
$\sqrt{(x + 3)^2} = \sqrt{1} = \pm 1$
Thus $x + 3 = \pm 1$
This means that:
$x = +1 - 3 = -2$ and $x = -1 - 3 = -4$

Answer 13.1
If $2x + y = 8$, then
$y = 8 - 2x$
Substituting this value of y into $3x + 2y = 14$ gives:
$3x + 2(8 - 2x) = 3x - 4x + 16 = -x + 16 = 14$
So $x = 2$
Putting this value of x into $2x + y = 8$ gives:
$4 + y = 8$
Thus $y = 4$
Therefore, $x = 2$, $y = 4$.

Answer 13.2
$2x + y = 8$ (1)
and
$3x + 2y = 14$ (2)
We need to multiply both sides of equation (1) by 2:
$2(2x + y) = 2 \times 8$
Multiplying this out gives:
$4x + 2y = 16$
Both equations can now be manipulated to equal $2y$.
Equation (1) becomes:
$2y = 16 - 4x$
Equation (2) becomes:
$2y = 14 - 3x$
Now that both equations are equal,
$16 - 4x = 2y = 14 - 3x$
so, $16 - 4x = 14 - 3x$
This simplifies to: $x = 2$
Substituting into either of the original equations gives: $y = 4$
Therefore, $x = 2$, $y = 4$.

Answer 13.3
The equation that describes the population of the first village is:
$y = 50x + 1000$
where y = population size, and x = time in years.
The population of the second village can be described by:
$y = -50x + 1600$
The graph of these equations is shown on the following page:

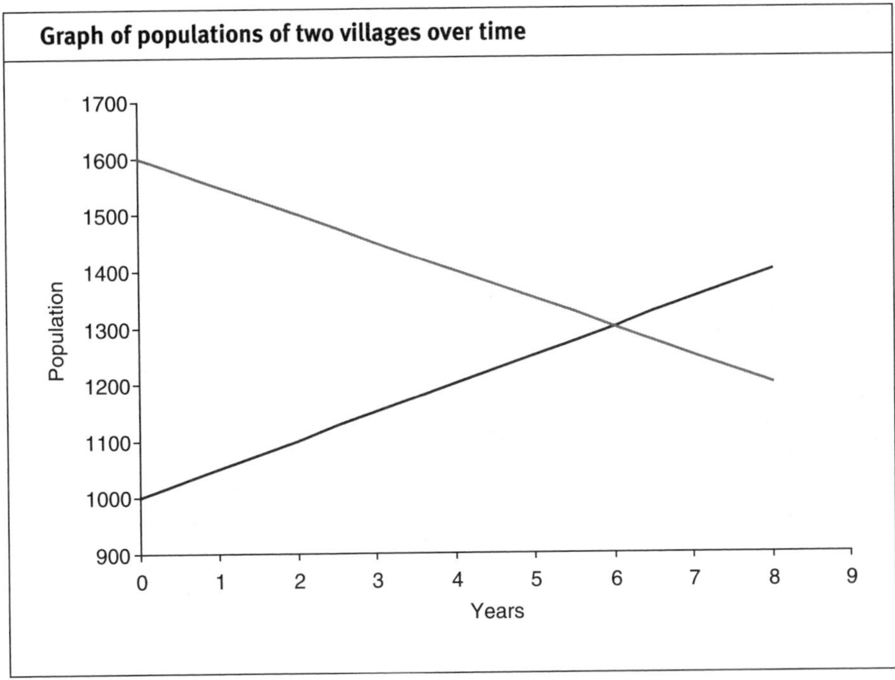

The intersection is at 6 years, when the population will be 1300.

To solve $y = 50x + 1000$ and $y = -50x + 1600$ by elimination, as both equations equal y, we can write:

$50x + 1000 = -50x + 1600$

This can be manipulated to:

$50x + 50x = 1600 - 1000$

So,

$100x = 600$

$x = 6$

So, the population size will be the same in 6 years.

Substituting this value of x into $y = 50x + 1000$ gives:

$y = (50 \times 6) + 1000 = 1300$

So in 6 years both villages will have 1300 inhabitants.

Answer 14.1

In the equation $a_n = a + (n - 1)d$,

$a = 12$ as the first count is 12; $d = 4$ as the common difference is 4; $n = 10$.

Thus we get

$12 + (10 - 1)4 = 12 + 36 = 48$

In the 10th year, 48 robins would be expected.

Answer 14.2

Again, $a = 12$, $d = 4$ and $n = 10$.

Putting these into the formula

$$S_n = \frac{n}{2}[2a + (n - 1)d]$$

gives:

$$S_{10} = \frac{10}{2}[(2 \times 12) + (10 - 1)4] = 5 \times (24 + 36) =$$

300

Answer 14.3

The first term is 250, so $a = 250$. The common ratio is 3, so $r = 3$, and $n = 7$.

Substituting these into the formula $a\,r^{n-1}$ gives

$250 \times 3^{7-1} = 250 \times 3^6 = 182\,250$

Answer 14.4

Once again, $a = 250$, $r = 3$, and $n = 7$.

Putting these into the formula

$$S_n = \frac{a(r^n - 1)}{r - 1} \text{ gives:}$$

$$S_n = \frac{250(3^7 - 1)}{3 - 1} = \frac{250 \times 2186}{2} =$$

273 250 plants

Answer 15.1

$$\frac{(a+2)^7}{(a+2)^5} = (a+2)^{7-5} = (a+2)^2 = a^2 + 4a + 4$$

Answer 15.2

$\sqrt[3]{(2a-1)^6} = (2a-1)^{6/3} = (2a-1)^2 = 4a^2 - 4a + 1$

Answer 16.1

1) $\log(1.2 \times 10^{-5}) = \log 0.000012 = -4.92$
$pH = -\log[H^+] = -(-4.92) = 4.92$
So the pH is 4.9 to two significant figures.
2) $pH = -\log[H^+] = 6.3$
$\log[H^+] = -6.3$
$[H^+] = 5.0 \times 10^{-7}$

Answer 16.2

$\ln e^4 = 4$

Answer 17.1

Initial number of patients $= N_0 = 5$; 3 weeks later, i.e. at $t = 3$ weeks, there are 25 patients, so $N_3 = 25$.
Plugging these into the growth-decay formula, $N = N_0\, e^{kt}$, we get:
$25 = 5e^{3k}$
This manipulates to:

$$\frac{25}{5} = 5 = e^{3k}$$

$\ln 5 = 3k$

$$\frac{\ln 5}{3} = k$$

$k = 0.54$ per week

Answer 17.2

Using our calculated infectivity growth constant of $k = 0.54$ per week, we can calculate how many patients would be affected in another 4 weeks, i.e. 7 weeks after initial diagnosis ($t = 7$ weeks). $N_0 = 5$ as before.
$N_7 = N_0\, e^{kt} = 5e^{0.54\times7} = 5e^{3.78} = 219$ patients

Answer 17.3

We can use any initial value for N_0, say $N_0 = 1$. After 8 days ($t = 8$ days), we have half the initial value, $N_8 = 0.5$.
Substituting these values into $N = N_0\, e^{kt}$, we get
$0.5 = 1e^{8k}$
Taking the natural logarithm of both sides gives:
$\ln 0.5 = 8k$

$$\frac{\ln 0.5}{8} = k$$

$k = -0.087$ per day

Answer 18.1

The volume of a sphere is $= \frac{4}{3}\pi r^3$.

The yolk diameter is 24 mm, so its radius is 12 mm.

$$V = \frac{4}{3}\, 3.1416(12^3) = 7238.2464\,\text{mm}^3$$

However, the yolk diameter was measured to two significant figures, so the volume should also be given to two significant figures:
Yolk volume $= 7200\,\text{mm}^3$

Answer 19.1

The rate of increase in heat loss at that point will be:

$$\frac{dy}{dx} = 2(50x) = 2 \times 50 \times 1.2 = 120\,\text{Wm}^{-1}$$

So, at 1.2 m in length, the heat loss is increasing at the rate of $120\,\text{Wm}^{-1}$ growth.

Answer 19.2

When $x = 30$, the rate of increase, or gradient, is

$$\frac{dy}{dx} = \frac{3x^2}{60} = \frac{x^2}{20} = \frac{30^2}{20} = 45$$

Thus, when the radius of the colony reaches 30 mm, the rate of increase is 45 wasps per mm increase in radius.

Answer 19.3

The differential of $y = 3x^{20} - 8$ is

$$\frac{dy}{dx} = 20(3x^{20-1}) - 0 = 60x^{19}$$

Answer 19.4

$y = \frac{3}{x^5}$ is the same as $y = 3x^{-5}$.

So,

$$\frac{dy}{dx} = 3(-5x^{-5-1}) = -15x^{-6}.$$

Answer 20.1

$$\int 7x^4\, dx = 7\frac{x^{4+1}}{4+1} + C = \frac{7x^5}{5} + C$$

The integral is thus $\frac{7x^5}{5} + C$

Answer 20.2

Subsituting $n = 3$ into the formula

$$\int x^n\, dx = \frac{x^{n+1}}{n+1} + C$$

we get the definite integral

$$A = \int_3^5 2x^3\, dx = \left[\frac{2x^4}{4} + C\right]_3^5$$

$$= \left(\frac{2 \times 5^4}{4} + C\right) - \left(\frac{2 \times 3^4}{4} + C\right)$$

$$= (312.5 + C) - (40.5 + C) = 272.$$

Answer 22.1

$5.5\,\text{pg} = 5.5 \times 10^{-12}\,\text{g}$

Answer 22.2

$(5 \times 10^3)(6 \times 10^3) = 30 \times 10^6 = 3 \times 10^7\,\text{m}^2.$

Answer 22.3
0°C = 273.15 K so
−40.5°C = −40.5 + 273.15 = 232.65K
The temperature in °C was given to one decimal place, so the temperature in K also needs to be given to one decimal place:
−40.5°C = 232.7 K.

Answer 23.1

$$\text{Number of moles} = \frac{\text{mass}}{\text{relative molecular mass}} =$$

$$\frac{450.45\,\text{g}}{180.18} = 2.5\ \text{mol}$$

Answer 23.2

C_5	$5 \times 12.01 =$	60.05 Da
H_4	$4 \times 1.01 =$	4.04 Da
N_4	$4 \times 14.01 =$	56.04 Da
O_3	$3 \times 16.00 =$	48.00 Da

Molecular mass of uric acid = 168.13 Da.

Answer 23.3
Mass of glucose = $180.18 \times 0.5\,\text{M} \times 0.2\,\text{l} =$ 18.018 g

Answer 23.4

$$\frac{0.25\,\text{M}}{1\,\text{M}} = \frac{500\,\text{ml}}{x\,\text{ml}}$$

$$x = \frac{1 \times 500}{0.25} = 2000\,\text{ml}$$

So, 2 l of 0.25 M solution can be prepared.

Answer 23.5
A 5% w/v solution contains 5 g per 100 ml, thus 50 g l^{-1}.

$$50\,\text{g}\,\text{l}^{-1} = \frac{50}{180.18} \approx 0.28\,\text{M}.$$

Answer 24.1
In 0.1 M HCl, as there is almost complete dissociation, the concentration of H^+ is effectively also 0.1 M, so $[H^+] = 0.1$
$pH = -\log[H^+] = -\log 0.1 = 1$
The pH of 0.1 M HCl is 1.

Answer 24.2
$[H^+][OH^-] = 10^{-14}$
So $[H^+][OH^-] = [H^+] \times 0.1 = 10^{-14}$

$$\text{Thus } [H^+] = \frac{10^{-14}}{0.1} = 10^{-13}$$

$pH = -\log[H^+] = -\log 10^{-13} = 13$
So the pH of 0.1 M NaOH is 13.

Answer 24.3
$pK_a = -\log K_a = 9.25$
so
$\log K_a = -9.25$
$K_a = 5.62 \times 10^{-10}$.

Answer 25.1
First calculate the resulting volume and then the molarities. The resulting volume would be 500 ml.
So, the resulting molarity of Tris–HCl would be:

$$\frac{300}{500} = 0.6\,\text{M}.$$

The molarity of Tris base would be:

$$\frac{200}{500} = 0.4\,\text{M}.$$

$$pH = pK_a + \log\frac{[A^-]}{[HA]} = 8.3 + \log\frac{0.4}{0.6}$$

$$= 8.3 + (-0.1761) = 8.1239$$

The pH is 8.1, to two significant figures.

Answer 25.2
Desired pH = 7.4

$$= 7.2 + \log\frac{[A^-]}{[HA]}$$

$$\text{So, } \log\frac{[A^-]}{[HA]} = 7.4 - 7.2 = 0.2$$

$$\frac{[A^-]}{[HA]} = 1.585$$

We therefore need the ratio of 1 M conjugate base and diethylmalonic acid to be 1.585 to 1.

$$\text{Amount of conjugate base needed} = \frac{1.585}{1 + 1.585} \approx$$

0.61 litres.

$$\text{Amount of acid needed} = \frac{1}{1 + 1.585} \approx 0.39\ \text{litres.}$$

Answer 28.1

$$\bar{x} = \frac{\sum x}{n} = \frac{11.7 + 11.9 + 12.2 + 12.7 + 13.0}{5} =$$

$$\frac{61.5}{5} = 12.3\,\text{g}\,\text{dl}^{-1}$$

Answer 28.2
The median is 12.5 g dl^{-1}, half-way between the middle two values.

Answer 29.1
Answer 28.1 showed the calculation that gives the sample mean, 12.3 g dl^{-1}.

Haemoglobin x	Sample mean $\bar{x}$	Deviation $x - \bar{x}$	Square of deviation $(x - \bar{x})^2$
11.7	12.3	-0.6	0.36
11.9	12.3	-0.4	0.16
12.2	12.3	-0.1	0.01
12.7	12.3	0.4	0.16
13.0	12.3	0.7	0.49
	Sum of deviations $\Sigma(x - \bar{x})^2$		1.18
	Dividing by n-1 gives the variance, V		$\frac{1.18}{(5-1)} = 0.295$
	The square root of the variance gives the SD, σ		$\sqrt{0.295} = 0.543$

Thus SD = 0.543 g dl^{-1}.

Answer 30.1

$$z = \frac{(x - \mu)}{\sigma} = \frac{(15.5 - 12.5)}{1.2} = 2.5$$

Answer 33.1

$$\text{SEM} = \frac{\text{SD}}{\sqrt{n}} = \frac{0.4}{\sqrt{16}} = 0.1$$

Answer 34.1

SEM × 1.96 = 0.8 × 1.96 = 1.57
$\bar{x}$ − 1.57 = 12.8 − 1.57 = 11.23
$\bar{x}$ + 1.57 = 12.8 + 1.57 = 14.37
so the 95% CI is 11.23 to 14.37

SEM × 2.58 = 0.8 × 2.58 = 2.06
$\bar{x}$ − 2.06 = 12.8 − 2.06 = 10.74
$\bar{x}$ + 2.06 = 12.8 + 2.06 = 14.86
so the 99% CI is 10.74 to 14.86

SEM × 3.29 = 0.8 × 3.29 = 2.63
$\bar{x}$ − 2.63 = 12.8 − 2.63 = 10.17
$\bar{x}$ + 2.63 = 12.8 + 2.63 = 15.43
so the 99.9% CI is 10.17 to 15.43

Answer 35.1

The chance of throwing two sixes is:

- a 1 in 36 chance
- a 1/36 chance
- a 0.028 probability
- P = 0.028
- a 2.8% probability.

Answer 36.1

The null hypothesis is that there is no difference between the effects of the two temperatures on the germination rate of the wheat seeds.

Answer 36.2

P = 0.25 means that the probability of the difference having happened by chance is 0.25 in 1, i.e. 1 in 4.
It is not statistically significant.

Answer 38.1

$$t = \frac{\bar{x} - E}{\text{SEM}} = \frac{62 - 70}{8} = -1$$

The t distribution is symmetric, but usually only tabulated for positive values, so for $t = -1$, look up $t = +1$.
So, $t = 1$, df 11
The table in Appendix 2 shows that, for df 11, the critical value for 5% is 2.20. As $t < 2.20$, the result is not significant.

Answer 38.2

$$t = \frac{\bar{d}}{\text{SE}_d} = \frac{(36.9 - 36.8)}{0.04} = 2.5$$

So, $t = 2.5$, df 19
Using the table of t values in Appendix 2, for 19 df the critical value of t for 5% is 2.09 and the critical value of t for 1% is 2.86. Our calculated value of 2.5 is therefore greater than needed for the 5% significance level but less than needed for the 1% level. Therefore the P value is less than 0.5 but greater than 0.01.

Answer 38.3

$$t = \frac{\bar{x}_a - \bar{x}_b}{\text{SE}_d} = \frac{1.5 - 1.2}{0.1} = 3$$

So, $t = 3$, df 58 (as an approximation use df = 60 from table)
Using the table of t values in Appendix 2, for 58 df the critical value of t for 1% is 2.66 and the critical value of t for 0.1% is 3.46. Our calculated value of 3.0 is therefore greater than needed for the 1% significance level but less than needed for the 0.1% level. Therefore the P value is less than 0.01 but greater than 0.001.

Answer 40.1

The expected frequency, E, of cultures with colonies for the standard medium is given by:
E = total number of cultures with bacterial colony present × total number of standard culture Petri dishes/Total number of Petri dishes
So,

$$E = \frac{304 \times 240}{480} = 152 \text{ for standard medium,}$$

colony present;

$$E = \frac{176 \times 240}{480} = 88 \text{ for standard medium, no}$$

colony present;

$$E = \frac{304 \times 240}{480} = 152 \text{ for new medium, colony}$$

present;

$$E = \frac{176 \times 240}{480} = 88 \text{ for new medium, no colony}$$

present.
The following table shows the calculation of

$$\Sigma \frac{(O - E)^2}{E}$$

Table calculating χ^2 for different culture media					
	Observed frequency, O	Expected frequency, E	$O - E$	$(O - E)^2$	$\frac{(O - E)^2}{E}$
Standard medium, colony	144	152	−8	64	0.4211
Standard medium, no colony	96	88	8	64	0.7273
New medium, colony	160	152	8	64	0.4211
New medium, no colony	80	88	−8	64	0.7273
				$\sum \frac{(O - E)^2}{E} = 2.297$	

So, $\chi^2 = 2.297$

Answer 40.2

$df = (2 - 1)(2 - 1) = 1$

The critical value for the 5% significance level is 3.84 for 1 df.

But in our example, χ^2 is only 2.297, so the result is not significant and the null hypothesis stands.

Answer 41.1

$$\rho = \frac{\sum (x - \bar{x})(y - \bar{y})}{\sqrt{\sum (x - \bar{x})^2 \sum (y - \bar{y})^2}}$$

$$= \frac{-284}{\sqrt{536 \times 160}} = -0.97$$

Answer 42.1

	Mass (kg)	log of (mass)	Resting heart rate (beats min^{-1})	log of heart rate
Mouse	0.02	−1.70	700	2.85
Rat	0.2	−0.70	400	2.60
Cat	5	0.70	150	2.18
Dog	10	1.00	120	2.08
Man	70	1.85	70	1.85
Horse	450	2.65	40	1.60
Mean		0.63		2.19

After calculating the various values of $(x - \bar{x})$ and $(y - \bar{y})$ we can calculate:

$$m = \frac{\sum (x - \bar{x})(y - \bar{y})}{\sum (x - \bar{x})^2} = \frac{-3.72}{12.90} = -0.288$$

Substituting into the regression line formula $\bar{y} = m\bar{x} + c$ gives:

$2.19 = (-0.288 \times 0.63) + c$

Thus, $c = 2.37$

So, $y = -0.288x + 2.37$

For a 15-kg animal, $\log(15) = 1.176$

$y = (-0.288 \times 1.176) + 2.37 = 2.03$

This is a logarithm of the expected resting heart rate, so

Expected resting heart rate $= 10^{2.03} =$ 107 beats min^{-1}

Appendix 1: Flow chart for choosing statistical tests

The key tests for each category are given in the lozenges.
The tests in *italics* are the non-parametric tests for each category.

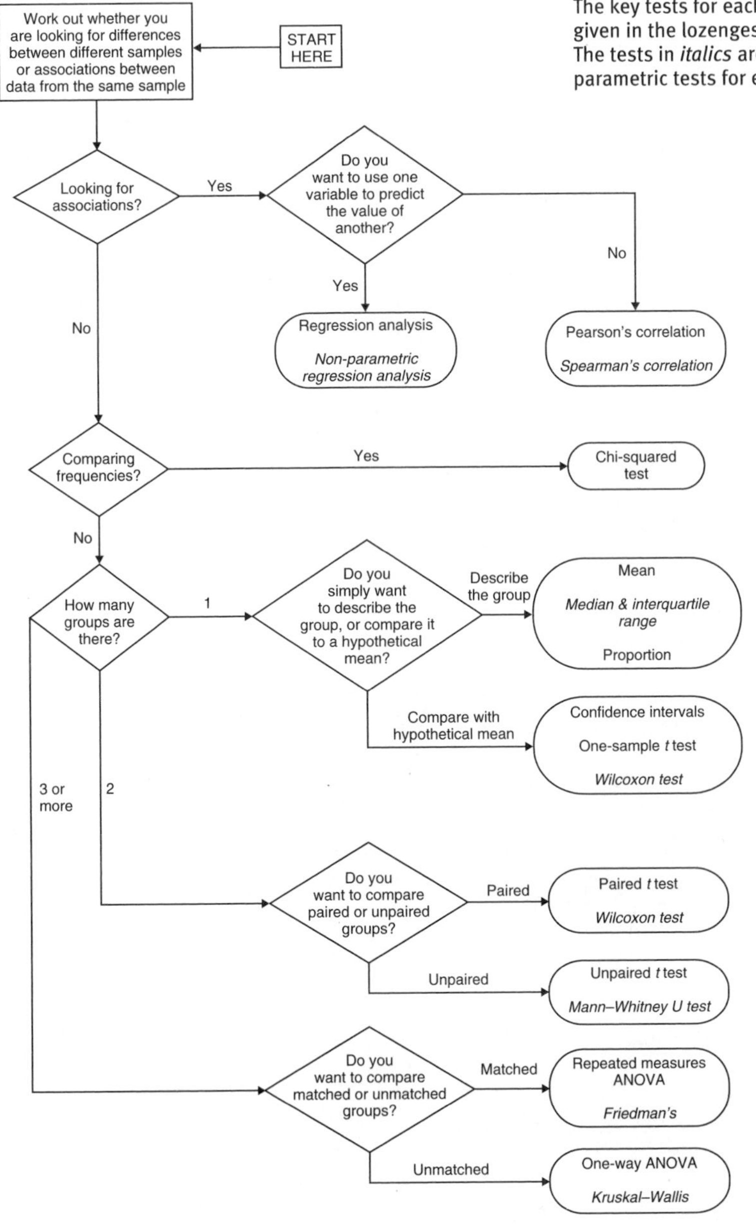

Appendix 2: Critical values for the *t* distribution

This table gives us the critical *t* values for different degrees of freedom and commonly used significance levels.

We normally need to use the two-tailed significance levels. See Section 38.2 for an explanation of one- and two-tailed testing.

We reject the null hypothesis if the calculated value of *t* is larger than the value on the table for a chosen significance level.

	Significance level for two-tailed test		
	5%	1%	0.1%
Degrees of freedom (df)	Significance level for one-tailed test 2.5%	0.5%	0.05%
1	12.71	63.66	636.58
2	4.30	9.92	31.60
3	3.18	5.84	12.92
4	2.78	4.60	8.61
5	2.57	4.03	6.87
6	2.45	3.71	5.96
7	2.36	3.50	5.41
8	2.31	3.36	5.04
9	2.26	3.25	4.78
10	2.23	3.17	4.59
11	2.20	3.11	4.44
12	2.18	3.05	4.32
13	2.16	3.01	4.22
14	2.14	2.98	4.14
15	2.13	2.95	4.07
16	2.12	2.92	4.01
17	2.11	2.90	3.97
18	2.10	2.88	3.92
19	2.09	2.86	3.88
20	2.09	2.85	3.85
25	2.06	2.79	3.73
30	2.04	2.75	3.65
40	2.02	2.70	3.55
50	2.01	2.68	3.50
60	2.00	2.66	3.46
70	1.99	2.65	3.43
80	1.99	2.64	3.42
90	1.99	2.63	3.40
100	1.98	2.63	3.39
Infinity	1.96	2.58	3.29

Note: A *t* distribution with infinite degrees of freedom is equivalent to the values of the normal distribution.

Appendix 3: Critical values for the chi-squared distribution

This table gives us the critical chi-squared values for different degrees of freedom and commonly used significance levels.

We reject the null hypothesis if our calculated value of chi-squared is larger than the value on the table for a chosen significance level.

Degrees of freedom (df)	Significance level 5%	1%	0.1%
1	3.84	6.63	10.83
2	5.99	9.21	13.82
3	7.81	11.34	16.27
4	9.49	13.28	18.47
5	11.07	15.09	20.51
6	12.59	16.81	22.46
7	14.07	18.48	24.32
8	15.51	20.09	26.12
9	16.92	21.67	27.88
10	18.31	23.21	29.59
11	19.68	24.73	31.26
12	21.03	26.22	32.91
13	22.36	27.69	34.53
14	23.68	29.14	36.12
15	25.00	30.58	37.70
16	26.30	32.00	39.25
17	27.59	33.41	40.79
18	28.87	34.81	42.31
19	30.14	36.19	43.82
20	31.41	37.57	45.31
25	37.65	44.31	52.62
30	43.77	50.89	59.70
40	55.76	63.69	73.40
50	67.50	76.15	86.66
60	79.08	88.38	99.61
70	90.53	100.43	112.32
80	101.88	112.33	124.84
90	113.15	124.12	137.21
100	124.34	135.81	149.45

Index

Where multiple page numbers are given, the bold page numbers indicates the key pages.